W9-DGD-063

THE NATURE OF GEOGRAPHY

A CRITICAL SURVEY OF CURRENT THOUGHT
IN THE LIGHT OF THE PAST

By
RICHARD HARTSHORNE

Reprinted, 1961, with corrections by the author, from the
Annals of the Association of American Geographers
Volume XXIX, Numbers 3 and 4
DERWENT WHITTLESEY, *Editor*

Published by the Association
Lancaster, Pennsylvania
1939

Second Printing, 1946
Third Printing, 1949
Fourth Printing, 1951
Fifth Printing, 1956
Sixth Printing, 1958

Reprinted with corrections, 1961

THE SCIENCE PRESS PRINTING COMPANY
LANCASTER, PENNSYLVANIA
Manufactured in the United States of America

Photo-Lithoprint Reproduction
EDWARDS BROTHERS, INC.
Lithoprinters
ANN ARBOR, MICHIGAN

New developments in geographic thought in the twenty years since the first publication of this work have led me to reconsider a number of its conclusions. This reconsideration is presented in my *Perspective on the Nature of Geography,* published in 1959 by Rand McNally and Company as the first of the monograph series of the Association of American Geographers (subsequently referred to simply as *Perspective. . . .*). In each case in which the view expressed in these pages has been changed in the later discussion, a note has been inserted to inform the reader of the change. The major changes are indicated in the following Supplementary Notes: 32, 35, 36, 39, 41, 50, 54.

In view of the wide-spread use of this volume in the training of graduate students, it seems desirable to explain, somewhat more fully than is stated in the Introduction, the principles under which the work was written, and in the light of which it should be understood.

It was never the author's intention to present a pronouncement as to the scope and purpose of geography or to set up a dictum by which specific works could be judged as admissible or inadmissible in the field of geography. On this topic the reader is referred to the paper "On the Mores of Methodological Discussion in American Geography," included in the back of this volume.

Rather the purpose, as stated in the sub-title, was to examine the issues which were—and in many cases still are—in dispute among American geographers, in the light of what geographers had written on those questions throughout the history of development of the field. Not that the past should or can dictate to the present and future; every individual geographer contributes, through his expression of views and the character of his substantive work, to the ever-developing nature of the field. But the influence of any individual or group is necessarily small in comparison to the total of all previous development. Methodological writing therefore will be most effective in the long run if it is based on thorough understanding of the development on which it seeks to build.

It is true that the development of methodological thought results from viewpoints and concepts developed in the course of substantive research and is demonstrated in substantive publications, so that the most reliable view of how geographers have considered the methodology of their work might conceivably best be attained by thorough examination of the substantive literature. Certainly students should be encouraged to do this as widely and deeply as they are able. But to examine all the significant literature of substantive work in geography to induce the methodological concepts implicit in them seemed to me a task far beyond the competence of any one student. To the extent that the methodological aspects of the work of individual outstanding geographers—such as Humboldt, Ritter or Ratzel—had been analyzed by qualified scholars familiar with their work, these studies were used in my writing. In the main however,

24722

the study was limited to examination of explicitly methodological writings. Thus limited, the literature includes writings of may scores of geographers over the past two centuries. This should provide, I believed, an adequate and yet manageable body of material from which to construct a reliable statement of the nature and scope of geography as its professional students have studied it.

Because this material is widely scattered in many publications, more commonly in journals than in books, and in large part in foreign languages, it was a special purpose of this work to bring together in organized form the essential arguments and conclusions of all such studies which had been found to have had significant influence on the thinking of geographers on questions of current concern. To the degree that this purpose was accomplished, it should be unnecessary for others to repeat it. Whatever difficulties result from the somewhat encyclopedic character of the work are far less than if the readers were forced to cover the literature for themselves.

One consequence of this method, for which a caveat is in order, is that the work may appear to give undue prominence to methodological writings as compared with substantive works and thereby to give an unbalanced picture of the relative importance of the two kinds of contributions in the overall work of individual geographers and likewise an unbalanced picture of the comparative importance of different geographers in the development of the field. The purpose of the study was not to appraise the work of different geographers, but rather to test the validity of different methodological ideas, regardless of their source.

The principle that the reader should be able to test the validity of the findings imposed on the author, as it does on any student who publishes work purporting to represent methodological research, the obligation to present the evidence found in the works of previous writers with meticulous adherence to established standards of scholarship. This principle is expounded in my paper of 1948 on "Mores in Methodological Discussion," included in the back of this volume, and still further in the preface to a later paper, " 'Exceptionalism in Geography' Re-Examined," *Annals*, Association of American Geographers, XLX (1955), pages 205-208.

A work constructed from more than a thousand individual references can hardly be entirely free of misstatement or misunderstanding. While no such cases have been demonstrated in formal criticisms, several have been brought to my attention in correspondence and more have been discovered in work with my graduate students. Each such case is corrected in brief notes added in this edition. Further, in order more nearly to accomplish the original purpose of providing a coverage of the methodological literature available at the time of writing, references have been added to writings which were in print in 1939 but only later came to my attention. Particular mention may be made of those referred to in Supplementary Notes 2, 13, 14, 18, 22, 28, 45 and 50.

Richard Hartshorne
University of Wisconsin
November 7, 1960

A FOREWORD BY THE EDITOR

In this issue and the next, the Association takes the unprecedented step of presenting a monograph of book length. This appears to be warranted by the merit of the paper. It is, moreover, peculiarly appropriate because the work is a study in the methodology of geography, a phase of the science that has found repeated expression in this periodical from its inception. As is customary in learned societies, a goodly number of its presidents have presented before the Association their views of the theory and the technique of geography in addresses afterwards published in the ANNALS. In addition there have been occasional contributions by members in unofficial capacity, including the Association's founder, W. M. Davis. Some of these are devoted exclusively to technical aspects of the topic. Others, more numerous, are concerned with methodology incidentally to presentation of matter on areas. The fitness of discussing theory and technique in the publication maintained by professional American geographers for the interchange of their views and the presentation of their researches, can hardly be questioned.

The perennial issues of methodology have alternately lain dormant and put forth fresh shoots. The past few years have constituted a period of active thinking about these issues. This has been voiced in vigorous debate at every recent gathering of American geographers. Besides, no less than six papers taking cognizance of the issues have been published in the ANNALS within two years, and others have appeared elsewhere, both in America and in European countries. To those debates and articles the monograph offered here is in a sense the capsheaf. This it becomes because it far transcends the scope of any of its predecessors.

A word as to its origin is in order, to account for publication in the ANNALS of a paper of such extraordinary length. A year ago the author submitted for publication a rather long paper on the nature of geography, but not longer than some which have appeared in earlier volumes of the ANNALS. Pursuant to suggestions made by the editor and other critical readers of that manuscript, the author undertook to make emendations. Meanwhile he had settled temporarily in Central Europe. The changes suggested, as well as others prompted by studies published by several students during the period of reworking, resulted in a complete revision, thanks in part to the atmosphere of the native heath of geographic methodology. Indeed, many of the gleanings from the printed word have been verified and amplified in discussions between the author and German authorities.

The increase in length entailed by the enlargement in scope gave pause to the author, who repeatedly expressed his misgivings in letters to the editor. Convinced that the work needed doing, the editor encouraged its completion. When finally the revised manuscript was received it was accompanied by the following statement:

"Looking at the product that has grown out of what I started to do over a year ago, it seems to me that for anyone in my position to have planned to produce such a study, in which the work of so many colleagues is critically discussed, would have been presumptuous. To have sent such a thing, unasked, to the ANNALS, would certainly have required more boldness than I could command.

"For me, therefore, it is important that the work was not planned from beforehand, but grew of itself out of a very much smaller idea. This does not mean that it is planless; it has been revamped many times to form an organized whole, but its nature developed from itself rather than from any intention on my part.

"Finally, it is only fair to remind the editor that the original suggestion that I write something on the general subject came from him (though this is not intended to make him responsible for the result)."

The editor, believing that the monograph is both timely and timeless, seizes this opportunity to claim for himself the degree of responsibility for its appearance which the author so generously accords him.

The Nature of Geography
A Critical Survey of Current Thought in the Light of the Past

RICHARD HARTSHORNE

> "Lassen Sie uns mit vorsichtiger Kritik, aber nicht zaghaften Sinnes, sondern in mutigem Denken und in froher Zuversicht an dem weiteren Ausbau . . . unserer Wissenschaft arbeiten."
>
> ALFRED HETTNER, 1907.

CONTENTS

[iii]

* Also, Abstract [vi]

ACKNOWLEDGMENTS

The study originally prepared at the writer's own institution was revised and greatly enlarged on the basis of materials available only in European libraries. That the writer was able to use these materials resulted from the fact that he was studying another project in Europe during the year 1938–39, enabled to do so, in part, by a grant from the Social Science Research Funds of the University of Minnesota.

Appreciation should be expressed for the cooperation of the library staffs at the Universities of Vienna, Zurich, and Clark, of the National Library at Vienna, and particularly of the Geographical Institute of the University of Vienna. At each of these libraries publications were found that were not available at any of the others, nor in the library of the writer's own institution.

The privilege of attending the "Oberdeutschen Geographentag" at Isle Reichenau, April 1–5, 1939, enabled the author to receive helpful suggestions from German and Swiss geographers attending those meetings. Similarly the author is indebted to a number of other European geographers, most especially to Professor Johann Sölch. He wishes also to express his gratitude to the editor and to the anonymous critics of the ANNALS whose assistance far exceeded the requirements of their positions.

In the last analysis, the value of the study rests on the works, both those expressly methodologic and those illustrative of methods and philosophies, of the geographers, numbering nearly two hundred, whose publications provided the material from which my monograph is constructed.   R. H.

ABSTRACT

Although the concluding chapter (XII) provides a summary
of the positive conclusions reached in the text study, experience has demonstrated the need for a preliminary abstract to
serve as a guide to readers interested in some, but not all,
of the problems discussed.

Introduction

I. (A) This study was undertaken because American geographers,
though given to frequent discussions of the nature and
scope of their field have been unfamiliar with the past
discussions and more serious studies of the problems in
question, particularly the studies in the foreign literature. (B) Lack of understanding of the consistent characteristics of the field during its modern development as
a discipline has led to dissatisfaction and recurrent attempts to reform. (C) In place of presenting a personal
view of what geography ought to be, or what one might desire it to be, the problem is conceived of as a research
problem, to examine the field as scholars have worked in
it and conceived it [Section II] and thereby to determine
what geography is [Section IV] and what are its characteristics and qualifications as a branch of knowledge [Sec-
tion XI]. (Note the fuller statement of purpose on pages
31-32.) Certain proposals for change or for emphasis of
certain concepts that have recently been urged among
American geographers but have long been studied by our
European colleagues, are examined and tested. [Sections
V-X]. Since it was desirable to give full consideration
to ideas seriously urged by competent geographers, these
discussions are necessarily lengthy and detailed. Some
readers may wish to omit those sections, totalling nearly
one third of the total text, that arrive at negative conclusions. These are Sections III, part A of V, VII, IX
excepting part F, and parts C and E of X.

II. Although geography has its roots in Classical Antiquity,
its development as a modern discipline crystallized in
Europe, and primarily in Germany, during the period 1750
to 1900. The examination of the historical development
of concepts concerning geography, and substantive work in
it leads to the following conclusions: (1) Geography
studies the areas of the earth according to their causally
related differences, in other words, the areal differentiation of the earth. (2) With few exceptions geographers
have recognized the need for two different methods of approach - the systematic studies concentrated on areal

differences of specific elements over the whole earth, or
major parts of it, and regional studies of the complete
geography of specific areas. (3) A different form of
"dualism," the division between physical (in the sense of
natural) geography and human geography became a problem
during the mid-nineteenth century, but was largely overcome in the latter part of the period. (4) Special emphasis on the study of land forms in the actual research
work of geographers has firmly established geomorphology
(or physiography) as a part of the field in Germany and
perhaps also in America and other countries. (5) The concept of unity or Ganzheit of the earth, conceived even as
an organism, though very important in the early development of the subject has almost completely dropped out.
(6) On the other hand, the similar concept of a specific
region as a unit in itself, a "whole" or an "organism,"
though effectively criticized in the early nineteenth
century, has re-emerged in Germany in recent decades and
been transplated and vigorously supported by a number of
students. (7) The related problem of the concept of the
"natural region," though critically attacked early in the
nineteenth century, remains as a current problem today.

III. At various times in the past certain geographers, or
groups of geographers, have conceived of the field in
quite different terms from those found to represent the
main lines of development of the field. Since these devia-
tions from the course of historical development have some
representation in recent thought among geographers, the
history of the concept is studied in each case to determine
the reasons that led to its being discarded.

These are: (A.) The attempt to make geography an
"exact" science, or an "essentially natural" science by
the arbitrary elimination of those kinds of phenomena not
regarded as capable of study by "scientific" methods;
(B.) the definition of geography as the science of the
planet earth, rather than of the earth surface; (C.) the
definition of geography as the study of relationships between the natural environment and man, or the adjustment
of human activities to the natural environment; (D.) the
definition of geography as the study of distributions on
the earth surface.

IV. A. The concept of geography as the study of the areal differentiation of the earth surface is justified in common
sense by the well-known fact that things are different in
different areas of the world and that these variations are
somehow causally related to each other. There is a constant need, both in intellectual thought and for practical

purposes, to know and understand what these differences
are and how they are related, in order to understand the
character of different areas.

B. The logical position of such a field in relation to
other fields of knowledge has been expounded in much the
same terms by Immanuel Kant, Humboldt, and Hettner.   In
contrast to the systematic sciences that study particular
kinds of phenomena, wherever and whenever found, two other
groups of studies are necessary to interpret the complexes
of phenomena as found associated together in terms of
space and of time. The historical sciences study the association in segments of time. Astronomy is concerned
with the association of phenomena in celestial space,
geography in the space of the earth surface.

C. The nature of geography is best understood in comparison with history; the character of geographic spaces, regions, is best understood in comparison with historical
units or periods.

V. A. The concept of "landscape," as based on the German term
of double meaning <u>Landschaft</u>, has led inevitably to lack
of clear thinking on a number of major problems in the
field. In particular it is unsound and unnecessary to attempt to define the field of geography in terms of this
ambiguous concept.

B. The normal connotation of the English word "landscape"
can be of value in geography if the term is defined
clearly to mean the external, visible, (or touchable) surface of the earth. This surface is formed by the outer
surfaces, those in immediate contact with the atmosphere,
of vegetation, bare earth, snow, ice, or water bodies or
the features made by man. Minor penetrations of the atmosphere into this surface are confusing only in theory:
for practical purposes they are readily ignored.

C. The terms "natural landscape" and "cultural landscape"
are not properly used to indicate separate components of
a total landscape. At any one time there is only one
landscape and only in areas untouched by man can it properly be called "natural." In place of the use of these
terms for a wide variety of different concepts that need
to be carefully distinguished, the following solutions
are suggested: (1) For the sum of all the natural factors in an area, the term "natural environment" is well-
established and clearly understood. (2) "Natural landscape" should be used only to indicate the original landscape of an area as it existed before the entry of man,
because of past corruption of the term, clarity may require

the redundant phrase "original natural landscape," or
that may be avoided by using the term "primeval landscape." (3) The concept of the theoretical landscape
that would now exist in an inhabited area if that area
had never been touched by man is not a concept frequently
needed and therefore had best be spelled out in full if
used at all. (4) The landscape of areas of primitive development, prior to the entry of civilized man, is not a
natural landscape, since even primitive peoples may cause
notable alterations, but may be called a "primitive landscape." (5) Likewise the general landscapes of such
primitive areas, as well as the patches of uncontrolled
areas in the midst of well-developed lands may be called
"wild landscapes" in contrast to the "cultivated" or
"tamed" landscapes of fields, farmsteads, roads and cities.

VI. The distinction between history and geography - like any
other division of the sciences - is in conflict with reality. Phenomena are actually associated together both in
respect to time and space. The separation can be justified only in terms of the limits of the human intellect
in examining reality. In each field many topics can be
studied adequately only if the methods of the other field
are utilized.

A. The use of the geographic approach in historical studies,
in particular the attempt to determine the significance to
history of specific geographic features is logically a
part of history - whether pursued by historian or geographers.

B. Adequate interpretation of many of the individual features of a region may require consideration of past conditions, that have led to the situation under examination.
Geography focuses its attention on things as they are in
any cross-section of time, considering developments for
the purpose of interpretation rather than because of concern for the processes themselves.

C. Historical geography is properly the geography of any
past period, that period considered as though it were
present. In comparative historical geography, the geographies of successive periods in the same area, are
studied to bring out the differences at successive times.

VII. Because of the multiplicity of heterogenous phenomena
that are associated in areas, geographers concerned with
the scientific development of the field have long been
concerned over the problem of selection of data that
should be included in their studies. Beginning at the
turn of the century, a small number of European

geographers have maintained that geography is properly
limited to material features, both natural and cultural,
excluding thereby the non-material cultural features
(frequently expressed as the limitation to "visible features" or to "features perceptible by the senses)." This
thesis has recently been vigorously urged, or stated as a
settled principle, by a considerable number of American
geographers. Consequently it is subjected to exhaustive
examination before being dismissed as illogical, historically inconsistent, disruptive, and impractical [see Summary on page 235 f].

VIII. A. If geography is considered as the study of the areal
differentiation of the world, the logical basis for selection of data to be considered is to select those that contribute significantly, both in themselves and in their
causal relations to other variables, to the total complex of areal differentiation.

B. The essential criteria therefore, as expressed by
Hettner and accepted by a large body of German geographers
are: (1) the feature concerned varies from place to place;
(2) the variations form a system, or systems, in which
there is spatial association of the phenomena in terms of
their location in reference to each other forming an areal
expression; and (3) there is a causal connection between
the variations of the feature or element and those of
other elements, and their different phenomena are united
at one place.

C. The criteria are applied to specific cases, by way of
illustration.

D. Any body of data that fulfills these criteria can be
presented on a map that will show significant comparisons
with maps of other elements. Cartographic presentation
is the most distinctive technique in geographic work. As
a rule of thumb, therefore, a simple test of the geographic quality of any study is whether it can be studied
fundamentally by maps.

IX. A-E. Whenever regional geography has been vigorously
studied by geographers, disagreements have developed over
the nature of the areal units into which they divide the
world. In earlier periods, and again in this century,
various students have claimed that the areal units -
whether called regions, natural regions, geographic regions, Landschaften, or landscapes - represent individual
concrete objects, or Wholes, or even organisms that can
be studied like any other individual objects. The world
therefore consists of a mosaic of these individual units

and can be studied in terms of the relations of each unit,
as a whole, to the others. Since this concept, in one form
or another, has entered also into the American literature,
including textbooks, its claims for acceptance are considered in full detail before rejecting it in all its manifestations. The region is simply a more or less convenient, because more or less intelligently delimited, arbitrary division of the earth surface necessary for regional
study.

F. Certain areal units however do conform to the terms
listed. A farmer's field or a city block is a distinct
individual unit. A farm, or a factory, or even a city is
in many respects properly considered to be a Whole. Finally, the work of man in creating the cultural landscape does
produce a mosaic, though one that is far from perfect.

G. The emphasis on regions as distinct objects, as though
complete each in itself, has apparently led to neglect of
one of the most fundamental of geographic factors - the
significance of location of phenomena on the earth surface in relation to each other.

X. A. The conclusion that regions are not definite concrete
objects but merely arbitrary divisions of the earth surface
made by the student does not cast out the problem of dividing the world, or any large part of it, into regions
nor reduce to unimportance the basis for such a division.
It is important to find the most intelligent and useful
method, or methods, of divising the world into regions.

B. There are two principal types of systems of regional
division, both of utility in different ways. A realistic
system of specific regions is based on consideration of
all the factors involved, including relative location in
reference to land and sea. A comparative system of generic regions considers only the internal characteristics of
the areas without respect to relative location; strictly
speaking, it does not establish regions but simply areas
of particular types.

C. In either kind of system the term "natural regions" is
misleading. Strict analysis shows that such regions are
actually regions based on a combination of certain natural
factors determined according to their importance to men
of a particular culture and technology.

D. The analysis of one system of specific regions dis-
closes a number of major difficulties inherent in the
problem and discusses the arbitrary solutions that must
be made. The thesis that such a division must be thoroughly genetic in foundation is analyzed and rejected as impractical.

E. Various attempts to construct comparative systems of generic regions based on combinations of natural elements are analyzed and found to be unsatisfactory. In most cases they are little more than systems of climatic or vegetation types.

F.-G. Greater success is to be expected from systems of generic regions based on the actual syntheses of cultural features constructed by man. Two such systems, one based on the present landscape cover of the world, the other on the synthesis of features involved in land use, are analyzed in detail with conclusions concerning their advantages and limitations.

H. A full summary of the chapter is provided in the concluding section H. [Pages 361-365].

XI. A. The purpose of this concluding chapter is to determine the characteristics of geography, as found in the preceeding chapters, in comparison with other branches of that type of knowledge that for convenience is called science.

B. Certain characteristics of geography result from the fact that it is a subject that penetrates into both the natural sciences and the social sciences and partakes of characteristics of both groups. More significant is the conclusion that geography, like history, does not have a distinctive group of phenomena at the center of its interest, as do the systematic sciences, but has the distinctive function of studying the integration of heterogenous phenomena in sections of space, the areas of the earth surface. Geography, like history, examines reality as it is, naively looking at things as they are actually arranged.

C. The character of geography is to be tested in terms of its adherence to the ideals of certainty, exactness, universality and system. Geography seeks to make its knowledge as certain and accurate as possible. Its attainments, in comparison with other sciences, are not to be measured merely in terms of the degree of success in reaching these ideals, but also in terms of the relative difficulties in the tasks undertaken.

D. Geography seeks for universality of its knowledge by the development of a sound system of generic concepts and the construction thereby of general principles of interrelationships. There remains however, as in any branch of science, a large number of significant phenomena that can be studied only in terms of the unique. In geography a relatively large proportion of the work, though not as large as that in history, is necessarily concerned with unique cases.

In systematic geography, however, there is increasing emphasis on the development of generic concepts and general principles, though the multiplicity of factors commonly involved in a geographic problem makes extremely difficult the application of general principles or laws. The attempt to establish generic concepts of areas, as units, with the hope of finding general principles or laws in regional geography, is found to be a pursuit of what is logically impossible since the area is not an object or a phenomenon. Generic concepts of systematic geography however are used in regional geography, and generic descriptions of types of areas are of assistance in reaching a partial understanding of the character of specific areas.

E. Knowledge in geography is organized into systems in two different ways. Specialized branches of the field divide the phenomena of areal differentiation into major groups each consisting of closely related phenomena; these include the various parts of physical geography - such as climatology, the geography of landforms, soil geography, etc.; and the several branches of human or cultural geography - economic geography, political geography, and social geography. Knowledge in all these branches is also organized in terms of systematic geography and regional geography.

F. This division of geography into these two forms of organization is compared with the situation in other integrating sciences, namely astronomy, historical geology and history.

G. In analyzing the character of systematic geography particular attention is paid to the following topics: The problem of establishing clear distinction between the purposes of systematic geography on the one hand and of the related systematic sciences on the other; special techniques; the study of element-complexes; the capacity for prediction; and the comprehensive range of natural and cultural phenomena studied in systematic geography.

H. Three major steps are involved in a study in regional geography. In order to comprehend the actual interrelation of phenomena in specific places it is necessary to consider small subdivisions within each of which the local variations of the various factors are arbitrarily ignored. The second step is to relate the unit areas to each other to discover the structural and functional formation of

the larger region. Finally it is necessary to study the
arrangement of the regions to each other and the interrelations of phenomena in one region to those in another.
The several arbitrary devices that must be used in this
process present problems that are discussed in detail.

Other specific questions discussed include: the problem
of transition areas; the kind of knowledge to be included
in a regional study; "the genesis of an area:" scientific
laws or principles concerning regions; comparative regional geography; and the question of the size of region appropriate for intensive study in particular, the value of
studies in "microgeography."

I. Systematic and regional geography do not represent
separate branches of the field, distinct in substance, but
rather two different approaches that are mutually dependent on each other and must be combined in specific studies.

The detailed conclusions in this chapter will be found
summarized in the final chapter, pages 464-468.

XII. The summary of conclusions omits all discussions that arrived at negative conclusions, but recapitulates the positive conclusions reached in the previous chapters concerning the nature of geography.

This study of the nature of geography is derived from writings by students of the subject. In the text, citations of these writings occur on nearly every page, where they stand, abbreviated, in brackets. To facilitate reference to the titles thus cited, the bibliography is unconventionally placed *before* the text.

In each bracketed citation in the text the number in italics gives the number of the title in this list. Succeeding numbers in Roman type cite the pages to which reference is made.

The works are arranged in groups such that the list, in itself, may be of most service to the student. In each group (excepting the miscellaneous group listed last) the titles are arranged in approximately chronological order. The first part is concerned with the historical development of thought concerning geography prior to the present century, including historical and biographical studies as well as examples of studies in geography. The second and larger part is concerned with the development of thought in the present century. The following outline indicates the arrangement.

I. History of geographic thought, prior to 1900 [*1–82*]
   General historical studies [*1–11*]
   Biographical and critical studies of the work of individual geographers [*12–37*]
   Methodology, with illustrative works [*38–82*]
II. Geographic thought in the twentieth century [*83–400*]
   General surveys of geographic work [*83–110*]
   Philosophy of science and of non-geographic fields [*111–120*]
   The methodology of geography, in general [*121–224*]
      in Germany*[*121–181*]
      in France [*182–187*]
      in other continental countries [*188–191*]
      in Great Britain [*192–202*]
      in the United States [*203–224*]
   The theory of regions, *Landschaften,* landscapes, and boundaries [*225–297*]
      in Europe [*225–280*]
      in the United States [*281–297*]
   Systems of regional division [*298–328*]
   Studies cited as illustrations (arranged alphabetically, by authors), [*329–400*]

[1]

* and other German-speaking countries

The word "Bibliography," in parentheses indicates that the study includes a particularly useful list of references, whether in the form of a separate list or in footnote references.

A. HISTORY OF GEOGRAPHIC THOUGHT, PRIOR TO 1900

GENERAL HISTORICAL STUDIES

1. Wisotzki, Emil: *Zeitströmungen in der Geographie*. Leipzig, 1897. (Bibliography.)
2. Hettner, Alfred: "Die Entwicklung der Geographie im 19.Jahrhundert," *Geogr. Ztschr.*, 4 (1898), 305–320. [Expanded in *161*, 1–109.]
3. Richthofen, Ferdinand Frh. von: "Triebkräfte und Richtungen der Erdkunde im neunzehnten Jahrhundert" (Rektoratsrede, University of Berlin, 1903), *Ztschr. d. Ges. f. Erdkunde, Berlin*, 38 (1903), 655–692.
4. Günther, Siegmund: "Entwicklung der Erdkunde als Wissenschaft: Teil- und Hilfswissenschaften derselben," in Rothe, K. C., and E. Weyrich, *Der moderne Erdkunde-Unterricht*, Wien, Leipzig, 1912, 7–40.
5. Becker, Anton: "Entwicklung der Methodik des Erdkundeunterrichtes," in Rothe, K. C., and E. Weyrich, *Der moderne Erdkunde-Unterricht*, Wien, Leipzig, 1912, 41–55.
6. Wagner, Hermann: Lehrbuch der Geographie.* Hannover, 1920. "Einleitung," 17–36.
7. Schmidt, Peter Heinrich: *Wirtschaftsforschung und Geographie*. Jena, 1925. (Bibliography.)
8. Plewe, Ernst: *Untersuchung über den Begriff der "vergleichenden" Erdkunde und seine Anwendung in der neueren Geographie*. Ztschr. d. Ges. f. Erdkunde, Berlin, Erg. Heft 4 (1932). (Bibliography.)
9. Wright, John Kirtland: "A Plea for the History of Geography," *Isis* (Bruxelles), 8 (1925), 477–491. (Excellent bibliography on history of geography.)
10. Dickinson, Robert E., and O. J. R. Howarth: *The Making of Geography*. Oxford, 1933.
11. Bürger, Kurt: *Der Landschaftsbegriff: ein Beitrag zur geographischen Erdraumauffassung*. Dresdener geograph. Studien 7, 1935. (Bibliography). Abstract by K. H. Huggins, "Landscape and Landschaft," *Geography* 21 (1936), 225 f.

BIOGRAPHICAL AND CRITICAL STUDIES OF THE WORK OF INDIVIDUAL GEOGRAPHERS

12. Gerland, Georg: "Immanuel Kant, seine geographischen und anthropologischen Arbeiten," *Kant-Studien*, 19 (1905), 1–43, 417–547. Also published separately, Berlin, 1906.
13. Adickes, Erich: *Kant's Ansichten über Geschichte und Bau der Erde*. Tübingen, 1911.
14. *idem: Untersuchungen zu Kants physicher Geographie*. Tübingen, 1911.
15. *idem: Ein neu aufgefundenes Kollegheft nach Kants Vorlesung über physische Geographie*. University, Tübingen, 1913. Reviews of this and two preceding titles, by O. Schlüter, *Geogr. Ztschr.*, 19 (1913), 115; 20 (1914), 415.

[2]

* Title should be in italics

16. Dove, Alfred: *"Johann Reinhold Forster,"* and *"George F. Forster,"* *Allg. deutsche Biographie,* 7 (1878), 166–181.

17. *Fünf Briefe der Gebrüder von Humboldt an Johann Reinhold Forster,* edited by F. Jonas. Berlin, 1899.

18. *Goethes Briefwechsel mit Wilhelm und Alexander v. Humboldt,* edited by L. Geiger. Berlin, 1919.

19. *Briefwechsel Alexander v. Humboldt's mit Heinrich Berghaus,* edited by Berghaus. 3 vols., Jena 1862, 1869.

20. Bruhns, Karl, editor: *Alexander v. Humboldt: eine wissenschaftliche Biographie.* 3 vols., Leipzig, 1872; English edition, vols. 1–2 only, London, 1873. (Bibliography.) (The references in this paper are taken from the biographical sections of Vol. I, by J. Löwenberg, and Vol. II, by A. Dove, and from the critical essay in Vol. III, by A. H. R. Grisebach; Peschel's essay is reprinted in his *Abhandlung* [66].)

21. Dove, Alfred: "Alexander v. Humboldt," *Allg. deutsche Biographie,* 13 (1881), 358–383.

22. Döring, Lothar: *Wesen und Aufgaben der Geographie bei Alexander von Humboldt.* Diss., Univ. Frankfurt, 1930. Also published in *Frankfurter Geograph. Hefte,* 1931. (Bibliography.)

23. Rehder, Helmut: *Die Philosophie der unendlichen Landschaft: Ihr Ursprung und ihre Vollendung.* Diss., Heidelberg, 1929.

24. Kramer, Gustav: *Carl Ritter: ein Lebensbild nach seinem handschriftlichen Nachlass.* 2 vols., Halle, 1864, 1870.

25. Marthe, F.: *Was bedeutet Carl Ritter für die Geographie.* Berlin, 1880. Brochure, reprinted with additions, from *Ztschr. d. Ges. f. Erdkunde, Berlin,* 1879, 374–400.

26. Ratzel, Friedrich: "Zu Carl Ritters hundertjährigem Geburtstage," in *Kleine Schriften,* Munich, 1906, I, 377–428. Originally published in newspaper, 1879, in part also in *Allg. Dtsch. Biographie,* 28, 679–697.

27. Hözel, Emil: "Das geographische Individuum bei Carl Ritter und seine Bedeutung für den Begriff des Naturgebietes und der Naturgrenze," *Geogr. Ztschr.,* 2 (1896), 378–96, 433–444.

28. Fröbel, Julius: *Ein Lebenslauf: Aufzeichnungen, Erinnerungen und Bekenntnisse.* 2 vols., Stuttgart, 1890–91. Abstract in *Allg. deutsche Biographie,* 49 (1904), 163–172.

29. Hantzsch, Viktor: "Ernst Kapp," *Allg. deutsche Biographie,* 51 (1906), 31–33.

30. Girardin, Paul, and Jean Brunhes: "Elisée Reclus' Leben und Wirken," *Geogr. Ztschr.,* 12 (1906), 65–79.

31. Ratzel, Friedrich: "Oscar Peschel," *Allg. deutsche Biographie,* 25 (1887), 416–430. Repub. in Ratzel, *Kleine Schriften,* Munich, 1906, I, 429–447.

32. Hettner, Alfred: "Ferdinand von Richthofens Bedeutung für die Geographie," *Geogr. Ztschr.,* 12 (1906), 1–11.

33. Ule, Willi: "Alfred Kirchhoff," *Geogr. Ztschr.,* 13 (1907), 537–552.

34. Steffen, Hans: "Erinnerungen an Alfred Kirchoff als Methodiker und Universitätslehrer," *Geogr. Ztschr.,* 25 (1919), 289–302.

35. Helmodt, Hans: "Friedrich Ratzel, ein Lebensabriss von ihm selbst und vom Herausgeber," in Ratzel, *Kleine Schriften,* Munich, 1906 (H. Helmodt, editor), I, xxi–xxiii.

36. Hassert, Kurt: "Friedrich Ratzel, Sein Leben und Wirken," *Geogr. Ztschr.*, 11 (1905), 305–325, 361–380.
37. Sapper, Karl: "Georg Gerland," *Geogr. Ztschr.*, 25 (1919), 329–340.

METHODOLOGY, WITH ILLUSTRATIVE WORKS *

38. *The Geography of Strabo.* Trans. by H. C. Hamilton and W. Falconer, London, 1892.
39. Kant, Immanuel: *Vorkritische Schriften,* in Kant's Gesammelte Schriften, Berlin Academy of Sciences edition, Bd. I, II (1902, 1905). (Includes various short studies on the earth and its movements, earthquakes, and winds, and statements of his program for the course in "Physical Geography" for various years, originally published 1754–1765, Bd. I, 183–204, 417–472, 489–503; II, 1–12, 312 f., 443. See also *Reflexionen zur physischen Geographie,* edited by E. Adickes from manuscripts, Bd. XIV (1911), 539–635.)
40. *Immanuel Kant's physische Geographie,* edited by F. T. Rink. First published Königsberg, 1802, evidently from manuscripts of 1775 and 1759 (see footnote 3 in text). Republished in various editions of Kant's works, particularly in edition pub. by Berlin Academy of Sciences, Kant's Gesammelte Schriften, Bd. IX (1923), 151–436, with notes by Paul Gedan, 509–568.
41. Forster, John Reinhold: *Observations made during A Voyage Round the World, on Physical Geography, Natural History and Ethic Philosophy.* London, 1778.
42. Humboldt, Alexander v.: *Florae Fribergensis Specimen.* Berlin, 1793. (Long footnote on the division of the sciences concerned with nature, ix–x.)
43. *idem: Ideen zu einer Physiognomik der Gewächse.* (Brochure), Tübingen, 1806.
44. *idem: Ideen zu einer Geographie der Pflanzen nebst einem Naturgemälde der Tropenländer.* Tübingen, 1807.
45. *idem: Ansichten der Natur: mit wissenschaftlichen Erläuterungen.* Stuttgart, 1808, 1849.
46. *idem:Essai politique sur le royaume de la nouvelle Espagne.* Paris, 1811.
47. *idem: Relation historique du Voyage aux régions équinoxiales du Nouveau Continent.* 3 vols.; Paris, 1814–25.
48. *Alexander von Humboldts Natur- und Kulturschilderungen,* ausgewählt und eingeleitet von Karl H. Dietzel. Leipzig, 1923. [Selections chiefly from the three preceding works.]
49. Ritter, Carl: *Die Erdkunde, im Verhältniss zur Natur und zur Geschichte des Menschen, oder allgemeine, vergleichende Geographie als sichere Grundlage des Studiums und Unterrichts in physikalischen und historischen Wissenschaften.* 19 vols. (2nd. ed. of Vols. I. II), Berlin, 1822–59. [Introduction, I, 1–88, republished in *50.*]
50. *idem: Einleitung zur allgemeinen vergleichenden Geographie, und Abhandlungen zur Begründung einer mehr wissenschaftlichen Behandlung der Erdkunde.* Berlin, 1852. Republication of "Einleitung," and "Allgemeine Vorbemerkungen über die festen Formen der Erdrinde," both from the *Erdkunde* (1817); and five lectures given before the Acad. d. Wissenschaften, Berlin, 1826, 1828, 1833, 1836, 1850, and previously published in *Abhandlung d. kgl. Akad. . . . (hist.-phil. Kl.).*

[4]

* See Supplementary Note 1

51. Bucher, August Leopold: *Von den Hindernissen, welche der Einführung eines besseren Ganges beym Vortrage der Erdkunde auf Schulen im Wege stehen.* Cöslin, 1827.

52. Humboldt, Alexander v.: *Vorlesungen über physikalische Geographie nebst Prolegomenen über Stellung der Gestirne, Berlin im Winter 1827–28.* Ed. by Miron Goldstein, Berlin, 1934.

53. Fröbel, Julius: *Geographisch-statistische Beschreibung von Ober-und Nieder-Peru, Argentinien, Uruguay und Paraguay.* Vol. 20 of Handbuch der neuesten Erdbeschreibung, ed. by A. C. Gaspari, GuthsMuts and others. Weimar, 1831–32.

54. *idem:* "Einige Blicke auf den jetzigen formellen Zustand der Erdkunde," *Annalen der Erd-, Völker- und Staatenkunde* (Berghaus Annalen), 4 (1831), 493–506.

55. Ritter, Carl: "Carl Ritter's Schreiben an Heinrich Berghaus, in Beziehung auf den vorstehenden Aufsatz des Herrn Julius Fröbel," *Annalen der Erd-, Völker- und Staatenkunde* (Berghaus Annalen), 4 (1831), 506–520.

56. Fröbel, Julius: "Ueber die Unterscheidung einer Erdkunde als eigentlicher Naturwissenschaft und einer historischen Erdkunde," *Annalen der Erd-, Völker- und Staatenkunde* (Berghaus Annalen), 6 (1832), 1–10.

57. *idem:* "Entwurf eines Systemes der geographischen Wissenschaften," *Mittheil. aus d. Gebiete d. Theoretischen Erdkunde* (Zurich), 1 (1836), 1–35, 121–132. Discussed by H. Wagner, *Geogr. Jahrb.* 7 (1878), 621 f.

58. *idem:* "Ueber den orographischen Begriff des Gebirges, mit Andeutungen zu einer reinen Hypsographie," *Mittheil. aus d. Gebiete d. Theoretischen Erdkunde* (Zurich), 1 (1836), 469–481,

59. Humboldt, Alexander v.: *Asie Centrale: Recherches sur les Chaines de Montagnes et la climatologie comparée.* 2 vols., Paris, 1843.

60. *idem: Kosmos: Entwurf einer physischen Weltbeschreibung.* 5 vols., Stuttgart, 1845–62.

61. Ritter, Carl: *Allgemeine Erdkunde.* Vorlesungen an d. Univ. Berlin, hrsg. v. H. A. Daniel, Berlin, 1862.

62. Fröbel, Julius: *Aus Amerika: Erfahrungen, Reisen und Studien.* 2 vols., Leipzig, 1857–58. Rewritten as *Seven Years Travel in Central America, Northern Mexico, and the Far West of the United States.* London, 1859. Review of former by Neumann, in *Ztschr. f. allg. Erdkunde,* 1858, 83 ff.; see also review by same of Fröbel's study of German immigration in America, *op. cit.,* 1860.

63. *idem: Amerika, Europa und die politischen Gesichtspunkte der Gegenwart.* Berlin, 1859.

64. Guyot, Arnold H.: *The Earth and Man.* (Trans. from French), Boston, New York, 1863.

65. Brown, Ralph H.: "Arnold Guyot's Notes on the Southern Appalachians," *Geogr. Rev.,* 29 (1939), 157 f.

66. Peschel, Oscar: *Abhandlungen zur Erd- und Völkerkunde.* 3 vols., Leipzig, 1877. Republication, after his death, of articles published 1854–75.

67. *idem: Neue Probleme der Vergleichenden Erdkunde als Versuch einer Morphologie der Erdoberfläche.* Leipzig, 1870, 1878. In part previously published in *Ausland,* 1866 and following years.

68. Spörer, Julius: "Zur historischen Erdkunde: Ein Streifzug durch das Gebiet der geographischen Literatur," *Geogr. Jahrb.,* 3 (1870), 326–420. (Bibliography.)

69. Richthofen, Ferdinand Frh. v.: *China.* Vol. I, Berlin, 1877.
70. Marthe, F.: "Begriff, Ziel und Methode der Geographie," *Ztschr. d. Ges. f. Erdk., Berlin,* 1877, 422–467. Abstract in *Geogr. Jahrb.,* 7 (1878), 628.
71. Wagner, Hermann: "Bericht über die Methodik der Erdkunde," *Geogr. Jahrb.,* 9 (1882), 651–700.
72. Ratzel, Friedrich: *Anthropogeographie.* Stuttgart, 1. Teil, 1882, 1899; 2. Teil, 1891, 1912.
73. Richthofen, Ferdinand Frh. v.: *Aufgaben und Methoden der heutigen Geographie.* Akad. Antrittsrede, Leipzig, 1883.
74. Wimmer, J.: *Historische Landschaftskunde.* Innsbruck, 1885.
75. Wagner, Hermann: "Bericht über die Methodik der Erdkunde (1883–1885)," *Geogr. Jahrb.,* 10 (1885), 607–610.
76. Gerland, Georg: "Vorwort des Herausgebers," ("Die wissenschaftliche Aufgabe der Geographie, ihre Methode und ihre Stellung im praktischen Leben," title according to Ratzel), *Beiträge zur Geophysik,* 1 (1887), i–liv.
77. Wagner, Herman: "Bericht über die Entwickelung der Methodik und des Studiums der Erdkunde (1885–1888)," *Geogr. Jahrb.,* 12 (1888), 418–444.
78. Supan, Alexander: "Über die Aufgaben der Spezialgeographie und ihre gegenwärtige Stellung in der geographischen Literatur," *Peterm. Mitt.,* 35 (1889), 153–157.
79. Vidal de la Blache, Paul: *États et nations de l'Europe autour de France.* Paris, 1889.
80. Wagner, Hermann: "Bericht über die Methodik der Erdkunde," *Geogr. Jahrb.,* 14 (1890), 371–399. (The last of a series of critical surveys of methodological literature, by Wagner, in Vols. 7, 8, 9, 10, 12, and 14.)
81. Richthofen, Ferdinand Frh. v.: "Antrittsrede"; *Sitzg. Ber., Akad. d. Wissensch. (phys.-math. Kl.)* Berlin, 1899, xxxii (603–607).
82. Richter, Edward: *Die Grenzen der Geographie.* Rektoratsrede, Graz Univ., 1900.

B. GEOGRAPHIC THOUGHT IN THE TWENTIETH CENTURY

GENERAL SURVEYS OF GEOGRAPHIC WORK

83. Brunhes, Jean: "Human Geography"; chapter in *History and Prospects of the Social Sciences,* H. E. Barnes, ed., New York, 1925.
84. Sauer, Carl: "Recent Developments in Cultural Geography," Chap. 4 of *Recent Developments in the Social Sciences,* E. C. Hayes, ed., Philadelphia, 1927, pp. 154–212. (Bibliography.)
85. *idem:* "Cultural Geography," in *Encycl. of Social Sciences,* 6 (1931), 621–623.
86. Vallaux, Camille: "Human Geography," in *Encycl. of Social Sciences,* 6 (1931), 624–626.
87. Sapper, Karl: "Economic Geography," in *Encycl. of Social Sciences,* 6 (1931), 626–628.
88. Joerg, W. L. G.: "Recent Geographical Work in Europe," *Geogr. Rev.,* 12 (1922), 431–483. (Bibliography.)
89. Vogel, Walther: "Stand und Aufgaben der historisch-geographischen Forschung in Deutschland," *Peterm. Mitteil.,* Erg., H. 209 (1930), 346–360. Bibliography.)

90. Penck, Albrecht: on Alfred Hettner's importance in the development of geography, "Review" of Hettner's *Vergleichende Länderkunde*, in *Deutsche Literaturzeitung*, (I), 1935, 38–43; (II), 1936, 31–39.
91. Krebs, Norbert: "Der Stand der deutschen Geographie," *Geogr. Ztschr.*, 44 (1938), 241–249.
92. Oestreich, Karl: "Die neueren Strömungen in der niederländischen Geographie," *Geogr. Ztschr.*, 44 (1938), 289–297.
93. Musset, R.: "Der Stand der Geographie und ihre neueren wissenschaftlichen Strömungen in den Ländern französischer Zunge," *Geogr. Ztschr.*, 44 (1938), 269–277. (Bibligraphy.)
94. Migliorini, Elio: "Die heutigen neuen Strömungen in der italienischen Geographie," *Geogr. Ztschr.*, 44 (1938), 277–283.
95. Galon, Rajmund: "Die Geographie in Polen, ihre Fortschritte und Ziele," *Geogr. Ztschr.*, 44 (1938), 297–306. (Bibliography.)
96. Ahlmann, Hans W., and Nils Friberg: "Neue Strömungen in der nordischen geographischen Forschung," *Geogr. Ztschr.*, 44 (1938), 307–315. (Bibliography.)
97. Berg, Lev Simonovich: *Outline of the History of Geography in Russia* (in Russian). Reference based on review in *Geogr. Ztschr.*, 36 (1930), 103.
98. Sölch, Johann: "Der geographische Unterricht in England," *Geogr. Ztschr.*, 31 (1925), 26 ff.
99. *idem:* "Die Verknüpfung von Geographie und Gesellschaftslehre in England," *Geogr. Ztschr.*, 36 (1930), 145–157.
100. Huender, W. J.: *De Englesche Geographie en de 20ste Euw*. Utrecht, 1934. Synopsis in two pages in English. Review by J. Sölch in *Geogr. Ztschr.*, 42 (1936), 27.
101. Dickinson, Robert E.: "Die gegenwärtigen Strömungen der britischen Geographie," *Geogr. Ztschr.*, 44 (1938), 258–269. (Bibliography.)
102. Davis, William Morris: "The Progress of Geography in the United States," *Ann. Assn. Am. Geogr.*, 14 (1924), 159–215.
103. Johnson, Douglas: "The Geographic Prospect," *Ann. Assn. Am. Geogrs.*, 19 (1929), 167–231.
104. Davis, William Morris: "A Retrospect of Geography," *Ann. Assn. Am. Geogrs.*, 22 (1932), 211–230.
105. Parkins, Almon E.: "The Geography of American Geographers," *Journal of Geog.*, 33 (1934), 221–230.
106. Bowman, Isaiah: *Geography in Relation to the Social Sciences.* New York, 1934. (Bibliography.)
107. Colby, Charles C.: "Changing Currents of Geographic Thought," *Ann. Assn. Am. Geogrs.*, 26 (1936), 1–37. (Bibliography.)
108. Broek, J. O. M.: "Neuere Strömungen in der amerikanischen Geographie," *Geogr. Ztschr.*, 44 (1938), 249–258. (Bibliography.)
109. Pfeifer, Gottfried: "Entwicklungstendenzen in Theorie und Methode der regionalen Geographie in den Vereinigten Staaten, nach dem Kriege," *Ztsch. d. Ges. f. Erdk., Berlin*, 1938, 93–125. (Bibliography.) Abstract by J. Leighly, *Geogr. Rev.*, 28 (1938), 679. (A complete translation by Leighly has been distributed in mimeograph form by the American Geographical Society.)
110. Inouyé, Syuzi: "Die japanische Geographie der letzten zehn Jahre," *Geogr. Ztschr.*, 44 (1938), 284–289.

PHILOSOPHY OF SCIENCE, AND OF NON-GEOGRAPHIC FIELDS

111. Hettner, Alfred: Das System der Wissenschaften," *Preuss. Jahrbücher,* 122 (1905), 251–277 [in part in *161,* 110–117].

112. Wundt, Wilhelm: *Logik.* Stuttgart, 1907. Vol. II. "Logik der exakten Wissenschaften."

113. Lehmann, Otto: "Über die Stellung der Geographie in der Wissenschaft," *Vierteljahrsschr. d. Naturforsch. Ges. in Zürich,* 81 (1936), 217–239.

114. Barry, Frederick: *The Scientific Habit of Thought: An Informal Discussion of the Source and Character of Dependable Knowledge.* New York, 1927.

115. Cohen, Morris: *Reason and Nature: An Essay on the Meaning of the Scientific Method.* New York, 1931.

116. Kroeber, Alfred Louis: "History and Science in Anthropology," *American Anthropologist,* 37 (1935), 539–569.

117. Johnson, Allen: *The Historian and Historical Evidence.* New York, 1934.

118. Steefel, Lawrence D.: "History," Chap. vii in *Man and Society: A Substantive Introduction to the Social Sciences,* E. P. Schmidt, ed., New York, 1937, 305–322.

119. Shotwell, James T.: "History," Encyc. Britt., 14th ed.

120. Barnes, Harry Elmer: "History," Encycl. Amer., rev. ed., 1938.

THE METHODOLOGY OF GEOGRAPHY, IN GENERAL

IN GERMAN-SPEAKING COUNTRIES *

121. Hettner, Alfred: "Geographische Forschung und Bildung," *Geogr. Ztschr.,* 1 (1895), 1–19. [Concepts later developed in *126* and *161.*]

122. Schlüter, Otto: "Bemerkungen zur Siedlungsgeographie," *Geogr. Ztschr.,* 5 (1899), 65–84, especially, 65–68.

123. Hettner, Alfred: "Grundbegriffe und Grundsätze der physischen Geographie," *Geogr. Ztschr.,* 9 (1903), 21–40, 121–139, 193–213. [Largely repeated in *161,* 215–343.]

124. Oberhummer, Eugen: *Die Stellung der Geographie zu den historischen Wissenschaften.* (Brochure), 1904.

125. Penck, Albrecht: "Die Physiographie als Physiogeographie in ihren Beziehungen zu anderen Wissenschaften," *Geogr. Ztschr.,* 11 (1905), 1–20.

126. Hettner, Alfred: "Das Wesen und die Methoden der Geographie," *Geogr. Ztschr.,* 11 (1905), 545–564, 615–629, 671–686. [Included in *161,* 110–132, etc.]

127. Schlüter, Otto: *Die Ziele der Georgraphie des Menschen.* Antrittsrede, Munich, 1906.

128. Penck, Albrecht: *Beobachtung als Grundlagen der Geographie.* Two addresses, Berlin, 1906.

129. *idem:* "Antrittsrede," *Sitzg. Ber., Akad. d. Wissensch. (phys.-math. Kl.),* Berlin, 1907: xxxiii (634–641).

130. Hettner, Alfred: "Die Geographie des Menschen," *Geogr. Ztschr.,* 13 (1907), 401–425. [Included in *161,* 266–273.]

131. Schlüter, Otto: "Über das Verhältnis von Natur und Mensch in der Anthropogeographie," *Geogr. Ztschr.,* 13 (1907), 505–517. Discussion by A. Hettner, 580–583.

132. Hettner, Alfred: "Methodische Streifzüge," *Geogr. Ztschr.*, 13 (1907), 627–632, 694–699; 14 (1908), 561–568.

133. Banse, Ewald: "Geographie," *Peterm. Mitt.*, 58 (1912), 1–4, 69–74, 128–131.

134. Schlüter, Otto: "Die Erde als Wohnraum des Menschen," in Rothe, K. C., and E. Weyrich, *Der moderne Erdkunde-Unterricht*, Wien, Leipzig, 1912, 379–430.

135. Hahn, Friedrich: "Methodische Untersuchungen über die Grenzen der Geographie (Erdbeschreibung) gegen die Nachbarwissenschaften," *Peterm. Mitt.*, 60 (1914), 1–4, 65–68, 121–124.

136. Gradmann, Robert: "Geographie und Landeskunde," *Geogr. Ztschr.*, 21 (1915), 700–704.

137. Penck, Albrecht: "Der Krieg und das Studium der Geographie," *Ztschr. d. Ges. f. Erdk.*, Berlin, 1916; 158–176, 222–248.

138. Merz, Al.: "Die Heidelberger Tagung deutscher Hochschullehrer der Geographie, 26–27..April 1916," *Ztschr. d. Ges. f. Erdk.*, Berlin, 1916, 392–408.

139. Hellpach, Willy: *Die geopsychischen Erscheinungen.* Leipzig, 1917, 1923. Revised, 1935, as *Geopsyche.*

140. Hettner, Alfred: "Die allgemeine Geographie und ihre Stellung im Unterricht," *Geogr. Ztschr.*, 24 (1918), 172–178. [Included in *161*, 122–132].

141. Hassinger, Hugo: "Über einige Aufgaben geographischer Forschung und Lehre," *Kartogr. u. Schulgeogr. Ztschr.*, 1919, 1–12.

142. Hettner, Alfred: "Die Einheit der Geographie in Wissenschaft und Unterricht," *Geogr. Abende*, im Zentralinst. f. Erzhg. u. Unterr., Heft 1, Berlin, 1919.

143. Philippson, Alfred: "Die Lehre vom Formenschatz der Erdoberfläche als Grundlage für die geographische Wissenschaft," *Geogr. Abende*, Heft 2, Berlin 1919.

144. Gradmann, Robert: "Pflanzen und Tiere im Lehrgebäude der Geographie," *Geogr. Abende*, Heft 4, Berlin 1919.

145. Schlüter, Otto: "Stellung der Geographie des Menschen in der erdkundlichen Wissenschaft," *Geogr. Abende*, Heft 5, Berlin 1919.

146. Branca, W., and Em. Kayser: "Zu welchen schweren Schäden führt eine übertriebene Betonung der Geologie in der Geographie?" *Ztschr. d. dtsch. geol. Ges.*, 71 (1919), B 30–44.

147. Penck, Albrecht: "Zu welchen schweren Schäden führt eine übertriebene Betonung der Geologie in der Geographie?" *Ztschr. d. dtsch. geol. Ges.*, 72 (1920), 124–138. [Reply to *146*.]

148. Schlüter, Otto: "Die Erdkunde in ihrem Verhältnis zu den Natur- und Geisteswissenschaften," *Geogr. Anzeig.*, 21 (1920), 145–152, 213–218.

149. Philippson, Alfred: *Grundzüge der allgemeinen Geographie.* Leipzig, 1920, 1933. Especially pp. 1–16.

150. Leutenegger, Albert: *Begriff, Stellung und Einteilung der Geographie.* 1922.

151. Volz, Wilhelm: "Das Wesen der Geographie in Forschung und Darstellung," Antrittsrede, Leipzig 1923, *Schles. Jahrb. f. Geistes- u. Naturwissensch.*, 1 (1923), 239–272.

152. Hettner, Alfred: "Methodische Zeit- und Streitfragen," *Geogr. Ztschr.*, 29 (1923), 37–59.

153. Heiderich, Franz: "Geographisch-methodische Streiflichter," in *Sieger-Festschrift*, Wien, 1924, 212–222.

154. Schlüter, Otto: "Staat, Wirtschaft, Volk, Religion in ihrem Verhältnis zur Erdoberfläche," *Ztschr. f. Geopolitik*, 1 (1924), 378–383, 432–443.

155. Braun, Gustav: *Zur Methode der Geographie als Wissenschaft.* (Brochure), Greifswald, 1925. Also pub. as Erg. H. z. 17/38 *Jahresbericht d. geogr. Ges., Greifswald.*

156. Graf, Otto: *Vom Begriff der Geographie.* Munich and Berlin, 1925 (Bibliography). Review by A. Hettner, *Geogr. Ztschr.*, 32 (1926), 304–306, with reply by Graf and counter-reply by Hettner, *Geogr. Ztschr.*, 33 (1927), 341–344.

157. Maull, Otto: *Politische Geographie.* Berlin, 1925. (Bibliography.) Review by O. Schlüter, *Geogr. Anz.*, 27 (1926), 62–66; reply by Maull, 245–253; and review by R. Sieger, *Geogr. Ztschr.*, 32 (1926), 379 f.

158. Penck, Albrecht: "Geographie und Geschichte," *Neue Jahrbücher f. Wissensch. u. Jugendbildg.*, 2 (1926), 47–54. Abstract in *Ztschr. d. Ges. f. Erdk.*, Berlin, 60 (1925), 384 f.

159. *idem:* "Geography among the Earth's Sciences," *Proceed. Amer. Philos. Soc.*, 66 (1927), 621–644.

160. Tiessen, Ernst: "Die Eingrenzung der Geographie," *Peterm. Mitt.*, 73 (1927), 1–9.

161. Hettner, Alfred: *Die Geographie, ihre Geschichte, ihr Wesen und ihre Methoden.* Breslau, 1927. (Contains most of the material published in his numerous articles in the *Geogr. Ztschr.* from 1895 to 1927.)

162. Penck, Albrecht: "Die Geographie unter den erdkundlichen Wissenschaften," *Die Naturwissenschaften*, 16 (1928), 33–41.

163. *idem:* "Neuere Geographie," *Ztschr. d. Ges. f. Erdkunde*, Berlin, Sonderband 1928, 31–56.

164. Pfeifer, Gottfried: "Uber raumwirtschaftliche Begriffe und Vorstellungen und ihre bisherige Anwendung in der Geographie und Wirtschaftswissenschaft," *Geogr. Ztschr.*, 34 (1928), 321–340, 411–425.

165. Hassinger, Hugo: "Über einige Beziehungen der Geographie zu den Geschichtswissenschaften," *Jahrb. f. Landeskunde v. Niederösterreich*, 21 (1928), Festschrift f. O. Redlich, 3–29.

166. Kraft, Victor: "Die Geographie als Wissenschaft," in *Enzykl. d. Erdk.*, Teil: Methoden d. Geographie, Leipzig, Vienna, 1929, 1–22.

167. Hettner, Alfred: "Methodische Zeit- und Streitfragen: Neue Folge," *Geogr. Ztschr.*, 35 (1929), 264–286, 332–345.

168. *idem:* "Unsere Auffassung von der Geographie" (addressed to Alfred Philippson), *Geogr. Ztschr.*, 35 (1929), 486–491.

169. Schlüter, Otto: "Über die Aufgaben der Verkehrsgeographie im Rahmen der 'reinen' Geographie," *Peterm. Mitt.*, Erg. Heft 209 (1930), 298–309.

170. Ule, Willi: "Methoden der geographischen Forschung," in *Handbuch d. biolog. Arbeitsmethoden*, Abt. X (1930), 485–528.

171. Hettner, Alfred: "Die Geographie als Wissenschaft und als Lehrfach," *Geogr. Anz.*, 32 (1931), 107–117.

172. Passarge, Siegfried: "Aufgaben und Methoden der politischen Geographie," *Ztschr. f. Politik*, 21 (1931), 443–460.

173. Lautensach, Hermann: "Wesen und Methoden der geographischen Wissenschaft," in *Klute's Handbuch der geographischen Wissenschaft*, Potsdam, 1933, I, 23–56.

174. Schrepfer, Hans: "Einheit und Aufgaben der Geographie als Wissenschaft," in J. Peterson and H. Schrepfer, *Die Geographie vor neuen Aufgaben*. Frankfort a. M., 1934.

175. Hettner, Alfred: "Neue Angriffe auf die heutige Geographie," *Geogr. Ztschr.*, 40 (1934), 341–343, 380–383.

176. *idem*: "Gesetzmässigkeit und Zufall in der Geographie," *Geogr. Ztschr.*, 41 (1935), 2–15.

177. Plewe, Ernst: "Randbemerkungen zur geographischen Methodik," *Geogr. Ztschr.*, 41 (1935), 226–237.

178. Obst, Erich: "Zur Auseinandersetzung über die zukünftige Gestaltung der Geographie," *Geogr. Wochenschr.*, 3 (1935), 1–16.

179. Maull, Otto: "Allgemeine vergleichende Länderkunde in *Länderkundliche Forschung*," *Krebsfestschrift*, Stuttgart, 1936, 172–186.

180. Schmidt, Peter Heinrich: *Philosophische Erdkunde: Die Gedankenwelt der Geographie unde ihre nationalen Aufgaben*. Stuttgart, 1937.

181. Lehmann, Otto: *Der Zerfall der Kausalität und die Geographie*. (Brochure published by author), Zurich, 1937.

IN FRANCE

182. Brunhes, Jean: *La géographie humaine*. Paris, 1910, 1925. Translated as *Human Geography*. Chicago, 1920.

183. Vidal de la Blache, Paul: "Les caractères distinctifs de la géographie," *Ann. de Géogr.*, 22 (1913), 289–299.

184. *idem: Principes de géographie humaine*. Ed. by Emm. de Martonne, Paris, 1921. Translated as *Principles of Human Geography*, New York, 1926.

185. Febvre, Lucien: *Le terre et l'évolution humaine*, in the series, "L'Evolution de l'Humanité." Paris 1924. Translated as *A Geographical Introduction to History*, in the series, The History of Civilization. New York, 1925. (Bibliography.)

186. Vallaux, Camille: *Les sciences géographiques*. Paris, 1925.

187. Ancel, Jacques: *Geopolitique*. (Brochure), Paris, 1938.

IN OTHER CONTINENTAL COUNTRIES

188. Almagià, Roberto: "Problemi e indirizzi attuali della Geografia," *Atti d. Soc. Italiana per il Progresso d. Scienze* (Pavia), 1929, 3–33.

189. Michotte, P.: "L'Orienation nouvelle en géographie"; *Bull. de la Soc. R. Belge de Géogr.*, 1921, 5–43.

190. DeGeer, Sten: "On the Definition, Method, and Classification of Geography," *Geografiska Annaler*, 1923, 1–37.

191. Markus, Eduard: *Geographische Kausalität*. Tartu, 1936. (Bibliography.)

IN GREAT BRITAIN

192. Chisholm, George G.: "The Meaning and Scope of Geography" (Inaugural address, Edinburgh), *Scott. Geogr. Mag.*, 24 (1908), 561–575.

193. Unstead, John Frederick: "Geography and Historical Geography," *Geogr. Journal*, 49 (1922), 55–59; a review of Carl Sauer, *Geography of the Ozark Highland of Missouri*.

194. Fairgrieve, James: *Geography in School.* London, 1926, 1933.
195. Roxby, Percy M.: "The Scope and Aims of Human Geography," *Scott. Geog. Mag.,* 46 (1930), 276–289.
196. Mackinder, Halford John: "The Human Habitat," *Scott. Geog. Mag.,* 47 (1931), 321–335.
197. "What is Historical Geography?" report of a joint meeting of British geographers and historians, *Geography,* 17 (1932), 39–45.
198. Gilbert, Edmund William: "What is Historical Geography?" *Scott. Geogr. Mag.,* 48 (1932), 129–136.
199. East, William Gordon: "The Nature of Political Geography," *Politica,* 2 (1937), 259–286.
200. Stamp, Josiah C.: "Geography and Economic Theory," *Geography,* 22 (1937), 1–14.
201. Crowe, P. R.: "On Progress in Geography," *Scott. Geogr. Mag.,* 54 (1938), 1–19.
202. Dickinson, Robert E.: "Landscape and Society," *Scott. Geogr. Mag.,* 55 (1939), 1–14, with comment by P. R. Crowe, 14 f.

IN THE UNITED STATES

203. Davis, William Morris: "An Inductive Study of the Content of Geography," *Bull. Am. Geogr. Soc.,* 38 (1906), 67–84.
204. Semple, Ellen Churchill: *Influences of Geographical Environment,* New York, 1911. (Bibliography.)
205. Brigham, Albert Perry: "Problems of Geographical Influence," *Ann. Assn. Am. Geogrs.,* 5 (1915), 3–25.
206. Fenneman, Nevin: "The Circumference of Geography," *Ann. Assn. Am. Geogrs.,* 9 (1919), 3–12.
207. Dryer, Charles Redway: "Genetic geography: The Development of the Geographic Sense and Concept," *Ann. Assn. Am. Geogrs.,* 10 (1920), 3–16.
208. Barrows, Harlan H.: "Geography as Human Ecology," *Ann. Assn. Am. Geogrs.,* 13 (1923), 1–14.
209. Sauer, Carl: "The Survey Method in Geography and its Objectives," *Ann. Assn. Am. Geogrs.,* 14 (1924), 17–33.
210. Thomas, Franklin: *The Environmental Basis of Society: A Study in the History of Sociological Thought.* New York, 1925. (Bibliography.)
211. Sauer, Carl: "The Morphology of Landscape," *Univ. Calif. Publs. in Geog.,* 2 (1925), 19–53.
212. Whitbeck, Ray H.: "Adjustments to Environment in South America: An Interplay of Influences," (long introductory footnote), *Ann. Assn. Am. Geogrs.,* 16 (1926), 1 ff.
213. Huntington, Ellsworth: "The Terrestrial Canvas," Chap. 1 of his *The Human Habitat.* New York, 1927.
214. Turner, Frederick J.: *The Significance of Sections in American History.* New York, 1932. "Introduction."
215. Whitbeck, Ray H., and Thomas, Olive J.: *The Geographic Factor: Its Role in Life and Civilization.* New York, 1932. (Bibliography.)
216. Hartshorne, Richard: "Recent Developments in Political Geography," *Amer. Pol. Sci. Rev.,* 29 (1935), 785–804, 943–966. (Bibliography.)

217. Whittlesey, Derwent: "Geography," *Thirty-sixth Yearbook, Nt. Soc. for the Study of Educ.*, (1937), II, 119–125.
218. Hartshorne, Richard: "Geography for What?" (editor's title), *Social Education*, 1 (1937), 166–172.
219. Huntington, Ellsworth: "Geography and History," *Canadian Journal of Econ. and Pol. Science*, 3 (1937), 565–572 [a review of *389*].
220. Leighly, John: "Some Comments on Contemporary Geographic Methods," *Ann. Assn. Am. Geogrs.*, 27 (1937), 125–141.
221. Platt, Robert S.: "Items in the Regional Geography of Panamá: With Some Comments on Some Comments on Contemporary Geographic Method," *Ann. Assn. Am. Geogrs.*, 28 (1938), 13–36.
222. Leighly, John: "Methodologic Controversy in Nineteenth Century German Geography," *Ann. Assn. Am. Geogrs.*, 28 (1938), 238–258.
223. Finch, Vernor C.: "Geographical Science and Social Philosophy," *Ann. Assn. Am. Geogrs.*, 29 (1939), 1–28.
224. Platt, Robert S.: "Reconnaissance in British Guiana, with Comments on Microgeography," *Ann. Assn. Am. Geogrs.*, 29 (1939), 105–126.

THE THEORY OF REGIONS, LANDSCHAFTEN, LANDSCAPES, AND BOUNDARIES

IN EUROPE

225. Hassinger, Hugo: "Das geographische Wesen Mitteleuropas nebst einigen grundsätzlichen Bemerkungen über die geographischen Naturgebiete Europas und ihre Begrenzung," *Mitt. d. k. k. Geogr. Ges., Wien*, 60 (1917), 437–493.
226. Penck, Albrecht: *Über politische Grenzen.* Rektoratsrede, Berlin, 1917.
227. Sieger, Robert: "Zur politisch-geographischen Terminologie," *Ztschr. d. Ges. f. Erdk., Berlin*, 52 (1917), 497–529; 53 (1918), 48–69.
228. Maull, Otto: "Geographische Staatsstruktur und Staatsgrenzen," *Kartogr. u. Schulgeogr. Ztschr.*, 8 (1919), 129–136.
229. Passarge, Siegfried: *Vergleichende Landschaftskunde*, Heft 1. Berlin, 1921.
230. Friedrichsen, Max: "Die geographische Landschaft," *Geogr. Anz.*, 22 (1921), 154–161, 233–240.
231. Bluntschli, Hans: "Die Amazonasniederung als harmonischer Organismus," *Geogr. Ztschr.*, 27 (1921), 49–67.
232. Sieger, Robert: "Natürliche Räume und Lebensräume," *Peterm. Mitt.*, 69 (1923), 252–256.
233. Sidaritsch, Marian: "Landschaftseinheiten und Lebensräume in den Ostalpen," *Peterm. Mitt.*, 69 (1923), 256–261.
234. Krebs, Norbert: "Natur- und Kulturlandschaft," *Ztschr. d. Ges. f. Erdk., Berlin*, 58 (1923), 81–94.
235. Younghusband, Francis: *Das Herz der Natur.* Leipzig, 1923.
236. Gradmann, Robert: "Das harmonische Landschaftsbild," *Ztschr. d. Ges. f. Erdk., Berlin*, 59 (1924), 129–147. Reply by Passarge, and counter-reply, 331–337.
237. Sölch, Johann: *Die Auffassung der "natürlichen Grenzen" in der wissenschaftlichen Geographie.* (Brochure), Innsbruck, 1924. (Bibliography.) Review by R. Sieger, *Peterm. Mitt.*, 71 (1925), 57–59.
238. Passarge, Siegfried: *Vergleichende Landschaftskunde*, Heft 4. Berlin, 1924. Review by Sapper, *Geogr. Ztschr.*, 31 (1925), 179.

239. Markus, Eduard: "Naturkomplexe," *Sitzungsber. d. Naturf.-Ges. bei d. Univ. Dorpat*, 32 (1925), 79–94.

240. Passarge, Siegfried: "Harmonie und Rhythmus in der Landschaft," *Peterm. Mitt.*, 71 (1925), 250–252.

241. Maull, Otto: "Zur Geographie der Kulturlandschaft," in *Freie Wege vergleichender Erdkunde, Festgabe für E. von Drygalski*, München, Berlin, 1926, 11–30.

242. Hettner, Alfred: "Methodische Zeit- und Streitfragen: Passarges Landschaftskunde!," *Geogr. Ztschr.*, 31 (1925), 162–164.

243. Volz, Wilhelm: "Der Begriff des "Rhythmus" in der Geographie," *Mitt. d. Ges. f. Erdk., Leipzig*, 1923–5 (1926), 8–41.

244. Vogel, Walther: "Zur Lehre von den Grenzen und Räumen," *Geogr. Ztschr.*, 32 (1926), 191–198.

245. Granö, Johannes G.: "Die Forschungsgegenstände der Geographie," *Acta Geographica*, 1, 1–15 (Helsinki, 1927).

246. Banse, Ewald: *Landschaft und Seele: Neue Wege der Untersuchung und Gestaltung.* Munich, 1928.

247. Schlüter, Otto: "Die analytische Geographie der Kulturlandschaft erläutert am Beispiel der Brücken," *Ztschr. d. Ges. f. Erdk., Berlin*, Sonderband, 1928, 388–411.

248. Creutzburg, Nikolaus: "Über den Werdegang von Kulturlandschaften," *Ztschr. d. Ges. f. Erdk., Berlin*, Sonderband, 1928, 412–425.

249. Penck, Albrecht: "Deutschland als geographische Gestalt," in *Deutschland: die natürlichen Grundlagen seiner Kultur*, pub. by K. L. Deutsch, Akad. d. Naturfr. z. Halle, Leipzig, 1928.

250. Waibel, Leo: "Beitrag zur Landschaftskunde," *Geogr. Ztschr.*, 34 (1928), 475–486. Reply by Rudolf Ahrens and counter-reply by Weibel, *Geogr. Ztschr.*, 35 (1929), 166–170.

251. Spethmann, Hans: *Dynamische Länderkunde.* Breslau, 1928. Review by R. Gradmann, *Geogr. Ztschr.*, 34 (1928), 551–553; [discussions by Hettner, *167*, Philippson, *260;* reply by Spethmann, and counter discussion by Gradmann, *261*].

252. Granö, Johannes G.: "Reine Geographie: eine methodologische Studie beleuchtet mit Beispielen aus Finnland und Estland," *Acta Geographica*, 2, No. 2 (Helsinki, 1929). (Bibliography.) Review by H. Hassinger, *Geogr. Ztschr.*, 36 (1930), 293–296.

253. Hassinger, Hugo: Discussion of Passarge school; in a review in *Peterm. Mitt.*, 74 (1928), 371 f. (§ 435) and in discussion with Passarge, *op. cit.*, 75 (1929), 86 f.

254. Lehmann, Otto: "Länderkunde und – – – Länderkunde," *Mitt. d. Geogr Ges., Wien*, 72 (1929), 292–334.

255. Krebs, Norbert: "Revolution und Evolution in der Geographie," *Mitt. d. Geogr. Ges., Wien*, 72 (1929), 334–345. [Reply to *254*.]

256. Hassinger, Hugo: "Zum Darstellungsproblem in der Geographie," *Geogr. Ztschr.* 35 (1929), 541–546.

257. Passarge, Siegfried: *Beschreibende Landschaftskunde.* Hamburg, 1929.

258. *idem:* "Wesen, Aufgaben und Grenzen der Landschaftskunde," *Peterm. Mitt.*, Erg. Heft 209 (1930), 29–44.

259. *idem:* "Das Problem des logischen Systems der Landschaftstypen," *Naturwissensch.,* 19 (1931), 702–704.

260. Philippson, Alfred: "Methodologische Bemerkungen zu Spethmann's Länderkunde," *Geogr. Ztschr.,* 36 (1930), 1–16.

261. Spethmann, Hans: *Das länderkundliche Schema in der deutschen Geographie: Kämpfe um Fortschritt und Freiheit.* Berlin, 1931. Review by E. v. Drygalski, *Peterm. Mitt.,* 1932, 5–6; discussion by R. Gradmann, *Geogr. Ztschr.,* 37 (1931), 540–548.

262. Volz, Wilhelm: "Geographische Ganzheitlichkeit," *Ber. d. math.-phys. Kl. d. sächs. Akad. d. Wiss.,* 84 (1931), 91–113.

263. Lautensach, Hermann: "Die länderkundliche Gliederung Portugals," *Geogr. Ztschr.,* 38 (1932), 193–205, 271–284; especially 193–198.

264. Hassinger, Hugo: "Der Staat als Landschaftsgestalter," *Ztschr. f. Geopolitik,* 9 (1932), 117–122, 182–187.

265. Hettner, Alfred: "Zur aesthetischen Landschaftskunde," *Geogr. Ztschr.,* 39 (1933), 93–98.

266. Waibel, Leo: "Was verstehen wir unter Landschaftskunde?," *Geogr. Anz.,* 34 (1933), 197–207.

267. Passarge, Siegfried: "Das Problem der kulturgeographischen Räume," *Peterm. Mitt.,* 79 (1933), 1–6.

268. *idem: Einführung in die Landschaftskunde.* Leipzig, 1933. Review by C. Troll, and discussion by Passarge and Troll, *Geogr. Ztschr.,* 40 (1934), 109, 464–468.

269. Hettner, Alfred: "Der Begriff der Ganzheit in der Geographie," *Geogr. Ztschr.,* 40 (1934), 141–144.

270. Granö, Johannes G.: "Geographische Ganzheiten," *Peterm. Mitt.,* 81 (1935), 295–302.

271. Vogel, Walther: "Landschaft und Land als Raumeinheiten der Geographie," *Mitt. d. Ver. d. Ges. d. Geogr. a. d. Univ. Leipzig,* H. 14/15, 1936, pp. 1–8.

272. Passarge, Siegfried: "Versuch einer Darlegung der eigenen wissenschaftlichen Tätigkeit," *Ztschr. f. Erdk.,* 1 (1936), 49–62.

273. Hassinger, Hugo: "Die Landschaft als Forschungsgegenstand," *Schriften d. Ver. z. Verbreitg. naturwissensch. Kenntnisse in Wien,* 1937, 76–95.

274. Wörner, Rolf: "Das geographische Ganzheitsproblem vom Standpunkt der Psychologie aus," *Geogr. Ztschr.,* 44 (1938), 340–347.

275. James, Preston E.: "The Geography of the Oceans: A Review of the Work of Gerhard Schott," *Geogr. Rev.,* 26 (1936), 664–669.

276. Pawlowsky, Stanislaw: "Inwieweit kann in der Anthropogeographie von einer Landschaft die Rede sein," *Comptes rendus d. Congr. Intern. d. Géogr., Amsterdam, 1938,* Tome 2, Sec. 3a, 202–208.

277. Geisler, Walter: "Die Bedeutung der kulturmorphologischen Strukturelemente bei der Bildung des Landschaftsbegriffes," *Comptes rendus du Congrès Intern. d. Géogr. Amsterdam, 1938,* Tome 2, Sec. 5, 4–11.

278. Lautensach, Hermann: "Über die Erfassung und Abgrenzung von Landschaftsräumen," *Comptes rendus du Congrès Intern. d. Géogr. Amsterdam, 1938,* Tome 2, Sec. 5, 12–26.

279. Krebs, Norbert: "Question: La Concept Paysage dans la Géographie Humaine" (report, in German, of 8 papers, particularly of 4 papers concerned with the concept of Landschaft or landscape [*276–8* and *297*]). *Comptes rendus d. Congr. Intern. d. Géogr. Amsterdam, 1938,* Tome 2, Rapports, 207–213.

280. Bryan, Patrick W.: *Man's Adaptation of Nature: Studies of the Cultural Landscape.* London, 1933. Especially Chaps. 2–6.

IN THE UNITED STATES

281. Jones, Wellington D., and Vernor C. Finch: "Detailed Field Mapping in the study of the Economic Geography of an Agricultural Area" ("the joint conclusions of a group composed of Charles C. Colby, D. H. Davis, V. C. Finch, William H. Haas, Wellington D. Jones, A. K. Lobeck, Kenneth C. McMurry, A. E. Parkins, Robert S. Platt, and Derwent S. Whittlesey"), *Ann. Assn. Am. Geogrs.* 15 (1925), 148–157.
282. Whittlesey, Derwent: "Sequent Occupance," *Ann. Assn. Am. Geogrs.,* 19 (1929), 162–165.
283. Jones, Wellington D.: "Ratios and Isopleth Maps in Regional Investigation," *Ann. Assn. Am. Geogrs.,* 20 (1930), 177–195.
284. Whitaker, Russell: "Regional Interdependence," *Journ. of Geogr.,* 31 (1932), 164 f.
285. Finch, Vernor C.: "Montfort, a Study in Landscape Types in Southwestern Wisconsin," *Geogr. Soc. of Chicago, Bull. 9* (1933).
   "Conventionalizing Geographic Investigation and Presentation; a Symposium," three papers listed below, with discussion; *Ann. Assn. Am. Geogrs.,* 24 (1934), 77–122.
286. James, Preston E.: "The Terminology of Regional Description," *Ann. Assn. Am. Geogrs.,* 24 (1934), 77–86.
287. Jones, Wellington D.: "Procedures in Investigating Human Occupance of a Region," *Ann. Assn. Am. Geogrs.,* 24 (1934), 93–107.
288. Finch, Vernor C.: "Written Structures for Presenting the Geography of Regions," *Ann. Assn. Am. Geogrs.,* 24 (1934), 113–120.
289. Whittlesey, Derwent: "The Impress of Effective Central Authority upon the Landscape," *Ann. Assn. Am. Geogrs.,* 25 (1935), 85–97.
   "A Conference on Regions," Papers by R. B. Hall, George T. Remer, Samuel Van Valkenburg (not pub.), and Robert S. Platt, with discussion, *Ann. Assn. Am. Geogrs.,* 25 (1935), 121–174.
290. Hall, Robert Burnett: "The Geographic Region: A Resumé," *Ann. Assn. Am. Geogrs.,* 25 (1935), 122–136.
291. National Resources Committee: *Regional Factors in National Planning.* Govt. Printing Off., Washington, 1935. Especially Parts 4, 5.
292. Joerg, W. L. G.: "The Geography of North America: A History of its Regional Exposition," *Geogr. Rev.,* 26 (1936), 640–663. (Bibliography only on fairly complete works.)
293. Wright, John Kirtland: "Some Measures of Distributions," *Ann. Assn. Am. Geogrs.,* 27 (1937), 177–212.
294. James, Preston E.: "On the Treatment of Surface Features in Regional Studies," *Ann. Assn. Am. Geogrs.,* 27 (1937), 213–228. (Bibliography.)
295. "Round Table on Problems in Cultural Geography," Discussion led by W. W. Atwood, Ralph Brown, F. B. Kniffen, R. B. Hall, and Derwent Whittlesey; S. D. Dodge, Chairman. *Ann. Assn. Am. Geogrs.,* 27 (1937), 155–175.
296. Dodge, Richard E.: "The Interpretation of Sequent Occupance," *Ann. Assn. Am. Geogrs.,* 28 (1938), 233–237.

297. Broek, J. O. M.: "The Concept Landscape in Human Geography," *Comptes rendus d. Congr. Intern. d. Géogr. Amsterdam, 1938,* Tome 2, Sec. 3a, 103–109.

SYSTEMS OF REGIONAL DIVISION

298. Hahn, Eduard: "Die Wirtschaftsformen der Erde," *Peterm. Mitt.,* 1892, 8–12.
299. *idem: Von der Hacke zum Pflug, Garten und Feld: Bauern und Hirten in unserer Wirtschaft und Geschichte.* Leipzig, 1919.
300. Hettner, Alfred: "Die geographische Einteilung der Erdoberfläche," *Geogr. Ztschr.,* 14 (1908), 1–13, 94–110, 137–150. [Principles also in *161,* 118–132; system in detail in *301.*]
301. *idem: Grundzüge der Länderkunde: I. Europa.* Leipzig, 1907, completely revised, 1923, 1932. Review by J. Sölch, *Geograf. Annaler,* 1923, 323 ff. II. *Die Aussereuropäischen Erdteile.* Leipzig, 1923, 1926. (First published, in briefer form as text of *Spamer's Handatlas,* 1897.)
302. *idem: Der Gang der Kultur über die Erde.* Leipzig, 1923, 1929.
303. Gradmann, Robert: "Wüste und Steppe," *Geogr. Ztschr.,* 22 (1916), 418–441, 489–509.
304. Sieger, Robert: "Die wirtschaftsgeographische Einteilung der Erde," in *Karl Andree's Geographie des Welthandels,* Vol. 4, 2nd ed., Wien, 1921, 3–128.
305. Passarge, Siegfried: *Die Landschaftsgürtel der Erde.* Natur und Kultur, Breslau, 1923, 1929.
306. Busch, Wilhelm: *Die Landbauzonen im deutschen Lebensraum.* Stuttgart, 1936.
307. Herbertson, Andrew John: "The Major Natural Regions: An Essay in Systematic Geography," *Geogr. Journ.,* 25 (1905), 300–312.
308. *idem:* "The Higher Units: A Geographical Essay," *Scientia,* 14 (1913), 203–212.
309. Unstead, John Frederick: "A System of Regional Geography," *Geography,* 18 (1933), 175–187.
310. Committee of the Geographical Association (J. F. Unstead, J. L. Myres, P. M. Roxby, and L. D. Stamp): Report on "Classification of Regions of the World"; *Geography,* 22 (1937), 253–282.
311. Smith, Middleton, O. E. Baker, and R. G. Hainsworth: "A Graphic Summary of American Agriculture," in *Yearbook of U. S. Dept. of Agric.,* 1915. Washington, Govt. Ptg. Off., 1915.
312. Baker, Oliver E.: "Agricultural Regions of North America," *Econ. Geogr.,* intermittently 2–10 (Oct. 1926–1934). Basis in 2 (Oct. 1926), 459–493.
313. Jonasson, Olof: "Agricultural Regions of Europe," *Econ. Geogr.,* 1 (Oct., 1925), 277–314; 2 (Jan., 1926), 19–48.
314. Huntington, Ellsworth, Frank E. Williams, and Samuel Van Valkenburg: *Economic and Social Geography.* New York, 1933.
315. Jones, Wellington D., and Derwent Whittlesey: "Types of Agricultural Land Occupance." (Map.) Univ. of Chicago Bookstore, 1932.
316. *idem:* "Nomadic Herding Regions," *Econ. Geogr.,* 8 (1932), 378–385.
317. Whittlesey, Derwent: "Shifting Cultivation," *Econ. Geogr.,* 13 (1937), 35–52.
318. *idem:* "Fixation of Shifting Cultivation," *Econ. Geogr.,* 13 (1937), 139–154.
319. *idem:* "Major Agricultural Regions of the Earth," *Ann. Assn. Am. Geogrs.,* 26 (1936), 199–240. (The three preceding studies represent amplifications of parts of this study.)

320. U. S. Bureau of the Census (in co-operation with Bur. Agric. Econ.): *Types of Farming in the United States;* text by Foster F. Elliott. Govt. Print. Off., Washington, 1933.

321. James, Preston E.: *An Outline of Geography.* New York, 1935.

322. Finch, Vernor C., and Trewartha, Glenn T.: *Elements of Geography.* New York, 1936.

323. Dodge, Richard Elwood, and Stanley Dalton Dodge: *Foundations of Geography.* New York, 1937.

324. Hartshorne, Richard, and Samuel N. Dicken: "A Classification of the Agricultural Regions of Europe and North America on a Uniform Statistical Basis," *Ann. Assn. Am. Geogrs.,* 25 (1935), 99–120.

325. Hartshorne, Richard: "A New Map of the Dairy Areas of the United States," *Econ. Geogr.,* 11 (1935), 347–355.

326. *idem:* "A New Map of the Manufacturing Belt of North America," *Econ. Geogr.,* 12 (1936), 45–53.

327. *idem:* "Human Geography," Chap. viii in Man and Society: A Substantive Introduction to the Social Sciences, E. P. Schmidt, ed., New York, 1937, 323–379.

328. Hartshorne, Richard, and Samuel N. Dicken: *Syllabus in Economic Geography.* (Lithoprint, Edwards Bros.) Ann Arbor, Mich., 1933, 1938.

STUDIES FROM THE TWENTIETH CENTURY CITED AS ILLUSTRATIONS

(arranged alphabetically, by authors)

329. Allix, André: *Un pays de haute montagne: L'Oisans. Étude géographique.* Paris, 1929. Review by Demangeon, *Ann. de Géogr.,* 1930, 91.

330. Banse, Ewald: *Buch der Länder: Landschaft und Seele der Erde. I. Das Buch Abendland; II. Das Buch Fremdland.* Berlin, 1929, 1930. Review by A. Hettner, *Geogr. Ztschr.,* 34 (1928), 626 ff.

331. Blache, Jules: *Les Massifs de la Grande Chartreuse et du Vercors.* Grenoble, 1931.

332. Bowman, Isaiah: *The New World: Problems in Political Geography.* Chicago, 1921, 4th ed., 1928. (Bibliography.)

333. Broek, J. O. M.: *The Santa Clara Valley, California: A Study in Landscape Changes.* University, Utrecht, 1932.

334. Brown, Ralph H.: "Materials Bearing upon the Geography of the Atlantic Seaboard, 1790 to 1810," *Ann. Assn. Am. Geogrs.,* 28 (1938), 201–231.

335. Brunhes, Jean, and C. Vallaux: *La géographie de l'histoire: Géographie de la Paix et de la Guerre sur terre et sur Mer.* Paris, 1921.

336. Brunhes, Jean, and P. Deffontaines: "Géographie politique," Pt. 3 of J. Brunhes, *Géographie humaine de la France;* Vol. 1 of G. Hanotaux: Histoire de la Nation Francaise. Paris, 1926.

337. Colby, Charles C.: "The California Raisin Industry: A Study in Geographic Interpretation," *Ann. Assn. Am. Geogrs.,* 14 (1924), 49–108.

338. Cressey, George B.: "The Land Forms of Chekiang, China," *Ann. Assn. Am. Geogrs.,* 28 (1938), 259–276.

339. Darby, Henry Clifford: *An Historical Geography of England before 1800.* Cambridge, 1936. Review by J. Sölch, *Ztschr. d. Ges. f. Erdk.,* Berlin, 72 (1937), 212 ff.

340. Dicken, Samuel N.: "Galeana: A Mexican Highland Community," *Journal of Geogr.,* 34 (1935), 140–147.

341. Dodge, Stanley D.: "Bureau and the Princeton Community," *Ann. Assn. Am. Geogrs.,* 22 (1932), 159–209.

342. *idem:* "The Chorology of the Claremont-Springfield Region in the Upper Connecticut Valley in New Hampshire and Vermont," *Papers of the Mich. Acad. of Sci., Arts, and Letters,* 22 (1936), 335–353.

343. Finch, Vernor C., and O. E. Baker: *Geography of the World's Agriculture.* Washington, Govt. Ptg. Office, 1917.

344. *Friedrichsen Denkschrift: Vom deutschen Osten.* Hubert Knothe, ed., Berlin, 1934. Especially articles by W. Geisler and H. Schlenger. Review by R. Hartshorne, *Geogr. Rev.,* 25 (1935), 518.

345. Geisler, Walter: *Schlesien als Raumorganismus.* (Brochure), Breslau, 1932.

346. *idem: Oberschlesien-Atlas.* Berlin, 1938.

347. Garver, Frederic, F. M. Boddy, and A. J. Nixon: *The Location of Manufacturing in the United States, 1899–1929.* Univ. of Minnesota Press, 1933.

348. Hall, Robert Burnett: "Some Rural Settlement Forms in Japan," *Geogr. Rev.,* 21 (1931), 93–123.

349. *idem:* "The Yamato Basin, Japan," *Ann. Assn. Am. Geogrs.,* 22 (1932), 211–230.

350. *idem:* "Tokaido Road and Region," *Geogr. Rev.,* 27 (1937), 353–377.

351. *idem:* "A Map of Settlement Agglomeration and Dissemination in Japan," *Papers of the Mich. Acad. of Science, Arts and Letters,* 22 (1936), 365–367.

352. Hartshorne, Richard: "Location Factors in the Iron and Steel Industry," *Econ. Geogr.,* 4 (1928), 241–252.

353. *idem:* "The Iron and Steel Industry of the United States," *Journ. of Geogr.,* 28 (1929), 133–153.

354. *idem:* "The Twin City District: A Unique Form of Urban Landscape," *Geogr. Rev.,* 22 (1932), 431–442.

355. *idem:* "Geographic and Political Boundaries in Upper Silesia," *Ann. Assn. Am. Geogrs.,* 23 (1933), 195–228.

356. *idem:* "The Upper Silesian Industrial District," *Geogr. Rev.,* 24 (1934), 423–438.

357. *idem:* "Suggestions on the Terminology of Political Boundaries," *Mitt. d. Ver. d. Geogr. a. d. Univ. Leipzig,* Heft 14/15, (1936) 180–192.

358. *idem:* "The Tragedy of Austria-Hungary: A Post-Mortem in Political Geography" (abstract), *Ann. Assn. Am. Geogrs.,* 28 (1938), 49.

359. *idem:* "Racial Maps of the United States," *Geogr. Rev.,* 28 (1938), 276–288.

360. Hassinger, Hugo: *Die Geographie des Menschen.* Vol. II of Klute: Handbuch geographischer Wissenschaft. Potsdam, 1937.

361. Hettner, Alfred: *Die Oberflächenformen des Festlandes: Probleme und Methoden der Morphologie.* Leipzig, Berlin, 1921, 1928.

362. *idem: Die Klimate der Erde.* Leipzig, Berlin, 1930.

363. *idem: Vergleichende Länderkunde.* 4 vols. Leipzig, Berlin, 1933–35. Review by Penck [*90*].

364. Hoover, Edgar Malone: *Location Theory and the Shoe and Leather Industry.* Cambridge, 1937.

365. James, Preston E.: "The Coffee Lands of Southeastern Brazil," *Geogr. Rev.,* 22 (1932), 225–244.

366. Jefferson, Mark: "The Civilizing Rails," *Econ. Geogr.,* 4 (1928), 217–231.

367. Jones, Stephen B.: "The Forty-Ninth Parallel in the Great Plains: The Historical Geography of a Boundary," *Journ. of Geogr.*, 31 (1932), 357–368.
368. Kniffen, Fred B.: "Louisiana House Types," *Ann. Assn. Am. Geogrs.*, 26 (1936), 179–193.
369. Langhans, Manfred: "Karte des Selbstbestimmungsrechtes der Völker," *Peterm. Mitt.*, 72 (1926), 1–9.
370. Milojevic, Borivoji Z.: "The Kingdom of the Serbs, Croats, and Slovenes: Administrative Divisions in Relation to Natural Regions," *Geogr. Rev.*, 15 (1925), 70–83.
371. Overbeck, Hermann, and George Wilhelm Sante, edits.: *Saar-Atlas*. Botha, 1934. Review by R. Hartshorne, *Geogr. Rev.*, 24 (1934), 680 ff.
372. Palander, Tord: *Beiträge zur Standortstheorie*. Diss., Uppsala, 1935. (Bibliography.)
373. Passarge, Siegfried: "Die politische Erdkunde Afrikas vor dem Eingreifen der europäischen Kolonisation," *Peterm. Mitt.*, 70 (1924), 253–261.
374. *idem:* "Madrid: Das Werden einer Grosstadt in einer Steppenlandschaft," *Ztschr. f. Geopolitik*, 1 (1924), 688–694.
375. *idem: Das Judentum als landschaftskundliches-ethnologisches Problem.* Munich, 1929.
376. *idem: Aegypten und der arabische Orient: eine politisch-geographische Studie.* Weltpol. Büch., Berlin, 1931.
377. *idem:* "Politisch-Geographische Betrachtungen über Alt-Hellas," *Geogr. Wochensch.*, 1935, 926–932.
378. *idem:* "Byzanz: eine politisch-geographische Betrachtung," *Geogr. Anz.*, 36 (1935), 484–488.
379. *idem:* "Die grosse geopolitische Gefahrenzone Europas und ihre Raumbedingtheit," *Ztschr. f. Geopolitik*, 13 (1936), 137–145.
380. Platt, Robert S.: "Conflicting Territorial Claims in the Upper Amazon," in *Geographic Aspects of International Affairs* (Harris Institute Lectures for 1937), C. C. Colby, ed. Chicago, 1938.
381. Ripley, William Z.: "Geographical Limitations of Consolidated Systems," *Amer. Econ. Rev.*, 14 (1924), Supplement, 52–64.
382. Sauer, Carl, and Peveril Meigs: "Site and Culture at San Fernando de Velicatá," *Univ. Calif. Pubs. in Geogr.*, 2 (1927), 271–302.
383. Sauer, Carl, and Donald Brand: "Pueblo Sites in Southeastern Arizona," *Univ. Calif. Publ. in Geogr.*, 3 (1930), 415–457.
384. *idem:* "Prehistoric Settlements of Sonora, with Special Reference to Cerros de Trincheras," *Univ. Calif. Publ. in Geogr.*, 5 (1931), 67–148.
385. Schlüter, Otto: "Nation und Nationalität" (review), *Geogr. Ztschr.*, 12 (1906), 528 f.
386. Schmidt, Peter Heinrich: *Einführung in die allgemeine Geographie der Wirtschaft.* Jena, 1932.
387. Scofield, Edna: "The Evolution and Development of Tennessee Houses," *Journ. Tenn. Acad. Sci.*, 11 (1936), 229–240.
388. Spethmann, Hans: *Das Ruhrgebiet im Wechselspiel von Land und Leuten, Wirtschaft, Technik und Politik.* Berlin, 1933. Review by R. Hartshorne, *Geogr. Rev.*, 26 (1936), 343.
389. Taylor, Griffith: *Environment and Nation: Geographical Factors in the Cultural*

and Political History of Europe. Toronto and Chicago, 1936. Reviews by Ellsworth Huntington [*219*]; by J. O. M. Broek, *Canad. Hist. Rev.,* 18 (1937), 69; by J. K. Wright, *Am. Hist. Rev.,* 42 (1937), 700; and by R. Hartshorne, *Journ. Mod. Hist.,* 9 (1937), 372.

390. Tiessen, Ernst: "Der Friedensvertrag von Versailles und die Politische Geographie," *Ztschr. f. Geopolitik,* 1 (1924), 203–220. In extended form: *Versailles und Fortsetzung: eine geopolitische Studie.* (Brochure), Berlin, 1924.

391. Thornthwaite, Warren: "The Climates of the Earth," *Geogr. Rev.,* 23 (1933), 433–440.

392. Trewartha, Glenn T.: "Ratio Maps of China's Farms and Crops," *Geogr. Rev.,* 28 (1938), 102–111.

393. *idem:* "French Settlement in the Driftless Hill Land," *Ann. Assn. Am. Geogrs.,* 28 (1938), 179–200.

394. Troll, Karl: "Die Landbauzonen Europas in ihrer Beziehung zur natürlichen Vegetation," *Geogr. Ztschr.,* 31 (1925), 265–280.

395. Waibel, Leo: *Probleme der Landwirtschaftsgeographie.* Breslau, 1933.

396. Weber, Alfred: *Über den Standort der Industrien: reine Theorie des Standortes.* Tübingen, 1909. Trans. by C. J. Friedrich, *Theory of the Location of Industries.* Chicago, 1928.

397. Whitbeck, Ray H.: "Geographical Influences in the Development of New Jersey," *Journ. of Geogr.,* 7 (1908), 177–182. Adapted in C. C. Colby, *Source-Book for the Economic Geography of North America,* Chicago, 1922, 205–210.

398. Whittlesey, Derwent: "Early Geography in Northern Illinois," *Science,* 81 (1935), 227–229.

399. *idem:* "Environment and the Student of Human Geography," *Sci. Monthly,* 35 (1932), 265–267.

400. Wright, John Kirtland: "Voting Habits in the United States," *Geogr. Rev.,* 22 (1932), 666–672.

ADDITIONAL WORKS CITED
(Numbers 1-400 are in the original list, pages 1-21.)

A. HISTORY OF GEOGRAPHIC THOUGHT, PRIOR TO 1900

GENERAL HISTORICAL STUDIES

401. Peschel, Oscar: *Geschichte der Erdkunde bis auf Alexander von Humboldt und Carl Ritter.* 2d edition, edited by S. Ruge, Munich, 1877. (Bibliography.)
402. Bunbury, Edward H.: *History of Ancient Geography.* 2 vols., London 1879, 1883.
403. Tozer, Henry F.: *History of Ancient Geography.* 2d edition, Cambridge, 1935.
404. Berger, Ernst Hugo: *Geschichte der wissenschaftlichen Erdkunde der Griechen.* 2d edition, Leipzig 1903.
405. Wright, John Kirtland: "The History of Geography: A Point of View," *Annals Assn. Am. Geogrs.*, 15 (1925), 192-201. (Bibliography.)
406. Wood, Ella Lucile: "History of the Modern Concepts of Geography," Diss., 1927, manuscript in University of Wisconsin Library.
407. Kühn, Arthur: "Die neugestaltung der deutsche Geographie in 18. Jahrhundert." *Quellen u. Forsch. z. Gesch. d. Geogr. u. Volkerkunde,* Band V. Leipzig, 1939.

BIOGRAPHICAL STUDIES

408. Gottmann, Jean: "Vauban and Modern Geography," *Geogr. Rev.,* 34 (1944), 120-128.
409. Brown, Ralph H.: "American Geographies of Jedidiah Morse," *Annals Assn. Am. Geogrs.*, 31 (1941), 145-217.
410. Dana, James D.: "Memoir of Arnold Henry Guyot," *Smithson. Inst. Ann. Report* 1887, Pt. 1, 693-722; also in *Biographical Memoirs of the National Academy of Science,* 2 (1886).

B. GEOGRAPHIC THOUGHT IN THE TWENTIETH CENTURY

GENERAL SURVEYS OF GEOGRAPHIC WORK

411. Keltie, Sir John Scott: "The Position of Geography in British Universities," *Amer. Geogr. Soc. Research Series No. 4.* New York, 1921.
412. DeMartonne, Emmanuel: "Geography in France," *Amer. Geogr. Soc. Research Series No. 4a.,* New York, 1924. (Bibliography.)
413. Winkler, Ernst: "Internationale Geographie: Gedanken zum Ausbau der erdkundlichen Kongresse," *Schweiz. Hochschulzeitung,* 1939 (8 p.).

414.   *idem:* "Fortschritte und Probleme der Erdkunde in der Schweiz,"
       Ztschr. f. Erdkunde, 7 (1939), Heft 1, 1-18. (Bibliography.)
415.   Martonne, Emmanuel de. "Evolution de la Géographie," in his
       Traite de Geographie Physique. Paris, 1919 and 1948, pp. 3-
       26.

THE METHODOLOGY OF GEOGRAPHY, IN GENERAL

IN GERMANY AND OTHER GERMAN-SPEAKING COUNTRIES

416.   Krebs, Norbert: "Die Geographie in ihrer Stellung zu anderen
       Wissenschaften," *Ztschr. f. Schulgeogr.* (Vienna), 27 (1906),
       129-137.
417.   Philippson, Alfred: "Inhalt, Einheitlichkeit und Umgrenzung der
       Erdkunde und der erdkundlichen Unterrichts," *Mitt. d. Haupt-
       stelle f. naturwiss. Unterricht,* Berlin, 1919, 2 Heft, 22-43.
418.   Winkler, Ernst: "Geographie als Zeitwissenschaft," *Ztschr. f.
       Erdkunde,* 5 (1937), 49-58.
419.   idem: "Kulturlandschaftsgeschichte," *Ztschr. f. Schweiz.
       Geschichte,* 19 (1939), 54-76. (Bibliography.)

IN FRANCE

420.   Vidal de la Blache, Paul. "Le principe de la géographie générale,"
       Annales de Geographie, V (1896), 129-42.
421.   Demangeon, Albert: *Problemes de Geographie Humaine.* 2d edi-
       tion, Paris, 1943. Review by H. J. Fleure, Geogr. Rev., 36
       (1946), 172 f.

IN GREAT BRITAIN

422.   Mackinder, Halford J. O. "On the Scope and Methods of Geogra-
       phy," *Proceedings of the Royal Geographical Society,* N.S. IX
       (1887), 141-60.
423.   Mackinder, Halford J. O. "Modern Geography, German and
       English," *Geographical Journal,* VI (1895), 367-79.
424.   Herbertson, Andrew J.: "Regional Environment, Heredity and
       Consciousness," *Geogr. Teacher,* 8 (1915), 147-153. Preceded
       by a memoir to "Andrew John Herbertson," by H. J. Mackinder,
       N. E. MacMunn, and E. F. Elton, 143-146.
425.   Forde, C. Daryll: "Values in Human Geography," *Geogr. Teacher,*
       13 (1925), 215-220.
426.   Mackinder, Sir Halford J.: Comments on the relation of history to
       geography, in *Geogr. Journ.,* 78 (1931), 268 f.
427.   East, William G.: "A Note on Historical Geography," *Geography,*
       18 (1933), 282-292. (Bibliography.)
428.   Holmes, J. Macdonald: "The Content of Geographic Study," *Re-
       port, Austral. and New Zealand Assoc, Adv. Sci.,* 1935, 401-
       433.

429. Taylor, E.R.G.: "Whither Geography? A Review of Some Recent Geographical Texts," *Geogr. Rev.*, 27 (1937), 129-135.

430. Ogilvie, Alan G. "The Relations of Geology and Geography," *Geography*, XXIII (1938), 75-82.

431. Forde, C. Daryll: "Human Geography, History and Sociology," *Scott. Geogr. Mag.*, 55 (1939), 217-235.

IN THE UNITED STATES AND CANADA

432. Taylor, Griffith: "Geography the Correlative Science," *Canad. Journ. of Econ. and Pol. Sci.*, 1 (1935), 535-550.

433. VanCleef, Eugene: "A Stratigraphic View of Geography," *Science*, 83 (1936), 313-317.

434. Taylor, Griffith: "Correlations and Culture: A Study in Technique," *Proc. Brit. Assn. Adv. Sci.*, 1938, 103-138.

435. Adams, Charles C.: "A Note for Social-Minded Ecologists and Geographers," *Ecology*, 19 (1938), 500-502.

436. Renner, George T.: "Human Ecology: A New Social Science," *Teachers College Record* (New York), 39 (1938), 488-493.

THE THEORY OF REGIONS, LANDSCHAFTEN, LANDSCAPES,

AND BOUNDARIES
(including studies of special techniques)

IN EUROPE

437. Gallois, Lucien: *Régions Naturelle et Noms de Pays: Étude sur la Region Parisienne*. Paris, 1908. (especially Chapters I and XII.)

438. Hettner, Alfred: "Die Terminologie der Oberflächenformen," *Geogr. Ztschr.*, 17 (1911), 135-144.

439. Creutzburg, Nikolaus: "Wirtschaft und Landschaft," *Hermann Wagner Gedänknisschr. Pet. Mitt.*, Erg. Heft 209 (1930), 275-286.

440. Jessen, Otto: "Der Vergleich als ein Mittel geographischer Schilderung und Forschung," *Hermann Wagner Gedanknisschr. Pet. Mitt.*, Erg. Heft 209 (1930), 17-28.

441. Grano, Johannes: "Die geographischen Gebiete Finnlands: Eine vergleichende Übersicht nebst methodischen Erörtungen," *Publs. Inst. Univ. Aboensis*, No. 6, 1931. Review by Eugene VanCleef in *Geogr. Rev.*, 22 (1932), 497 f.

442. Winkler, Ernst: "Zur Frage der allgemeinen Geographie," *Athenaeums-Schriften* (Zurich), 1938, Heft 2, 1-24. (Bibliography.)

IN THE UNITED STATES AND CANADA

443. Joerg, W.L.G.: "The Subdivisions of North America into Natural Regions: A Preliminary Inquiry," *Anals Assn. Am. Geogrs.*, 4 (1914), 55-83.

444. Mathes, F.S.: "The Conference on the Delineation of Physiographic Provinces in the United States," *Annals Assn. Am. Geogrs.*, 5 (1915), 127-129.
445. Davis, William Morris: "The Principles of Geographical Description," *Annals Assn. Am. Geogrs.*, 5 (1915), 61-105.
446. Jefferson, Mark: "Some Considerations on the Geographical Provinces of the United States," *Annals Assn. Am. Geogrs.*, 7 (1917), 3-15, Plates I, II.
447. Campbell, Marius R.: "Geographic Terminology," *Annals Assn. Am. Geogrs.*, 18 (1928), 25-40.
448. Jones, Wellington D.: "A Method of Determining the Degree of Coincidence in Distribution of Agricultural Uses of Land with Slope-Soil-Drainage Complexes," *Transact. Illinois State Acad. Sci.*, 22 (1930), 549-554.
449. Lösch, August: "The Nature of Economic Regions," *South. Econ. Journ.*, 5 (1938), 71-78.
450. Meigs, Peveril, 3d: "A New Index for the Analysis of Regional Trends," *Scott. Geogr. Mag.*, 55 (1939), 161-170.

METHODOLOGY OF APPLIED GEOGRAPHY

451. Taylor, Griffith: "The Geographer's Aid in Nation-Planning," *Scott. Geogr. Mag.*, 48 (1932), 1-20, 65-78.
452. Bowman, Isaiah: "Planning in Pioneer Settlement," *Annals Assn. Am. Geogrs.*, 22 (1932), 93-107.
453. Sauer, Carl O.: "Land Resource and Land Use in Relation to Public Policy," *Report of the Science Advisory Board, July 31, 1933, to September 1, 1934.* Washington, D.C., 1934, Appendix 9, 165-260. (Bibliography.)
454. Joerg, W.L.G.: "Geography and National Land Planning," *Geogr. Rev.*, 25 (1935), 177-208. (Bibliography.)
455. McMurry, K.C.: "Geographic Contributions to Land-Use Planning," *Annals Assn. Am. Geogrs.*, 26 (1936), 91-98.
456. Zuber, Leo J.: "A Comparative View of Regional Planning," *Journ. Tenn. Acad. Sci.*, 12 (1937), 267-272.
457. Bowman, Isaiah: "Geography in the Creative Experiment," *Geogr. Rev.*, 28 (1938), 1-19.
458. Hudson, G. Donald: "The Unit Area Method of Land Classification," *Annals Assn. Am. Geogrs.*, 26 (1936), 99-112.
459. idem: "Methods Employed by Geographers in Regional Surveys," *Econ. Geogr.*, 12 (1936), 98-104.
460. Trefethen, Joseph M.: "A Method for Geographic Surveying," *Amer. Journ. Sci.*, 32 (1936), 454-464.

I. Introduction

A. historical background of american geography

Geographers are wont to boast of their subject as a very old one, extending, even as an organized science, far back to antiquity. But often when geographers in this country discuss the nature of their subject, whether in symposia or in published articles, one has the impression that geography was founded by a group of American scholars at the beginning of the twentieth century. Likewise many such discussions in the past have had only slight relation to similar discussions by European geographers, in spite of the well known fact that geography has had a greater development in the universities of Germany alone than in those of this country.

European geographers, notably the Germans, have been much more generally interested in the study of the nature of geography than those in this country. Many of the questions raised here only in recent years have been very effectively discussed by certain German geographers since the beginning of the century. It seems desirable therefore to examine these discussions at some length. If satisfactory answers can be found there to objections raised by our critics we may be spared the disturbance created by apparently new suggestions for radical departures from established lines of work, and those working in established lines may be assured of the value of their efforts. On the other hand if these many studies of the problem indicate that in certain respects we have been following false hopes, further loss of effort may be spared.

The studies of the nature of geography on which this paper is based are listed, together with illustrative studies, in the foregoing bibliography. Some effort was made to include in the former all of the more significant studies of recent years—though it is no doubt deficient in respect to the French literature, not to mention those in languages other than English, French and German. No such effort was made in connection with the illustrative literature however, the writer merely using whatever studies happened to come to his mind.

In a critical survey of this character, based almost entirely on the work of living colleagues in his field, a writer might prefer to omit specific references, but this easier choice is forbidden him because of his responsibility not only to the reader but also to those colleagues whose work has provided the material for the study. (See index of authors at end.) In particular, the critical position assumed places an imperative responsibility upon the writer to provide a means by which quotations, paraphrases, and critical comments

[22]

may easily be checked against the writings concerned. Consequently a voluminous number of individual references are required. To avoid the disruption of the reader's thought created by so many footnotes, a simple reference system, suggested by the editor, is followed. The numbers in brackets provide the necessary basis for reference without, it is hoped, unduly interrupting the reading.

It would of course be absurd to suggest that American geographers were entirely unfamiliar with the concepts of geography developed in Europe.* Arnold Guyot of Princeton, the first professor of geography in an American university, was a follower of Carl Ritter, but he had no successors. Far more important was the missionary work of Ellen C. Semple, whose presentation of Ratzel's concepts of anthropogeography set the course of human geography in this country for a quarter of a century. During the same period, William Morris Davis and Professor Albrecht Penck of the University of Berlin did more than any others before or since to establish mutual relations between geography in this country and Germany. Not only was there a close connection in their own work, but each taught for a year or so in the other country—Penck at Yale and Columbia, Davis at Berlin—and each published to some extent in the other language. Their field of mutual interest, however, was almost exclusively limited to physiography. The emphasis that Davis gave to this part of geography combined with Semple's exclusive consideration of human geography developed much the same form of dualism that was characteristic of an earlier period in Germany. Davis' attempt to bridge the gap was purely theoretical [203], as he himself made little effort to make the connections in practice.

For nearly two decades these two points of view dominated most of the thought of American geographers, both in their research and in their concepts of their subject as presented, particularly, in the presidential addresses of Brigham, Dryer, Barrows, and Whitbeck [305; 207; 208; 212]. The notable exception offered by Fenneman [206], influenced by the work of Hettner, appears to have made little impression until some years later. With that exception, geography was conceived of as a science of relationships between the natural environment and human activities. This "environmentalist concept" of geography was given its most careful statement by Barrows, who, reversing the usual form, speaks of geography, or "human ecology," as the study of man's adjustments to the natural environment.

A somewhat different point of view was introduced into this country from France by the publication in English translation of works of Vidal de la Blache and Jean Brunhes. Although both of these writers retained in theory the concept of geography as the study of relationships between man

[23]

* Early in the nineteenth century Jedidiah Morse worked in close relationship with European geographers [409].

and the earth, in detail their work pointed in a different direction, and particularly tended to emphasize regional studies [Brunhes, *182*, 4, 13–27, 552 ff., *83*, 55; Vidal, *184*, 3–24; *cf*. Sauer, *84*, 171, 180–1].

The greatest change in point of view, however, has been caused by the work of Carl Sauer in bringing the attention of American geographers to concepts developed in Germany nearly a generation ago. Through his three somewhat similar studies, published between 1925 and 1931, as well as by his marked personal influence, he has led many, particularly among the younger workers, to think of geography in terms of the study of material landscape features, both natural and cultural, and to consider these features according to their chorographic, or regional, interrelations [*211; 84; 85*].

American geography has therefore been markedly influenced by the work of European geographers, but it is notable that this influence has come through but very few contacts—for the most part through Davis, Semple, and Sauer. A few individuals, like Fenneman, or Wellington Jones, have been independently influenced, but have had little effect on the concepts of geography held by the rest. To be sure, many American geographers have followed the work of British geographers, but unfortunately the latter have been as provincial as ourselves if indeed not more so. The exceptions appear conspicuous in their lack of influence. Thus Chisholm, in his inaugural address at Edinburgh in 1908 [*192*], abstracted at some length "the most comprehensive and to my mind the most illuminating recent series of papers on the scope of geography," namely Hettner's classic statement of 1905. But this view of the field, essentially, as we shall see, that of most German geographers, appears to be known to but few in England today. Similarly the views that Herbertson brought back from his studies in Germany made little impression in England until relatively recently. The great majority of English geographers appear completely unaware of the developments in German geography since Ratzel (note, for example, Bryan's frank admission of this fact in a footnote report of a conversation with one German geographer—whose name is misspelled and whose views are misstated [*280, 7*]). Most English geographers, depending largely on Semple for their knowledge of Ratzel, have tended greatly to overemphasize his importance in Germany [*cf*. Roxby, *195*, 280; and Dickinson and Howarth, *10*, 195–202]. In more recent discussions, however, Dickinson has strongly emphasized the concepts of German geographers of this century [*101; 202*]; further, a number of English geographers have been influenced by the work of Passarge.

The situation in France has been somewhat similar. Vidal de la Blache and Vallaux gave great attention to Ratzel's work, but in the later methodological discussions of Vallaux and of Brunhes one finds little understanding of what German geographers since Ratzel have thought or achieved. Consequently the several publications in English of these French geographers have done nothing to lessen American ignorance of modern German geography.

In return, we may add, German geography has been but little affected by the methodological views developed in other countries. As exceptions, we may note the influence of the regional monographs produced under the direction of Vidal de la Blache, which undoubtedly helped to further the recent emphasis on regional geography in Germany; and Herbertson's efforts to establish "natural regions of the world," which for a time gained more attention in Germany than in England. The publication in German of lectures by Sir Francis Younghusband has influenced a number of present German geographers, notably Banse and Volz. Apparently the only American geographer who has notably influenced German geography to date was W. M. Davis, whose developmental system of land forms classification was taken up by Penck, and is also represented in one of the standard German texts of geomorphology of which Davis and Braun were co-authors. German students of this subject continue to argue over the relative merits of the *"Davische"* system. On the more general aspects of methodology of geography as a whole, however, Davis appears to have had little influence in Germany.

These relatively limited associations in the methodology of the subject are in marked contrast to the increasing degree to which American geographers are utilizing the research products of European geography. Koeppen has become almost a household word in our considerations of climatology; the work of Russian and German students has revolutionized our classification of soils; *Siedlungsgeographie* has added new words like *Strassendorf* to our terminology; Passarge's concepts of *Landschaft* find expression in at least one recent American text; those interested in political geography have been made familiar with the developments in this field in Europe [*216*]; and Joerg provided us with a survey of the whole field of European geography during the World War and post-War periods [*88*].

In spite of this development, however, the tradition apparently remains that any question as to the nature of the subject is a matter of personal opinion rather than one requiring study. Even the animated discussions that followed Sauer's challenging statements appear to have led few to examine the original studies on which his conclusions were based. With the exception of Hall's résumé of the concept of regions [*290*] and a short discussion by the writer [*216*, 795–804], most geographers have been content with second-hand quotations, or have depended on their own consideration as to

[25]

what geography ought to be. The most important exception is Finch's recent presidential address, published as this paper was nearly completed [*223*].

B. ATTEMPTS TO REFORM THE NATURE OF GEOGRAPHY

It has often been suggested that the question of the nature of geography is of minor importance—that what is needed is simply for geographers to work. *"Es ist viel herumgedokiert worden an der Geographie"* Penck complains, and states that "progress in a science is not made by writing about methods, but by working methodically" [*163, 50*]. But we may presume that Penck, who has published at least eight methodological discussions, does *not* mean that it doesn't matter whether we know where we are going so long as we are on the move. In a field in which the members who are officially recognized as "geographers" may, at a single meeting, read papers which to some of the hearers appear to belong in geology, climatology, soil science, economics, history, or political science, it is particularly necessary that we know what our field is, in order that we may hope to understand what its individual students are trying to do.

Even more important is the need for each individual who proposes to devote his professional life to the field of geography to have a clear picture of the scope and nature of that field. Not that any prescribed limits should be outlined within which every one who is called a geographer must confine his work; no individual or group is given any authority in science to prescribe such limits nor to determine the character of the work to be carried on within them. But experience has all too clearly demonstrated that those who enter the field of geography with preconceived ideas of what it is or should be are afflicted with a feeling of dissatisfaction, if these ideas prove to be in conflict with what geography actually is. Whether this feeling of disatisfaction be directed at themselves or at their colleagues, it represents an unhappy waste of intellectual energy deleterious to the effective progress of both the individual and the field as a whole. It is therefore one of the primary duties of those who attract young students into any particular field to give them so accurate a picture of the nature of that field that any later disillusionment will be unnecessary. But we cannot do this unless we ourselves know the nature of our field.

No one can read or listen to many of the discussions on the nature of geography by American geographers during recent decades without observing a marked motivation of reform. Evidently there is a certain general dissatisfaction with the field of geography, at least in its present state. To some degree at least these symptoms are echoes of similar symptoms that may be observed in the literature of European geographers, notably of the Germans.

Needless to say, a feeling of dissatisfaction with the present state of any field of knowledge—as distinct from a dissatisfaction with its essential nature—is a symptom of healthy vitality. But violent or arbitrary reactions to that feeling are not evidences of maturity. A bright-eyed youngster may observe errors in the way in which the pieces of a stained-glass window have been put together, and if he demonstrates his point by aiming stones accurately at the points in question he will create a notable disturbance but he will have done little for the improvement of the window. In many such cases the following statement of Humboldt seems appropriate: "It is far from my practice to find fault with endeavors, in which I have not myself made any efforts, because their success so far appears very doubtful" [60, I, 68].

In discussing one of the more violent reformers among German geographers, Gradmann has asked: "Is it really completely unavoidable that our science should be exposed naked before all the world, in that every new year a new 'reformer' arises who condemns the whole previous works to the foundations and feels himself called upon to erect something entirely new out of the ruins" [251, review, 552 f.].

We have already suggested—and the later consideration of one or two of the more famous attempts to reform geography radically will demonstrate the suggestion—that a thorough dissatisfaction with the state of the field is a phenomenon whose explanation involves the critic as well as the object of his criticism. Although we are justified in keeping this possibility in mind there is no intention of dismissing in any such manner any proposals for reform in a field of study for which none of its students would claim perfection. Each of the important proposals of recent years will be analyzed on its own merits, in its relation to the historical development of geographic thought and in its relation to a logical concept of the field.

Several different tendencies may be observed in the various proposals that have been urged for the reform of geography. The current *Weltanschauung* in German thought finds expression both in the demand that geographers should give greater emphasis to the importance of individual men responsible for the historical development of the features of any region, and in the demand that geography must serve the national interest [*e.g.,* Schrepfer, *174*]. To anyone familiar with the German geographical literature of the World War and post-War periods the suggestion that German geographers have been deficient in studying geography in Germany's interest is at least as surprising as to the elder German geographers themselves. On the assumption that this movement is not likely to make a positive impression in this country, we need not consider it further [the interested reader will find pertinent discussions in Hettner, *175,* 341–3; and Plewe, *177,* 226 ff.].

A second tendency, of greater importance in this country, seeks to add to geography some element that provides "problems." The "mere description" of the character of a region, even with the addition of interpretation, apparently, is not a problem, or at least not a problem in the sense in which that term is used in some other sciences. Consequently, it is thought necessary to add the element of change in time, to compare the situation of today with that of a previous time, to study the effects of some particular change in one factor in a region upon the others, or, in general, to make geography "dynamic" rather than "static."

The remaining efforts to reform geography that need to be considered appear to be much more definitely motivated by a desire to make it what apparently it has not been, a proper science. In the first place, since science must deal with observable phenomena, geography, in order to be a science, should limit itself to observable phenomena, and these it finds in the visible "landscape"—or, if it is claimed that some things are observable but not visible, geography must limit itself to material objects. These, in contrast with immaterial phenomena, we may observe and measure with some degree of accuracy and certainty.

Another suggestion is based on the assumption that each particular science has its own distinctive objects or phenomena to study. A science of geography must therefore somehow discover its distinctive objects of study. Most of the visible or material objects that it finds in areas have already been claimed, if not thoroughly exploited, as properties of other sciences. One possibility, therefore, is for the geographer to pick up whatever objects he can find that no one else has claimed, and mark these for his own.

Other geographers insist however that the proper objects of study for geography are the pieces of land which he calls regions. In order that these may be studied as other scientists study their objects, arranging them in classifications of types and developing scientific principles or laws about them, they must be concrete unitary objects. If one questions this assumption but accepts the major premise as to the nature of science, then the study of regions is not science but perhaps a form of art, which must be excluded from a "science of geography."

At this point one is reminded of a remark of Professor Whitehead when confronted by one of his students with a discontinuity in his logical reasoning: "What we need at this point is a beautiful phrase." The escape from the dilemma in regard to areas as objects is aided by one of those blessed words of double-meaning—"landscape." Since the landscape is the visible scene, including within it visible material objects, and may be limited in various undefined ways, it is called a unitary concrete object. But the landscape—in its German form *"Landschaft,"* if not in English—is also the limited area of land, province or region. Hence the region or landscape is a concrete unitary object having form and structure and therefore subject to classification and the development of scientific principles concerning its relations.

C. THE PURPOSE OF THIS STUDY

The detailed examination of the nature of geography which this paper endeavors to present is not based on any assumption that geography is or ought to be a science—or that it ought to be anything other than it is. Assuming only that geography is some kind of knowledge concerned with the earth, we will endeavor to discover exactly what kind of knowledge it is. Whether science or an art, or in what particular sense a science or an art, or both, are questions which we must face free of any value concepts of titles. We will, therefore, not ask ourselves whether geography, as Douglas Johnson puts it, can "expect to enjoy the high prestige of the sciences," whether it "should pretend to a position of equality among them," or should accept a more humble though "honorable place along with history, economics, and sociology." Such questions naturally produce only the reaction that Johnson himself expressed: a refusal to "surrender geography's claim to a place of equality among the sciences until the scientific possibilities of the subject are more thoroughly explored than they have as yet been" [*103*, 220]. Geography is thereby put on the defensive, ever attempting to be something that perhaps it cannot be.

On one aspect of this question we may clear the ground at the start. That geography could be "a science in the sense that geology and botany are sciences" is as impossible as it would be for botany to be a science in the sense that geology is a science, or vice versa, or for either to be a science in the sense that physics and chemistry are sciences. Whatever geography is, its venerable if not honored age would nullify the most enthusiastic efforts of any students to make it over into something entirely different. Geography is not an infant subject, born out of the womb of American geology a few decades ago, which each new generation of American students may change around at will. Granted that one must keep an open mind on any subject in science, we are hardly free, as one unnamed geographer claimed, to change our concept of geography each year as the spirit moves us [see quotations in Parkins, *105*, 222].

In any established field, the problem of arriving at its proper definition is not so much a creative task, a problem in logical reasoning from *a priori* assumptions, as a research problem in knowledge itself. The philosopher, Kraft, in subjecting the field of geography to a critical examination—to which we will refer in detail later—claims that the problem of the position of any subject within the totality of sciences is a work that extends beyond the particular scientific knowledge of the department concerned, is properly a problem "in scientific theory, *i.e.*, in the theory of cognition." But the function of such a research in the theory of science cannot be to prescribe what geography should be, but rather to ascertain what it contains [*166*, 1, 3]. Similarly Hettner wrote, in 1905: "The system of the sciences has been an historical growth; abstract designations of the sciences that tend to take no account of the historical development—unfortunately the methodological literature of geography has been particularly rich in such *a priori* conceptions—are foredoomed to unfruitfulness" [*126*, 545].

This point of view has recently been characterized by Leighly as blind adherence to "tradition" in the face of "strict logic" [*222*]. Even if the concept of geography that he defends could be shown to be logically sound—a question to which we will come later—the argument is not to be won by devitalizing the historical evolution of a field of study as mere "tradition." When one considers the cultural origins of the inhabitants of the United States, one might claim that logically they should speak either a variety of European and African languages or a mixture in which each was represented in proportion to its part in the total population; it is merely a "tradition" of our schools that all should be taught something approximating "the King's English."

The growth of knowledge in any branch of science depends less on the particular scope for which it may find a logical justification, than on the continuity of its life, from the past through the present to the future. In the history of science—as in all history—common knowledge among educated people so emphasizes the achievements of a few outstanding "masters" as to lead one to suppose that each of them, starting from scratch, had singlehanded accomplished his wonders. In geography, we have a striking example in Alexander v. Humboldt. He is frequently presented as a pioneer without predecessors, a lone figure who explored America and returned to establish the science of physical geography. In reality, as we shall shortly observe, it may be questioned whether Humboldt would have become a geographer at all, had it not been for the younger Forster, and the latter's geographic work was very largely dependent on the elder Forster; his antecedents we do not know. Furthermore Humboldt throughout his work depended on a veritable host of other workers, both of earlier times and of his own time, as the multitudinous references in his writings indicate.

If, in contrast, each new generation of geographers, or each individual geographer presents new suggestions, untested either by example or by study of the previous development of the subject, there is introduced, as Hettner observed, a feeling of unrest and insecurity among geographers that handicaps the successful cultivation of the field and, in addition, causes workers in other sciences to become sceptical of the whole subject [*152, 52* f.; *167, 265*; *168,* 490 f.]. American geographers are all too familiar with such results. If we are to continue to experience violent shifts of the helm—formerly toward physiography, then toward environmentalism, now toward landscape studies, tomorrow to "the topography of art" and thereafter who knows whither[1]—our ship will beat around with ever-changing aim, hence aimlessly, and will arrive nowhere.

Throughout the preparation of this study the writer has assumed that few if any readers are interested in the particular view that he as a single geographer may have of the field. The privilege of presenting statements of that kind is reserved, one presumes, for the masters of our field whom this Association elects to its presidency. The writer's concern, on the contrary, is to present geography as other geographers see it—or have seen it in the past. Since the geographers of any particular time and country represent but a small part of the whole field, it will be necessary to extend our examination both into other countries and into the past. If we wish to keep on the track—or return to the proper track, as was suggested in a recent symposium—we must first look back of us to see in what direction that track has led. Our first task will be to learn what geography has been in its historical development.

Reversing the usual procedure, therefore, we will not start with any particular concept of the nature of science, but rather with the field of geography as it has been produced. Having ascertained its essential characteristics we will endeavor to find what logical position it holds among the different branches of knowledge [*cf.* Kraft, *166,* 3]. The first of three major portions of the paper is therefore concerned with the question "What is geography?" (Secs. II–IV).

On the foundation thus established we will be in a position to examine various suggestions that have been made for improving geography. This examination of major problems in geography today constitutes approxi-

---

[1] Since the above sentence was first written, a new answer has already been given: "geophysics" [*222,* 252 ff.]; although I am later informed that that was not the intention.

mately half of the total study (Secs. V–X). The reader will understand that the brief statement of suggestions given above (Sec. B) does not do justice, even in outline, to the ideas of the students proposing them. Because many of these students have made distinguished contributions to geographical research their theoretical suggestions for its improvement demand our careful attention. On the other hand we may presume that they themselves will wish that their ideas should not to be taken on authority but should be subjected to critical examination. Since in many cases the significance of a theoretical suggestion can most clearly be seen where it has been applied in practice, our examination may also require a critical survey of the material products. In both cases the reader is particularly requested to keep in mind that it is only the validity of the ideas of the nature of geography that concerns us here. The informed reader will recognize from the bibliography that I have confined myself largely to the writings of geographers of established distinction. Since the standing of these writers is in nearly every case based on their non-methodological works, it cannot be affected by any discussions in this paper. It is the methodological ideas of these writers that we are concerned to question: are these, either in terms of logic or of fact, valid statements of the nature of geography or acceptable suggestions for improvement within geography? If these questions cannot be answered in the positive, it will be necessary to consider a sound logical basis on which geography may develop as a broad but unified field.

Finally, by way of conclusion, we may seek to relate the field of geography, in its essential characteristics, to the general nature of knowledge, to determine, that is, what kind of a field of study geography is (Secs. XI–XII). Whether it is properly to be called a science or not, we will leave to those more interested in terminology; if we know definitely what kind of a study it is, we need not concern ourselves over titles.

The writer wishes particularly to stress, that in considering the suggestions of his colleagues, his purpose is not to engage in dialectic arguments, seizing any opportunity that lack of clarity of statements might offer, but rather to understand the intention of the suggestions made. In particular he has no wish to introduce into American geography a form of personal argumentation that has unhappily marked the methodological discussions of German geographers in recent times. With Obst, his purpose, is to seek for the maximum degree of understanding and agreement possible, not to establish any particular direction in geography but to make possible a common basis for geographic work, so that the geographic work itself—the only real purpose—may be furthered to the highest degree [*178*, 1–3].

D. THE NEED FOR AN HISTORICAL SURVEY

In a brief examination of published studies on the history of geography, John K. Wright commented on the lack of any adequate treatment of this subject in English [9]. We have numerous excellent treatments of the history of American geography, notably those by W. M. Davis [102; 104], Douglas Johnson [103], and Colby [107]. Attention may also be called to the recent publication in German literature of two separate surveys of recent developments in American geography, by Broek, of the University of California, and by Pfeifer, who has also worked in the same institution [108; 109; but note Platt's critical comment, 224, 125]. The development of geography in this country, however, can only be understood in the light of earlier and contemporary developments in Europe; for a thorough treatment of this subject we still look in vain in our literature. Since Wright first noted this lack, in 1925, Dickinson and Howarth have published a useful survey of the field, but one that unfortunately depends largely on foreign interpretations of the work in Germany, and gives but the briefest sketch of the developments in that country since Ratzel [10].

There can be no question that the foundation of geography as a modern science was primarily the work of German students. In whatever country one starts, the study of the development of geography leads backward to the work of Humboldt and Ritter in the early part of the nineteenth century. Further, though in many countries geographic work has become more or less independent, major changes in geographic thought are found to be expressions, frequently very belated, of the developments in German geography. It is particularly unfortunate therefore that we have in English hardly even an outline of the development of geographic thought in Germany during the past century, so that nearly all American geographers have been trained in complete ignorance of the methodological background of their subject. German students, in contrast, have readily available not only the original writings of the period, but also many historical studies of the development of geographic thought.

To present this history adequately would require a book in itself. For our present purpose we must confine ourselves to the consideration of the major characteristics of geography as a field of study at each of several successive stages in its development. Our concern is not with the development of geographic work, nor with the relative importance of different writers or groups in the production of geographic material, but rather with the establishment and changes in thought concerning the nature of geography, as represented however in geographic works as well as in direct methodological studies.

[33]

Our examination of earlier work in geography has a two-fold purpose. We have already emphasized its importance in providing us with an understanding of the character of the field in which we work. The second purpose will be to observe what earlier writers have concluded in regard to problems which are of current concern in the methodology of geography, for there are few of these that have not claimed the attention of geographers in the past—in most cases, repeatedly in the past. As Braun observed, "there will be no improvement in the methodological confusion as long as we geographers continue the habit of completely ignoring the writings of our predecessors in this problem . . . and attempt to construct a building with stones that the neighbor has perhaps long ago recognized as useless or insufficient for a better arrangement" [155, 17 f.]. "Any progress in methodology," wrote Obst, "must be built on the past. No methodological change therefore without knowledge and consideration of past understandings and concepts" [178, 3].

These two purposes of our historical study cannot be attained if one enters the study with a pre-established concept of what geography should be, and on these terms searches among the writers of the past for those whose ideas are in agreement with ours, whom one then places in opposition to those who appear to have taken a different view of geography. In any methodological debate, to be sure, one may look for supporting arguments wherever he may find them—whether in the present or the past—but such a presentation is not to be confused with an historical study which aims at an objective presentation of the past; on the contrary it leads easily to a distortion of the past development of geography and of the ideas of earlier geographers in order to fit the purposes of the particular debate. We will therefore attempt to consider the concepts of previous geographers, so far as is humanly possible, independent of our own particular views.

II. The Nature of Geography According to Its Historical Development

A. THE PRE-CLASSICAL PERIOD OF MODERN GEOGRAPHY

Although the roots of geography, as a field of study, reach back to Classical Antiquity,* its establishment as a modern science was essentially the work of the century from 1750 to 1850. The second half of this period, the time of Humboldt and Ritter, is commonly spoken of as the "classical period" of geography. Undoubtedly the extraordinary accomplishment of each of these men, working at the same time but in very different ways, and the influence of their work on all subsequent geography justifies our regarding them as the first masters of modern geography—in that sense as the "founders." But in applying such titles one is easily led to ignore the importance of previous workers—those, we might say, who laid the foundations for the founders. *"Carl Ritter hat auch seine Vorgänger,"* Peschel is quoted as saying, and the statement applies as well to Humboldt. Among these predecessors we are not here concerned with the geographers of Classical Antiquity nor with the majority of the writers of geographical works during the two or three centuries preceding 1800. Our interest is with that smaller, but still considerable, number of students of the second half of the eighteenth century who consciously strove to convert a more or less miscellaneous and useful study into an independent science. Although one could hardly affirm that they had accomplished that purpose, nevertheless it is clear that they had put down the main outlines of the science of geography, as we know it to-day, before the appearance of geographical publications by Ritter and Humboldt at the turn of the century. They had not *established* those outlines, however. It is for that work, the establishment of the modern science of geography, that we are indebted to Ritter and Humboldt. Nevertheless, most of the fundamental concepts of geography—including almost all of those which we will have to examine in this paper—may be found in the writings of the German geographers immediately preceding Ritter and Humboldt. It hardly seems too much to say that, had neither of these men lived, the development of geography after 1800 would have led ultimately, even though far more slowly, to something like that which we know.

As the inheritors of the geography of Classical Antiquity and that of the Renaissance, the geographers of the eighteenth century as a matter of course considered all kinds of phenomena whose differences in different parts of the world made them appear significant to a knowledge of the world. They echoed the famous statement of Strabo: "Geography, in addition to its vast

[35]

* See Supplementary Note 2.

importance to social life and the art of government . . . acquaints us with the occupants of the land and ocean and the vegetation, fruits, and peculiarities of the various quarters of the earth, a knowledge of which marks him who cultivates it as a man earnest in the great problem of life and happiness" [38, introduction].[2] The "Physical Geography" of Immanuel Kant was concerned with all the elements included by Strabo; on the basis of this physical geography, Kant would found "other geographies" including political, commercial, moral (in the sense of *mores*), and theological geography, as well as the descriptions of each particular land [40, § 5]. Physical geography is the essential propaedeutic for an understanding of our perceptions of the world, whether those received directly by travel or indirectly by reading [§ 2, end of § 4]. In brief, "it serves as a suitable arrangement for our perceptions, contributes to our intellectual pleasure, and provides rich material for social discourse" [§ 5; though the ideas are presumably those of Kant, the words are in the voice of Rink, see footnote 3].

These statements of Strabo and Kant are susceptible of translation in terms of modern science. What Leighly refers to as "the unspecialized curiosity in the minds of stay-at home readers" was not considered by Kant as in opposition to science. On the contrary the curiosity which desires a knowledge of the world is the stimulus that requires science to pursue that knowledge and seek to organize it. For this purpose, to be sure, it is necessary that individual students and groups of students should specialize, but Kant recognized that there is more than one form of specialization. While he recognized the importance of the specialization according to kinds of objects (which was later to dominate the field of the natural sciences), he saw that that form of specialization was not possible in geography, since its single object, the earth, is for us unique in the universe. We shall consider later the form of specialization which he outlined for geography (Sec. IV B).

The great majority of the writers of geographical works of the seventeenth and eighteenth centuries, however, did not attempt to translate the purpose of geography in terms of scientific interest. Rather, they were concerned with its practical utility. Over and over again they emphasized the value of the study of geography as a means for other purposes—for an understanding of history and as a practical aid to government. Wisotzki has shown how this utilitarian point of view prevented scientific progress in geography [1, 96–130].

Overlooking earlier but unsuccessful efforts, we need to consider a small but increasing number of writers in the second half of the eighteenth century

---

[2] References are given only to sources that I have directly examined, in order that it may be clear that in other cases I have depended on secondary commentators.

who worked to elevate geography from its subordinate position to that of an independent science, *"reine Geographie,"* or "pure geography," as they came to call it. (In contrast with later uses of this term, both in Europe and America, the reader will note the suitability of its use in this connection.)

This movement found its first expression in the demand that geography should consider its object, the world, in terms not of political divisions but of more real and lasting divisions of nature. Divisions of area can be established only by boundaries, so that this concept found its first expression in a demand that geography should divide the world by "natural boundaries" of "lands" rather than by political boundaries of states. This movement, which represented an echo from Strabo, was expressed by many writers, both geographers and students of law (notably Grotius), as early as the 16th century [*1*, 193 ff.; *5*, 44 ff.]. It could make little headway, however, so long as geography was considered to be merely the handmaid of history and government. Although it was argued that the frequent changes in political boundaries made the geographer's work seem of but transient value, the argument was irrelevant so long as it was the major business of the geography teacher to provide information about political areas. To accomplish this task at a time when the political map of central Europe—Germany and Italy—was a crazy-quilt of hundreds of states, the geography teacher could hardly spare time to consider "natural regions."

On the other hand it seemed obvious to those who were attempting to develop a scientific geography, that the conventional organization of the knowledge of different parts of the world by political divisions was anything but satisfactory. Whether in terms of the political map of Europe of that day or this, or of the political map of the United States today, it hardly seems too much to say that no student who is seriously concerned with the attempt to develop regional geography could regard political divisions as even "generally satisfactory."

Then as now, however, the effort to consider the areas of the world in divisions other than political presented the difficult problem of finding an equally definite method of dividing areas of the lands. The answer to this long-felt need appeared to be offered by a theory put forward by various students in the middle of the eighteenth century, notably Buache in France— namely, the theory of the continuous network of mountain systems [see Wisotzki's chapter on *"Der Zusammenhang der Gebirge," 1*]. The theory was immediately taken up in Germany, and Gatterer made it the basis of a physical division of the world into lands and regions (*Gebiete*). Whereas the similar attempts made much earlier—in Holland and Italy as well as in Germany—had been largely ignored, Gatterer's work had a marked influence on his contemporaries and on later students, including Humboldt [cf. Richthofen, 3, 671]. We may therefore mark the beginning of a continuous reformation of geography with the appearance of his publications in 1773-75 [1, 201 ff.].

A somewhat different stimulus to the development of scientific geography throughout this period came from the lecture course on "physical geography" that the philosopher Immanuel Kant presented at the University of Königsberg, repeated some forty-eight semesters during the period 1756 to 1796 [14, 9 f.]. For Kant himself, as Gerland has noted, the study of geography represented only an approach to empirical knowledge necessary for his philosophical considerations [12]. But finding the subject inadequately developed and organized, he devoted a great deal of attention to the assembly and organization of materials from a wide variety of sources, and also to the consideration of a number of specific problems—for example, the deflection of wind direction resulting from the earth's rotation [39; 13].

It is difficult to estimate the historical importance of Kant's work in our field. His course in physical geography, one of the most popular of his courses, was heard by a very large number of students during its forty-year period, and long before the publication of the lectures, after 1800, there were many fairly complete handwritten copies in circulation. Furthermore, former students, like Herder, had given wider circulation to his ideas. Whether by indirect means, or only after their publication in 1801-05, the lectures had a significant influence, as we shall see, on both Humboldt and Ritter.

The greater part of Kant's course followed the outline of "general geography" that Varenius had laid down a century before and which Lulofs had followed much later (in German translation, 1755). In addition to the large amount of material taken from Lulofs, Kant later drew from the work of Bergman (published in Swedish, 1766, in German, 1769), as well as Buache, Buffon, and especially Büsching [40, notes, 551 ff.; 14, 285 ff.]. In his relatively briefer consideration of individual countries, however, he does not appear to have been stimulated either by Buache or Buffon to have altered Büsching's conventional division by political units; certainly the outline and material of the course were well established before the appearance of Gatterer's work [39, II, 1-12; 14, 27, 31 f.; 15].[3]

---

[3] For over a century there was uncertainty as to the authenticity of the various published versions of Kant's lectures on physical geography, but the extraordinarily painstaking research of the Kantian philosopher, Adickes, appears to have finally established the facts, which we may briefly summarize [14; 15; cf. Gedan in 40, 509 ff.]. Although Kant himself never prepared his lectures for publication, he had at least one full manuscript copy, from which however he evidently departed frequently in "dictating" his lectures. Further, there have been preserved a few original records by students as well as a larger number of handwritten copies prepared in what Adickes

describes as "a special branch of industry" in Königsberg—copies put together from various students' records, often those of different years mixed together, to be sold to other students. From such sources, a certain Gottfr. Vollmer issued in 1801, three years before Kant's death, the first volume of a six-volume set purporting to be Kant's geography. Since it has been shown that, though this was based on Kant's lectures, those form not more than a fifth of the total material presented, without differentiation, this edition must be dismissed (unfortunately Peschel based his consideration of Kant's work on Vollmer's production). Kant himself denounced it and requested F. T. Rink to prepare an "authorized edition" from various manuscripts that Kant had on hand. Unfortunately he had become too senile to examine what Rink produced and published in 1802. By comparing this edition with more than twenty other handwritten manuscripts, of the types described above, Adickes concludes that its authenticity is in general beyond doubt, but that it is based on sources separated nearly twenty years in time. The greater part of the first volume (§ 1–52, omitting §§ 11 and 14) was based on a full and fairly reliable record of Kant's lectures of 1775, on which Kant himself had made marginal corrections. The remainder of the first volume (*Teil*), and all of the second were taken from Kant's own manuscript, which however dates from not later than 1759. Further, all of the notes and references, as well as sections 11 and 14, were added by Rink, who also presumed to improve at many points on the literary style; although in most cases he accomplished the reverse, he does not appear to have changed the meaning significantly.

Adickes feels that, had Kant been able to realize what was being done, he would not have permitted Rink to publish the second portion, based as it was on outdated material which Kant evidently omitted or corrected in his verbal lectures. The first part however (§§ 1–52), represents approximately Kant's lectures in the latter years. (All the more significant references used in this paper are taken from this first part; quotations are corrected in language, without altering the meaning, according to suggestions from Adickes, taken from an original manuscript—probably, he thinks, the very manuscript that Rink had used.)

It has long been known, from the publication of Kant's program for his course for the year 1765–6, that he planned to revise his outline materially, to contract greatly the "physical geography" in order to devote approximately two thirds of the time to "moral" and "political" geography, but Adickes concludes, from the study of students' manuscripts of later years, that he did not find the time to do this, that, in general, the outline remained much the same as it had been.

In the various editions in which Rink's version of Kant's lectures have been republished, editors have corrected a number of his more obvious errors; without significantly altering the meaning in most cases. Probably the most nearly satisfactory edition is to be found in the publication by the Academy of Sciences in Berlin [40]. Adickes' suggestions for the preparation of a more authentic edition of Kant's lectures from the available manuscripts have apparently not been taken up.

For geography today, Adickes is no doubt correct in assuming that Kant's work is of little more than historical interest. Among geographers it has been examined in detail particularly by Gerland, whose course of twelve lectures on Kant's work in geography and anthropology, given in 1901, was later published [12]. One section of Kant's study, however—namely his concept of the relation of geography to other sciences—is of fundamental interest to us today, as will be seen later in this paper. We may note in particular, therefore, that Rink's presentation of this section is substantiated in the manuscript sources that Adickes studied.

We cannot here consider the various steps in the subsequent development of the "pre-classical" period, and can only briefly suggest the underlying factors that stimulated its progress. Of great importance was the rapid growth in various physical sciences. In particular the development of reasonably accurate barometric methods, notably by de Luc, whose work was first published in 1772, made it possible to measure the "vertical" dimension of the earth's surface [Marthe, 25, 30; Peschel, 66, I, 329 f.]. The fundamental approach of many geographers of this and the following period was notably influenced by the great reformer of education of the time, Pestalozzi, of Zurich, who brought into German schools the philosophical viewpoint of Rousseau [25, 30; 1, 260 ff.]. Closely related to the change in geographic thought—whether as cause or effect—was the so-called "natural division" of France into departments according to rivers and mountains. Somewhat later the radical and repeated changes in political boundaries during the Napoleonic Wars helped to discredit geography texts based on such fluctuating conditions, and thus helped to make way for a geography based on the physical description of the world [1, 257–66].

Our present interest is in the nature of the field of geography as it was conceived by a small but active minority of writers in the late eighteenth century who, before Humboldt and Ritter, were endeavoring to establish geography as a science. We need not attempt to set a definite limit as the end of this period, but may consider it as overlapping perhaps a decade or so into the nineteenth century. Although Humboldt published at least one geographical article before his all-important journey to tropical America in 1799, his major publications in our field did not begin until after his return in 1804, the year in which Ritter's name first appears in geographical literature. According to Marthe, however, it was not until nearly 1820 that each of these two figures assumed a dominant position in geography [25, 6].

The geography which the late eighteenth century had inherited from the past was essentially limited to phenomena of the world as known to man, that is, of the earth's surface, in the broad sense of a surface of some thickness extending into the atmosphere and into the solid earth wherever man had penetrated. Kant had described the field of physical geography as the world (*Welt*), so far as we can come in relation to it, the scene of our experiences. The study of the earth body belongs to mathematical geography, but is considered in physical geography insofar as it causes differences in its different parts [40, I, §2, 7]. In most of the geographies of this period the treatment of the earth-body was confined to an introductory chapter— "*Mathematische Vorbegriffe*," according to Kant—and the actual study of geography confined to the earth-surface. The specific term *"Erdober-fläche"* was frequently used, and in 1820 Wilhelmi specifically limited the field of geography to the earth-surface.[4]

The geography of this period included both "general" studies of particular kinds of phenomena of the earth surface, and descriptions of many kinds of phenomena found in particular areas. The inclusion of these two forms of study within the single field has no doubt been the cause of more controversy than any other single problem in the methodology of geography. It is significant therefore that it was not introduced into modern geography as a result of the chance combination of Humboldt and Ritter. On the contrary the same difference is found in the work of the geographers of antiquity, as Hettner observed in his first brief treatment of the history of geography [*2*, 306 f.; more fully, *161*, 33 f.]. The tendency of these two directions to come in conflict with each other and to interchange positions within geography at different times is evidence, he suggests, that they do not represent separate sciences but merely different directions within the same science. The distinction between the two directions was perhaps first clearly stated by Varenius (a German, Bernard Varen, living in Amsterdam) in 1650 [*10*, 100 ff.]. In contrast with most of the writers of his time who limited geography largely to a bare description of the several countries—including therein much of their political constitutions—Varenius divided the field into "general or universal geography" and "special geography" or "chorography" [note Humboldt's discussion of Varenius, *60*, I, 60]. His untimely death, at the age of twenty-eight, in 1650 prevented him from studying the second part, though his previous work on the geography and history of Japan would indicate that he did not lack interest in that part of the field.

The terms which Varenius had used, "general" and "special" geography, later became the standard terms in Europe for these two aspects of the field, though many later writers have found them unsatisfactory. As we will note in a later section, the frequent use by German writers of the adjective *"systematisch"* in describing "general geography" supports the common use in this country of the term "systematic geography." The term "special geography" was largely replaced in German literature by *"Länderkunde,"* which in spite of obvious disadvantages is commonly favored over non-Germanic

---

[4] Unless otherwise indicated the historical data concerning this period are taken from Wisotzki's chapter on *"Die reine Geographie,"* which is arranged in chronological order [*1*, 193–266]. In no case is the reader to assume the adjective "first." Conclusions concerning historical priority are not only dangerous but do not concern us here; we need only establish what ideas had been developed before the end of this period.

terms, "special," or—the term now nearly universal outside of Germany—"regional geography" (see Sec. XI E).

Following Varenius, Lulofs and Bergman had concentrated on systematic geography. Kant, however, though depending in part on them, likewise suggested the place of *Länderkunde,* but his beginning of such a regional study is not firmly based on his systematic studies [*40,* I, §3, II, Abs. 3]. The distinction between the two aspects of geography was more explicitly stated by Gatterer, 1773–75, by Krug, 1800, and particularly by Bucher, 1812.

Although the systematic studies of this period undoubtedly were of major importance in the progress of geography toward scientific standing, Plewe finds that many of them, notably those of Bergman, tended to fall out of geography and become studies in other sciences. In contrast, John Reinhold Forster—the first of the great scientific geographic travelers—consistently maintained the "macroscopic" sense suitable to geography and endeavored to explain the regional relations of each type of phenomena studied. The work of J. R. Forster, and particularly that of his son, George, was regarded as of great importance by Humboldt, upon whom, as we shall see, they had a determining effect. Plewe, among others, counts the elder Forster as the first—in order of time—of the great "classical geographers" [*8,* 22–26].[5]

The division between general or systematic geography and *Länderkunde,* or regional geography, represented therefore a form of dualism that was characteristic of geography throughout its initial period of development as a modern science. This form of dualism in geography is not to be confused, as is often the case, with a dualism in terms of content, as Wagner warned, in 1890 [*80,* 375].

In terms of content, the eighteenth century writers commonly distinguished three divisions: "mathematical geography," "physical geography,"

---

[5] It is hardly possible to separate the works of the two Forsters, Johann Reinhold (1729–98) and Georg F. (1754–94), since their most important journeys were made together, much of the work of the father was published in Germany by the son, and the latter depended in his methods of study on the father, who was still living when George Forster died at the age of forty. Though Germans, they had lived for some years in England, where the father held a professorship in natural history for two years, and both accompanied Captain Cook on his voyage of 1772–75, but they returned to Germany a few years later. J. R. Forster's *Observations* made on the voyage with Cook were written in English and published in England, but the German translation made by his son proved far more influential [*41;* the English writers Dickinson and Howarth mention only the German edition, *10,* 172]. Most students have concluded that the elder Forster was the more important pioneer [*cf.* A. Dove, *16*]; Plewe feels that personal feelings motivated Humboldt to refer repeatedly to the work of his friend, George Forster, at the expense of his less agreeable father [*8,* 22–26; but note also Humboldt's letters to J. R. Forster, *17*].

and what was variously called "historical" or "political geography," but which is not to be confused with the meanings of those terms as used today (or as used by Kant). Mathematical geography consisted in large part of the study of the earth as an astronomical body. As we have seen, it was always of minor importance and in practice its study was left largely to astronomers. Zeune, in 1808, and Butte, in 1811, definitely excluded it from their outlines of geography.

The major contrast in content was that between what was included in physical geography and that contained in what we might call social geography. With the much more rapid increase in development of scientific methods of studying physical, as distinct from social, phenomena the contrast between the two became more marked. Attempts to develop a unified science included suggestions of confining the field to what was then called "physical geography." But if the present-day student considers this period only in terms of the titles which its writers used, he will have an erroneous conception of their works. As used by nearly all the writers of this period—including Kant, Forster, and later, Humboldt—the term "physical geography" was not limited to that which later was to be called the "physical" or "natural" environment, but included races of men and, commonly, their physical works on the earth. Thus Kant not only included man as one of the features "encompassed in the earth surface (*Erdboden*)" but also considered man as one of the five principal agents affecting changes on the earth [*40*, II, §1–7, I, §74]. Indeed, Kant's "physical geography," both in purpose and in content, might be considered as "anthropocentric," a point of view which Ritter inherited from Kant [according to Becker, *5*, 53].

Very few writers of this period made the distinction which is most familiar to us, between a geography of natural, or non-human features, and a human geography, though it may be found in the work of J. M. F. Schulze, 1787, and of Rühle, 1811. For most writers of this and the following period —*i.e*, until perhaps the middle of the nineteenth century—the term "nature" (*Natur*) was not used in contradistinction to human, but to indicate that which was perceived externally in contrast to one's internal feelings and thoughts. So far as I can find, no geographer of this period questioned the place of either human or organic features of the earth in the field of geography. One might note, however, a trend in that direction, in the interpretation which Hommeyer, in 1810—and perhaps others—gave to *"reine Geographie,"* namely, as limited to the conditions of the terrain; the study of climates, minerals, and organic life was left to *"Naturkunde," "Naturbeschreibung"* and *"Länderkunde"* [*1*, 221].

Although the writers of this period recognized that the scientific geography which they were attempting to develop included somewhat separate parts—"physical" and social geography—they considered these to be associated not merely because the phenomena were to be found in the same places, but because they were causally connected by mutual relationships. No doubt the interest in relationships between different groups of men and the particular character of their natural environment has ever been considered appropriate in geography. Kant, for example, noted that "in the mountains, men are actively and continuously bold lovers of freedom and their homeland" [40, II, §4], and similar relations were discussed by Gatterer and his followers [5, 48–53]. Kant specifically based the social branches of geography on their relations to physical geography: "theological principles in many cases undergo important changes as a result of differences in the land (Boden)" [40, I, 19]. In this he was followed by Müller, in 1785, and Fr. Schultz, in 1803, and by many others in the following decade. So far as I can find, however, it did not occur to any of the students of this period to consider these relationships as in themselves the direct object of geography study.

During the latter part of this period, it became common to regard the intricate interlacing of relationships between all the phenomena found in the earth surface, organic and inorganic, as functional relations of parts of a single whole. The concept was early expressed by Kant, for whom the systematic arrangement of objects by classes, as in the system of Linnaeus, divided nature into parts without organizing a system,[6] whereas physical geography "gives an idea of the whole, in terms of area" (Raum) [cf. Plewe, 8, 39]. At the beginning of the following century this concept was apparently so strongly supported by the philosophical views of the day that Butte stated, in 1811: "no scientist doubts the reality of an earth organism" [1, 230]. Similarly, within any particular area, the combination of all interrelated phenomena is not a mere aggregate but an interrelated "Whole," according to Krause as well as Butte, in 1811.

Few of the writers of that period—or indeed of any later period—distinguish clearly between the concept of unity of all the phenomena at any particular place or area, in what we may call a vertical totality or unity, and the horizontal unity of the area as an individual unit distinct from neighboring units. This latter concept was, as Bucher noted in 1827, in a sense forced upon the promoters of the new school of geography in competition with the old [51, 86 f.]. In place of the definite areal units of states, sharply defined by political boundaries, the new geography required equally definite "nat-

---

[6] Kant was writing, of course, long before Darwin provided biological science with a unified system. Gerland however thinks Kant's consideration of Linnaeus in error [12].

ural" units, somehow defined in nature. For a time, such definite natural boundaries appeared to be provided by drainage basins sharply separated by the "network of mountains," and, of course, by the seas. As increased knowledge of the actual conditions of the earth's surface made this theory untenable, the problem of finding "natural boundaries" for such "natural units" of area became much more difficult. We need not here concern ourselves with the long conflict between the supporters of "dry boundaries" (watersheds) and "wet boundaries" (the rivers), nor with the attempts to combine them. In large part, the general theory seems to have been supported by the concepts of natural philosophy which had a marked influence on nearly all the geographers at the end of this period. Indeed the expression of this philosophical concept in the geography of the time was taken up by political leaders—at least when it suited their particular purposes. On this basis Dalton justified the expansion of revolutionary France to the Rhine [*1*, 258 f.; *cf.* Spörer's long footnote on similar political arguments of a later period, *68*, 363].

While the concept of natural divisions of the lands was effectively introduced as early as 1773, by Gatterer, the consideration of each of the divisions as in itself a natural unit does not appear clearly until 1805, in the work of Hommeyer [Bürger, *11*, 7–12]. With Hommeyer, as with his predecessors, this unity represented perhaps little more than the unity of the landforms. The concept of a composite unity—the integrated total of all the phenomena of an area into an individual unit, distinct from those of neighboring areas was stated emphatically by Zeune, 1808–11, and by Butte in 1811. For Butte the individual lands and districts were "organisms" which like any organism included a physical side—inanimate nature, and a psychical side—animate nature including man. "The unit areas (*Räume*) assimilate their inhabitants" and "the inhabitants strive no less constantly to assimilate their areas" [*1*, 231].

Other writers of the period, however, did not accept these concepts without question. Rühle von Lilienstern immediately, 1811, opposed Zeune's theory with the impracticability of establishing definite "natural regions." A decade later both Wilhelmi and Selten urged that the boundaries of unit areas were not to be determined on the basis of any one kind of phenomena and recognized the difficulty of establishing definite limits where the boundaries of many different kinds were each gradual rather than sharp and, taken together, failed to coincide. "Nevertheless," said Wilhelmi, "the forms of nature are clear and distinctly separated as soon as we regard the particular constitution (of the total of all factors in any area) in its full form, rather than the boundary or transition zone" [from Bucher, *51*, 89; also *1*, 245].

The most thorough-going criticism of these concepts was made by A. L. Bucher in 1827. In a discussion fifteen years earlier, he had, with some hesitation, adhered to the concept of natural boundaries and apparently also —though this point is not clear—to that of regions as natural units. Rühle's critical discussion of Zeune, however, had made a strong impression upon him, his own efforts to establish natural divisions between parts of the continents had made him still more sceptical of the theory, and the discussions of Selten and particular Wilhelmi revealed, he thought, its fundamental fallacies—even though neither of them had arrived at that final conclusion.[7] His concern over this and similar problems in geography led him to make a critical examination of nearly forty geography texts published in Germany in the first quarter of the nineteenth century. This remarkable study, whose value was at once perceived by Berghaus [55, editor's footnote, p. 516], merits a high rank among critical works in geographic methodology. Wisotzki rightly gives its discussion a prominent place at the conclusion of his survey of the pre-classical period. The arguments with which Bucher attacked the concepts of natural boundaries and natural regions leave little for the present-day writer to add.[8] Indeed it would hardly seem too much to say that if Bucher's study had not been so generally overlooked by recent German as well as American geographers, much of the discussion of regions as natural concrete units in a later section of this paper would be superfluous.[9]

As the culminating expression of methodology of regional geography of the pre-classical period, Bucher's study represents in a sense a dead end. The long search for the true "natural divisions" he convicted as futile from the start, but he would not return to the use of political divisions. Rather he suggested that it was not necessary to establish either boundaries or divisions; regional studies were needed only for special purposes for which the areas concerned could be arbitrarily bounded in any convenient way [84–93]. The essential work in geography was concerned with the syste-

---

[7] Bucher [51, 34, 37, 44, 89 f., 116 ff.]; Wilhelmi, who had published his work anonymously, was unknown by name to Bucher; the authorship is taken from Wisotzki [1, 243–7].

[8] On the basis of Bucher's earlier discussion, Bürger lists him with Zeune as one of the two forerunners of Humboldt and Ritter to present the concept of unity of areas, but fails even to mention his later complete and emphatic reversal, even though that is clearly presented by Wisotzki, on whom he appears in part to depend [11, 11 f.].

[9] Bucher's critique also included, among other points, a vigorous attack on the use of the Mercator projection both for world maps and for those of areas in mid-latitudes, as found in a number of texts of the period [51, 53–56]. Likewise, we may mention as timely, his arguments against combining history and geography in a common course of instruction [237–42].

matic studies of individual categories of phenomena, each however in relation to the earth. As Bucher's final study falls well within the "classical
period" we will have occasion to consider these conclusions again in that
connection.

Certain other features of the geography of the pre-classical period may be
more briefly noted. The impression which the view of an area makes upon
the spirit of the beholder was emphasized by Hommeyer, who considered the
study of the aesthetic-geographic character of the *Landschaft* as a part of
geography. Hommeyer, however, was one of the first to introduce confusion
in geography regarding this term, as he specifically defined it as a portion
of territory, intermediate in size between a *Gegend* and a *Land* [*1*, 211 f.,
220]. More important in the development of the aesthetic description of
nature, according to Humboldt, were Rousseau, Buffon, Bernardin de St.
Pierre (in *Paul et Virginie*), Chateaubriand, Playfair, and Goethe, and,
among geographers, particularly George Forster [*43*, 13; *52*, 27; and *60*, II,
65–75].

It was a natural consequence of the manner of development of the new
science of geography that its proponents should see it closely related to history, but they insisted that geography should not be "the handmaid of history," but rather that the two were "sister subjects." The comparison of
the two fields was most clearly stated by Kant, as we shall see in a later section. The extent to which geographical studies should include historical
treatment was discussed by Rühle, in 1811, and, notably, by Bucher in 1827
[*51*, 237 ff.].

Though the new geography found its basis in physical rather than political features, it did not exclude the specific problems of political geography,
as we understand the term. On the contrary, one of its claims was that it
provided a firmer foundation for the study of that field than had a geography
which constituted little more than political and historical geography. In
particular, the drastic changes in the areas of states during this period constantly engaged the attention of geographers, and as we have noted, the concept of "natural boundaries," first developed as a framework for physical
geography, was carried over—all too simply—both into political geography
and political practice [*1*, 258–9]. One byproduct of this was the further
confusion of the term "natural boundary" to include linguistic boundaries
[240], a concept which recently has again been emphasized—not without
inconsistency, to be sure—by one of the rulers of Europe.

Although the new science of geography sought its foundations in the
more permanent physical features of the earth, it was characteristic of its

time in considering these primarily in relation to man. Throughout the period most writers, consciously or unconsciously, followed Kant in regarding physical features in terms of their ultimate importance to man. Müller, in 1785, and Kayser, in 1810, specifically emphasized the study of the earth as "the dwelling place (*Wohnsitz*) of man." On the other hand, one can observe a swing away from this viewpoint in those writers like Hommeyer who emphasized the description of the earth forms—in terms of pure description, without explanation—as the sole content of *"reine Geographie"*; all other features might be considered in applied geography [Plewe, *8*, 19].

One other concept of geography expressed at this time was that it was a study concerned with the "Where" of things. Stated by Lindner in 1806, it was vigorously opposed by Ritter in the same year.

Finally, it may be added, the students who were consciously endeavoring to construct a science of geography, repeatedly emphasized the importance of two cardinal principles of any scientific work. In contrast with the blind dependence on authority, characteristic of earlier periods, and the *a priori* construction of systems of supposed facts about the earth that characterized much of the work of the eighteenth century, many students toward the end of this period emphasized the importance of determining first the actual facts. Likewise they wrote repeatedly of the need to indicate specifically the sources of information [*1*, 255 ff.].

In large part, however, the new science of *"reine Geographie"* represented the expression of purposes and hopes rather than of accomplishment. In spite of the insistence on establishing facts, the geographers of this period, with few exceptions, continued to set up *a priori* systems of facts without putting them to the test, and many of these concepts we still inherit—or have constructed anew. In the century and a quarter since Butte spoke of areas as organisms no geographer, so far as I can find, has seriously attempted to establish the organic character of any single region, and yet, as we will see, current textbooks of the most modern type repeat this *a priori* assumption. In the century since Bucher's scientific integrity led him to announce his failure to establish a division of lands into natural regions, no one, so far as I can find, has seriously attempted to show the error of his methods or argument, or to produce what he found impossible; nevertheless the concept of the region as a definite unitary entity is prominent in the writings of a large number of geographers today.

B. THE CLASSICAL PERIOD: HUMBOLDT AND RITTER

If the geographers of the late eighteenth century developed the greater part of the theoretical concepts of the new science of geography, the transformation of their "ideas, demands, and wishes into facts" was largely the work of Alexander von Humboldt and Carl Ritter.[10] It is an extraordinary fact that modern geography in all lands should owe so much to two men living at the same time in the same country—for over thirty years in the same city.

This situation was not merely the result of a coincidence. Both of the "founders" of geography depended in large part on their predecessors whom, of course, they had in common. Furthermore, Ritter repeatedly asserted— and no student of his work has questioned the fact—that in many respects Humboldt, the elder by ten years, was his teacher. Peschel rather emphatically endorses Ritter's statement that, without the work of Humboldt, his own work could never have been produced [66, I, 324; 49, I, 54 f.]. Though we may find some exaggeration in the statement, its essential truth is testified by the innumerable places where Ritter echoes the concepts of Humboldt,[11] as well as by his hundreds of specific references to the latter's writings. That these are not merely the meticulous notations of a pedantic scholar is testified by Humboldt himself. In the introduction to his final work on Central Asia, he refers to the earlier publication of a series of itineraries and observes that "the terminology that I proposed for the mountain systems of Asia, and the geologic views on their mutual independence, have been discussed with great superiority by my old and illustrious friend, M. Carl Ritter, in his great work on Asia." [59, I, xlviii].

Although Humboldt made repeated use of Ritter's work in his own writings—the citations of Ritter in *Asie Centrale* must run well over a hundred—his general concepts of geography were evidently well formed before he met the younger student and before the appearance of Ritter's first geographic writings. Humboldt had been trained neither for geography nor for science, but rather for governmental service—through courses in technology, economics, history, philology, etc. His dominant interest however was in the study of nature—as a mine inspector he was concerned to

---

[10] The expression is taken from Wisotzki, who however speaks only of Ritter [1, 257]. In the judgment of most other writers, including Peschel, Richthofen, Hettner, and Penck, Humboldt's importance to geography was equal to or greater than that of Ritter; see particularly Penck [137, 158–76].

[11] Compare, for example, Ritter's statement of the comparison of history and geography in his lecture of 1833 [50, 152 f.], with the statement to be found in Humboldt's lectures given at Berlin, 1827–8 [52, 14]; it is possible however that both statements may be independently derived from Kant. It would make an extremely interesting subject for a dissertation to investigate in detail the common elements in the work of the two founders of geography. Döring's brief statement is based on an exhaustive study of Humboldt but on only a brief examination of Ritter's writings [22, 160–63].

study subterranean vegetation—and a close relationship with George Forster turned this interest in the direction of geography. In particular a trip which they made together in the Rhineland and to England, in 1790, stimulated Humboldt's ambition to undertake scientific travel and introduced him to Forster's methods of geographic study [47, I, 41; 60, II, 72; cf. 20, I, 95–108; 8, 23–4; see footnote 5]. It may well be that Humboldt's concept of geography had already been influenced, indirectly, by Kant [20, I, 46, 50 ff.]

During his short period in governmental service Humboldt published a number of scientific papers, including one or two in geography, but his great work, of course, came as a result of his travels in tropical America, in 1799 to 1804, as well as the later trip to Central Asia, in 1829. It would be an error, however, to suppose that his geographic work was limited to his own observations, or even to the lands which he had observed. Of over sixty years of active geographic study, his explorations hardly totalled five years. (To be sure much of Humboldt's time in the latter part of his life, when he was a very active privy councilor to the king of Prussia, was occupied with state problems.) Many of Humboldt's studies, even before the *Kosmos,* extended over the whole world and depended on the utilization of the work of countless other explorers and scientists. Indeed, during the greater part of his career Humboldt was as much a geographer of the study and library as was Ritter.

Although Humboldt's work had an important influence on Ritter throughout his career, Peschel was unquestionably exaggerating when he stated that their first personal contact, during several weeks in 1807, was the critical experience that definitely turned the younger student to geography. Ratzel has shown that neither this statement—frequently repeated by later writers after Peschel—nor the more general inference that Ritter came into geography from history, is supported by the facts of his life [26, 423, 389 ff.]. Probably no student of his time—and few since—have been so specifically trained for geography as was Ritter.[12] From very early childhood until he went to the university, he lived in a school, in Schnepfenthal, in which education followed the precepts of Rousseau and Pestalozzi: knowledge of the world was acquired by direct observation of nature on walks and longer trips. Throughout this period the youth was under

---

[12] Unless otherwise indicated, the biographical data are taken from the authoritative work by Kramer, based in part on Ritter's diary [24]; it has been abstracted by Fr. Ratzel in his critical sketch of Ritter's life and work [26]. The account given here is disproportionately long in order to correct erroneous impressions common in nearly all English accounts. Dickinson and Howarth [10, 150–61], though recognizing Ritter's importance, have apparently been misled in various important respects, possibly by the biography by the American writer, Gage [cf. Kramer, 24, II, 35].

the particular personal care of J. C. F. GutsMuths, a teacher whose major interests were in nature study and in geography, and who himself made some contribution to the development of geography [Ratzel *26*, 383, 388 ff.; Plewe *8*, 39]. At this time, geography was Carl Ritter's favorite study, and in the following university years the natural sciences commanded his chief attention together with history and theology. These studies did not end when he left the university: for nearly twenty years as tutor in a private family in Frankfurt, he had both the need and the opportunity to continue the study of natural sciences, geography, and history, and to make numerous field trips in the Rhineland, Switzerland, etc., making careful, detailed observations. Likewise he was able to come into close relations with many eminent scholars of the day, of whom the most significant for him were the anatomist, Sömmering, the geologist, Ebel, and the leader of educational reform, Pestalozzi.

In sum, we may conclude that Ritter's preparation for research covered an extremely wide field, in which on the one hand natural sciences, and particularly the observational methods of nature study, predominated, but in which, on the other hand, his interest was increasingly in human problems, *i.e.*, history. Geography, as he read it in the writings of Herder, a follower of Kant, maintained the connection between his earlier field of studies and his ultimate interests [*cf.* Schmidt, *7*, 78 f., and 41–4]. Although his meeting with Humboldt, in 1807, undoubtedly stimulated his interest in geography, for the very reason that Humboldt had demonstrated so clearly the importance of earth conditions to man, for at least ten years thereafter Ritter was still uncertain whether to continue his geographic work or to shift to history.

His first publications—overlooking a series of geographic studies for elementary students—dealt with the geography of Europe. These appeared between 1804 and 1807 and made a considerable impression in Germany. One of these, which was also enthusiastically reviewed in France, was a brief study in the systematic geography of Europe, based on six maps, with explanatory text, of individual categories of phenomena. Appearing some years before Humboldt published similar maps, it constituted, according to Plewe, "the first physical atlas" [*8*, 30; *cf.* Kramer, *24*, I, 255]. Though all these geographical studies appeared before the celebrated meeting with Humboldt, one may presume that Ritter had already been influenced by the latter's writings, since he wrote at that time that he "had devoured all his published works with a kind of ravenous appetite" [*24*, I, 167].

During the next few years Ritter prepared a more complete systematic geography of the world, a "Handbuch der physischen Geographie," which he intended to be the completion of his geographic work and a bridge to studies in history, to which he wished then to devote the rest of his life. For some reason however—possibly because of the criticism offered by the geologist, von Buch, who read the manuscript at Ritter's request—he decided not to publish it. Parts of it however were utilized by various colleagues to whom he freely offered it, so that the work appeared in a number of texts of the time, in some cases with the original authorship indicated, in others not [24, I, 205–7, 258–68].

During the same period Ritter's studies on Asia led him into the historical field, in which he prepared a lengthy study in ancient history (published later, 1820)—according to Ratzel "the only work of Ritter's of exclusively historical character" [26, 410]. Apparently, however, he contemplated shifting entirely to history. To fulfil his geographical work he outlined a project which should cover the world by regions—originally in a four volume Erdkunde. This would then supply the sound foundation for historical studies. Working at the University of Göttingen—where at the same time he took further courses in history, but also in mineralogy—he completed the first volume, on Africa, according to plan, but soon found that the enormous amount of material that he collected on Asia could not be handled in one volume, so that he issued the second volume as but a part of the study of Asia. The appearance of these two volumes, in 1817 and 1818, made a tremendous impression in the academic world; almost at once, says Marthe, Ritter was recognized "without question as the reformer of geography, as the master who first made that field into a science" [25, 6; see also Kramer, 24, I, 377 f., 404 f.; Ratzel, 26, 689 f.; Schmidt, 7, 86].

Ritter's position as the master of academic geography was therefore definitely established before he had held a single academic post. After one year as professor of history and geography at the gymnasium in Frankfurt, the efforts of various officials in Berlin—including Wilhelm v. Humboldt, the brother of Alexander [24, I, 436, 447]—were successful in arranging for a double position for Ritter, in the military college and at the university. In 1820, Ritter went to Berlin as the first university professor of geography. At just what time he recognized that his plan of "finishing up" on his geographic work in order to go on to history was unattainable is not clear; in any case he continued to work in geography in Berlin during the remaining 39 years of his life, although it may be said that in the latter volumes of the Erdkunde, his interest in history became more and more pronounced.

Following Humboldt's final shift of residence from Paris to Berlin, in 1827, the professional relations between the two increased steadily. According to Kramer, who lived for some years in Ritter's house, few days passed, at least in the latter years, in which—if they were both in Berlin—communications of greater or less importance did not pass between them [24, II, 85].[13]

This is not the place to attempt an appraisal of Ritter's contribution to geography, nor should such an attempt be necessary. No geographer in history had his work discussed so repeatedly, and often critically, as Ritter. The studies referred to in the following paragraphs are no mere obeisances to a great "traditional" figure. Except for the quotations from Humboldt, they are taken from critical examinations of Ritter's work by authors who did not hesitate to point out its weaknesses and the extent to which it failed to live up to his stated purposes. Nevertheless those students who have thoroughly examined his work appear to agree unanimously with Humboldt in recognizing its "masterly" character. The more important critical studies of Ritter's work are those of Peschel, Marthe, Ratzel, Hözel, and Wisotzki [66, I, 336 ff.; 25; 26, 405–28; 27; 1, 267–323].

In view of the real differences between Ritter and Humboldt, both in personal character and, as a result, in their work, but particularly because of the supposed opposition between their points of view in geography, it is appropriate to note Humboldt's opinion more fully. The earliest expression that I have found is in the lecture course that Humboldt gave at the University of Berlin, 1827–28. Humboldt did not limit himself to the laudatory phrases which the occasion called for when he described Ritter's *Erdkunde*—at that time only the first two volumes had appeared—as "the most inspired work of this kind (comparative geography), that our century has yielded; it is the first work in which is presented the influence which the surface-view has had on the peoples and their fates" [52, 14 f.]. In a personal letter to Ritter a few years later, 1832, he expressed his great admiration of the work on Asia, adding that it gave him great pleasure to sing its praises to the king and all at the court, telling them that "such an important work had not appeared in thirty years" [24, II, 120]. Humboldt used his

---

[13] Many of the secondary sources present a very different picture of the relation between Humboldt and Ritter, even indicating that there was essentially no personal relationships. I have found, however, no refutation of Kramer's statement; and possibly even some support in the references of each of the principals to the other in private letters to third parties [cf. 20, II, 131; 19, III, 40, 62, 89]. One gathers from these however that the relationship was very largely, if not entirely, limited to their single field of common interest, geography. Undoubtedly, as Ratzel suggests, the praise which each rendered to the work of the other was expressed in exaggerated tones—"as though both were hard of hearing"—but this was considered in good form at that time, and there appears to be no basis for questioning their essential sincerity. On the other hand, the great difference in temperament, social and religious background, and non-professional activities made it hardly possible, as Ratzel notes, for a close personal friendship to develop between them [26, 423 ff.].

influence at the court to have Ritter relieved of some of his heavy duties, presumably at the military college, in order that he might complete the work on the *Erdkunde* [*24*, II, 32 f.; *20*, II, 127]. That Humboldt did not alter his opinion is indicated by numerous statements published in later writings. In the introduction to his Central Asia he finds Guigniaut's verdict correct: "one of the greatest and most magnificent monuments raised to science in our day" [*59*, I, xlviii f.]. In the first volume of the *Kosmos* (1845), he speaks of "Carl Ritter's great and inspired work" which "has shown that the comparative geography attains thoroughness only when the whole mass of facts that have been gathered from various zones, is comprehended in one view, is placed at the disposal of the integrating (*combinierenden*) intelligence" [*60*, I, 18]. Later in the same volume he refers to Ritter's work as the fulfillment of a part of the plan of Varenius, which—in the earlier period of scientific development—Varenius himself had not been able to accomplish: "It was reserved for our time, to see the *comparative geography* cultivated in masterly fashion, in its widest compass, indeed in its reflex on human history, on the relations of the form of the earth to the direction of the characteristics of peoples and the progress of civilization" [*60*, I, 60, with footnote reference to Ritter].

The new direction which Ritter was endeavoring to bring into geography is explicitly stated in many writings from 1804 on. His first principle, frequently repeated, was that geography must be an empirical science, rather than one deduced either from rational principles—from philosophy— or from *a priori* theories of "general geography." "The fundamental rule which should assure truth to the whole work," he wrote of his *Erdkunde*, "is to proceed from observation to observation, not from opinion or hypothesis to observation" [*49*, I, 23]. *

The first major plank in Ritter's reform of geography therefore was to call a halt on all attempts that led from theoretical considerations to systems of earth forms. He was among the first to show that the theory of a continuous network of mountains was opposed by the records of observed facts and that, likewise, there was no general correspondence, as often assumed, between the crest-lines of the mountains and the divides of drainage basins. Although the teleological view of the universe that Ritter had received through Herder from Kant provided the background of purpose of his study, he was not, like Kant, a philosopher who required geography in order to establish theories of world knowledge [*cf.* Adickes, *13*, 75 f., 190 f.; Gerland, *12*]. Convinced that there were laws governing the relation of human and non-human phenomena on the earth, Ritter was in no great hurry to establish them, but rather felt that, if he could bring together all the facts and relationships observed in areas, these would make it possible

[54]

---

* In contrast to deductive reasoning from a priori theories, as described on page 48.

to state such laws. Attempts to develop these laws first—his own, perhaps, as well as others—had proven unsuccessful: "we must ask the earth itself for its laws" [*49*, I, 4].

Ritter was therefore the first great opponent of what may properly be called "arm-chair geography." On the other hand, as a scholar and teacher, rather than explorer or field worker, as Hettner observes [*2*, 310], Ritter represented the geographer in library and study: the observations used in his writings were not obtained directly, but rather from the works of others who did observe at first hand. Not that he did not travel and observe, as is often erroneously supposed; on the contrary he traveled much throughout his life, making detailed observations in many parts of Europe [*24*, I, 271–331, II, 64–83]. But since he unfortunately became so deeply involved in his studies of Asia as never to commence the volumes on Europe, the detailed observations which he made on his own trips had only the value of enabling him to "understand what others had observed."

In basing his work on observations of other students, Ritter was not satisfied with the limited amount of material which had sufficed for most of his predecessors. In order to achieve the highest degree of accuracy, his plan was to bring together on each point in question the greatest number and variety of trustworthy witnesses, of all times and from all peoples, so that their testimony might be seen together, whether in agreement or for comparison [*49*, I, 23]. The material was by no means limited to the works of travelers, whether scientific or otherwise, but was drawn also from the works of a large number of specialized scientists, as one may observe by glancing through the introduction or by noting the footnotes in any of the volumes.

It seems more than probable that Humboldt had Ritter's methods of work, as well as his own, in mind, when he discussed the use of sources in his lecture course given during the winter of 1827–28, at the University of Berlin. "What an individual observer can see, is naturally small in comparison with that which has been observed during so many centuries. If it is therefore very important to gather the observations, it is likewise necessary to be engaged in some part of the sciences of nature, for only so can one learn to understand that which others have observed. For those who can spend the great amount of time and energy necessary, the most important method is the study of all travel descriptions, of all the individual treatises, for only out of the special cases can the general be recognized . . . " [*52*, 26].

One cannot glance through the pages of Ritter's presentation of Asia without being impressed by the great number and variety of the sources upon which he depended: the works of contemporary scientists like Buch and Humboldt—whose works are no less honored by use than by praise—the reports of earlier travelers, whether non-scientific travelers or those of whom Humboldt complained that, while "well prepared in the isolated branches of science," they rarely had "a sufficiently varied knowledge" to study phenomena of different categories in interrelation [47, I, 4]; and in addition a great host of works of all kinds from the pens of medieval and classical writers. In large degree, these were the same kinds of sources, in many cases the same works, used by Humboldt in his writings. Although the very quantity involved made it impossible for Ritter to complete his original purpose of covering the whole world—and indeed deflected him from his particular aim of correlating the human and physical facts established—he at any rate could claim to have assembled for comparison all the evidence available to establish the facts. The first of many to testify to his success in this regard was Humboldt, with particular reference to the area with which he himself was personally familiar, Central Asia. Shortly after Ritter sent Humboldt the first volume on Asia, in its second edition of 1832, Humboldt wrote to him that he could find no words in which "to express the real admiration with which I am filled by your gigantic work on Asia. For the past two years I have been most earnestly engaged with Inner-Asia, with the use of all sources, but how much has become clear to me only in the three days in which, without break, I have been reading in this work. You know *every-thing* that has been observed for centuries, with your particular sagacity you arrange everything together, from much-used material you gain ever new and grand views, and you reproduce the whole in the most desirable clarity" [24, II, 120].

It was not, however, Ritter's purpose merely to assemble a mountainous accumulation of information. The enormous multiplicity of observations must be organized according to the chorological (*räumliche,* or spatial) principle, which he considered both in a horizontal, and—with reference to elevation—in a vertical sense [49, I, 24]. This did not mean, however, simply an accumulation of facts concerning each area; rather he wished to show the "coherent relation" (*Zusammenhang*) in terms of cause and effect, of the different features [24, I, 250 f.] and the "formation of the multiplicity of features into the essential character of an area" [8, 32]. By proceeding "from the simple to the compound," he wished to establish the totality of interrelated features as the distinctive character of each area [49, I, 24]. We are but paraphrasing any number of his statements if we say that he was concerned with reducing to order observations made in nature [49, II, xv ff.].

In these respects Ritter's purposes were not so sharply in contrast with those of his predecessors as he believed, but the great difference was in the extent to which he put these into practice—even though this was by no means complete. In certain other respects his position was in clearer opposition. Though he followed the movement of *"reine Geographie"* in using "natural lands" rather than political states, he objected to the overly-simple, and in fact impossible, procedure of dividing only by mountain ridges, or the equally simple division by drainage systems. According to his theoretical statements his divisions were made in terms of relief (in particular he divided the upper, middle and lower portions of stream basins), but in practice he also recognized other factors, as in his separation of the Sahara and in its subdivisions in terms of climate, vegetation, etc. [49, I, 959 ff.]. In this, he was consistent with his fundamental principle, that we should not evolve a theory of natural divisions of the lands, but must always look to nature for its principles [49, I, 4].

He reminded his contemporaries, in their intense argument over the relative values of "wet" and "dry" boundaries, that the mere drawing of boundaries was but a means toward the real purpose of geography, the understanding of the content of areas. This view he emphasized by a phrase which, though not clear in itself, was perhaps sufficiently clear in the context in which he used it as to justify its frequent repetition in later decades, so long as the context was not overlooked. Geography, he suggested, is the study of *"der irdisch erfüllten Räume der Erdoberfläche"*; i.e., the areas of the earth are not to be studied in themselves, as mere divisions of the earth's surface, but neither are the objects that are found on the earth surface to be studied in themselves—in geography—but rather the areas of the earth surface are to be studied in terms of the particular character resulting from the phenomena, interrelated to each other and to the earth, which fill the areas [50, 152].[14]

[14] In Leighly's recent discussion of "Methodological Controversy in Nineteenth Century German Geography," this phrase is mistranslated to describe geography as "the science of space-filling terrestrial objects" [222, 251]. The phrase, to be sure, is not ascribed to Ritter, but appears in the discussion of Gerland's thesis, as "a windy definition that had long been current." As no specific page reference is given, it is only after searching through Gerland's essay—more than fifty pages of fine type—that one finds that Gerland had quoted Ritter's phrase correctly [76, xvi], though without mentioning its authorship, supposing, no doubt, that any reader would recognize it as one of the best known quotations from Ritter. On the other hand, it is only fair to add, Gerland is in part responsible for the misunderstanding: his supposed "disposal" of Ritter's concept is based on the misconstruction represented by Leighly's translation—i.e., that translation correctly represents Gerland's misinterpretation of Ritter's meaning.

Although Leighly's paper appears in general well-documented, the general absence of specific page references makes it very difficult to find the original statements given in translation or paraphrase. As a number of errors in interpretation have been made, it seems necessary to note these, in footnotes, at the appropriate places in our historical survey. *

* See Supplementary Note 3.

In all of these respects there was no essential difference between Ritter's view of geography and that of Humboldt. Since Ritter had expressed many of these views as early as 1804 (*24*, I, 250], one may question Peschel's statement that it was in their first personal contact in 1807, that Ritter "was brought to see clearly the great function of geography—to show the unison of human social phenomena with the complex of natural forces in the locality" [*66*, I, 324, 341; see also Döring, *22*, 160–3]. Ritter may however have acquired his ideas from Humboldt's writings, with which he was already familiar, and the lengthy conversations with Humboldt over several weeks may well have emphasized and clarified his concepts [*24*, I, 165–7]. Furthermore Peschel is simply echoing many of Ritter's statements when he says that the latter's work in relating historical phenomena to physical phenomena would have been impossible had not physical geography been advanced so notably by Humboldt: "a Carl Ritter could not have preceded an Alexander v. Humboldt, but could only follow him" [*66*, I, 324].[15]

On the other hand, the great difference in temperament and in general outlook on life gave a very different color to the work of the two founders of modern geography, and in the works of their followers this difference became fundamental. It is natural enough therefore that later students should tend to emphasize the differences between the two men and even to find an opposition between them of which neither of them were conscious—familiar though they were with each other's work. Interesting and dramatic as such a contrast may be, historical objectivity does not permit us to exaggerate differences that may be superficial into a fundamental contrast. *

(Subsequent to writing this conclusion, I find that Ratzel had made much the same statement, emphasizing the many respects in which Humboldt's geographical work, even in its philosophy, was similar to that of Ritter [*26*, A. d. Biog., 690 f.].)

In fairness to all varieties of critics of Ritter it must be said that in his voluminous and variegated writngs one may find apparent justification for almost any possible judgment. If at times he wrote as a scientist, at other times he wrote as a philosopher—indeed it did not occur either to Ritter or to Humboldt that science and philosophy could be completely separated—and at other times Ritter wrote as a theologian. Furthermore, Ritter was unfortunately far from careful in his selection of terms, "he did nothing to confine the fluid and ambiguous quality of words," "strict logic was not his

---

[15] Ritter's most complete tribute to Humboldt is to be found in the address he gave at the celebration of the fortieth anniversary of Humboldt's return from America [*Zeitschr. f. allegm. Erdk.*, Berlin, 1844, 384 ff.; republished in part in *20*, I, 469 ff.]. The festive occasion may have excused the somewhat excessive strain of the laudation.

* See Supplementary Note 4

forte" [Plewe, *8*, 28]. Any number of writers have debated what Ritter intended in calling his *Erdkunde*, *"allgemeine vergleichende"* (general comparative) and the net conclusion appears to be that we are not to look for any specific meaning in these terms, but merely to understand that he had certain general purposes in mind that he wished to contrast with those of his predecessors [see Hettner, *2*, 310; and Plewe, *8*, 28 ff.].

Ritter is therefore a writer who cannot be judged, either in his work or in his methodology, on the basis of a few striking quotations pulled out of their context and out of their relation to the ideas of his contemporaries. A view of Ritter constructed in this manner has the effectiveness, but only the degree of truth, of a caricature. We need not however make an exhaustive study of Ritter's work, but may base our appraisal on the many thorough studies of that kind made by previous students, in each case, however, referring their findings to the original work. The latter must include his actual geographic work as well as his methodological studies, since the methodology is expressed no less in the actual work itself.

In particular, one must avoid the error of attempting to determine the character of the geographic work of either Ritter or Humboldt—or of any other writer—from their own statements of ultimate purposes. The fact that Columbus set out across the Atlantic for the purpose of finding a shorter route to the Indies, and apparently died still believing that he had found it, does not cause us to judge his importance in history in terms of the establishment of connections with the Indies—to that he contributed little more than a permanent confusion in geographic and ethnographic terminology.

Writers of a later period, living in a different philosophical atmosphere, attempted to discredit the importance of Ritter's work by attacking his philosophical concepts, in particular his pietistic, teleological view of the universe. Commonly this is done under the naïve assumption that the writer has a purely "scientific" concept, in contrast with a philosophical concept of the universe. We cannot here discuss this elementary philosophical problem, but merely point out that every scientist has his philosophy of science, and that that philosophy is neither science itself nor is it the product of science. That Ritter recognized this distinction may be seen particularly clearly in his discussion of Fröbel's criticism of his work. To Fröbel's objection to his teleological view—which was not an argument but simply a different philosophical assumption of science asserted without foundation—Ritter replied properly in philosophical terms [*55*, 517 f.].[16] Though we may

---

[16] Leighly's comment that Fröbel's attack on Ritter's teleology "evidently touched a tender spot" [*222*, 245], is a conclusion for which this writer finds no evidence. Any, who care to, may readily judge for themselves, as Leighly quotes most of Ritter's reply on this point, omitting only the meat of it, namely, the direct reply to Fröbel's analogy of cows and men which he had quoted in full [244]. It is appropriate to call attention

reject his philosophy we can respect his statement of it. On the other hand, in reply to the attack on his scientific methods and concepts, he limits himself to scientific terms.

The proper question to ask in criticising the scientific works of any writer—as Plewe among others has noted—is not: what particular philosophy of science is exposed by these writings, but rather: to what extent does the writer's particular philosophy affect his work. One may have a teleological philosophy—as did Leibnitz, for example—and nevertheless still produce genuinely impressive research [8, 70 f.]. Though many of Ritter's followers, like Guyot [64], depended on the teleological factor to explain geographic details, competent critics agree that Ritter seldom was guilty of this procedure, which in fact he vigorously condemned [Marthe, 25, 21 f.; Wisotzki, 1, 297–304].

In general, we may say that the teleology in Ritter's geography was an attempt to interpret philosophically that which science could not explain. Many of the statements at which the scientific critics of the late nineteenth century pointed in holy horror would stand comparison with current utterances of a number of eminent physical scientists who find themselves in a similar position. In Ritter's case, there were three fundamental facts of geography for which science had no explanation—and presumably still has none—namely, the uniqueness of the earth in the universe, so far as we know it; the earth as the home of that unique creature, man; and finally— the fundamental explanation of a host of geographic facts—the differentiation in character among the major land units of the world.

In our day, to be sure, it is more common for scientists to express their philosophical views of problems presented by their scientific work in separate publication; the intellectual world of a century ago did not call for this physical separation, nor can we regard it in itself as a major difference. "Even though a spirit turned toward the eternal breathes through Ritter's presentation," concludes a modern student, Schmidt, "even though his highest enthusiasm is fired by his religious *Weltanschauung,* nevertheless in his research, he in no way proceeds from preconceived opinions; his scientific procedure was directed throughout on temperate, purely factual comprehension of the facts and their relations. . . . Ritter strove in the knowledge of

---

to a lengthy statement that the same Fröbel wrote a quarter of a century later: "If nature—and history . . . —must be studied according to physical forms of judgment, it does not follow that one may not also regard nature, as well as the moral world, according to ethical forms of judgment. It would put a ridiculous restraint on reason to deny it this right. . . . There is of course nothing new for philosophy in all this; but many of our younger natural scientists, who take pride in not being philosophers, commit the error of believing that that against which they close the eyes of the mind therefore does not exist" [62, I, 79 f.].

the earth for a comprehension of the divine world plan in no other way than the natural scientists pursue the thought of evolution" [7, 85].

Although most critics would add some degree of qualification to Schmidt's judgment, it seems fair to conclude that the disturbance caused in the modern scientist's mind by Ritter's teleological expressions is far out of proportion to their importance in his work. An illustration is offered in the case of his treatment of the continents, specifically in two lectures given before the Berlin Academy of Sciences, one in 1826, the other in 1850, republished later together [50, 103–28, 206–46]. They constitute some 65 pages in which the author describes the size, form, construction, and climatic conditions of the different continents, seeks to construct "a law of the arrangement" of the parts of each continent to the whole [220], and particularly endeavors to show how these conditions have determined the development of the peoples of the different continents at different periods of history. One must look with a very careful and critical eye to find any statement that would disturb the scientific determinists of the late nineteenth century. Near the beginning of the first lecture, to be sure, Ritter expressed his belief that the earth had been planned as "the temporary nursery of the human species" [104] and a sentence at the end of the second lecture suggests that a part of this plan was the particular form and position of each of the continents, which had led—as his detailed discussion had demonstrated—to the particular function that each had played in the course of world history [243].[17] Those who can overlook such expressions of belief will, like Humboldt, test the essays in terms of their scientific content.

Humboldt's comment in this connection also throws light on his interest in human geography. Noting that he himself had long insisted on the great

---

[17] Leighly's translations of these two statements are likely to mislead the reader [222, 242 f.]. The omission of a major phrase in the first quotation—an omission not indicated—seriously changes the meaning, as may be seen by examining the following quotation of the original in full (the parentheses indicate the parts omitted).

"In den Gesamterscheinungen der Natur und der Geschichte treten die Einwirkungen dieser tellurischen Anordnung des Planeten und seiner Verhältnisse überall hervor, da er (zum Schauplatz der Natur und ihre Kräfte wie) zum Träger der Völker von Anfang an eingerichtet ward, als Heimath, Wohnort und temporäre Entwicklungsanstalt für das Menschengeschlecht, (das ohne diese Bedingung nicht gedacht werden kann)." In other words if anyone should say "geography studies nature as well as man," he could be quoted as saying: "geography studies man."

In the second quotation the phrase "from the beginning of time" is transferred from the middle of the sentence to the position of emphasis at the beginning. Since this construction is no less common in German than in English we must assume that Ritter did not intend to give the phrase that emphasis. (The originals may most readily be found in Ritter's republication of his essays [50, 104 and 243], unchanged from their original form as they appeared in the less readily available publications to which Leighly refers.)

influence that the form of a continent exerts on its climate and vegetation, he discusses Strabo's effort to show the relation of the articulate and peninsular form of Europe to the development of its civilization, and adds: "In our time M. Ritter has developed with great wisdom the analogies, in physical and political characteristics, that the three peninsulas of Asia offer to the three peninsulas of Europe. The two groups offer centers of culture of very distinct physiognomy" [59, I, 67 f., footnote].

Nevertheless there were certain respects in which Ritter's philosophy did affect his work, and even more that of his followers. His purpose led always from the individual facts toward a "Whole" of all phenomena; rather than to trace the inner relations of phenomena to a "last cause," his purpose was to find the coherence of forces in a Whole, and thus ultimately to indicate the purpose of the Whole [Wisotzki, 1, 304]. This purpose must surely be found most particularly in the highest of earth creatures, the only one who could conceive of an organization of the Whole, namely man (it was logically false, he replied to Fröbel, to suggest that it could just as well be found in cows [55, 518]). His consideration of the earth therefore logically centered on man.

This view, which may be found in his first geographical publication, in 1804, is expressed particularly in the statement of purpose with which he introduced his *Erdkunde:* "to present the generally most important geographic-physical conditions of the earth surface in their [or possibly "its"] natural coherent interrelation (*Naturzusammenhang*), and that (the earth-surface) in terms of its most essential characters and main outlines, especially as the fatherland of the peoples in its most manifold influence on humanity developing in body and mind" [49, I, v].[18]

---

[18] Ritter's statement is not entirely clear in the original, and minor differences in translation into English are possible. The translation given above was checked by two German students. The omissions in Leighly's quotation [222, 243] make the translation grammatically impossible, as is shown by the following reproduction of the original in which the parts he omits are enclosed in parentheses. Ritter stated that his purpose in the *Erdkunde* was: "*die (allgemein) wichtigsten, geographisch-physikalischen Verhältnisse der Erd (oberfläche in ihrem Naturzusammenhange, und zwar ihren wesentlichen Zügen und Hauptumrissen nach) darzustellen, (insbesondere als Vaterland der Völker) in dessen mannigfaltigstem Einflusse auf körperlich und geistig sich entwickelnde Menschheit*" [49, I, v]. Granted that the original statement is long-winded and somewhat difficult to translate, the omissions reduce it to a grammatical ruin, from which one can form a complete statement only by ignoring grammatical forms and cases. The result is a simpler statement than Ritter's, but it is not Ritter's. Grammatically it is clear that the main part of the statement is given by "*die . . . geographisch-physikalischen Verhältnisse der Erdoberfläche in ihrem Naturzusammenhange darzustellen.*" Even if one may overlook the dictates of grammar, no one familiar with the writings of Ritter would suppose that he could state his purpose in geography without the use of some form of the word "*Zusammenhang*"—

While this concept of geography unquestionably places a major emphasis on the earth as the home of man, it does not restrict it to that particular view. "Independent of man, the earth is also without him, and before him, the scene of the natural phenomena; the law of its formation cannot proceed from man. In a science of the earth, the earth itself must be asked for its laws" [*49*, I, 4]. The first volume of the *Erdkunde* gives detailed consideration to areas in Africa of apparently no concern to man [*cf.* Plewe, 8, 70 f.]. In the later volumes, however, Hettner finds that Ritter concentrated his attention on man to such a degree as to forget his purpose of establishing the total association of nature and man [*161*, 87; *cf.* Marthe, *25*, 18; Ratzel, *26*]. Nevertheless one of his major accomplishments, according to Marthe, was to establish the physical foundations of geography in detailed study of what we would term regional landforms [*25*, 24 ff.].

Further, Ritter's particular philosophical view—but also perhaps his failure to submit concepts to rigid logical examination—permitted him to further the traditional concept of regions as "natural divisions" of the earth-surface. That Ritter's natural philosophy has continued to affect, in this respect, the views of geographers who would deny that philosophy has been repeatedly suggested by Hettner [in 1908, *300*, 7–13; repeated, *161*, 299–306].

Finally one may conclude that in his teaching, in contrast to his writing, Ritter's teleological views were more strongly impressed upon his students. There is some evidence for that in the lectures published shortly after his death [*61*], and far more in the performance of his students.

Unquestionably the modern scientist feels more at home in the writings of Humboldt than in those of Ritter. Peschel, though deducing from certain of Humboldt's statements that he could not have been a materialist, honors his avoidance of the confusion "not uncommon with the materialists as well as with their opponents, of presenting scientific investigations in some form of religious light" [*66*, I, 305]. On one point, to be sure, Peschel finds that "the noble heart of Humboldt appears to have a bit corrupted his critical sense," namely in regard to race questions. Humboldt vigorously maintained the theory of the "unity of the human race," a concept that "came to prevail first through Christianity" and could be ignored on the slave-market only because of "the degeneration of Christianity by great wealth." The opposite "hypothesis of racial gradations among men" he characterized as

---

in later parts of his introduction it is repeated two or three times per page on consecutive pages; it was no less important, as we will see, in Humboldt's conception of geography.

"not only unkind (*lieblos*) but also false" [*52*, 183–4; *60*, I, 385 f. and foot-note]. Although Humboldt's thesis would find more support among anthropologists today than that on which Fröbel as well as Peschel based their criticism, the criticism itself was essentially justified: his views on this question were strongly influenced by non-scientific considerations. In his youth, Humboldt had first found intellectual stimulus in what seemed to Goethe, Forster, and many others as the "barren environment of Berlin," almost exclusively in a small circle of Jewish intellectuals. From them he had learned of Lessing and Kant [according to both Löwenberg and Dove, *20*, I, 40–49; II, 292 f.]. He himself described the impressions made upon him by the degrading scenes of the slave-market, which he first witnessed in tropical America [*47*, I, Bk. II, Chap. 4]. Both in political circles and in his scientific writings he repeatedly and vigorously expressed his views on the race question. In his master work, the *Kosmos,* he chose to "close the general presentation of the natural phenomena of the universe" with the following quotation from his brother, Wilhelm v. Humboldt: "We wish to note one idea which is visible in ever increasing validity through the whole of history . . ., the idea of humanity, . . . to treat the whole of humanity, without consideration of religion, nationality, and color, as One great closely related race, as one Whole existing for the attainment of one purpose, the free development of inner powers" [*60*, I, 385 f.].[19]

In common with most other students, Peschel fails to note the extent to which Humboldt's philosophical views find repeated expression in his scientific writings, nor does he note a certain similarity between those views

---

[19] Humboldt's active work on this question is of some significance in the history of both Europe and America. Though a nobleman and privy councilor to two successive kings of Prussia, Humboldt retained through life "the ideas of 1789" acquired in his young manhood from the French Revolution, and stimulated again in 1804 by his six weeks' visit to the United States, nearly half of which was spent with Thomas Jefferson, then president [*20*, I, 393, ff., II, 293, 295]. Fifty years later, discouraged with the political situation in Europe, and looking on the United States as "the bulwark of a rational freedom," but fearful that the curse of slavery might destroy it [*19*, I, 16], he permitted his name to be used in support of Fremont's presidential campaign of 1856 [*20*, II, 295 ff.]. Greatly disappointed in the outcome of that election, he was on the other hand successful in the following year in introducing a law prohibiting slavery in Prussia; further, he courageously continued the general attack on negro slavery by publishing an open letter essentially supporting the views of Fröbel, at that time still politically suspect. (Their disagreement was over the scientific arguments; on the issue itself they agreed, [*cf.* Fröbel, *28*, I, 303].)

In the 1840's Humboldt took a vigorous and effective part in opposing the enactment of restrictive laws against the Jews. The present commentator, writing at the moment in the Germany of 1938, can only regret that Humboldt's statement of nearly a century ago is still pertinent: "the history of the Dark Ages shows to what aberrations such interpretations give strength" [*20*, II, 291 ff.].

and the concepts which Ritter expressed, commonly in a more religious tone. Geography was not, for Humboldt, a field studied as an end in itself, but rather as the means of comprehending "the harmonious unity of the cosmos" as a "living whole," "a unity in multiplicity" [*60*, I, 4 ff.]. "Insight into the cosmic organism (*Weltorganismus*) creates a spiritual enjoyment and an inner freedom which even under fate's hard blows can not be destroyed by any external power [*44, 32*; *cf. 60*, II, 89 ff.].

For Humboldt, this unity of nature was no teleological, anthropocentric unity—he wrote of the comprehension of "the inner, *secret* play of natural forces" [*44*, 32], where Ritter spoke of discovering the "divine secrets" [*50*, 228]—but was a balanced unity of the whole of nature, of which man was a part [note the many quotations in Döring, *22*, 18, 37 ff., 59]. "In the forests of the Amazon as on the ridges of the High Andes, I perceived how, animated by one breath from pole to pole, only One Life is poured out in stones, plants, and animals, and in the swelling breast of man" [from a letter quoted by Rehder, *23*, 136]. An important factor in the evolution of Humboldt's thoughts was his close personal relationship with Goethe [*20*, I, 187–201]. With him, and with the entire generation of the "romanticists" (in the widest sense), Humboldt shared the idea of an organic coherence of all phenomena; this was a common characteristic of both of the founders of modern geography [Plewe, *8*, 49–51].

In other words both Humboldt and Ritter, in their philosophical point of view, were products of their time—in particular both were influenced by the thought of Kant and of Rousseau. To Ritter's religious nature the concept of a universe of order and law in which all phenomena of nature and man were interrelated required the assumption of some divine purpose and plan which the scientist should attempt to establish. Granted this assumption it appeared obvious that the ultimate purpose involved could not be simply the production of an intricate universal mechanism; the purpose could only be found in the life of the highest of the creatures of the universe, the only thing in the universe that was capable of recognizing either the order of the universe or a purpose behind it—namely, man. In Humboldt, on the other hand, the same philosophical concept found a responsive chord in his feeling for the aesthetic rather than the religious. In the writings of Rousseau and St. Pierre, it was the descriptions of the "harmonies of nature" that appealed to him; through Goethe he received the concept of the *"Landschaft"* as conceived by the "romantic" movement [*cf.* Rehder, *23*, 134–38].[20]

---

[20] Even before his travels to America, Humboldt was drawn into close personal relationship with Goethe thanks to the close friendship between the latter and Humboldt's brother Wilhelm. The publication of some twenty letters that passed between Alexander v. Humboldt and Goethe during the period 1795 to 1827 gives some indica-

Interesting as this comparison may be, between the philosophical viewpoints of the two founders of modern geography, it does not directly concern us here. The essential point is that both of them differed from their philosophical and literary predecessors in that both strove to demonstrate their philosophical concepts not by deductive logic, as in the case of Kant, nor by sentimental descriptions of subjective impressions of nature, as in the case of St. Pierre, but by objective descriptions of observations of nature. If there are exceptions in Ritter's work that we must overlook, the same is true, if in lesser extent, in the work of Humboldt [cf. Dove, 21]. Throughout the writings of both students, the modern scientist finds repeated expressions of philosophical concepts that are now commonly excluded from such works. But modern science, which has had bitter fights with religious philosophy, has never come into serious conflict with the aesthetic view of nature; perhaps for this reason Humboldt's expressions of his *Weltanschauung* are less disturbing to scientific readers today than are those of Ritter.

In general, neither Ritter nor Humboldt saw any conflict between science and philosophy. Ritter found that both geography and history "are directed toward the integration (*die Combination und das Mass*) of ideas and are therefore forced to philosophize" [50, 152]. Likewise Humboldt, a student of the works of Kant and Fichte, felt a need for "something better and higher, to which everything could be referred" and therefore greeted with enthusiasm the natural philosophy which Schelling had founded. In a letter to Schelling, in 1805, and in publication two years later, he expressed his conviction that "a true natural philosophy could not harm empirical research"; on the contrary "such a philosophy leads findings back to principles and is likewise the foundation for new findings" [20, I, 228 f.; 41, v]. Some years later, to be sure, when many students gave "pure thought" the preference over empirical studies and seemed to think that science could be produced by thought without research, Humboldt scathingly attacked this "mad" form of natural philosophy [20, I, 230].

That Humboldt did not associate Ritter's work with the *"bal en masque"* of a philosophy that despised observations and experiments and taught "a physics that one can pursue without wetting one's hands" is indicated by a discussion immediately following his praise of Ritter's work in one of his lectures at the University of Berlin: "The description of the universe provides materials for a proper natural philosophy, the foundations for which

---

tion of the interchange of ideas that took place during Humboldt's various visits with Goethe [18, 289–314]. Humboldt dedicated the first part of his travel descriptions, *Das Naturgemälde der Tropenwelt*, to Goethe; in his letter of Feb. 6, 1806, he speaks of a long-held plan to express his "respect and gratitude" in this way.

are sought in many different ways. I can not find fault with these efforts, although I would be inclined myself to go to work more empirically. In this natural philosophy we need only fear and avoid with difficulty: false facts. Empiricists and philosophers should not mutually despise each other, for only bound together can we attain the highest goal" [52, 15].

For both Humboldt and Ritter, the concept of unity of nature presumed a causal interrelation of all the individual features in nature. The phenomena of nature were studied in order to establish this coherence and unity. For both it was axiomatic that the unity of "nature" included organic as well as inorganic, human as well as non-human, immaterial as well as material. The exclusion of any part would be not only arbitrary but would destroy the coherence and unity of the whole. In his first publication, in 1804, Ritter vigorously opposed the trend in "pure geography" that tended to over-emphasize, he thought, the "natürliche Landschaft," insisting that geography must describe and explain all the present conditions of an area [24, I, 250 ff.; Plewe, 8, 32 f.]. Ultimately, to be sure, his predominant interest in man tended to emphasize the human element to such an extent that he actually failed to accomplish his fundamental purpose [Marthe, 25, 18; Ratzel, 26].

Humboldt's view of science was limited only by the universe outside his own mind, "Natur." Goethe, as well as Ritter, called him "an academy in himself" [20, I, 198; 24, I, 154]. Even if we eliminate those parts of his work that he did not regard as geography (Erdbeschreibung), so extensive and rich was his view of the field that almost every movement for some particular form of geography—from the geophysics of Gerland to the aesthetic geography of Banse can find its antecedents in Humboldt. None, however, may look to him as the precedent for excluding other parts of the field. "In the great enchainment of causes and effects, no material and no activity may be studied in isolation" [44, 39]. "The highest goal of all observation of nature is the knowledge of our own nature: and therefore we conclude our description with a consideration of the races of men" [52, 13, 182–90]. Likewise the outline of systematic geography in the Kosmos—called "physical geography" according to the usage of the time—culminates in the section on man [60, I, 378–86]. The consideration of man includes spiritual aspects as well as material, "as though the spiritual were not also contained within the natural whole (Naturganzen)" [60, I, 69]. Moral and aesthetic problems therefore formed a part of his considerations [see 44, 24; * and 47, I, 348–52].

It is true that in many of Humboldt's regional descriptions one will find little concerning man and his works, but he himself explained that his travels

[67]

* See Supplementary Note 5

in equatorial America had taken him chiefly to areas where "man and his products disappear, so to speak, in the midst of a wild and gigantic nature" [47, I, 32]. His purpose, in presenting "the general results that interest all enlightened men" is to describe not only the climate and "the appearance of the countryside (*paysage*), varied according to the nature of the soil and its vegetable cover," but also the "influence of climate on organic life," "the direction of the mountains and rivers that separate the races of men as well as the plant societies" and "the modifications found in the condition of peoples placed in different latitudes and in circumstances more or less favorable to the development of their faculties. I do not fear that I have increased too much the number of objects so worthy of attention: for one of the fine characteristics that distinguish present civilization from that of past times is that it has enlarged the mass of our conceptions so that we realize more clearly the relationships between the physical and the intellectual world; hence there has developed a wider interest in objects that previously were of concern only to a small number of savants because they had been considered isolated from each other according to an overly narrow point of view." [47, I, 14.]

In his descriptions of populated areas, Humboldt consistently strove to fulfill this purpose. Man, his culture and his works, are included as integral parts of his description and interpretation of nature. (In his studies of Central Asia, he was limited by specific restrictions imposed by the Russian government.)

Both Humboldt and Ritter carried the concept of unity of all nature into the consideration of individual areas, at least in what I have termed the "vertical" sense—*i.e.*, that all the features of an area in their interconnections form a naturally unified complex, whether or not that is to be considered as a unit Whole or merely a part of the one natural Whole of the world. * For Humboldt, every part of the world is a reflection of the unity of the Whole [60, II, 89]. Ritter's statements on this point are more emphatic; both however, were, as we have seen, following paths already suggested by their predecessors.

How was this unity to be studied? It hardly seems necessary to say that neither of these "founders" of a science supposed that the study of an area was to be confined to a consideration of it as a whole, or that it should begin with such a consideration of the whole. While both assumed from the start that all the objects and forces of nature (nature in the sense of the world outside our minds) were interconnected to form a Whole, both recognized, in theory as well as in practice, that this could be established scientifically only by investigating the individual, single features in their relation to each other, and building these up in their actual relationships to form the whole.

[68]

* See Supplementary Note 6

The function of the scientific geographer was to perceive these features, not separately, but in their interrelations so that he could thereby reproduce intellectually the unified whole that was nature [*60*, I, 4 ff., 51 ff.; *cf.* Plewe, *8*, 39].

That the previous statements correctly represent Ritter's purpose and method is supported not merely by the views of many other students more familiar with his work, but appears clear from even a brief consideration of either his methodological discussions or of his work. To be sure, one may find sentences in his writings which appear to say the opposite. Thus a century ago the young student, Fröbel, extracted one or two of Ritter's sentences from their context so that it appeared as though Ritter attempted to use synthesis without preliminary analysis [*54*, 502, 504]—the "holistic" principle, as Leighly puts it, that "forbade the investigator to dissect a whole into its parts for fear of destroying its coherence and so its essential character" [*222*, 257]. That Ritter was not so naïve as not to recognize that synthesis presumed analysis can be read from the paragraphs that follow the quoted sentence on which Fröbel has based his argument [*49*, I, 3]; indeed every section of his work demonstrates the principle which he had definitely stated in his introduction, "that the procedure must lead from the simple to the compound, from the individual parts to the unit" [*49*, I, 24]. It was consistent with Ritter's personal character that he should have some hesitation in replying to the charge that his method "which builds up and combines" was one "in which analysis is unconsciously presumed" [Fröbel, *54*, 504]. Many thought that he might have asked Fröbel to read again the pages referred to, but he contented himself with the suggestion that it might have occurred to his critic that it would not be news to the author of a synthetic system that synthesis must be developed from analysis [*55*, 515]. Later critics, indeed, questioned whether Ritter had actually succeeded in getting beyond analysis to synthesis.

The "horizontal" concept of unity of particular areas, as individual wholes, was also furthered by Ritter, though apparently with some qualifications. He did, to be sure, stress the individuality of the continents, which at times he referred to as "organs," but in his reply to Fröbel he insists that the terms are not to be taken literally, but in the form of analogies (just as, one may note, Fröbel had written of the systems of like phenomena on the earth as "organs" [*54*, 495]. He was not entirely satisfied with the term "individual" and would drop it whenever he learned a better one. Nevertheless he conceived of each continent as a "natural Whole"—in terms of all its characteristics [*55*, 518 f.]. Similarly, he wrote, in some cases at least, of the divisions of the continents, *die Länder*, as "individuals" that are

"members," rather than mere parts of the "organisms" of the continent [*cf.*
Bürger, *11*, 14–19]. Although these concepts fit Ritter's teleological philosophy, they are not necessary consequences of that philosophy, nor is such a
philosophy necessary to those concepts. In any case we are not to take the
terms literally [*cf.* Hözel, *27*].

Humboldt does not appear to have considered this concept. Döring concludes that "whereas for Ritter the individual area had its own particular
determined value, for Humboldt it was only a variation of the great cosmic
theme of law and causality" [*22*, 162]. "Nature in every corner of the
earth is a reflection of the Whole. The forms of the organism repeat themselves in ever different combinations" [*60*, II, 89]. In a sense Humboldt
was not faced with the question of regional division since he made no attempt
to organize the regions of the world, or of any major part, into a system.
That he recognized the need for such an organization of geographical study
we will note later.

If we put aside the question of the philosophical viewpoints of the two
founders of geography as irrelevant except as it affected their work—and in
that respect by no means so important as one has been led to believe—what
are the major differences between the directions which they gave to geographic work?

Although we find it erroneous to draw a fundamental contrast between
them in regard to the proper content of geography, it is nevertheless true
that the difference in character of their work led to an important difference
in later interpretations. Though Humboldt, at various places in his writings
presents brief systematic considerations in anthropological geography
[Peschel, *66*, I, 351–5], his studies in systematic geography were primarily
limited to non-human features. In Ritter, on the other hand, the interest in
the study of non-human elements was subordinated to the interest in man,
and his great work is presented in the form of a regional geography; further
his methodological studies repeatedly emphasized the importance of regional
organization of geography in contrast to that by classes of phenomena. In
consequence of this double contrast we find the confusion between the two
forms of dualism in geography to which we have previously referred—
namely, that the distinction between human and natural (non-human) geography is contained in the distinction between regional and systematic geography.

If we separate these two issues, we find, as already noted, that the inclusion of human as well as non-human features in geography did not raise any
question of dualism in the thought of either Humboldt or Ritter, but rather
was absolutely essential to their fundamental concept of the unity of nature.

[70]

The second form of dualism, however, was unquestionably recognized by both men as a contrast in form of study and organization of geographical material. Furthermore, this dualism in methods has presented an issue that has recurred repeatedly in the history of geography, including the present time, and geographers have not been lacking who wished to solve it by the simple method of excluding either one or the other aspect—that is, by limiting geography either completely to systematic studies or completely to regional studies. What was the position of the two founders of modern geography with reference to this issue—as expressed both by their theoretical statements and in their practice?

There can be no question that Ritter's emphasis on regional organization, both in his work and in his methodological statements, tends to give the impression that he had little interest in systematic studies of each analyzed element over the world. But if our comprehension of a student, generally regarded as of major importance in the development of geography, is to consist of more than mere impressions based on the outline of his work and a few sentences extracted from his methodology, we must examine this question more carefully than did the young critic of a century ago [Fröbel, 54].

The present issue is not to be confused with the related question, previously discussed—the argument that Ritter attempted to synthesize the factors associated in an area without preliminary analysis. That question might be dismissed with the mere statement that no matter what he may have appeared to say, in a few sentences removed from their context, the man whom successive generations of eminent scholars regard as a master of geography was no fool. The present question, however, is a genuine one.

Ritter's methodological discussions are to be regarded less as balanced treatments of a field of science than as the program of a reformer attempting to reconstruct a previous geography into a science. In the systematic studies of certain of his predecessors it seemed to him that the intellectual ("subjective") process of dividing the interrelated phenomena of nature into separate classes and studying these separately over the world was a disruption of the actual coherence of nature [49, I, 20]—a conclusion, we may note, that requires no teleological basis. He observed that, when each of these kinds of phenomena is studied in its forms and processes, the result is not only comparable with that of other sciences, but is rather a part of some other science. A "compendium" of these separate studies would not form a geography and the aggregation of areal sections of each of them for any particular area would be no science at all.[21]

[21] Ritter repeatedly used the word "compendium" (in the same form in German) to indicate the antithesis of his work—notably in his reply to Fröbel (both in

Ritter therefore insisted repeatedly that the proper method for a science of the earth was "to ask the earth itself for its laws"; it must use the "objective" (we would say "empirical") method, "calling attention to the main types of formations of nature"; by investigating the relationship grounded in nature itself, this method leads to a "natural system" [49, I, 20; cf. Wisotzki, 1, 273 ff.; Bürger 11, 15].

On the basis of such statements, then, one might, like Fröbel, accuse Ritter of ignoring the importance of systematic studies of individual features. To be sure, one finds statements which indicate the opposite, but one must look fairly carefully to find these. For example: "the earth's surface, its depths and its heights, must be measured, its forms arranged according to their important characters" [49, I, 4]. Further in the lectures given before the Academy of Sciences in Berlin, in 1828 and 1833, one will find considerable discussions of the need for studies of individual categories of phenomena [50, 129 ff., 152 ff.].

Nevertheless, most later critics have felt that Ritter's regional work would have been more effective had it been based more thoroughly on systematic studies, even though the resultant delay would have limited its scope. This criticism was never effectively presented during Ritter's lifetime. It might be inferred in Bucher's remarkable critique of 1827—to which we have already referred—though that is certainly not directed at Ritter. Whereas the *Erdkunde* is barely mentioned, Ritter's earlier systematic geography of Europe is treated at length as the principal example of the correct systematic approach [51, 93 f.]. But whether Bucher intended to imply a criticism of Ritter's later work, or whether Ritter was conscious of that, I do not know; the work was certainly called to his attention by an editorial footnote to his own article in *Berghaus Annalen*, in 1831 [55, 516]. Neither Marthe nor Wisotzki, both of whom recognized the importance of Bucher's study, appear to have inferred any criticism of the *Erdkunde*.

Much better known is the critical essay by Julius Fröbel, at that time a young student with two years' experience in academic geography (on Fröbel's own program, see Sec. III, A). On Ritter's recommendation, Berghaus published the article in his *Annalen* and the fact that Ritter commented on it at length in the same issue focussed the attention of the academic world upon it [54; 55]. Describing the episode some fifty years later, Fröbel speaks of his surprise and embarrassment at finding himself, as a young student of no particular attainments at the time, thrust suddenly

German and in Latin) [55, 508, 513]. In paraphrasing Ritter's statement of his purpose in that article, Leighly not merely omits two of the three different but coordinate purposes listed in a single sentence, but actually has Ritter call his own work by the very word that Ritter had used to indicate what his work was *not* [222, 245].

into the academic limelight as an opponent of the master. It is an interesting indication of Ritter's character that in his published remarks about Fröbel's critique, he ignored the easy opportunity to make Fröbel look foolish, and on the contrary commended his critical efforts as an expression of an "earnest thought for the furtherance" of geography from an "unmistakably keen mind" [55, 506 f.]. Further, when Fröbel came to Berlin shortly after, Ritter received him cordially and the young student received much needed employment working on the maps for the *Erdkunde*. Fröbel's account of his reception by Humboldt is also interesting, but his statement of the latter's apparently favorable reaction to what he himself termed his "impertinent" criticism (of Ritter) is not specific and has only the degree of credibility granted to personal reminiscences written more than fifty years after the event [28, I, 60 ff.].

In spite of the attention which Fröbel aroused, the net effect on the geographic thought of either Ritter or his contemporaries, appears to have been nil. Whatever degree of truth there may have been in Fröbel's criticism was lost sight of because his attack, based on but limited knowledge of Ritter's work, so widely overshot its mark—in striking contrast, we may note, to Bucher's critique which had been based on very careful analysis of the works criticized [*cf.* Marthe, 25, 29 f.; Fröbel does not mention Bucher's study, not even in his succeeding article, after Berghaus had rather pointedly called attention to it].

Although it was Ritter's "point of view" and "method of treatment" in geography, that Fröbel subjected to criticism, his consideration ignored Ritter's methodological essays and confined itself to the two volumes of the *Erdkunde* that had been published up to that time. On the basis of a few sentences in the introduction of the first volume, some of which he misunderstood, Fröbel arrived at a critical conclusion concerning the value of the work as a whole. In one case, as previously noted, the criticism is valid only if one considers the single sentences removed from their context. Fröbel's discussion of Ritter's use of the term "comparative," together with the latter's comments in reply, did little to clarify that question, since, as Plewe notes, each party proceeded from different assumptions and ignored those of the other [8, 38]. Undoubtedly Ritter's lack of precision in choice of words contributed further confusion to a problem that had previously bothered geographers, and over which many later students have wrestled; the most thorough and illuminating discussion of the whole question is presented in Plewe's "*Untersuchungen über den Begriff der 'vergleichenden' Erdkunde*" [8].

On the issue with which we are here concerned—the relation of systematic studies to regional geography—Ritter's theoretical statements offered clear grounds for argument. In reaction to those who had specialized on systematic studies and never reached the actual formations of interrelated factors in areas, he had called first for the study of the areas filled with interrelated phenomena, asserting that, when this had been done for the whole world, the material would then be available for a more successful development of the general principles of systematic geography [49, I, 75]. Ritter maintained his belief in this methodological principle in many later writings, notably in his essay of 1833 [50, 181].

In assuming, however, that these theoretical principles constituted the foundation for Ritter's *Erdkunde,* his critic ignored, or was ignorant of, the fact that elsewhere in his methodological discussions Ritter had shown the need for the opposite method, and that in practice he had in large part followed it. We have already noted his early study of the major geographic features of Europe that Bucher had cited as the model for systematic studies, and the handbook of general (systematic) geography. Although the latter had not been published, various sections had been used in different school texts, in some cases with the original authorship indicated. Ritter could therefore point to these, in his observations on Fröbel's critique, as well as to the systematic discussion of individual categories of features that fills more than fifty pages of the introduction to the *Erdkunde.* Whereas Fröbel had overlooked these, Humboldt and other writers had found in them general concepts and principles that could be used in their own geographic writings—as Ritter observed with detailed references [55, 511]. In subsequent volumes, we may add, Marthe has counted no less than twenty-four systematic studies, some running over a hundred pages each, dealing particularly with minerals and cultivated plants; in many other cases, however, Ritter refers to studies to appear in later volumes that never appeared [25, footnote 14].*

Further, the essays read before the Academy of Sciences in 1826 and 1828, include methodological discussions of systematic studies [50, 103–28, 129–51]. In the latter, for example, there is a discussion of the study of geometric forms of areas of different categories of phenomena, an early example, we may note, of studies of "patterns." Finally, Ritter could point to his lectures in general (systematic) geography that had been heard by hundreds of his students during the preceding decade. Although he had not published these, they had actually appeared in modified form as Berghaus' *Elementen der Erdbeschreibung*—the very work that Fröbel had praised as the model of what should be done in geography! [54, 500, 505]. Berghaus himself, as editor of the *Annalen,* confirmed this statement in

* See Supplementary Note 7.

two emphatic footnotes—not without implying that the fact should have been sufficiently clear to any reader of the *Elementen* [513, 516].[22] The irony goes even deeper than Ritter recognized. In repeating the statement more clearly in the introduction to the first volume on Asia, the following year, Ritter claimed that the inner organization of Berghaus' work was the result of his lectures, but gave his friend credit for having improved the style of presentation for elementary students [*49*, II, xv and 20 f.]. Lüdde however found, by comparing the originals, that Berghaus had published straight dictation of Ritter's course—on many pages, word for word [Plewe, *8*, 59].

It is interesting to note that Ritter, who in his youth had been scathingly ridiculed by one reviewer for presuming to criticize his elders—in a work that at the same time offered positive materials—should have limited his personal criticism of Fröbel to the polite but pointed request that he should make himself more familiar with a man's work before attempting to subject it to a critical judgment [*55*, 520]. It is not surprising, therefore, that Ritter's contemporaries should largely have ignored the degree of truth included in Fröbel's immature and misinformed critique. On the contrary many appear to have followed the more zealously the methodology that Fröbel had attempted to criticize. Undoubtedly this was impressed upon them by the constant emphasis in Ritter's methodological treatments on the full studies of areas, as well as by the contrast between the nineteen volumes of the *Erdkunde* and the small amount of distinctly systematic studies published under his name. With the notable exception of Reclus, they did little to further the systematic studies that Ritter had made, and in their regional studies they depended even less than he on systematic materials already available. The fact that so much of the work of Ritter's followers later came to be regarded as of little value might be taken as a demonstration of the argument that Fröbel had directed at the *Erdkunde:* that it could not attain scientific value because it had not been preceded by systematic studies. Indeed, the argument against the *Erdkunde,* itself, though evidently exaggerated, was perhaps ultimately vindicated. Although scientific travelers, as well as geographers, for many decades—in Germany and especially in Russia—found in Ritter's work "the essential register of facts on Asia"

[22] In Leighly's presentation of Ritter's defence of his work, the sole reference to this list of exhibits is the following sentence: "In his reply, Ritter reminded Fröbel that he, Ritter, had long used in his lectures the analytic procedure Fröbel recommended" [*222*, 245]. The later reference to these "unpublished lectures" as having little influence in the post-Ritterian period [248] is an error; they were published three years after Ritter's death, as *Allgemeine Erdkunde*, a volume frequently referred to in German literature [*61*].

[Marthe, *25, 25*], yet for modern students it appears to offer little of importance even to the historical geographer (as distinct from the historian of geography). On the other hand, one may remember, Ritter had found that his earlier attempt to provide first a systematic geography of the world had not been regarded by Buch as successful. Consequently the relative success or failure of Ritter's work does not lead to a clear conclusion as to the proper order of study of systematic and regional geography [*cf.* Marthe, *25,* 23 ff.].

Humboldt's approach was undoubtedly different in many respects. Even less than Ritter, did he regard geography as an end in itself. In order to establish the unity of the total cosmos, it seemed more important—at least for the time—to make systematic studies of particular kinds of phenomena in their interrelations in areas, than to prepare complete studies of individual areas [*60,* I, 65; *47,* I, 2]. Though he made many of both types of studies, his effort to cover the whole of "physical geography" in the *Kosmos* demonstrated his earlier statement that he regarded this as the more important of his two purposes. In referring to that well-known quotation however, one must understand clearly what he meant by the term *"physische Erdbeschreibung"*—the expression which he finally adopted after trying various others, such as "physics of the world." We have already noted that in its scope it included all earth phenomena outside the individual student's mind: man and his culture, as well as other organic and inorganic phenomena. Further, it was not the individual categories of phenomena in themselves that primarily interested him, though he had been "passionately devoted to botany and parts of zoology," but rather the interconnections of different categories of phenomena—*i.e.,* of plants of different kinds with each other, and with the differences in climate, relief, and soil and in relation to animals and human life. In addition to the types into which everything that *is,* may be classified in terms of its intrinsic characteristics, one perceives types in the arrangement, distribution and mutual relations of things of different categories. "It is the great problem of physical geography (*physique de monde*) to determine the form of the types, the laws of these relationships, the eternal bonds that enchain the phenomena of life with those of inanimate nature" [*47,* I, 2–6, quotation on page 6; *44,* 2 ff., 33–6]. Throughout all his geographic work Humboldt remained true to the principle he had set for himself in a letter written on the day of his departure for America: "On the interrelation (*Zusammenhang*) of forces, the influence of inanimate creation on animate animal and plant world, on this harmony, my attention shall always be directed" [quoted by Richthofen, *81,* 605–7]. More specifically, in the first volume of the *Kosmos,* Humboldt described physical geography as the study of "that which exists arealy (*im Raume*) together," it studies phenomena as arranged in areas in their mutual relations to all other phenomena with which they form a natural whole [*60*, I, 49–72, quotation on page 50; *cf. 20*, I, 274; *22*, 129 ff.].

In contrast with those who attempt to construct geography after the pattern of other sciences, Humboldt regarded it as having a fundamentally different viewpoint from all the other sciences concerned in the study of nature. The individual natural sciences—what we would call the physical and biological sciences concerned with earth phenomena—"study the forms, construction and processes of individual animals, plants, solid objects or fossils" and seek to arrange these in classes and families "according to their internal analogies." Geography is concerned with these objects as they exist together, related to each other causally in an area (*Raum*). In contrast with physics or botany, in which "the objects of nature are divided according to kinds of objects," geography regards all the objects as a natural whole, as they stand in areal connection, in part with the earth body, in part with the universe. Cosmography (*Weltbeschreibung*—the term is taken from Kant) is "the study of all that has been created, all that is (the things and forces of reality) in the area, as a simultaneously existing *Naturganzen*." It includes both an astronomical part and a geographical part. Finally, geography is closely related to the third point of view in studying the earth— more closely than to that of the specialized sciences—namely, the historical, which is concerned with the changes in time of all the phenomena [*42; 52,* 14; *60*, I, 49 ff.].

Similar statements of the comparison of geography and history may be found in Ritter's writings, including his first publication of 1804. In particular he notes that as chronology provides the framework into which the multiplicity of historical facts are ordered, the area (*Raum*) provides the skeleton for geography; both fields are concerned with integrating different kinds of phenomena together, each in its respective frame [*24*, I, 250 f.; *50*, 152 f.]. To what extent he was indebted to Humboldt for these ideas we cannot tell; neither is it clear to what extent each may have depended on Kant, who had long before expressed similar ideas in the lectures on physical geography that were not published until 1801–2. (Humboldt refers to Kant in his lectures of 1827–28, but his own first statement of the concept of geography, on which later versions are based, was published in 1793 [*42*].)

Heterogeneity therefore is an inescapable characteristic of geography, as Humboldt understood it. On the other hand, as an eminently practical scientist (by which I do not mean an applied scientist) he recognized that to comprehend a whole consisting of a multitude of things in interrelation,

[77]

it was necessary to understand first the relations of some of the things to all the others. His studies in systematic geography, therefore, concentrated on particular classes of phenomena—particularly vegetation—in relation to all other phenomena in areas. Whereas "botany proper studies the character, organic formation and relations of plants, in the geography of plants descriptive botany is tied to climatology." The development of plant geography required the combination of means of measuring elevation and temperature with a knowledge of plants, and had been delayed until the methods of the physicist could be combined with those of the botanist. In plant and animal geography however "we speak not of the plants and animals, but of the earth-crust covered with plants and animals" [52, 168–71; cf. also 19, I, 64, 74].

In other words, Humboldt recognized that, though a systematic study in geography must be focussed on one category of objects, it was in no sense to be restricted to the objects of that category but was to consider those objects in their relationship to other geographic phenomena. His physical geography therefore cannot be completely divided into a list of highly specialized fields each separate from the other; rather, in each part, one group of features is the center of attention, but its relations with others carries the work into all the other divisions. Since these relationships are those based on areal position—rather than historical sequence—the fundamental unifying principle is chorological [cf. Döring, 22, 51–9].

This chorological point of view Humboldt consistently maintained in all studies that he entitled geography. Even in a textbook for geography that he and Berghaus projected (for schools in India, at the request of certain English authorities) he refused to insert material that properly belonged in botany or zoology. "A physical geography can concern itself neither with energy and material, nor with the physiology of the organic bodies; all that must be assumed as known" [quoted by Berghaus, 19, III, 94].

Humboldt's contribution to the development of systematic geography is therefore to be found not only in the value of the works which he produced but also in the fact that he first clearly portrayed the distinction between systematic, but chorological, studies in geography and systematic studies in the special sciences. That he was for a time regarded as an important figure in such sciences as botany or geology rested on a misunderstanding of his work, as he himself explained it. The judgment of later specialists in these fields, that Humboldt had done little of major importance in their sciences, was an echo of his own purpose [Peschel, 66, I, 310]. The fact that, rather than attempting systematic studies in the special fields, Humboldt, through the greater part of his work, maintained the chorological point of view, enabled him to found several branches of systematic geography, notably climatology and plant geography [Peschel, *66*, I, 316 ff., 328 ff.; Grisebach in *20*, III, 232–68; Hettner, *2*, 309; *161*, 85 f.].

On the other hand it would represent a complete misunderstanding of Humboldt's concept of the field of geography to suppose that he thought it limited to what he called "physical geography," *i.e.*, to chorological studies in systematic geography. Since his own work was based on his field observations—enriched by those of a great host of other writers, both scientists and travelers—he never prepared a regional geography of the world or even of a single continent. But his recognition of the need for such studies, co-ordinate with work in systematic geography, is shown by the outline which he drew up in 1848 for the geography text to be used in India, in which *"Specielle Geographie: Länderbeschreibung"* forms the second half, co-ordinate with the first, *Physikalische Erdbeschreibung [19*, III, 55–61].[23] The same view is indicated in his discussion of Varenius' outline of geography [*60*, I, 60; *52*, 14, 26], as well as in his many references to the work of his "great and celebrated friend," Ritter. More significant, however, is the large number of explanatory descriptions of individual regions which he studied in terms of the interrelated totality of their phenomena.

Many of these regional studies of Humboldt are interspersed in the record of his travels [*47*], others will be found in his *Ansichten der Natur* [*45*, note particularly *"Das Hochland von Caxamarca,"* 311–34]; an excellent collection of such regional descriptions has recently been published in a single volume of selections from many different writings by Humboldt [*48*]. If Humboldt considered these as less important than his systematic studies, he certainly did not regard them as unimportant, as is shown by his careful and detailed discussions of the difficulties involved in such studies and of his method of solving them [*45*, x f.; *47*, I, 137 ff., and, especially in the *Kosmos*, *60*, II, 1–134, *"Anregungsmittel zum Naturstudium"*].

For Humboldt, "there is a certain natural physiognomy that belongs exclusively to each climatic zone; every vegetation zone, in addition to its particular advantages, has also its particular (or specific) character." His purpose was to gain "a comprehension of the natural character of the different regions of the world" which "is most intimately related with the history of the human race, and with that of its culture" [*43*, 11–14; *Kosmos*, *60*, II, 90 ff.; *cf*. Grisebach in *20*, III, 267]. Although Humboldt's personal

---

[23] Humboldt outlined parts of the first section of the outline in detail; other parts. and the whole of the second section, he evidently left for Berghaus to work out. Since the English group which had requested the work—which they planned to have translated into Hindu—later cancelled the project, it was never completed [*19*, III, 34–204, *passim*].

inclinations may have led him to prefer to study areas in which man was of little importance, his interest was limited neither to the non-human features nor to the visible landscape features.   On the peninsula of Araya (now in Venezuela) his range of observations included astronomical and atmospheric conditions, plants and animals, the salt marsh and the salt works based upon it, the pearl industry, the physical characteristics of the various races represented, and particularly the changes in the character and customs of the native-born Spanish colonists resulting from the differences in natural conditions as compared with Spain.   This situation he contrasted with that found in the highlands, or in the United States, where European settlers had found climatic conditions more similar to those in Europe, and with the character of the ancient Phoenician and Greek colonies [47, I, Book II, Chap. IV].

Similar considerations may be found in what might be regarded as Humboldt's more scientific works.   "How powerfully has the sky of Greece influenced its inhabitants!   Were not the people who settled in this beautiful and fortunate part of the earth, between the Oxus, the Tigris, and the Aegean Sea, first awakened to moral grace and sensitive feelings?   And did not . . . our ancestors bring anew gentle manners from those gentle valleys?"   "The influence of the physical world on the moral, the mysterious interrelation of the material and the immaterial, gives to the study of nature, when one raises it to a higher point of view, a too-seldom recognized charm of its own" [43, 13 f.].   Peschel appears to have overlooked such statements in Humboldt's writings while objecting to similar suggestions in the work of Ritter's followers, specifically of Kapp.   In particular, Peschel attacks the assumption of "a causal relation between the impression of the Landschaft and the mental expressions of the population" [66, I, 404].

On the other hand, Humboldt's descriptions of areas, Peschel found, had introduced a new period in regional geography (Länderbeschreibung).   He succeeded not merely in presenting the picture of the area, but also in "making it live through the play of forces;" thus he "made real the interconnection of the observations with a higher order of the Whole."   "The post-Humboldtian geographer is concerned not only with the interrelations of physical phenomena, but also with those of the historical [human] phenomena with their scene."   In particular, Peschel finds that geography was raised to scientific stature in the regional studies of Cuba and Mexico ["New Spain," 46].   In these studies, to be sure, Humboldt's interest in natural conditions is subordinate to that in social conditions and much of the material, according to Hettner, is not to be included in geography [161, 84].   Humboldt himself apparently regarded these as studies in political economy.   Nevertheless the manner in which he described the differences in economic, social, and political conditions of different areas in relation to the differences in natural conditions was in fact followed, not by economists nor political scientists, but by geographers, who thereby, according to Peschel, made this form of regional study one of the functions of geography [66, I, 336–44].

In his study of the development of economics and of economic geography, Schmidt entitles Humboldt's study of New Spain "the first scientific, political and economic geography founded on the nature of the land" [7, 74]. Paradoxically we may note, while Humboldt thus made a major contribution to regional economic geography, Ritter, in his incompleted studies for a *"geographische Produktenkunde,"* was one of the first to make systematic studies in economic geography [Schmidt, 7, 84, referring to the systematic studies in the *Erdkunde,* previously mentioned here, and to 50, 182–206].

In general, we may recognize two major differences between the two founders of geography in their actual development of regional geography. Ritter's ambition to produce a regional geography of the world—even though his long life did not permit him to carry it beyond Africa and Asia—led him to consider regions of relatively large size in which he seldom if ever completed a full study of all related phenomena.[24] In contrast, Humboldt described in great detail areas he had visited, some quite small, and so was able to provide—within the means of observation then available—a relatively complete picture.   In order to accomplish this he not only considered each of the significant features of an area but combined these individual characteristics for each distinctive part of the area, then showing the relation of the different parts of the area in the total [47, I, 137 ff.].   Indeed the great attention which he gave to certain small areas would justify the students of "microgeography" in proclaiming Humboldt as their leader.

The second, more important, difference we have already indicated. Ritter maintained in theory, and in part in practice, that geography should first study all the interrelated phenomena to be found in each of all the areas of the world, on the basis of which systematic studies could be made of the relations of individual types of phenomena.   The opposite view, that Fröbel proclaimed, called first for a complete study of the individual phenomena systematically over the whole world.   Humboldt, as a practical scientist, evidently felt he must do what he could do: he could more easily consider

---

[24] This statement can be made only tentatively, since the writer does not pretend to be familiar with the nineteen volumes of Ritter's *Erdkunde.* It appears to be justified by the outline and by a glance through parts of the text, as well as by the statements of other students.   But this does not mean that Ritter did not present small details concerning his regions; for example, note the descriptions of villages and farmhouses—a beginning of *Siedlungsgeographie*—in the mining district of the Altai [49, II, 847].

the relations of the two sets of phenomena—*e.g.,* plants and elevation—than those of all phenomena found together; furthermore he could not wait for a complete world regional survey. On the other hand he felt called upon to provide an explanatory description of the lands which he had studied as fully as he could study them. For this purpose he could not wait until systematic geography had been completed. But, by using the results of his systematic studies in his descriptions of areas, he produced "masterpieces" of regional interpretation [Hettner, *161,* 86].

The ultimate results of these differences are both interesting and paradoxical. Ritter's relative neglect of systematic studies not only detracted from the value of his regional work, but set a tradition which for a time limited geography in such a manner that regional geography itself could not progress. In the case of Humboldt, on the other hand, the great thesis of the cosmos, that was of first importance to him, is now of little moment either in science or philosophy. Though his studies in systematic geography were of major importance in the historical development of the field, they have long since become obsolete. In part the same is true of his studies in what has come to be called "comparative regional geography" in which he, rather than Ritter, was the pioneer [Hettner, *161,* 403; Plewe, *8,* 46–55]. But that part of his work which he once indicated as of lesser importance, the masterly explanatory descriptions, analytic and synthetic in form, of individual regions of tropical America and Mexico, remain of imperishable value in geography. As Lehmann noted of any successful studies in regional geography: they cannot become obsolete, they provide irreplacable material for historical geography [*113,* 239; *cf.* Sauer, *84,* 185].

Brief consideration may be given to certain other effects of the work of Humboldt and Ritter on the subsequent direction of geographic thought. Humboldt was always greatly interested in the aesthetic aspects of geography, "as a means of stimulating and widening scientific nature study." Though he found models of such description in the writings of Goethe, Rousseau, de St. Pierre, and various French and English travelers and literary writers, previously mentioned, it was particularly his "illustrious teacher and friend, George Forster," who had "most effectively pioneered in this direction" [*60,* II, 74]. Humboldt insisted that this aspect of geography was not to be distinguished from scientific geography; he was evidently unmoved by Fröbel's specific demand that such studies should be grouped with Ritter's historical-geographical studies in a non-scientific "historical-philosophical geography" [*56,* 7 ff.]. Although he does not mention Fröbel's essay, his statements in the second volume of the *Kosmos* present a specific reply to the former's suggestions. It was not the description of his own feelings that the author of nature descriptions was to present; on the contrary by describing objectively the external nature surrounding him he should leave complete freedom for the feelings of the reader. The geographer's descriptions of nature, we may conclude, are not to be artistic in the sense of expressionistic; neither are they to be artistic in the sense of form. The aesthetic aspect is not to be supplied by "poetic" phrases of the writer, but by the actual scene itself which it is merely his function to reproduce as well as his command of language may permit. "Descriptions of nature can be sharply limited and scientifically exact without thereby losing the living breath of the power of imagination. The aesthetic aspect must proceed from the presentiment of the interrelation of the sensual with the intellectual, from the feeling of the universality, reciprocal limitation, and unity of nature-life." It is not alone the description of the view naively beheld that is to provide the intellectual pleasure, but far more the understanding of the "harmonious interplay in the forces in the landscape" [60, I, 34; II, 72–4; cf. Döring, 22, 89; Peschel, 66, I, 336 f.; Grisebach, 20, III, 251 ff.; Dove, however, felt that Humboldt had himself been guilty of employing the poetic form of description to which he objected in theory, 21].

Neither Humboldt nor Ritter give a categorical answer to the question of the physical scope of geography—whether it includes the earth as a whole or is limited to the earth surface. No doubt one may find in Ritter's writings, as Gerland was able to do, statements that indicate the former, but it is clear that these are in conflict with the greater part of his methodological discussions as well as with all of his work: the "earth" which he described as the show place of the forces of nature, and particularly as the dwelling place of man, was not the earth body but its enveloping shell. It was charac- * teristic of Ritter that he should have been primarily responsible for establishing the term, *Erdkunde* or earth science—a term, as Hettner observes, that so poorly expressed his own concept. Whereas for Ritter the word was used as an alternative to the word *Geographie,* of foreign origin, and to the less scientifically sounding *Erdbeschreibung* which Humboldt favored, later students came to interpret it literally as the science of the earth.

Humboldt, in order to establish his thesis of the unity of the cosmos, took the entire universe in his view, but within this distinguished *"Erdbeschreibung"* as the "telluric or earthly (*irdisch*)" part [60, I, 51 f.]. His own research was largely limited to the earth, and indeed, to that part of it which he knew—the earth surface [Döring, 22, 55]. Consequently the tendency of the pre-classical geographers to confine their field to the earth surface was, in practice, continued through the classical period.

In one other respect, Humboldt's actual work extended beyond the field

* See Supplementary Note 8

that Ritter considered, namely in the study of the distribution of different types of phenomena—essentially one may say, "the Where of things" [Döring, *22*, 64 ff.]. He may, however, have regarded such distributional studies as merely preliminary investigations necessary for the systematic studies of interrelated phenomena; in any case they constitute but a part of his geography.

Finally, we may add, in the development of the methodology of geography during the middle of the nineteenth century—if not longer—two particular factors caused Ritter's influence to be of much greater importance. As the holder of the only university chair of geography in Germany—if not in the world—he exerted a great influence beyond the range of his immediate students, whereas Humboldt's influence was for some time greatest with men who did not consider themselves geographers. Secondly, whereas Ritter repeatedly expounded his views on the nature and problems of geography in methodological papers, Humboldt's numerous discussions of such questions were scattered through his general writings. For this reason, perhaps, his concept of the relation of geography to other sciences was lost sight of for nearly a century.[25]

C. SHIFTING VIEWPOINTS IN THE SECOND HALF OF THE NINETEENTH CENTURY

The "classical period" in the development of geography may conveniently be considered as terminating with the death of both Humboldt and Ritter in 1859. During the following decade academic geography was dominated by the school which Ritter had founded. Since his followers tended to emphasize the "historical" aspects of the field even more than had Ritter, one may speak of a drift away from systematic geography to a regional geography primarily concerned with man. According to Peschel, Ritter's "single laudable student" was Ernst Kapp, but Kapp's interest extended also into political problems, and following 1848, disciplinary action led him, like

---

[25] Previous to his full treatment of this concept in the chapter of the *Kosmos*, entitled *"Begrenzung und wissenschaftliche Behandlung einer physischen Weltbeschreibung"* [*60*, I, 49–78], Humboldt had stated it briefly in two studies published in Latin, the first in 1793 [*42*]. The most complete presentation of his geographical methodology is provided by Döring's dissertation, a painstaking, thorough, and well-arranged treatment, consisting in large part of quotations or accurate paraphrases that I have found to be reliable [*22*]. The first volume of the *Kosmos* presents Humboldt's outline of systematic geography. A brief survey of his geography is provided by the recent publication of his lectures given at the University of Berlin in the winter of 1827–28 [*52*]. Although I have no information as to the literalness of this publication, the concepts, expressions, and statements presented are unquestionably correct reproductions of Humboldt's views as found in his writings.

Fröbel, to come to the United States, where he settled as a cotton-farmer in a German community in Texas and apparently did not again engage in geographic studies until after his return to Germany in 1865 [see Peschel's review in *66*, I, 399–413, 418; see also Fröbel *28*, I, 477 f.; biographical facts from *29*]. Another follower, if not student, of Ritter to come to the United States, was the French Swiss, Arnold Guyot, who held perhaps the first chair of geography in this country, at Princeton. That he had but little influence may perhaps be accounted for by the manner in which the teleological view dominated even his interpretation of details [*64*].[26]

By far the most successful of Ritter's students and followers was the French geographer, Elisée Reclus. According to Girardin and Brunhes, it was his study under Ritter, in 1851, that made Reclus a geographer; from Ritter he derived his main principles and ideas concerning geography [*30*, 67–9, 71]. It is interesting to note that Reclus also followed Ritter's practice, rather than his theory, in first preparing limited studies in systematic geography and then a general systematic geography, before proceeding to a complete regional survey of the world. Unlike Ritter, however, he established his reputation by his systematic study, a major work in physical geography (*La Terre*, 1866–67; later translated into English). Although Girardin and Brunhes found that the influence of Ritter especially predominated in this work, other students have noted the dependence on many other writers, including the English student, Mary Sommerville. Spörer praised * this work as superior to any of its kind in the German literature, contrasting it with the work of the Ritterian school, apparently not recognizing Reclus as a student and follower of Ritter [*68*, 331 f.; presumably the same is true with regard to Peschel's statement quoted above]. Hettner however finds the work distinctly limited by Ritter's influence [*161*, 108]. Schmidt's statement that Reclus "became the Ritter of France" no doubt is based largely on his nineteen-volume regional survey, *Nouvelle géographie universelle*, 1875–94. Though this obviously followed Ritter's *Erdkunde*, it was far more successful, "both in the taut organization, which permitted him to complete the great work, and in the close coherence of nature and culture in every earth area" [*7*, 151–3].

Ritter's influence was also significant in certain other fields. One of his most interested students at the military college was Moltke, who published various geographic studies long before his military-geographic plans became

[26] Note however the recent republication of his "Notes on the Southern Appalachians" in which Ralph H. Brown finds material of value for the historical geography of that region [*65*]. Mention may also be made here of his memorial address for Ritter, which I have not seen: Carl Ritter. An address to the Amer. geogr. and ** statist. soc., Princeton, N. J., 1860.

* See Supplementary Note 9
** See Supplementary Note 10

of momentous importance [Schmidt, *7, 86*]. Likewise Ritter was influential in calling the attention of historians to the significant relation of geography to the course of human events; Marthe notes this influence particularly in the work of the celebrated historian E. Curtius, a former student of Ritter's [*25,* footnote 10].

Though Humboldt, who held no university post, had no immediate followers in academic ranks, his influence outside the universities—and outside of Germany—was vastly greater than that of Ritter. During the brief "post-Ritterian period" Hettner finds that "the real representatives of the true geographical science were the scientific travelers who took Humboldt for their model" [*2,* 313].

The post-Ritterian period however proved to be but a brief interlude before the last quarter or third of the century brought a very rapid development in academic geography in Germany. In many respects this period may be regarded as the critical period in the development of the field. The foundations which Humboldt and Ritter had established for geography did not provide, in appearance certainly, a clearly unified field. To the extent to which their followers exaggerated certain aspects of the views of each of the founders, or attempted to introduce new concepts of the nature of the field, geography was for a time split in several directions and its position as a branch of knowledge thereby brought into serious question. Following the death of Ritter there was no professor of geography in any German university and the return to university status and particularly the rapid subsequent growth was largely the work not of the "historical geographers" who followed Ritter, but of students who had been trained as geologists and tended to specialize in the study of non-human features of the earth—*i.e.,* physical geography as we understand the term [*cf.* Penck, *129,* 635 f.]. With the rise in academic status of geography and the productive work of this period we are not here concerned; in the development of geographic thought, its major problem was to overcome the apparent disunity in the methodology of the field and thus definitely to establish its position as a single field of science.

The general scientific atmosphere during the latter part of the nineteenth century was far from receptive to the philosophical concepts of the earlier "romantic" period, whether those of Ritter or of Humboldt. This change was evident even before the end of the classical period. Indeed the differences in geographic thought overlap our arbitrary divisions of its historical development to such an extent that we find certain of the concepts of the post-classical period expressed by the last of the pre-classical geographers, Bucher. We have already noted Bucher's thorough attack on the use of "natural boundaries" and "natural regions" in the works of his contemporaries. Utilizing anatomy and physiology as analogies—though recognizing that the analogies were not complete—he arrived at the negative conclusion that geographers need not attempt in any way to divide the earth into areal parts except for special purposes; rather that they should study it in terms of classified phenomena, *i.e.,* systematic geography [51, 90–94]. Although Bucher used Ritter's study of 1806 as an example of what should be done, and speaks in other connections in praise of Ritter's work, his criticism unquestionably was applicable to the volumes of the *Erdkunde,* whether he realized that or not.

Though Berghaus appears to have been influenced by Bucher's study, and Marthe was evidently familiar with it, it is not clear that it had any important significance in the development of geographic thought, possibly because of its publication by a small provincial press [*cf.* Spörer's discussion, in a similar connection, 68, 365]. On the other hand the very similar views which Fröbel presented a few years later in the principal geographic journal of the time created quite a stir in academic quarters, particularly because they were directed at Ritter and were accompanied by Ritter's comments.

The controversy however proved to be a flash in the pan; Fröbel's demands "died away without effect" and Hettner therefore omits them from his consideration of the development of geographic thought [2, 305]. That view may be somewhat exaggerated however, since Fröbel's articles, at least the two published in Berghaus' Annals, retained a fairly conspicuous place in geographic literature and may well have influenced later students whether they realized it or not (see also Sec. III A). Certainly the viewpoint which he was trying to express ultimately came to the fore and Plewe therefore regards him as a herald of the new period in geographic thought— preceded, we may add, by Bucher [51, 59 f.].

The new viewpoint in science which came to dominate scientific thought, Plewe characterizes as marked by the increasing specialization of the sciences, by an increasing emphasis on the development of "scientific laws," and by a conscious isolation of science—and specifically of geography—from any particular *Weltanschauung:* this latter point of view, however, often represented an equally definite philosophical presumption of a materialistic, mechanistic, universe in which man was to be studied as a "thing," like any other— "a sum of atomic movements" [Plewe, 8, 60, 66 f.].

The shift in geographic work is generally regarded by German geographers to have been due primarily to the work of Peschel and Richthofen. The movement was originated in the essays which Peschel published from 1866 on, a number of which were collected in 1870, in *"Neue Probleme der vergleichenden Erdkunde als Versuch einer Morphologie der Erdoberfläche'*

[67]. "It has been quite justly said," writes the Belgian geographer Michotte, "that with this work the scientific spirit re-entered geography" [189, 24]. Through this work, and through his teaching, Peschel led geographers to study primarily the morphology of landforms. His geographical work was by no means limited to this part of the field. On the contrary he was concerned also with the study of the influence of landforms on human history, and was successful, Marthe found, in limiting this more narrowly and sharply than Ritter had done [25, 22]. His early death, at the age of forty-nine, cut his work short just at the time when his university position should have enabled him to be most productive. (For a more adequate treatment of Peschel's contribution—as well as a biographical sketch—see the study by Ratzel [31], as well as briefer statements by Schmidt [7, 147–9], and Döring [22, 163–5]).

The scientific morphology for which Peschel had striven was given its foundations by geographers who had been trained as geologists—notably by Richthofen, in his studies of China, published from 1877 on [Plewe, 8, 74; Hettner, 161, 99, and especially 32; Penck, 128, 43–51, and 137; and Schmidt, 7, 153–6]. Among the followers of Richthofen in this particular direction we may mention two: Penck—also trained as a geologist—whose classical study of "Die Morphologie der Erdoberfläche" first appeared in 1894, and whose many students are still active in this field; and the American, W. M. Davis, part of whose work, as we have noted, is included in the German literature.

In consequence there was inaugurated in Germany—as later under Davis in America—a long period when geomorphology represented the major field of geography [Philippson, 143, 9 f.]. The inclusion of this subject within geography appears thereby to have been irrevocably established in Germany, whatever logical objections might be raised [Kraft, 166, 7], whereas in America, its position with reference to geography and geology is still uncertain. (For a further discussion of this question see Sec. XI G.)

Whatever views might be held in regard to geomorphology, it has been commonly recognized that under Peschel's leadership, geography for a time tended to expand in the natural sciences into fields that had long been claimed and cultivated by other sciences, so that it seemed for a time as though geography would claim all of physical science that was related to the earth [Hettner, 2, 314 f.]. Human geography on the other hand—including ethnography, agricultural land use, trade, and the movements of peoples, as in Richthofen's studies of California and of China [69]—was considered largely in relation to landforms, or was confined to studies in regional geography. In the development of regional studies particular mention may be made of Kirchhoff, as a connecting link from Ritter to Ratzel [33; 34].

One important result of this situation was the confusion, to which we have previously referred, of two forms of dualism in geography into one: a systematic, physical (now meaning non-human) geography, and a regional, human geography. In order to justify the inclusion of studies of man in geography, and at the same time to prevent the indefinite expansion of geography in that direction it seemed necessary to state the purpose of the two parts separately. Natural features were studied in their own right— presumably it mattered little how far the field expanded in that direction— but human features were studied only in terms of their relations to the natural features. This view was more or less definitely stated by Wagner, Kirchhoff, and Neumann [according to Hettner, 2, 316].

The outspoken emphasis on the division of the field into two sets of phenomena pointed logically to two opposite directions of development. That which came to have the more important effects can be traced from Ratzel through Semple to America where it had its logical ultimate expression in Barrow's statement of geography as the field of "human ecology," the mutual relations between man and the natural environment [208]. This view has been echoed in England [195], and also in Japan [110]. Because of its great importance in this country we shall need to consider it later in detail.

The opposite conclusion—that man should be excluded entirely from geography—was urged much earlier, namely by Gerland in a long essay in 1887 [76]. Although this statement came "at a time that was certainly open to the ways of thought of natural science" [Bürger, 11, 24], and might well have seemed the logical conclusion of the swing in geographic thought of the previous twenty years, it met with essentially no favorable response. It was clear to many, as Hermann Wagner observed in his lengthy critique, that if Gerland's apparently logical thesis were carried through to its necessary conclusions, so much would have to be excluded from geography that the remainder would not be recognizable as the field that had long carried that name [77]. Few thought it necessary to give much attention to a thesis that would have thrown overboard the greater part of the geography of Humboldt as well as that of Ritter [cf. Supan, 78, 153; Hettner, 2, 315 f.]. Indeed it would be difficult to point with certainty to any effect of the controversy which Gerland introduced on the subsequent development of geographic thought in Germany; hence, while we will have occasion to return to it in other connections (Sec. III A, B), we need not examine its claims here.

The historical significance of Gerland's proposal was suggested a few years later by Wagner: the announcement of geography as a pure natural science came at the high point in the swing toward more intensive cultivation of physical geography that Peschel had started nearly two decades before [80, 374–5]. Although Gerland, as Schmidt remarks, was "simply consistent" with the course of this movement [7, 156], none of the "morphologists" was willing to continue to the logical extreme. Indeed, it is possible that his arguments may have made some realize how far geography had been swung out of balance.

The counter-movement however had already set in, both in actual geographic work and in methodological considerations. A major event in the history of geography that was to have indirect influence on its methodological development was the publication, in 1882, of Ratzel's *"Anthropogeographie oder Grundzüge der Anwendung der Erdkunde auf die Geschichte"* [72]. The term, anthropogeography, which has ever since been associated with Ratzel's name, reflects his background in the natural sciences, particularly zoology, but is misleading. It suggests the geography of man in terms of individuals and races, anthropological geography; whereas the major objects of Ratzel's concern were the works of man, particularly—thanks perhaps to Moritz Wagner—the products of man's social life in relation to the earth [Schmidt, 7, 157-61].

Ratzel was therefore, like Ritter, a student trained primarily in the natural sciences for whom geography offered the connection between the natural sciences and the study of man. As we found in the case of Ritter, the fact that Ratzel's work was largely in human geography has led many to overlook his earlier background. Thus, Sauer's statement that he "got into geography through newspaper work" [84, 166] is less than half the story. According to Ratzel's own account it was his interest in nature that determined his career; he never departed from his early intention of "devoting himself to some kind of scientific study." He took his doctorate in zoology, geology, and comparative anatomy. Continuing his zoological research in southern France he sought to replenish his funds by sending travel letters to the *Kölnische Zeitung,* with the result that he was offered a position as scientific and travel reporter for the paper. Karl Andree advised him to use this opportunity to become a geographer, and though Ratzel at first planned to travel as zoologist he did ultimately, after six years of travel in Europe, the United States, and Mexico, make the shift. His first publications however were in zoology, and his work in geography included physical studies of the snow limit in the Alps [35].

Ratzel's purpose was to establish the study of human geography on a scientific basis. Far more successfully than Ritter, Schmidt concludes, he "remained true to the direction of the natural sciences." His great contribution was "to have brought this part of the geography of man, cultural geography, into a scientific system by organization of the phenomena, and establishment of concepts and significant connections of the results obtained"

[7, 158]. Although he is commonly considered as primarily a follower of Ritter, in a sense his *Anthropogeographie* represents the first major systematic study of the geography of man as suggested in the culminating section of Humboldt's "physical geography." The fire of criticism that has been directed against his work ever since, does not lessen its importance in having demonstrated that the human, as well as non-human, aspects of geography could be subjected to systematic study, thereby leading to more reliable interpretations in regional geography. (For a full appreciation of Ratzel's work, see particularly Hassert's study [36].)

In one major respect Ratzel's approach was different from that of Humboldt. Working purposely from the point of view of natural science, he organized his first *Anthropogeographie* largely in terms of the natural conditions of the earth, which he studied in their relations to human culture. Though this procedure was common among geographers concerned primarily with physical geography, others, such as Kirchhoff, had studied human geography by the reverse method—by considering human conditions in relation to natural conditions. In his second volume, Ratzel himself largely reversed the process, but many of his followers, notably Semple in this country, maintained his earlier orientation and thereby established a procedure that dominated human geography for some decades—at least in this country. German geographers, however, changed their procedure much earlier, thanks particularly to the efforts of Hettner and Schlüter [130; 131]. In the meantime, however, the fact that human geography was studied in terms of the influences of the natural environment on man led naturally to the concept of the field of geography as essentially the study of such relationships—a concept commonly, but probably erroneously, ascribed to Ratzel himself. (In particular it is misleading to place Schlüter in opposition to "von Richthofen, Ratzel, and their contemporaries" as Dickinson does [202, 2; cf. Schlüter, 131, 507 f.].)

In the year following the appearance of the first major study in systematic human geography, Richthofen presented, in his inaugural address at Leipzig, what came to be regarded as the programmatical statement of modern German geography [73]. His earlier statement, in the concluding pages of the first volume on China, 1877 [69, I, 729–32], had been an attempt to restate the concept of geography that was common with Ritter and Humboldt, but, as Hettner notes, it placed bounds on geography that were broken in the book itself [126, 560]. In the same year, however, Marthe had clearly restored the areal or chorological principle as the dominant criteria of geography, using the terms chorography and chorology which various

[91]

* See Supplementary Note 11

** The reference to Semple is in error; her work demonstrates both of these methods [204].

writers before him, including Peschel, had taken from the early Greek geographers. At the same time, he emphasized the study of distributions, which had formed a part of Humboldt's geography, and expressed geography in simple terms as the study of the Where of things [70, especially 426–9]. Marthe's concepts also found expression in a statement on the nature of geography—drawn up by a committee of which he was a member—adopted by the International Geographical Congress at Venice in 1882 [71, 679].

In his epoch-making address at Leipzig, then, Richthofen took over Marthe's chorological conception and made it a fundamental basis of his concept of geography [Hettner, 126, 552]. In this address, writes Plewe, Richthofen showed himself to be "the actual one to inherit and carry forward the ideas of Humboldt and Ritter, and also, in a consistent sense, to fulfill Peschel. In contrast with the latter, he had the sound, unprejudiced historical sense to fit himself into the course of development and to determine his position exactly" [8, 73; see also Penck, 128, 43–51].

Richthofen had been trained as a geologist, and as a geographer he was primarily interested in geomorphology. Upon his election to the Academy of Sciences, he chose to enter the "physical-mathematical class" in contrast to "the historian Ritter" who had belonged to the "historical-philosophical class" [81, 605]. Nevertheless, he recognized the value of Ritter's work in the development of geographic thought. If one reads his address against the background of the concepts of Humboldt and Ritter one observes how much less he was interested in the contrast between them than in the common elements of their concepts and work. He was able therefore, "following the precedent of Humboldt, to restore the close connection of geography to the natural sciences," and at the same time to restore Ritter's program to its place in geography [Schmidt, 7, 153–6; cf. Döring, 22, 165 f.; Hettner, 126, 552 f., and 32].

We need not concern ourselves with Richthofen's specific statements; as Hettner later noted, though he set the direction of geographic thought for the future, he had not been successful in finding a sharp formulation for his concept of the field. But the essential thought is clear from his discussion: geography studies the differences of phenomena causally related in different parts of the earth surface [161, 106 f.; 73, 25 ff.]. Likewise important, in the development of geographic thought, was Richthofen's exposition of the relation of systematic and regional geography to each other and to the field as a whole. The actual purpose of systematic geography is to lead to an understanding of the causal relations of phenomena in areas [42 f.], an understanding which may be expressed in principles that can be applied in the interpretation of individual regions, i.e., chorology. (Richthofen distinguished between a first step, chorography, which is non-explanatory description, providing material for systematic geography, and chorology, a final step, the explanatory study of regions, based on systematic geography; but this separation has not been followed.)

The major distinction in geography, between systematic and regional geography (chorology), is therefore not a difference in materials studied—Richthofen recognized that it is not possible to limit the materials to be studied in geography—but a difference in method of study. Because of the heterogeneous character of these materials, they must be studied by classes, in systematic geography. On this basis he recognizes not two but three major groups: physical phenomena, biological phenomena, and human phenomena. Though man is of course a biological object, and to a certain degree may therefore be included with the studies of other animals, the fact that his relations to the earth surface are governed by a host of factors not significant for other animals, requires an entirely different form of consideration. In the chorological studies, the causal relations between all the groups of phenomena unites them in a single unitary study. Not bound by any limited concept of science, Richthofen, like Humboldt, saw no objection to a single science considering different kinds of things that exist together and are bound together by causal connections.

D. THE IMMEDIATE BACKGROUND OF CURRENT GEOGRAPHY

Richthofen's program for geography and the fuller exposition of it which Hettner later contributed (first in 1895, but most fully in 1905) prepared the way for studies in regional geography interpreted in terms of the fruits of systematic geography. This was not new in geography, but rather a return to the method of Humboldt [2, 309 or 161, 86]. Though neglected, first by the followers of Ritter who studied regional geography with relatively little consideration of systematic geography, and later by the followers of Peschel who considered chiefly systematic geography, it had never been lost sight of. Supan however felt that Richthofen's address, in giving more detailed attention to systematic geography, had subordinated regional geography, and, in 1889, urged its more frequent cultivation [78].[27] Likewise Hettner vigorously encouraged the development of regional studies—by example, by his methodological statements, and through publication in the *Geographische Zeitschrift* which he founded in 1895. Nevertheless this

---

[27] Wagner discussed the importance of regional geography as early as his methodological report of 1882. In his report of 1891, he devoted some ten pages to the subject [80, 385–95], after first making a passing reference to it, with a footnote referring to his earlier discussion [375]. It is in reference to this report that Leighly states that "regional geography . . . was just coming into Wagner's view in 1891" and that he had noted it then "only in passing" [222, 256].

aspect of geography hardly reached a position comparable with systematic, notably morphological, studies, until relatively recent times, particularly after the World War led to a focussing of German interests on the full character of European areas, notably those of Germany.   The fact that Hettner is so often referred to as a promoter of regional geography should not lead one to suppose that he discouraged the development of systematic geography. Throughout the successive shifts in emphasis of the past forty-odd years he has consistently maintained the necessity for geographic research from both of these points of view.   In his first brief statement on the nature of geography, with which he introduced his *Geographische Zeitschrift*, in 1895, he commented on the neglect of systematic studies by the followers of Ritter, and credited the leadership of Peschel with having restored systematic physical studies in geography [*121*, 2; *cf*. *2*, 310].   His first detailed methodological study, of 1903, was limited to the fundamental concepts and principles of systematic physical geography [*123*]; in that of 1907 he considered systematic human geography [*130*], and in many others, since 1905, he has stressed the importance of systematic, or general, geography, as a co-ordinate part of the field [especially *140*].   Finally, in more recent years, he has himself presented detailed systematic treatises covering different categories of geographic features, first on surface forms [*361*], then on climate [*362*], and finally, a four–volume work covering the whole of systematic geography [*363*].   Human geography however is limited to less than a hundred pages of the fourth volume, since Hettner planned to devote a separate volume to this part of the field.   Geographers of all countries will hope that nothing will prevent the publication of this volume for which the manuscript is essentially completed.

In his first discussions of the relation of these two types of geographic work, Hettner introduced a somewhat unusual terminology designed to emphasize that there was no sharp separation between them.   In a regional study of any extensive area it is necessary to study the notable variations in the individual geographic features systematically.   On the other hand, the systematic study of a particular category of geographic features is not made with reference solely to that category, but rather in terms of its chorological relations to one or more other features—*i.e.*, their relations as each varies in different areas.   He wished to emphasize, that is, the distinction that Humboldt had clearly drawn, but which nevertheless had often been lost sight of, between studies in systematic geography and those of the special systematic sciences studying the same objects [see especially *60*, I, 48 ff.; on the relation of Hettner's views of geography to those of Humboldt, see Döring, *22*, 166–8].

[94]

In the German literature of the nineteenth century the terms introduced by Varenius, "general" and "special geography" had in considerable part been replaced—in the desire to eliminate words of foreign origin—by *"allgemeine Erdkunde"* and *"Länderkunde."* Not only did these suggest too great a cleavage between the two types of geographic study but the term "general science of the earth" expressed a very different concept from that commonly associated with the word "geography." It was forgotten, as Penck has recently observed, that *"Erdkunde"* was "simply the germanization of the word geography. One took it literally and derived therefrom the task of studying the earth as a whole." Gerland, under whom Hettner had studied, had attempted to follow this reasoning to its ultimate conclusion and even Richthofen was apparently confused by it [*90,* I, 39]. One consequence, Hettner felt, was that geographers were influenced to make systematic studies of the earth as a whole in which the chorological point of view was lost [*126,* 559]. Geography became even more dualistic in nature than Wagner had recognized; studies in systematic geography did not produce general concepts and principles concerning the interrelation of phenomena that were needed for regional geographic studies. To overcome this separation of the field into two entirely distinct types of work, Hettner regarded it as necessary to make the chorological concept dominate—as in Humboldt's work—in systematic as well as in regional studies. To emphasize this viewpoint he placed at the head of his outline of systematic or general geography —drawn up in 1889, two years after the publication of Gerland's proposal— the title *"Vergleichende Länderkunde."* The relation of this form of geographic study to *Länderkunde* proper, or regional geography, is clearly explained in his studies of 1895 and 1898. Although Hettner dropped this term in the methodological studies in which he most fully developed the relation of the two kinds of study [*161,* 398–404], later, when he ultimately fulfilled the outline written more than forty years before, he maintained the original title.[28]

---

[28] Penck found this to be an unfortunate "act of reverence" toward the original manuscript, but Hettner had the additional reason that his work is a "general geography" minus the study of the seas, and the term *"Vergleichende Länderkunde"* is therefore proper. [Penck, *90,* I, 38–40; II, 31–2; Hettner, *363,* IV, foreword.] That Hettner's terms should have led to complete misunderstanding by one American student [*cf.* 222, 256 f.] might be regarded as an additional point to Penck's objections, though, as the latter recognizes, there is no difficulty in understanding what Hettner intends if one reads beyond the title. In any case it is not for us to enter into a discussion of terminology in a foreign language.

The terms used in this paper, now fairly common in American geography, and readily adapted in all languages of Latin origin, do not supply suitable terms when translated into German. Fortunately the English language is not limited to words of any particular linguistic origin.

Before leaving this period of the late nineteenth century, we must note certain characteristics of the geography of the classical and pre-classical periods that, for a time at least, almost dropped out of sight. One of these is the concept of unity or *Ganzheit* in geography. In his survey of the development of this concept in the history of geography, Bürger is able to present but little acceptable evidence that it was important in German geography during this period. Relatively few German students appeared to have been concerned with Ritter's concept of the earth as an organic Whole, though Ratzel may have echoed this thought in certain connections, and it was a favorite concept of the great French geographer, Vidal de la Blache [*184,* 5]. Gerland's effective argument against this concept reflects an increasing demand in science for direct accurate description in place of either mystical assumptions or misleading analogies. The earth, he concluded, was not an "organism" but simply a "complex of cosmic matter" [*76,* vi].

To be sure, German geographers of the late nineteenth century continued to study the mutual relations between different phenomena in areas; Richthofen saw the unity of the field as resulting from the causal interconnections of different kinds of phenomena [*73,* 16 f., 67], and Hettner, as well as others, emphasized the importance in regional geography of presenting not merely some of the characteristics of an area, but the total character, as determined by the interrelated combination of all significant features [*161,* 217]. But this total character is not thought of as a "unity," an *"Einheit,"* or *"Ganzheit"* in the sense of the philosophy of early writers, including Humboldt and Ritter.

Likewise, one does not find during this period the concept of the individual region as a unit in itself, a "whole," or an "organism." Both of these aspects of Unity and Wholeness have been returned to geography in recent decades, chiefly since the World War. Its belated development makes it questionable to consider it as a result of the "return to Ritter," of which both Richthofen and Hettner spoke in a very limited sense.[29] Certainly neither Richthofen nor Hettner ever supported the concepts; the latter, as we shall see, has frequently challenged them.[30] He himself suggests that the concept

[29] Hettner wrote in 1898 that the position established by Richthofen might be called *"in gewisser Hinsicht eine Rückkehr zu Ritter"* [2, 317]. Leighly's quotation of this as simply "a return to Ritter," without the qualifying phrase [*222,* 258], is incomplete, particularly in view of Hettner's detailed explanations of what was involved in his qualification [2, 308 f., 313, 315]. What other evidence there is to be found in this or other writings by Hettner that he was responsible for a transfer of Ritter's "holistic view" to regional study is not indicated by Leighly, nor is this writer able to find any. If the holistic view of regions is present in American geography today, as Leighly implies, it could be traced through Sauer to Schlüter.

[30] Bürger is able to list Richthofen and Hettner as contributing to the develop-

might be traced from Ritter's teleological concepts [161, 306], presumably through Ratzel [126, 557; see also Plewe, 8, 72, and Bürger, 11, 76]. It was evidently from Ratzel that Vidal de la Blache took his concept of terrestrial unity, of the "earth organism"; though Vidal erred in supposing it was new in geography [184, 5]. But it is also possible that the notable development of these concepts in recent years is not so much a product of the ideas of Ritter and Humboldt, but rather a repetition of a similar cultural phenomenon—the introduction into geography of general philosophical concepts of a particular time and country. In the post-War atmosphere of Germany, * "Unity" and "Totality" are powerful concepts.[31]

In the scientific temper of the late nineteenth century there was apparently little place for the consideration of the aesthetic character of landscapes that Humboldt had so notably developed. Bürger cites as sole exceptions of note, studies by Ratzel, Oppel, and Wimmer [74]. Both of the latter continued Humboldt's usage of the term "Landschaft" in the sense of the visual scene.

In sum, we may say that in the latter half of the nineteenth century, under the influence of the development of the specialized natural sciences, geography for a time appeared to be changing into a field quite different in character from that which Humboldt and Ritter had inherited and developed. The emphasis on systematic studies appeared to divide geography into two halves, one a natural science, the other a social science, united only in a study of regions that hardly appeared to be a science at all, in the sense in which the time conceived that term. By the end of the period, however, reaction had set in, so that the direction of geography was again essentially that which it

---

ment of these concepts only because he does not distinguish clearly between different uses of these terms [11, 26, 76]. Both Richthofen and Hettner consider the sum total of interrelated factors at any one place as a total mechanism which will not clearly be comprehended if any important part is ignored. For Hettner, further, this total is unique at every spot of the earth (Erdstelle) and thereby that place "is stamped as an individual" [161, 217]; but I find no justification in the context there or elsewhere, for Bürger's addition of the phrase, "as a Landschaft." On the contrary, in his direct discussion of this concept, which Bürger lists in his bibliography, Hettner clearly indicates that only the spot, not an area, has individuality [269]. It may be added that in one of his earlier studies (not cited by either Bürger or Leighly) Hettner compares the relation of localities to the whole earth surface—with obvious hesitation—to the relation of organs to a great organism, but not in order to indicate that the locality was in fact a specific unit or that the earth was really an organism; since he drops thèse terms in his later treatment of the same theme, we may ignore them.

[31] It is an indication of the fact that German geography is entering a new period in which we may expect notable changes that due caution prevents me from indicating the sources of this suggestion. They include both geographers and non-geographers.

* I.e., the atmosphere of the 1930's.

had been before. At the turn of the century, the purposes of geography corresponded in major degree, as Richthofen stated in 1903, "in content and methodology, to the conception which Humboldt had given it" [3, 673, 689]. Geography then, as throughout most of its history, was concerned, as Hettner put it, to study the areas of the earth (*Erdräume*) according to their causally related differences—the science of areal differentiation of the earth surface [Hettner, 2, 320].

Although Hettner found in 1905 that German geographers were generally coming to accept Richthofen's statement of the field of geography, he, as well as others, felt that Richthofen "had not been entirely successful in finding the sharp methodological expression for his concept"; that he had not followed it through consistently, and, later, in his address of 1903, had somewhat obscured the issue [126, 552–3, 560; 161, 106]. Hettner therefore set himself the task of providing a sound methodological exposition, not of some concept of geography which he deduced in his own mind, but of the concept which seemed most closely to represent the field of geography as it had developed hitherto. He was therefore not merely standing on the shoulders of Richthofen, but likewise, both indirectly and directly, on those of Ritter and Humboldt and their predecessors of the pre-classical period. His dependence on Humboldt can be noted in countless passages; he himself has indicated it in many [cf. 161, 85 ff.]. Döring concluded that in Hettner's synthesis of the geographic thought of the two founders of modern geography, Humboldt stands not beside but before Ritter [22, 163].

Throughout the long series of methodological discussions which Hettner has published from 1905 to the present time—with further studies, one may add, ready for publication—he has consistently endeavored to express the concept of geography as the outgrowth of its historical development. In that sense they represent the culmination of our study of the historical development of geographic thought. Since his purpose in these studies, however, has been to provide the logical basis for the concept historically evolved, we will examine his views in the following section.

Our historical survey may be briefly concluded. The viewpoint which Richthofen had stated in 1883, as interpreted by Penck, Schlüter, and above all by Hettner, came to be widely accepted, so that German geography in the first part of the twentieth century has been marked by a greater degree of unity of fundamental concepts than ever before. In particular, Hettner's methodological discussions have come to be regarded as "classics" in geography which no German scholar would ignore in any consideration of the methodology of our field. Sölch wrote in 1924 that "almost all the scientific geographers of Germany have arrived at similar conceptions, thanks not least to Hettner's work" [237, 56; cf. also Braun, 155, 6–8; and the statement at the *Heidelberger Tagung* of representatives of 21 geography departments of university rank in Germany, 138].

Likewise the methodological studies of Hettner and others of his contemporaries have had a notable influence on the concepts of geographers of other countries. Among the geographers of continental European countries outside of Germany who have expressed similar views on the nature of geography we may list the following: Berg in Russia [97, 103] and his student Marcus in Estonia [191, 12 ff.]; Granö in Finland [270, 296]; Arstal in Norway (according to Braun) and Helge Nelson in Sweden (as quoted by DeGeer, who on the other hand expresses a concept similar to that of Marthe [190, 10]); Michotte in Belgium [189]; and in Italy, Giannitrapani and Marinelli (according to Sölch) and Almagià [188]. Both Chisholm [192] and Herbertson (see Sölch) attempted to introduce Hettner's views in Great Britain.   In Japan, a number of geographers today follow Hettner, according to Inouye—including Komaki and Watanuki and, evidently, the reporter himself [110, 287 f.].   The chorographical concept was first introduced into this country by Fenneman [206] and is to be found in one of the later discussions of W. M. Davis, in marked contrast to his earlier, better known statement [102, 209 f.; cf. 203]; further it may be discerned in the work of many others, notably of Wellington Jones.   Nevertheless it was largely overlooked by American geographers until expounded with great effectiveness by Sauer a little over a decade ago [211; 84]; since then it has become the general concept of the field for a large number, if not the majority, of the active research workers in the United States.

On the other hand, in countries outside of Germany, there is no such general agreement among geographers as to the general nature and scope of their field of study as has characterized German geography of the past generation.   In most countries indeed there is marked divergence and conflict of opinion.   It is significant to note therefore that geographers outside of Germany have shown but little interest in studying the methodology of their field, as Stamp observes in a recent survey [200]—though it should not be * supposed that they do not frequently talk about it and publish their individual views.

French geography presents an exception, in the relatively high degree of unity that was established by the predominating influence of Vidal de la Blache.   But important as the influence of Vidal, Brunhes, and Demangeon has been in the development of regional geography, it has resulted from their works, and those of their students, rather than from studies of the nature

[99]

* Josiah C. Stamp, the economist.

of geography [cf. Sauer, 84, 171, 180 f.]. Indeed if one read only what these writers say about geography, as distinct from what they have done in geography, one would have to place them simply as modifiers of Ratzel. The most thorough, and in many respects a highly valuable, analysis of geography from this point of view is Vallaux, *Les sciences géographiques,* which will be referred to frequently in later sections of this paper [186]. It must be expressly stated, however, that no adequate attempt has been made here to survey the current thought in French geography; the reader will find some references in the recent survey of work in that country by Musset [93].[32]

British geographers have shown perhaps the least interest in attempting to determine the nature and scope of their field. Chisholm's inaugural lecture of 1908, based largely on Hettner, appears to have had little influence. Both Sölch and Huender (of Utrecht) have noted that different English geographers express widely differing views, though nearly all are subject to the "environmentalist conception" [98; 99; 100]. At least one English geographer, Dickinson, has expressed distress at the lack of coherence in geographic work in England [101]. Roxby's presidential address in 1930 examined the concepts of earlier German geographers and the interpretation of them by Vidal and Brunhes, to arrive at a statement very similar to that of Barrows; but German geographers since Ratzel are confined to a footnote [195, 282 f.]. The following year Mackinder sought to state the unity of geography in terms of natural regions based primarily on the hydrosphere rather than on the lithosphere—a conception that no doubt owes much to Herbertson [196]. Bryan's book on "Cultural Landscapes" shows no acquaintance with the developments in Germany since Ratzel [280; cf. Dickinson's comment, 101]. On the other hand the work of the Passarge school is discussed at length in Unstead's study of regional geography [309] and Stamp's survey of economic geography considers American, French, and German points of view [200]. In the last two or three years both Crowe and Dickinson have given critical attention to more recent German views—in part, apparently, as a result of the stimulus from this country [201; 202].[33]

---

[32] Attention should be called to the series of studies by different authors of the current situation in geography in various countries published in a recent number of the *Geographische Zeitschrift* (1938, pp. 241–315). Some discussion of points of view on the nature of geography will be found in those dealing with the United States [108], Netherlands [92], France [93], and especially Great Britain [101] and Japan [110]. The other studies do not take up this question directly but present a general view of the geographic work in Germany, Italy, Poland, and the Scandinavian countries [91; 94; 95; 96].

[33] Dickinson's study purports to be based on investigation of "current trends in

* See Supplementary Note 13
** Josiah C. Stamp, the economist.
*** See Supplementary Note 14

That American geographers have been more given to discussing and arguing over the nature of their field than to studying the problem is well-known. Serious scholarly studies of this problem, including a due consideration of the findings of previous students, are limited chiefly to two or three presidential addresses, already referred to, and the well-known discussions of Carl Sauer. Insofar as Sauer's methodology is derived from German writers it depends largely on Schlüter. Though he recognizes Hettner's methodological studies as "at their best perhaps the most valuable appraisals of what geographers are trying to do" [*84*, 182] he appears to have otherwise largely ignored them, perhaps because of a misunderstanding of Hettner that he appears to have taken over from Schlüter (see footnote 48).

Though the majority of American geographers may not have written on this subject they nevertheless have pronounced, but divergent, convictions as to the proper definition and scope of geography [*cf.* Parkins's survey of "The Geography of American Geographers," *105*]. Categorical criticism of specific papers on the grounds that they are "not geography" are common, and highly debatable views of the nature of the subject are published, of all places, in college text-books [*cf.* Crowe's comments, *201*, 10 f.]. One feels justified in asking these critics to give us the fuller benefit of their judgment in thorough scholarly studies of the nature of geography.

The geographers in Germany and the large number in other countries who agree with them in considering geography as a chorological science concerned with studying the areal differentiation of the world, do not suppose that they present a new concept of an old field. On the contrary, as many, including Hettner and Sauer, have noted, this concept may be derived from the work of the earliest geographers, from Herodotus and Strabo [*161*, 122; *211*, 25]. Modern geography, Sauer concludes, is "the modern expression of the most ancient geography"; it is appropriate therefore to use the term common in ancient geography (and revived by Marthe and Richthofen), namely, chorology, the science of regions.

---

geography in the United States, Germany, and France." But he greatly over-emphasizes Schlüter's importance in Germany in presenting his concept of geography as representative of that country. It is hardly a sufficient correction to speak of "Schlüter's concept . . . together with that of Hettner," without pointing out clearly that Hettner has repeatedly opposed almost all the specific aspects of Schlüter's concept that he describes in detail—almost nothing is said concerning the points on which they agree. The consideration of geography in the United States is limited exclusively to the work and concepts of the geographers of one university, a group that could certainly not be considered as representative.

[101]

III. Deviations from the Course of Historical Development

A. attempts to construct a "scientific" geography

In the preceding survey of the development of geographic thought in Germany, we gave but brief attention to radical suggestions calling for major changes in the concept of geography that had little if any effect on subsequent thought. It is true however, as Leighly has recently commented, that methodological controversies that are in themselves of minor importance may suggest significant conclusions. The fundamental attitude that underlay the proposals of Fröbel in the 1830's and of Gerland a half century later is one to which expression is frequently given in discussions—particularly oral discussions—on the nature of geography. Since Fröbel and Gerland both expressed their views fully in print, and the subsequent works of both give us some basis for judging the significance of their methodology to their practice, it may repay us to examine their programs more closely, in the light both of the background which produced them and the results to which they led.

We have previously noted that Fröbel's program was based on a very limited knowledge of the field which he sought to reform. Prior to his coming to Berlin in 1832, his preparation for geography consisted of several years' experience in topographic mapping and map drawing, university studies in natural sciences other than geography, a general reading of travel books, and finally—work which chance threw in his way and which need impelled him to accept—the preparation of a geographic handbook on Peru, Bolivia and the Plata countries, needed to complete the twenty volume *"Handbuch der Erdbeschreibung,"* edited by Gaspari, GutsMuths, and others [*53*]. In many respects this series ("one of the traditional encyclopedic treatises," as Leighly correctly calls it, in another connection [*222*, footnote 1]) was of a type that antedated Gatterer as well as Ritter, and, though Fröbel attempted to introduce short treatments of the natural regions of each country in addition to the standardized division by minor political units, one does not wonder that he mentions it many years later with little pride. By the end of the year or two needed to complete this job, he knew what was wrong with regional geography, so that his program for reform in geography appeared in the same years as his first publications in the field [*28*, I, 40–66].

Fröbel was therefore perhaps the first of the many students of the past hundred years who have attempted to construct a science of geography based, not on a study of geography itself, but rather on a comparison with the "natural sciences." Starting with the two naive assumptions, namely that "geography" or "*Erdkunde*" must correspond to a literal translation of its name, and that it "can only be a natural science," Fröbel thought his way through in print—not without some confusion [compare the discussion, *54*, 499 f., with that on 505 f.]—to what may be regarded as a logical program [*56*]. This program divides geography into two parts. Geography as a natural science consists of separate studies of each of the features of the earth, classified by kind. Following, in part, Humboldt's point of view, as well as his terminology, he included the study of man as a feature of the earth in his "natural science," in terms of all objective facts of man and the earth that represent relations between them [*54*, 495, 504; *56*, 2–4, 6].[34]

On the other hand, those aspects of geography that could not be fitted into his frame of a "natural science"—including what we have called "aesthetic geography," the relation of historical phenomena to geography, and, apparently, regional geography—he recognized as a form of "applied geography" or "historical-philosophical geography." The two points of view are to be developed by their own methods, the former following the analytic. the latter the synthetic method. Each can be built into a separate system, but the two apparently cannot be combined [*56*, 10]. Geography is, therefore, not merely "dualistic" but rather, one would suppose, is completely separated into two kinds.[35]

---

[34] In view of the numerous arguments based on logical considerations that Fröbel offered in justification for the inclusion of human facts in his "natural science" part of geography, there are no objective grounds for assuming, as Leighly repeatedly remarks, that it was only because of the influence of "tradition" that Fröbel included "ethnography" in his scientific geography (in reality he included a much wider range of human phenomena) [*222, 247, 251*].

[35] This writer finds no evidence in any of Fröbel's writings to support Leighly's inference that Fröbel's separation of geography into two parts was a cynical device to save "his natural-scientific geography" while permitting "the devotees of the historical-philosophic geography . . . to discredit the discipline of their choice" [*247*]. It would indeed be ironic if Fröbel had had such a thought, since his subsequent publications can only be included—so far as they can be considered as geography at all— in this latter part of the field. But there is no reason to apply the iron to Fröbel: nothing that he wrote in either of the articles in question permits us to question his sincerity. Even if we can overlook his praise of Ritter, for having tied the study of man to the earth [*54*, 504], we can hardly suppose that he wished to consign a major part of the work of Humboldt to the scrap-heap. Unfortunately Fröbel's carelessness in not giving specific references contributed confusion to posterity. Presumably it did not occur to him that his words would be read by a generation that would not recognize "*der Meister in der Naturschilderung*" as Alexander v. Humboldt, **and** would likewise fail to note throughout the paragraph the many echoes of Hum-

To what consequences did this view of geography lead in Fröbel's work? Plewe presumes that there should have been no difficulty in pursuing a program so clearly laid out—*i.e.*, in the "scientific" half of geography—but Fröbel's later work was not known to him [*8*, 60]. Indeed in all the studies of the history of geography that I have seen, Fröbel's name disappears completely after its one sporadic entry. Although it recurs in periodical reviews in later decades, one hardly recognizes it as belonging to the same man; only his autobiography provides the connection.

During the decade following 1832, Fröbel taught geography and mineralogy in Zurich and apparently devoted a considerable part of his time to the preparation of a full study of systematic geography that was to lead to a theory of earth science, but this was never to be completed.[36] Overlooking a study in crystallography—a part of his work in his second field of instruction—the only distinctly geographic research publication of his life,

---

boldt's description of the "physiognomy of nature." The only specific quotation that Fröbel gives in discussing the non-scientific half of geography is taken, without reference, from Humboldt—and incidentally is used to arrive at a conclusion opposite of that of its author [*56*, 7; Humboldt's original in *43*, 17]. Any possible doubt on this point is removed by Fröbel's specific reference, in a paper written a few years later, to Humboldt's *"Ansichten der Natur"* as the masterpiece of this type of non-scientific geography [*57*].

[36] In Zurich, Fröbel and one of his colleagues founded and edited the *"Mittheilungen aus dem Gebiete der Theoretischen Erdkunde."* As this periodical, which lasted through but four issues, is available in but few places, I was unfortunately not able to see a copy until after this paper had been completed. The introductory article by Fröbel [*57*] is a presentation of his outline of geography that is more complete and more maturely considered than that of 1832. Geography, in the broadest sense, he defined as "the science of the phenomena of the earth (*Erdwelt*), insofar as these constitute this earth by their association in area." Instead of dividing geography into two parts, he recognized four different kinds of geography (1) "pure geography" in which the phenomena are studied "insofar as this association has a purely scientific interest"; (2) "political geography," or, possibly, "statistics," in which the interest is "ethical-practical"; (3) "historical-philosophical geography," in which the interest is "ethical-theoretical"; and (4) "physiognomic geography," in which the interest is aesthetic. Pure scientific geography included objective studies of human races and peoples and the geography of states as earth phenomena. Geography is not the entire science of the earth: as *Erdbeschreibung* it is to be kept distinct from earth history and the theory of the earth [based on Humboldt, *42*]. The dedication of the first volume to Humboldt is of interest: *"dem Ersten, welcher den grossen Zusammenhang der Naturerscheinungen an der Erde als besondere Aufgabe einer eigenen Wissenschaft aufgefasst hat."*

In a second article, an excellent discussion of the problem of defining relief forms leads to a conclusion that the present writer unwittingly echoed at a recent geographical meeting: "the necessity of emancipating the geographic study of the unevenness of the earth surface completely and absolutely from geology" [*58*, 476].

other than his initial *Handbook of Peru, etc.,* was a study of certain remote valleys in the Walliser Alps, in which he was chiefly concerned to map the topography and to study the ethnography of the population—a remnant of early Celtic tribes, as his linguistic comparisons led him to conclude [*28,* I, 71–90]. This was hardly, one may suppose, a study in the strict "natural science" sense of geography.

After a decade of teaching at Zurich, Fröbel resigned from his position to devote himself to revolutionary propaganda in the political field and never returned to academic life. After serving in the abortive Frankfurt parliament of 1848, and narrowly escaping the firing squad in Vienna, he went in exile to America, where he spent ten years in various activities, chiefly as a roving journalist. His two-volume story of his travels—several times across the continent, as well as in various Central American countries, but not in the South American countries of which he had written in his first geographic work—was favorably reviewed by Neumann in the periodical of the geographical society of Berlin and provides interesting reading today [*62*].

Although a large part of the work is concerned with social and political problems in the United States, and the author specifically indicates that it is not to be considered as a scientific work, he nevertheless has provided a large amount of geographic material. One short chapter is a systematic study of the mountain systems of western North America—the only study that Fröbel ever made that could incontestably be included in what he had defined as scientific geography. (Whether this was a more significant contribution to the physical geography of North America than Neumann thought, I am not in a position to judge; it is interesting to note that it was first published in the *California Chronicle,* Dec. 13 and 14, 1854.) More valuable are the many fine descriptions of areas, which could fit into his division of "applied geography"; these offer the historical geographer descriptions of our Southwest in the middle of the past century made by a trained geographer. The informed reader will note the student of Humboldt in the description of the scenery of the Limpias Valley as "a wonderful harmony and unity of physiognomic elements out of which the landscape is formed" [*62,* II, 382].

Returning to Germany, Fröbel published in the following years a large number of political studies, some of which are clearly based on a geographic foundation. Notable is his short study of the relation of America to the European political-geographic situation [*63*]—a study which at the end of his life he regarded as more important than all else that he had written, and which Humboldt found valuable [*28,* II, 28 f.]. Fröbel is therefore characterized by Spörer—in the same study to which Leighly has called our attention, because of its criticism of Ritter and his followers—as a "skillful historical-geographical chess player" . . . "the virtuoso in the art of combining history and geography" [*68*, 415]. Indeed one may say that the same Fröbel, who in his first geographic phase appears as a herald of the late nineteenth century "purely scientific" geography, in his later "most important" study stands as the forerunner of the twentieth century, post-War, school of *Geopolitik*.

The remainder of Fröbel's long and active life—ten years or more as publicist and special adviser to various German governments in Vienna, Stuttgart, and Munich, and some sixteen years in the consular service of the German Reich, in Smyrna and Algiers—does not appear to have produced any contributions to geography.

One cannot consider the career of this extraordinary man without wondering what position he might have attained in the field of geography had he remained in it. In view of his undoubted capacities and energy it seems probable that he might well have influenced its developments as much as any other student of his time, but that his predominant interest in human phenomena would have permitted him to confine himself to that part which he classified as "scientific" appears most doubtful. Whatever one may conjecture on that point, the fact is that his revolutionary program for geography, based neither on his own work nor on a study of the history of the field, had no more influence on his own subsequent publications than on the development of the field in general.

The radical program that Gerland proclaimed for geography bore even less relation than Fröbel's to the author's previous work; and likewise had but little influence on his subsequent studies; before and after, his major field of work was in ethnographical studies, which to some degree could be considered as ethnographic geography [Wagner, *80*, 384]. In this case, also, the biographical background is significant; it may be found in the sympathetic study by Sapper [*37*]. The fact that Gerland has had no university training in geography was not unusual. Most of the men who developed the field in the latter part of the century had come into geography after being trained in mathematics, geology, zoology, or history [Penck, *90*, I, 38]. In most cases, however, the shift resulted from the interest that they found in the field and was represented first by publications in the field of geography. Gerland's relation to geography, however, had apparently been limited to the surveys of ethnographic work that he wrote for the *Geographisches Jahrbuch* and the teaching of courses in geography that fell to his lot in various secondary schools, when, in 1875, the new German university of Strassburg called "the forty-two year old ethnologist and philologist" to the chair of geography.

Although Sapper regarded the experiment as successful, we have contrary evidence in Gerland's own statement. In explaining his proposals for radically changing the nature of the field to which he had so belatedly been called, he wrote that the unscientific nature of geography as commonly understood had been "embittering his life and career" [76, xliii]. If one reads the whole of Gerland's lengthy statement, one cannot but question whether he had been able to understand the field he was called upon to teach—whether the problem, in other words, was not simply the personal problem involved in the contrast between his particular way of thinking in science and that of the field of geography as Humboldt and Ritter had developed it.

Whatever may be the explanation of Gerland's personal problem, his programmatic essay of 1887 did not solve it. Throughout the next twenty years of teaching he continued to include the geography of man in his courses. Among his writings of this period but little can be found that would be included in the field to which he had proposed to limit geography. As a major work, Sapper notes his *"Geographische Schilderung des Reichslandes Elsass-Lothringen,"* published in 1894, a work that included consideration of the human population, but "on principle" excludes political geography—Hamlet without the Prince of Denmark one may say, since the only justification for considering this area as a unit is to be found in its political geography.

One can hardly read Gerland's examination of Kant's work in geography and anthropology—a lecture course given in 1901 and published a few years later—without wondering whether he had not actually given up his earlier thesis, even in theory. It seems but a weak echo to hear him urge "all geographers to take to heart the word of the great philosopher," namely that "physical geography, a general abstract of nature [including, as we have noted, man] is the foundation of history and all other possible geographies" [12, 504]. Ratzel might have said as much, and Richthofen in fact did. At about the same time, Sapper informs us, Gerland planned to write four great works: "geophysics as the geography of inorganic earth, the geography of plants, the geography of animals, and sociology as the geography of man" [37, 340].

The reader may well ask why we should further consider a program for radical limitation of the scope of geography which had no basis in the author's previous work and no influence on his subsequent work. Nevertheless, Gerland's program is significant as the most thorough-going attempt to outline a science of geography in terms of the concept of science developed by a particular group of sciences, from "purely logical" considerations, regardless of the historical development of the field. In this respect it is the logical successor to that of Fröbel; insofar as Gerland failed to be completely consistent, Leighly has appended the necessary modifications [222, 250 ff.].

To be sure, Gerland himself did not admit that he was opposing "logic" to "tradition," but claimed that he had given ample regard to the historical development of geography [xli]. But the only evidences for that claim are scattered instances where he reaches back to a few predecessors to find, in part of their work, that to which he would limit the whole of geography [as in the references to Humboldt, pp. xix, xxi–ii]. His explanation that he "considered only the main course of the development, not the momentary conditions that are of no historical value," reads somewhat ironically today.

We need not consider Gerland's thesis in detail, but may focus our interest on the two major premises on which it is based. In one he states the nature of a science, in the other the nature of the earth which geography's name requires it to study [v f.]. The greater part of his lengthy treatment is an attempt to fit these two premises together logically into a "science of geography."[37]

It is characteristic of this type of approach to the methodology of geography that the premise regarding the nature of a science is presented in a categorical statement devoid of any supporting discussion or references [v and xxix]. It seems unnecessary to repeat his somewhat lengthy definition, since students of such opposite viewpoints as Wagner and Leighly both recognize it as the description of "a physical science." If this premise be accepted, Wagner agreed that much that follows must logically be granted; but he did not feel compelled to accept a statement of science that would grant scientific standing only to the so-called exact physical sciences [77, 421 ff.]. If this major premise is not accepted, the entire subsequent discussion falls to the ground.

One sees at once the disadvantage of attempting to construct the concept of a particular field of knowledge on the basis of a general definition of science. The question of the nature of science is a philosophical rather than a scientific question; that scientists, as such, are not qualified to answer it is reflected in the wide range of answers that different scientists provide. Great as may be the disagreement among geographers in their views in regard to the proper scope of geography, the differences on this point are relatively small in comparison with those that appear when they attempt to state what science is in general. Furthermore, the viewpoint on this question

---

[37] Leighly's presentation of Gerland thesis [222, 250–3], is not to be taken, as the reader might suppose, as a full abstract. Thus the phrases, "Gerland directed his attack immediately . . .," and "Gerland turned next . . ." each skip over four or five pages in which Gerland had presented important arguments that are never mentioned, though they attempt to justify the inclusion of studies that Leighly rejects [76, viii–xiii, xxi–xxv].

changes much more radically in course of time. This is reflected in the comparison of Fröbel's program and Gerland's. As Leighly correctly notes: in both cases, "it is the voice of nineteenth-century natural science that finds utterance" [222, 250]. But whereas in Fröbel's time, "natural science" was thought to include the objective study of man, fifty years later the concept of "natural" had been narrowed to exclude man. Further, as Leighly indicates in unconscious agreement with Wagner, Gerland's definition of science is even more narrowly conceived: the basis of his argument is "the strict logic of physical science." It therefore requires fixed conclusions and laws of certainties; a human geography that could only deal with "probabilities" was no science at all [xxix; cf. Wagner, 77, 436 f.]. Naive as this idea may seem to us today in view of the change in thought forced upon physicists and chemists by the discovery of irreducable uncertainties in the study of electrons, in Gerland's time it no doubt seemed a fair assumption for physical science.

In his second basic assumption, Gerland deduces the field of geography from its name; whether "geography" or "Erdkunde" is immaterial, it is the science of the earth. (This etymological conclusion is also supported by reference to the concepts—in contrast, we may note, to the work—of Kant, Ritter, and one or two others [ix].) The essential question then is: what is the earth? This question, he evidently felt, is not to be answered in terms of some theory—i.e., the earth as an "organism"—nor in terms of a particular, anthropocentric point of view—i.e., the earth as the home of man. Looking at the earth from an objective, scientific point of view Gerland found that it was like millions of other units in the universe, a great complex of materials in state of change, bound together and interrelated by various forces to form a unit Whole, though acted upon also by external forces, notably from the sun. The problem and object of geography is therefore to study the interrelations of these forces and the resultant changes in earth materials. Vast though the problem is, it forms a complete unit [vi f.].[38]

We need not examine all the logical difficulties into which Gerland was brought by the incompatibility of his two major premises. Wagner discusses these in great detail, though without clearly labelling the fundamental cause.[39] Briefly, the premises are incompatible from two major points of

---

[38] One may feel that this statement of Gerland's is inconsistent with the objections which he later makes to any study of interrelation of heterogeneous phenomena [xvii, li]. It may be an improvement on his logic therefore to translate his phrase "Complex kosmischer Materie" into "assemblage of matter," as Leighly does [222, 250], but it is a change of the original statement.

[39] Although a number of Wagner's arguments will be found in the following pages, by no means all can be mentioned here. Leighly's statement that his own "paraphrase does not quite exhaust the grounds Wagner cited . . ." [222, 256] is an under-state-

view—one in regard to the content of "earth science," the other in regard to methods of studying it.

What is the content of geography according to Gerland's definition of the earth as its object of study? If the earth materials that are in a state of inter-related development include rock, water, and air, can they exclude these materials transformed into plants? Obviously the component forces effecting changes in rocks, water, and air include the life force of plants and animals. Equally obviously, the materials of which human animals are made, and those which they transform into cultural products, are earth materials, and the human energy that has altered the face of the earth over great areas is certainly one of the forces affecting changes in earth materials. Indeed it would seem a curious reversal to earlier religious viewpoints to argue that man and human energy were not of the earth, earthy. If, however, as Gerland stated in this premise, all the materials and forces of the earth represent a unit whole (*einheitliches Ganzes*) [v], we cannot exclude some of these materials and forces and still have a unit whole left.[40]

The reasoning of the preceding paragraph was essentially that followed by Fröbel in explaining the inclusion of man in his "scientific geography." Though Gerland recognized its applicability to plants and animals, and for that reason in part included them in his more limited "science of the earth," he could not follow it through logically because it comes into conflict with his other major premise, that of the nature of a science. Human phenomena could not be included in a science as he had defined it and must therefore be thrown out of geography. In other words, reality must be distorted to fit science, science is not to be constructed to fit reality. This viewpoint is illustrated, almost at the same time as this is being written, by an English radio song:

> "How does a hen know the size of an egg
> With no egg-cup beside her
> With no one there to guide her?"

Gerland's solution of the logical dilemma was a compromise. Human

---

ment. Wagner offered at least a dozen logical objections to Gerland's thesis, of which but three are to be found in the paraphrase. The argument concerning plant and animal geography, though of minor importance in Wagner's critique [77, 425 f.], is called "the only logical answer . . . that Wagner . . . offered" [222, 255]; if by this is meant "acceptable," the count is still short, as at least one other argument is accepted, though without mentioning it (see footnote 41, following).

[40] This is essentially the argument that Wagner based on Gerland's premises [77, 426 f.] and which Leighly describes as "an extremely thin logical thread" [222, 255 f.]. But the syllogism to which the latter reduces that argument cuts out essential parts and overlooks the difference between a "complex of interrelated material" and a mere assemblage of material.

phenomena are sacrificed to one premise, but non-human organisms, more amenable to "scientific laws," are included on the basis of the other [xxiv f.]. That this was inconsistent with his arguments for excluding man was all too obvious: the method of the biological sciences is not the same as that of the physical sciences. (That the physical sciences may all be included under one "method," as he implied, is another undemonstrated assumption that might be questioned.)

To Gerland's contemporaries it was therefore clear that to fit the earth to his concept of a science he must change his concept of the earth by stripping it of its plant and animal cover. A further amputation was also required. In contrast with Fröbel, who had wished to relegate regional studies to a very loosely defined "applied," but "unscientific" geography, Gerland included the study of areal combinations of phenomena (other than human); *Länderkunde* was one of the three major parts of his geography [xxx, xxxv]. Wagner, and later Hettner, observed that such studies could not logically be included under the strict definition of science assumed [77, 433; 2, 315 f.].[41]

Since it was clear to all who read Gerland's thesis that to be logical, he must exclude plant and animal geography as well as regional geography, few geographers considered it necessary to examine his thesis further. That is to say, they were not ready to throw overboard as surplus baggage not only all the work of Ritter and his followers but likewise the greater part of the work of Humboldt—in general, by far the greater part of the geographical literature of the past, not to mention the scientific organizations and periodicals, and the training of the current generation of geographers.[42] In his

---

[41] Leighly's discussion ignores Gerland's arguments on this point [76, xxx, xxxv]. Apparently he accepts Wagner's criticism—whether consciously or unconsciously—for he first reduces this major part of Gerland's field to a source of "observational data for geophysics" [222, 253], and then quietly drops it out entirely [254, footnote].

[42] Wagner's discussion included a consideration of certain practical effects that might be expected if Gerland's proposal were to be adopted, a discussion that no more than balances the ten pages that Gerland had devoted to similar considerations [xliii–liii]. While not mentioning the latter, Leighly comments on the former: "there is visible through this part of his (Wagner's) criticism a poorly disguised fear of disturbing the position geography had attained in the schools and universities" [222, 256]. The disguise was, in fact, that of Lady Godiva. Wagner introduced that part of his discussion with the phrase *"Ich wage zu behaupten . . ."* (I dare to maintain), and somewhat farther along stated *"Ich kann die Befürchtung nicht unterdrücken, dass man der Entwicklung unserer Disziplin auf den Schulen einen schweren Schlag . . . erteilen würde,"* (I cannot suppress the fear that it would be a heavy blow for the development of our discipline in the schools), and yet again referred to *"eine Gefahr"* [77, 442 f.]. Likewise, in referring to this discussion in his next report, Wagner wrote: *"Ich habe damals aus meiner Befürchtung kein Hehl gemacht"* [80, 395].

historical study, therefore, Hettner properly limited his consideration of the thesis to its lack of continuity with the historical growth of the field; its logical claims however he did consider in his first full methodological treatment, in 1905 [*126*, 546–9; also in *132*, 694–9]. Likewise Penck attacked the logic of the thesis in an address given in this country, at the Congress of Arts and Sciences at St. Louis in 1904 [as published in Germany, *125*, 3 ff.; whether the address was published in this country, I do not know].   *

We may say, therefore, that logically Gerland presented not one thesis but two conflicting theses. If one follows consistently his definition of the earth, then geography, as the study of the earth, would include all that had commonly been studied in geography and also all of geophysics—if not also of geology. It would not however be a science as he defined that term. If one follows consistently his other premise, as Leighly does, one may prove that, in the terms of "the strict logic of physical science," geography must be a physical science of the earth, *i.e.*, "geophysics." But one may well wonder why it required over 50 pages and 50 years to prove that proposition.

The historical fact is that Gerland's contemporaries saw clearly enough the ultimate conclusions to which either of these alternatives led. Although they recognized that the methodological statements and general outlines of earlier geographers had commonly included the study of the earth body, they were sufficiently realistic to recognize that the actual contributions of geographers to geophysics had been extremely slight and that the knowledge and training inherited in geography did not equip its students to work in that field [Wagner, *77*, 426 ff.]. Although a few still continued to include the study of the earth body in their definitions of geography—Richthofen, in particular, returning to this concept in his Rectoral address at Berlin, in 1903—in practice, the geographers in Germany as well as everywhere else, confined themselves to the study of the earth surface. To be sure, they recognized that a complete explanation of surface phenomena—at least of land forms—must ultimately be based on deductions concerning the interior, but they did not suppose that one branch of science must include all upon which it depends. As Bucher had observed in that connection, every science depends in part on other "help-sciences," and any science may serve as a "help-science" without thereby losing its independence [*51*, 239]. Properly speaking, no branch of science can be a complete unit in itself independent of other branches.

In reality, as Wagner explained in detail, the examples of geophysics presented in the volume of studies of which Gerland's thesis was an introduction, gave little promise that studies of the earth's interior could form the basis of detailed studies of the surface forms, or vice versa [424 f.]. Logi-

[112]

cally, as Gerland said, the study of relief forms should begin with an explanation of the phenomenon of the continents, but after a half century of work in geophysics what "certain" explanation do we have for the "particular form of the continents?" [xxxv].

Even more clearly did Gerland's contemporaries foresee the consequences that would result if geography were transformed into geophysics. As' followers of Humboldt they saw the particular value which geography offered to all science in forming a connection between the artificially separated portions of reality studied by the natural and the social sciences [cf. Penck, 158, 54]. They did not share Gerland's confidence that the social sciences could be trusted to make these connections themselves, that geography need only provide the raw material, in the form of knowledge of the physical facts arranged by categories, or—in Gerland's original thesis—combined within particular areas in their relations independent of man [xxxvi f.].

Specifically, Wagner asked [438 f.], if the geographer is to "present scientifically," for the benefit of "the sciences to which geography is auxiliary," "the nature and natural productivity of a country, that interest the geographer for their own sake" [xxxvii], what is this "natural productivity"—i.e., productivity independent of man (and according to the improved form of Gerland's thesis, independent of the vegetative cover)? [see also the discussion by Partsch, referred to in 80, 376]. This question has been raised in various symposia of American geographers; many readers may remember a discussion concerning the possibility of a "pure geography" of the Amazon Basin. Since the same combination of soil and climate may be highly productive for some crops and of low productivity for others, the very concept of "natural productivity" has no meaning excepting in reference to the particular crops that particular men wish to grow. We could map the natural productivity of an area only by making separate maps for every conceivable combination of agricultural crops and methods.

Further, various of Gerland's critics objected that even if we eliminate all these difficulties and reduce geography to the study of the physical aspects of the earth, literally in terms of the physics of the earth, such a science would have no more unity than the science of physics itself. It would represent, that is, not a part of physics, but a collection of all the parts of physics that can be applied to one particular object, the earth. This is most clearly shown by Leighly's reference to the several sections of the American Geophysical Union—an organization whose very title indicates that it does not represent a branch of science, but rather a collection of branches of science that have certain interrelations. If we add together meteorology, seismology, vulcanism, terrestrial magnetism, geodesy, and hydrology we have a sum that represents much of the field of physics as applied to various physical phenomena of the earth. It is united only as an applied field. In terms of pure science it represents widely separated branches of physics in which the kinds of phenomena studied and the methods necessary for studying them are as heterogeneous as could be found in all of physics. The sum of these branches does not form "a physical earth science" but rather many physical earth sciences—only one of which is commonly called "geophysics"—and if we attempt to put these many together we return to the field of physics as a whole.

Likewise, finally, Gerland's critics saw that, in trying to construct a logical edifice for the field of geography on the basis of a particular concept of science, Gerland in reality had not considered the problem from the point of view of science in general. He recognized that there were interrelations between human and non-human phenomena that should be studied in certain fields of science but at the same time denied scientific character to these fields [cf. Wagner, 77, 421]. Any field that studies the interrelations between these heterogeneous phenomena must use both of the "methods" that he found incompatible within a single science. Why should the social sciences have this duality of method imposed upon them, or on the other hand, if they must and can employ both methods, why may not geography?

For example, Gerland vigorously objected to Ratzel's attempt to develop a geographic law concerning the separation of peoples by mountains: "the appropriate law here is not found in the connection of men with mountains, but in certain psychophysical reactions that are released by the mountains, hence does not lie in the field of geographic thought, but completely in psychology and physiology [xxx].

Gerland is more consistent, however, in other parts of his discussion, indicating that such relationships are not to be studied in science at all—in what other form of knowledge they are to be studied he does not say. "A science of the heterogeneous things that fill area is impossible because of their heterogeneity" [xvii]. The significance of this statement is not entirely clear until we come to his illustration of it, almost at the end of the essay: "What sort of a connection is involved in the fact that Rome is located on the Tiber, Prague on the Moldau? River and city are heterogeneous concepts that geography can never logically unite." Since neither the particular river Moldau is logically necessary to the particular city Prague, nor vice versa, "an intrinsic connection does not exist" [li].

The answer of geographers since Herodotus and Strabo—including Humboldt and Ritter and the followers of both—is so obvious as to appear almost superfluous: in constructing the city of Prague man has produced a synthesis of city streets and buildings with the particular character of the river Moldau at that site. Though, in this case, the economic and political center of the Bohemian basin might perhaps just as well have developed on one of the other headwaters of the Upper Elbe, such a city—whether called Prague or not—would have been significantly different from that which actually exists. Further, if we look beyond the particular examples that Gerland selected as apparently suitable to his thesis, would he have maintained that the particular river Rhine is not essential to the city Cologne, that there is no significant relation between the city of New York and New York Bay?

Gerland's specific illustration of his thesis brings us clearly to a parting of the ways. Those who follow him to the logical conclusion will study the phenomena of the earth in clearly separated categories; they will not, like Ratzel, concern themselves with the contrasts in population in successive zones of elevation in Scandinavia nor with the zones of land use at different elevations in the Alps [72, I, 403-4]. Neither will they, like Humboldt, study the differences in natural vegetation at successive zones of elevation in the Andes, nor in the different climatic zones of the world [43; 44]. On the other hand is found that host of students who from earliest times have observed that these heterogeneous phenomena exist in reality (*Natur*, Humboldt would say) not merely beside each other, but in causal relation with one another, and they have devoted themselves to the study of areas of the earth filled by these things in their interrelations. This study, since the days of antiquity, they have called geography and the world of knowledge has recognized it under that name. Whether or not the study is a science is for them a secondary question, it is the field of study that has long since been claimed by, and acknowledged to, geography.

We arrive therefore at the ultimate conclusion—obvious enough, one may say, from the start—to which "the strict logic of physical science" leads us. Granted the need for a study of the physics of the earth, we need only recognize how completely different such a field is from that which scientists as well as laymen have for centuries called geography. Gerland's thesis, as completed in the discussions then and since, may well stand as the *reductio ad absurdum* of all attempts to make geography into either an "exact" or an "essentially natural" science.

B. GEOGRAPHY AS A SCIENCE OF THE PLANET EARTH

There is one further problem, on which Gerland's essay throws light— namely the question whether the field of geography includes the earth as a whole, as its name certainly implies, or whether it is limited to that part of it which we know directly—the earth surface. In most methodological discussions prior to Richthofen's, the former was assumed, and most text books opened with a section on the earth as a body—as is true of some recent texts. On the other hand, the actual studies of all but a very small number of geographers were limited to phenomena found within the thin shell of the earth surface. Geography, in practice, therefore was the study of the world, as that term is most commonly understood.

Unquestionably however it seemed arbitrary to attempt to restrict a science called the study of the earth to the outer shell of the earth. Richthofen, recognizing that the practice of geographers had in fact defined their field in this way, had attempted to defend the restriction logically in his inaugural address at Leipzig, in 1883. Although Penck found this "never to be forgotten address" completely convincing [128, 45 f.], other students, including Hettner, felt that Richthofen had not found the necessary logical basis for the thesis, and Richthofen himself departed from it later [3, 679 f., 689].[43]

Gerland's discussion of Richthofen's address does not help us; a certain lack of clarity in the original argument permitted Gerland to misconstrue it into the opposite of its intention.[44] But in his own manner of viewing the

---

[43] The reference in Sauer's chapter in *Recent Developments in the Social Sciences* [84, 177] to Richthofen's rectoral address at Berlin in 1903, as an able presentation of the position that geography is the science of the earth's surface, should presumably refer instead to the inaugural address at Leipzig of 1883 [73].

[44] The manner in which Gerland handles this question is instructive. Richthofen's discussion of the significance of the earth surface to various sciences occupies no less than thirteen pages [73, 7–8, 11–16, 18–24]. Gerland's interpretation [76, xiii–xiv] is based entirely on one paragraph—indeed essentially on a single phrase, *"gern gesehene Genossen auf dem Felde der Geographie"* [7]. By implying that *Genoss* (companion) is synonymous with *Gast* (guest) he interprets Richthofen to have claimed for geography exclusive rights to the earth surface as against geology or economics, although the entire thought—and many specific statements—in the following pages of the original address recognizes clearly that various other sciences must study within the earth surface. Indeed Gerland uses some of these arguments to attack the argument that he attributed to Richthofen! If we grant that there was some inconsistency, the only question was whether the phrase *"auf dem Felde der Geographie"* was appropriate; with that eliminated or altered, the issue disappears.

By accepting Gerland's misconstruction of Richthofen's argument, Leighly reports Richthofen as saying exactly the opposite of what he actually wrote. Far from attempting a "Solomonic assignment" between geography and geology [222, 251], Richthofen had declared that "to wish to draw a sharp boundary in this area (the surface of the solid land) between the objects of the two sciences must be described as both impractical and purposeless" [73, 18]. Ironically enough it was Gerland himself who claimed to have "sharply and definitely separated the problems and field of work of geologists and geographers" [76, xxii]. Although the reasoning by which he arrived at his separation formed an important part of his treatise [viii–xiii, xxi–xxii], neither that nor Wagner's objection are mentioned in the re-presentation. Wagner observed that few, if any, geologists would accept Gerland's distinction between the

earth as the object of study in geography, Gerland provided, to my mind, the key to the fundamental distinction in point of view between geography and geophysics. Although this presents a problem to few geographers today, there is evidently sufficient uncertainty on it to justify a brief consideration.

In order to give an objective, scientific answer to the fundamental question, "what is the earth," Gerland discarded not only the teleological and anthropocentric views, but also, by implication, the geocentric viewpoint. Looking at the earth from a cosmic point of view he saw it as but one of many similar units in the universe [v ff.]. There can be no question that from this point of view, the earth body would be included in an earth science, indeed the interior of the earth would appear vastly to outweigh the surface. One might then conclude that the actual concentration of our study on the surface is simply a result of our inability to make direct observations on the interior, and of the fact that the surface provides the expression of the changes in interior materials produced by interior, surface, and external forces.

While this point of view is logically sound, it is fair to ask whether it is as realistic and objective as it appears. Unless we are to determine the relative importance of different things in the universe solely on the basis of mass, it is just as arbitrary to consider the earth as no more than one of the smaller planets of one of the lesser of millions of stars in one of many galaxies, as it is to describe it as the home of man—both statements are equally true. The very fact that we have a science of geography, but no science of Mars-ography reflects the two objective facts that make the earth, for us, unique, and justify our distinguishing it out of all the units of the astronomer's universe for study in one or more earth-sciences.

One of these facts—and literally speaking we cannot get away from it— is that the earth is where we live. We may say that science is not in itself an abstract thing, but is rather man's pursuit of knowledge of his universe. In this universe the earth is, for science, unique above all other units, as the one on which man lives. If other satellites of our sun, or of any other, were known to be like ours in all important respects, we would no doubt have a special science concerned with studying them, but would at the same time require a separate science for the separate unit on which we not only live, but rather of which we are a part. In the comparative science of earth-like planets we would look at the earth from a more celestial point of view; but from that point of view there would be little justification for a separate

two fields—namely, that geology studies the materials of the earth's surface, as found, *"das Gewordene,"* whereas geography studies the forces of development, *"das Werden."* [*77*, 429 f.]

science of the earth. The justification for that science would rest only upon a geocentric view.

The previous consideration however is, in the fact, academic. In our actual science of the cosmos—however detached our point of view—the earth is unique. Whatever speculations one may make as to the possibility of the existence of similar planets—recently calculated by Jeans as a very minute possibility—our science knows no other planet like it. In other words, for a celestial scientist, attached to no planet or star, but knowing only what we know, this little planet Earth would be an object of unique and extraordinary importance. It is the one body in the universe on whose surface temperatures fall within that very narrow range in which different materials exist simultaneously in solid, liquid, and gaseous state; in consequence of these and other factors it is the body whose surface shows the greatest multiplicity of varying interactions of factors and differential results; further, on it alone can that extraordinary phenomenon called life be observed in millions of generic and individual variations; and finally, to our celestial scientist it would not be a matter of indifference, that on this planet alone (so far as science can say) creatures capable of considering science can exist. Indeed, to become real, our celestial scientist must come to earth—only on the earth can he find his home.

The reader will have observed in the previous paragraph that the uniqueness of the planet earth to science is expressed, entirely, in terms of conditions on or near its surface. Our unique interest in the earth is a unique interest in the earth surface. If it be argued later that the understanding of the surface requires a knowledge of the interior, it may be necessary to dig into the interior—so far as that is even intellectually possible. But the object of our study in a science devoted to the earth as a unique planet, is the earth surface—the thin shell of outer earth in which are found the unique phenomena of earth. In other words, the geocentric point of view, which is implicit in the existence of a science devoted exclusively to the tiny fraction of the universe called the earth, is centered not on the astronomical, planetary unit, but on that naively given section of reality that man, from earliest times, has thought of as the world of which he is a part, and which only recently he has discovered to be but the outer shell of an astronomical body which he has since called the earth [cf. Schmidt, 386, 2 f.].

The change in our knowledge of the earth, referred to in the preceding sentence, may well be the explanation for the fact that we have no single word of common speech that refers unquestionably to the field commonly ascribed to geography, thus forcing us to use a term that appears artificial and at the same time not clearly defined—namely "earth surface." On the other hand, as Lehmann has recently pointed out, every one—outside of learned circles at least—understands what we mean by the word "world," particularly in the many combinations in which it is commonly used: "world tour," "world map," "world trade," "world-wide," "World-War," etc. In all of these cases there is no uncertainty as to the physical extent of space included: they do not include the center of the earth, nor do they include the moon or other "heavenly" bodies, but simply that outer shell of our planet as high in the atmosphere and as deep in the ground as man can *experience* it [*113,* 218 f., 235; *cf.* Kant. Sec. II A].

If the physical scope of geography is considered as simply the world—as it has almost always been in practice—it seems unnecessary to attempt any sharp definition of the term "earth surface." Biologists have little difficulty in studying the field of "living bodies" even though they may find it extremely difficult to give an accurate definition of "life." Likewise geographers need not concern themselves with establishing an exact boundary between the "world" and the interior of the earth. If the explanation of surface phenomena requires deductions concerning conditions deeper in the interior, the geographer who is able to dig deeper will presumably not stop at any lower boundary that might be set; so long as he maintains the purpose of his study in the surface—the world—there is no need to set such a boundary [*cf.* also Marthe, *25,* 9].

If the earth surface be regarded as the domain of geographic study, forming in its entirety for geography alone the sole concern, it also forms at the same time, as Richthofen indicated, the meeting place for all sciences that are concerned with the earth: meteorology, geology, geophysics (in the narrowest sense), geography, economics, etc. [*73,* 11–24; *cf.* also Michotte, *189,* 7 ff.]. But it does not follow that each of the other sciences enters this area as the guest of geography—as Gerland's construction of one unfortunate expression of Richthofen's had him say—but rather that each of these other sciences is, with geography, equally at home in some part of the earth surface. But each of these is interested in the earth surface from a different point of view, just as mineralogy and paleontology may study the same rocks from entirely distinct points of view. Richthofen failed to make clear this distinction in point of view; apparently like all the geographers of his time he overlooked Humboldt's statements of the contrast between the geographic viewpoint and that of the systematic sciences concerned with the same objects. By the time this view was again clearly expressed, by Hettner in 1905, the particular argument concerning the place of the earth-body in the field of geography had become of little more than academic interest. Wagner's prediction that the highly specialized techniques required in geophysics

* See Supplementary Note 16

would prevent geographers from making important contributions in that study was apparently fulfilled and geophysics became definitely established as an independent field in close relation to physics and geology.[45]

C. GEOGRAPHY AS A SCIENCE OF RELATIONSHIPS   *

To many readers the outline of historical development of geographic thought may appear to give too slight attention to what was for half a century the dominating concept in American geography, namely the study of relationships between nature and man. In view of the importance of this concept, not only among research workers in geography in this country and in England, but particularly in the schools, it is necessary to give it special attention.

Throughout the previous discussion we noted that geographers seriously interested in developing their subject as an independent branch of science have constantly sought to understand the complex relationships existing among the vegetation, fruits, occupants, and peculiarities of the various quarters of the earth—to paraphrase the statement quoted from Strabo. In any science the study of its phenomena involves the study of relationships that may be found among them. In geography, in particular, such relationships form the connecting links between heterogeneous and otherwise unrelated phenomena, uniting them in the single study. But we give an entirely different orientation to a field if we focus the attention on the connecting links, rather than on the sum total of phenomena interconnected in areas.

We have seen that the idea of defining geography in terms of relationships developed only when the subject as a whole was thrown farthest out of balance—in the latter part of the last century—in order to retain the study of human phenomena in a field in which, according to its particular development at the time, they seemed out of place [cf. Hettner, 126, 548; 130, 413; and Schlüter, 127, 11]. With the restoration of a balanced view later in the century the concept gradually dropped out and is hardly to be found in methodological discussions in Germany in the past two decades.

It is an extraordinary fact that, while modern geography in all countries has been fundamentally dependent on that developed in Germany, the particular concept which influenced German thought for but a brief period, should have come to dominate the methodology, and much of the practice, of geographers in France, England, America, and apparently Japan. Though

---

[45] Wagner's statement of the distinction between geography and geophysics did not "hand over without a qualm . . . the physical phenomena of the earth's surface," as Leighly states [222, 256], but rather a narrowly and specifically limited group of problems concerning the earth-body as a whole [77, 432 f.].

* Title should read "Geography Defined as the Science of the Relationships between Nature and Man".

simple explanations of cultural developments are to be included among the less reliable of historical conclusions, this particular result appears to have been caused by the great interest which a small number of outstanding foreign students took in the anthropogeography of Friedrich Ratzel. Although this work was of great importance in establishing systematic (general) human geography, Ratzel's methodology—whether in his works or in specific methodological statements—does not appear to have had any important effect on the methodology of present–day geography in Germany.

Though this statement, which is in marked contrast to the assumptions of most English-speaking geographers, is not expressed directly by German writers, it may be inferred from the small attention paid to Ratzel's works in methodological discussions, and has been substantiated by personal statements. On the other hand, it should be added, parts of Ratzel's work of which Semple gave but little indication have also been important in German geography. His regional studies have been highly regarded, particularly by those interested in the aesthetic aspects of the subject. Relatively speaking, his importance was greatest in political geography. A sociologist, L. Gumplovicz, says: "Ratzel's works contain more, and more important, knowledge concerning the state, than the entire theoretical political science literature of the last hundred years" [quoted in 7, 160]. Granted that there may be some exaggeration in that statement, Ratzel is universally recognized as having established the foundations of this particular branch of modern geography [216, 789].

The close association, in France, of geography and history may account for the active interest in Ratzel's work taken by Vidal de la Blache, Vallaux, and Brunhes. At least one article by Ratzel was published in France, in French, and his works were given lengthy abstracts and reviews, so that, for a considerable time, French geographers were greatly concerned to criticize, modify, and correct his statements and point of view. As similar consideration was not paid to other German geographers, then or later, Brunhes, in a short outline of geography, published as late as 1925, gives more than four pages to Ratzel, less than half a page to Richthofen, and dismisses later German geographers as amplifiers and modifiers of Ratzel's work [83]. As Sauer correctly observes, Richthofen's influence has been far more important than Ratzel's in twentieth century German geography [84, 183].

Brunhes' methodological considerations are perhaps more suggestive than definitive, but his detailed portrayal of his concepts, in *La géographie humaine,* presents essentially a modification of the dualistic concept of geography, in which a large part of human geography is defined in terms or relationships [*182; cf.* Michotte's critique, *189*, 11–14, 31–3; see also Sec. VII C, following]. The most thorough methodological study in French geography, and perhaps the ablest presentation of the concept of human geography as the study of relationships (in which however physical geography is defined in terms of distribution) is Vallaux' *Les sciences géographiques* [*186*]. On the other hand, these works have been less important in determining the direction of French geography, than the concept of regional geography which Vidal and his school have developed to such a notable extent, particularly in monographic studies of particular regions of France.

Ratzel's most important disciple was an American, Ellen C. Semple. [*]
By her rendition of his ideas in her teaching, and in her publications, she had a marked influence not only in this country but also in Great Britain—where Chisholm's enthusiastic discussion of Hettner's methodology [*192*] appears to have passed unnoticed and Herbertson's ideas of regional geography long met with opposition or indifference. In this country the association of Semple's concept of "geographic influences" with the frankly dualistic concept which Davis had brought from Germany largely determined the methodological thought of at least a generation. It was expressed repeatedly in the presidential address of this Association from 1906 to 1926—with the notable exception of that of Fenneman—and is most fully developed in the works of Semple [*204*], and Whitbeck and Thomas [*215*]. Dominant [**] in the universities until recently, it is still the prevailing concept in the teachers colleges and secondary schools. It is therefore necessary for us to examine this "environmentalist concept" in some detail, even though the objections to it have been stated repeatedly in American publications during the past dozen years [notably by Sauer, *209*, 18 f.; *84*, 165–75; *85*, 622; and later by James, *286*, 81 f.; Hartshorne, *216*, 795–9; *218*, 168 ff.; and Hall, *290*, 125 f.; and recently in England, by Dickinson, *202*]. [***]

Until comparatively recently the environmentalist concept was commonly expressed in terms of the influences of natural conditions on human inhabitants, or human activities. The natural environment was studied in itself, with little question as to its limits; indeed it was suggested that that was "pure geography," and many geographers and, even more, social scientists came to speak of the natural environment as "the geographic factor." Human geography was the study of the influence of this factor on man. Geography was therefore dualistic not merely in the sense that it studied both natural and human phenomena, that it must use techniques of various natural sciences and of various social sciences—no science can limit itself to its own particular methods—but rather in the sense that its fundamental point of view was different in studying natural and human phenomena. "Dualis-

[122]

tic" therefore, Hettner suggested in 1905, implied the other meaning of its literal German translation, *"zwiespältig,"* that is, discordant. The "logical unitary structure of geography was thereby destroyed," whereas "if the conception of nature and man proceeds from the chorological viewpoint, it is homogeneous in all important points" [*126,* 548, 554].

In a lengthier discussion, published in 1907, Hettner noted that it was unsound to proceed from the consideration of natural factors to the explanation of human phenomena. The consideration of the natural factors can "arrive only at possibilities; the decision lies with man. . . . A sure scientific knowledge of the actually existing causal relationships is only possible if one proceeds from the human facts, classifies these and pursues them to their geographic roots" [*130,* 413–4; repeated in *161,* 143; see also Schlüter, *131,* 506 f.]. Much the same argument is to be found in Barrows' presidential address of 1922. He recommends that relationships in geography should be studied "from man's adjustment to environment, rather than the reverse. The former approach is more likely to result in the recognition of all factors involved; to minimize the danger of assigning to the environmental factors a determinative influence which they do not exert" [*208,* 3].

Barrows' modification of the environmentalist concept of geography has apparently the further effect of solving the problem of "dualism." But it does this by the elimination of the study of physical geography in itself. Both natural features and human features are studied in geography only in their relations to each other. Geography is therefore exclusively human geography, or as Barrows stated it, "human ecology;" it is therefore essentially a social science related to the natural sciences in the same manner that plant ecology as a biological science is related to the physical sciences [*cf.* Penck, *163,* 36]. In theory, therefore, this concept represented a departure from the previous history of geography which was very nearly the opposite pole from that suggested by Gerland. Its logical conclusions have been most definitely expressed by a group of geographers in Japan who hold that physical geography is no longer a part of geography, but is to be found in other sciences, that what remains to geography is exclusively human geography [*110*]. In practice, of course, there can be no clear separation, since a study of the adjustments of man to the environment requires a knowledge of the environment, but this knowledge is logically subordinate, not to be studied for its own sake. *

On closer consideration of this concept of geography as human geography, its apparent unity appears much less real. If considered, as formerly, in terms of geographic influences on man, geography, as anthropogeography,

[123]

was a collection of relationships which could be classified in categories according to each of the natural factors, but in each of these collections there was no basis for a unified organization of all sorts of relationships, and the collections together had likewise no unity. It is no wonder therefore that undergraduate students complained of the encyclopedic character of Semple's major work; probably most research students find it of value chiefly, if not solely, as a reference volume [204]. In other studies in which Semple succeeded in establishing an organized unity, the organization will be found to be primarily historical, but on the other hand, from the historical point of view, the work is only partial.

In a sense, therefore, Semple had in practice used the reverse form of the environmentalist concept—the relation of man to nature—in her study of *American History and its Geographic Influences*. The organization therefore is in terms of the social sciences, and geography on this basis has in one sense the same degree of unity as the whole of the social sciences, but in another far less: it organizes its collection of relationships in terms of the organization of the social sciences, but within each of these it presents but a part of the total. Dickinson has recently noted the dispersive effects of this view of geography in the work of British geographers [101, 260, 267, 269].

A further modification of the definition of geography, which describes the field as the study of relationships, or adjustments, *within regions,* provides an easier organization of the material, no doubt, but one which appears purely arbitrary. If the major concern of geography is with relationships, presumably the logical organization would be in terms of the character of the relationships; the particular places in which they were found, or the particular times in history in which they were found, would be irrelevant. In any case, even though the regional classification provided a convenient method of restricting the view, it would provide no basis for a comprehensive organization of the relationships found within the region.

Although the environmental concept, therefore, appears to provide no logical basis for unifying geography internally, we must consider whether it does not at least set off geography as a distinct unit from other fields. We have seen that Hettner stated, as early as 1907, the conclusion which many "environmentalists" in this country adopted in the 1920's—namely, that the procedure of study of relationships should lead from man to nature, rather than the reverse. Hettner however followed the argument farther, by asking how a science can come to "arrive at an answer to the question of the dependence of man on the nature of the earth. The function of answering this question cannot be the domain of one science. The systematic sciences that study man must establish connections with the nature of the earth surface if they are to understand the different social conditions of different lands and places. Historical writing must take account of the natural conditions if it wishes to explain and understand individual events, *e.g.*, the course of a campaign, the territorial changes of a peace treaty . . . etc., or if it wishes to explain and understand the whole course of man's historical development. Geographers are not to be jealous of such studies of other sciences; on the contrary, we should rejoice in them, accept their conclusions thankfully and come gladly to assist in them. But in such assistance, our considerations must retain their center of gravity in geography" [*130*, 422].

It can hardly be denied that many geographers have regarded as trespassers those students in other fields who presumed to study relationships between man and nature. Even more have students of other fields wondered whether geography had any field of its own. Defined not in terms of a particular circle of facts, but in terms of causal relationships presumed to exist, it could have but a parasitic character [Hettner, *130*, 423].

If we depart from theory for the moment and examine the character of the work produced in geography under this concept of its nature, we must recognize that there is ample evidence to support the accusation of our colleagues from other fields. No one, to be sure, would wish to lay down fixed boundaries patrolled by border guards between adjacent sciences, and it is both natural and proper that some of the workers in any field will be engaged near the borders and frequently overlap into a neighboring field, presumably to the benefit of all concerned. Further, geography unquestionably has close contacts with a large number of other fields. Nevertheless, if geography is an independent field of study, it would seem a fair assumption that most of its workers, most of the time, would be found working within that field. That appears to have been in fact the case among German geographers of the past generation (if we can overlook a possible argument in regard to geomorphology), and, in large part, among the followers of Vidal in France. In contrast, the geographers whose work has most clearly been governed by the concept of relationships have almost all carried on a large part of their studies in areas that must at least be regarded as transition zones, if not definitely parts of other fields. Semple, Brunhes, Huntington, and Taylor—to mention only the most prominent—have all published major works that were frankly indicated as combinations of geography and history, and each of the last two named have also explored in various other fields. If any should suppose that these are exceptions, he needs only to thumb through the past volumes of these *Annals* and note the large number of smaller studies that are as much studies in other fields as in geography. The attitude of students in the other fields towards these studies—whether, as in many cases, they find in them new ideas and new methods useful in their fields, or whether they find them lacking in the necessary background of their fields—need not concern us here. What does concern us, is that such studies offer no answer to the question as to how geography may be distinguished from other fields of science.

An even more unfortunate result of this situation has been that students in other social sciences have been brought to look upon geography as essentially an *a priori* thesis which, in the extreme form, is simply environmental determinism—the mechanistic form of predestinarianism [*cf.* Sauer, *84, 174*]. Overlooking the unfair attacks of many critics—based frequently on work by non-geographers, as Whitbeck and Thomas rightly note [*215*]—there is much that cannot be refuted in the detailed and critical, but by no means unfriendly, examination of geographical work by Febvre [*185; cf.* also, Thomas, *210*]. This unfortunate result is not to be eliminated merely by inserting qualifying terms, by speaking of "possibilities," etc. As Schlüter, Michotte, and Sauer have all stated, a definition of a field of science in terms of causal relationships *instructs* its student to seek and find such relationships, robs him of his impartiality, and easily leads to dogma; for "success lies most apparently or at least most easily, in the demonstration of an environmental adjustment" [*127*, 11 f.; *189*, 11–15; *84*, 171–3; *85*, 622]. If it were necessary to demonstrate this proposition our geographic literature would offer all too many examples. The most extreme forms tend to be promulgated just where the greatest reserve is desirable, in school geography; I have listed a number of examples of these elsewhere [*218*].[46]

Finally, the concept of geography as a study of relationships has failed to provide the subject "with sufficient tangible objectives," that is, with concrete phenomena to be studied, or with "a distinctive and sufficient method or discipline" [Sauer, *84*, 173 f.].

Because of these logical and practical difficulties, Hettner and Schlüter, and following them Michotte, Sauer, and many others, have concluded that "an independent science can never have for its object merely causal relations, but must rather apply itself to a particular circle of facts which it first establishes and describes in order then to search for causal relationships" [*130*, 423].[47]

---

[46] Since readers may have seen this article referred to in a publication, *School and Society*, which has much wider circulation than the original, I may be permitted to file a correction here. In paraphrasing one of my statements the authors have—quite unintentionally I am informed—altered the original in a way that appears minor but which results in essentially the opposite of the original statement [*218*, 172].

[47] It is of historical interest to note that Hettner objected, in passing, to the concept of geography as the study of relationships, at least as early as 1895 and 1897, and

D. GEOGRAPHY AS THE SCIENCE OF DISTRIBUTIONS

As early as the pre-classical period of modern geography, it was suggested that geography is essentially the science of distributions, a field whose purpose is to study the distribution of different phenomena, separately and in relation to each other, over the earth. Döring found that such studies represented a minor part of the geography of Humboldt [22, 64 ff.]. This view of the field was presented with greatest emphasis by Marthe, who described geography simply as the study of "the Where of things" [70, 426 ff.]. More recently it has been represented by De Geer, who in other respects parallels Hettner [190]. The latter, Hettner, was at one time accused by Schlüter of supporting this concept, but it was only necessary for Hettner to repeat his previous arguments (to be cited later) in order to demonstrate that he had been totally misunderstood.

If the study of distributions is fundamental to the character of geography, not merely incidental to other purposes, it must form a characteristic distinguishing geography from other fields. This necessary conclusion has led many to suppose that when a botanist determines the areal distribution of a particular plant, or a geologist determines the location of a volcano, or a sociologist maps the distribution of population in a country, instead of merely using a statistical table, the botanist, geologist, or sociologist becomes thereby a geographer—or, at least, is working in geography. But, as Michotte concludes from this series of examples, each of these students is carrying on work necessary to the understanding of the particular kind of phenomena that he is studying from the point of view, not of geography, but of his own science. The use of the inductive method, in any of the sciences dealing with phenomena located within the earth surface, in the effort to establish the relationships that govern the character and development of the phenomena of that particular science, will often require the determination of the location of the phenomena before any principles can be determined [189, 15–17].

Beginning with his earliest methodological discussion, in 1895, Hettner has repeatedly stated similar objections to this concept of geography. "Distribution by place," he wrote in 1905, "forms a characteristic of the objects . . . and must therefore necessarily be included by the systematic sciences [not geography] in the compass of their research and presentation." He emphasized the distinction that Wallace had made between a geographical zoology that studies the distribution of the individual genera and species of

---

that Schlüter, referring to Hettner's objections, criticized the concept more fully, in 1899, *Geogr. Zeitschr.* 1: 373 f.; 3: 625; 5: 66 f.

animals, and a zoological geography—or simply animal geography—that is concerned with the difference in faunal equipment of the different lands. (Michotte notes the importance of this distinction in titles [*189*, 41, footnote].) The same distinction in point of view, Hettner illustrated by several other examples of natural features. In each case, the view of the systematic science is focussed on the phenomena, which are studied in their distribution, that of geography on the *areas* that differ from each other in their mineral, floral, or faunal contents [*cf.* Michotte, 17–28]. This geographic view represents, we may note, a consistent derivation from Humboldt's description of geography as the study of "that which exists together in area" [*60*, I, 50]. It appears most clearly, perhaps, in Hettner's example with regard to cultural phenomena: "A study of the distribution of a particular tool, weapon, or in general of any particular object or custom is erroneously called anthropogeographic; it is much more ethnological, even though it may indirectly acquire anthropogeographical significance; for the point of interest here is not the land, but the object concerned or the people as possessors and bearers of this object" [*126*, 557 ff.].[48]

The comparison of geography and history, upon which Hettner bases his general concept of geography (to be considered in full in the following section), might suggest, as Sauer notes, that if history may be considered as the science of the When, geography would be logically the science of the

---

[48] The statements given above from Hettner are all taken from the essay (the conclusion is also repeated in the summary) to which Schlüter was replying when he asserted that Hettner supported the concept of geography as the science of *"Das Wo der Dinge"* [*127*, 53–6]. Whatever may have been the basis for this misconstruction, Hettner's reply should have left no possible grounds for misunderstanding [*132*, 628 f.]. Nevertheless, in introducing German methodological thought to American geographers a score of years later, Sauer revived the error. Without supplying any evidence for his statements, he speaks of Hettner as "vindicating it (geography) for the general study of terrestrial distributions as to their extent and explanation" [*84*, 182; repeated, 188], and indeed as defending the geographer who gives "his main attention to all kinds of distributional studies" [185]. Possibly he relied on Schlüter's critique of Hettner, which he specifically recommends, and ignored Hettner's reply, of which the reader is told nothing, though the article containing it is mentioned elsewhere, in a footnote recommending several of Hettner's methodological articles. The particular example cited by Sauer as not appropriate for geography, Langhans' "geographic study of the rights of self-determination" [*369*], would clearly fit in the list that Hettner gave of studies that are not within geography (in another of the articles mentioned in Sauer's footnote [*126*, 588 f.]). A third article of Hettner's included in that footnote also contains a refutation of the concept ascribed to him [*152*, 39 f.]. To complete the record we may note eight other publications, appearing between 1895 and 1934, in which Hettner stated his logical objections to this view, which, without explanation, has been attributed to him [*121*, 10; *123*, 23 f.; *140*, 173 f.; *142*, 11; *161*, 123–5; *167*, 339 f.; *175*, 382; *363*, *Einleitung*].

Where [*84*, 184]. Neither Hettner, nor Michotte, who accepts Hettner's comparison of the two fields, fall into this error. Though "When?" is one of the several questions that historians always seek to answer, just as the geographer is always concerned to know where his phenomena are, historians do not regard their field as defined simply in terms of the element of time. Not because of the limitation suggested by Sauer, which they do not accept (see Sec. VII B), but rather, as Kroeber puts it, because history, as a distinctive field of knowledge, is not concerned with analyzing the processes of particular kinds of phenomena—leaving that to the systematic sciences—but rather with the integration of different kinds of phenomena with respect to periods of time [*116*, 545 f.].

Finally we may add the very important practical objections that Sauer raises against the concept of geography as the science of distributions. Geographers would be influenced (as the present writer has found in his own experience, see Sec. XI G) to work "on the distributional expression of many facts wherein our competence is less than that of many others;" further, "there is known no discipline for the development of such studies;" and, finally, "distribution *per se* does not supply a common bond of interest" [*84*, 185].

Even though most geographers today would accept the conclusion that the special function of geography among the sciences is not to be found in the study of distributions, it is extremely difficult to avoid the apparently logical conclusion—often held unconsciously—that because the geographer cannot study any phenomena without knowing where they are, the study of the Where is in itself a part, at least, of our field. As this is a current problem, primarily in systematic geography, we will need to consider it again in examining that aspect of the field (Sec. XI G). For the present it is sufficient to note that all of the systematic sciences that are concerned with phenomena located within the earth surface must also answer the same question in at least part of their work. The problem of determining and explaining the Where of things is not the distinctive function of geography.

IV. The Justification for the Historical Concept of Geography as a Chorographic Science

A. the common-sense justification

The question is frequently raised whether the study of the areal differentiation of the earth's surface provides a sufficient function for geography, whether a "science" of geography does not need to seek more vital problems. Stated in these terms the question raises the problem of the nature of science in general, the discussion of which is postponed to the final section of this paper. For the present we will merely assume that whatever has engaged the intellectual curiosity of peoples of all times—including many of the most eminent thinkers—is a subject worthy of advanced study. Geography as a chorographic (or chorologic) study has always found its justification in the widespread desire of many people to know what other parts of the world are like, just as history finds ample justification in the common desire to know what happened and what things were like in past times. Further, since the ordinary person actually knows but partially the area in which he lives, just as he understands but little of what is happening in the world in his lifetime, "home geography" is as necessary as current history.

"The purpose and the irreplacable significance of geography," says Volz, "is that it teaches us to know the space we live in, the surface of the earth" [262, 93]. We do not need to prove that the study of the world in which we live is worthy of serious minds. "Geography assumes the responsibility for the study of areas," as Sauer says, "because there exists a common curiosity about that subject. The fact that every school child knows that geography provides information about different countries is enough to establish the validity of such a definition. No other subject has preempted the study of area. . . . If one were to establish a different discipline under the name of geography, the interest in the study of areas would not be destroyed thereby. The subject existed long before the name was coined. . . . The universality and persistence of the chorologic interest and the priority of claim which geography has to this field are the evidences on which the case for the popular definition may rest" [211, 21].

The character of the study that is presented to geography in the mandate from common intellectual curiosity is unquestionably a highly complicated one; from the point of view of the specialized sciences, each dealing with a restricted class of phenomena, it may well appear bewildering in its scope and character, in its heterogeneity of phenomena. But this mandate permits of no avoidance of the problem; to attempt to break the field up into separate, specialized parts is not to create various other sciences but to eliminate geography. For man, as Vidal observed, is interested above all in the totality of conditions in any area. "From the beginning of the era of voyages and explorations it is the spectacle of the social diversities associated with the diversities of places, that has aroused his attention." Consequently, "for most of the ancient writers from whom geography received its titles of origin, the concept of 'country' (contrée) is inseparable from its inhabitants; the particular character of a country is expressed no less by the means of nourishment and the physical aspect of its human population than by the mountains, deserts, and rivers that form their surroundings" [*184, 3*].

It should not be implied from the very general character of our statement of the function of geography—to learn what the different parts of the world are like—that we are to be satisfied with answers of equally general nature. On the contrary, to provide complete, detailed, accurate, sound, explanatory, and organized answers to such general questions will tax the very best abilities of the mature student.

Obviously there are many different ways of studying the world, but since men as individuals, and as individual groups, do not in any very full sense live in the entire world, but each in a relatively restricted area of the world, one of the most significant methods of studying the world is to study it by areas. Geography, in brief, can demand serious attention if it strives to provide complete, accurate, and organized knowledge to satisfy man's curiosity about how things differ in the different parts of the world, just as history in similar fashion strives to satisfy his curiosity about what things were like in the past; and just as history considers the past in terms of periods because men live and things happen together only within a limited space of time, so geography must consider the world in terms of limited areas within which things are closely associated.

Few, therefore, will question the value of geography in the education of man, by which I mean not merely in the schools, but rather in the full sense of widening of one's perception of the world. There exists, however, a remarkable but not uncommon misconception of the relation of education and research according to which a subject may be valuable enough for instructional purposes, but not be worthy of research by mature scholars. To be sure, the mere re-organization of information already known into an outline suitable for teaching purposes is a pedagogical, rather than a research task— though not necessarily unworthy of mature scholars. If, however, by regional geography were meant simply the arrangement of geographical data by regions, it is doubtful if a course of such character would be worthy to be taught at any level. The thorough regional study which aims at an adequate understanding of the spatial relations and interrelations of phenomena found together within any region is no mere problem of organization, but requires research calling for all the intellectual ability of mature scholars. If it is granted that the results are worthy to be taught to others, how can there be any question that the problem is worthy of research? For, while we cannot ask our ordinary students to learn all that has been learned by research, we can hardly justify asking them to learn anything that was not, at some time, worthy of research by mature scholars. All that we dare to teach, at any level, elementary as well as university, requires research—original research to ascertain if it be true, and repeated research to check its truth.

Another question often raised, though the answer might seem obvious, is whether geographers are the appropriate persons to secure and supply the knowledge of areas. Certainly it is true that, just as history is studied and written by many who are not by training historians, there are many non-geographers willing and able to supply knowledge of areas. Ralph Brown's recent study of the geographical material available for the Atlantic Seaboard of the United States in the period around 1800 shows that the world is not dependent alone on professional geographers for such information. If these be lacking, or fail to perform their central purpose, others may, in some fashion, take their place. But his study likewise illustrates the vast amount of work necessary simply to test the dependability of material gathered by untrained travelers, naturalists, agricultural students, literary writers or statesmen, and, particularly to organize it effectively into regional study [*334*].

From its beginning, geography has had to depend on the effective descriptions of travelers and literary writers as well as upon the more scientific, quantitatively measured and logically related studies of trained students. The development of the subject, Hettner notes, has brought these two divisions together and simple description has been replaced by an interpretative description which looks for causal connections [*2, 320*].

Nevertheless the regional geographer must still assume a task which is in some ways similar to that of the artistic (or literary) describer—that of presenting a synthetic description of an area. Undoubtedly he will envy the superior ability of the artist to give us the feeling of an area—I know of no better description of the feeling of the Spanish meseta than that of Washington Irving, in the introduction to *The Alhambra*. But before a geographer hands over his task to the artistic worker he would do well to consult artists and writers, or students of art and literature, to be sure that they will accept his assignment. The subjective impression which the artistic worker (including both painters and literary writers) receives from a landscape or region, and which he desires to convey to others, is something very different from the objective description which the geographer must attempt to provide. So far as pure description is concerned the method of the geographer is photographic in character, with the distinctive personal reactions of the observer reduced to a minimum [cf. Wörner, 274, 344 f.], a limitation which the artist cannot accept.[49] For this purpose the geographer has techniques that are unknown to the artistic worker. To a large extent he may depend on cartographical presentation. Further, by explaining the interrelations between the phenomena which he describes he should be able to give the reader a much fuller picture of the character of the area he is describing than any non-geographer can hope to do, for its character consists not merely of what is seen at first sight, but includes the complex of relationships of the things seen [cf. Schlüter, 148, 146–8; Hettner, 132, 561–8].

Finally, however effective may be the descriptions of artistic writers in presenting the character of an area, these descriptions cannot be expected to satisfy scientific standards of knowledge. One does not expect the writings of a Walter Scott or a Francis Parkman to supplant methodical historical narratives. Similarly, only trained geographers can provide an objective, quantitatively measured, scientifically interpretative, and dependable presentation of an area. If, at the same time, this requires a high degree of skill in the use of maps, diagrams, and the written word, geographical literature is not lacking in such skillful presentations. That is a question of style, and style in the presentation of knowledge does not remove it from the sphere of science. But if geography is to remain always objective, as Penck insists, the describer is not to express his feelings, but to impart how he has comprehended the things objectively [163, 50].[50]

---

[49] Since those who have proposed that geography, or a part of geography, be considered as a form of art have given no indication of having consulted either artists, students of art, or students of the theory of knowledge, it hardly seems necessary to list here the particular individuals whom I have consulted on this point. Among many available works may be mentioned two studies by philosophers, Cohen and Kraft, in which the distinction between art and science as different forms of knowledge is clearly presented [115; 166, 20 f.].

[50] The quite opposite views of Younghusband, which have been published in Germany in translation [235], and particularly those of Banse [246; 330] have aroused so much discussion that almost every German writer on methodology of the past ten or twenty years has taken the opportunity to criticise them. Since we do not consider this question in detail, we may list the principal references here: Friedrichsen [230, 233–7]; Krebs [234, 82 ff.]; Hettner [152, 53–7; 161, 151–3; 167, 276; and, particu-

B. THE LOGICAL JUSTIFICATION; THE POSITION OF GEOGRAPHY
IN RELATION TO OTHER SCIENCES

Although geography in view of its long history can be expected to remain essentially that which it has been, if it is to have a productive future its nature must bear logical analysis. In particular any clarification or limitation of its nature will depend on the place which reason assigns to it in some logical arrangement of the divisions of science—"science" that is, we repeat, in the broadest sense of organized, objective knowledge.

Although relatively few of the students who have attempted to determine the field of geography from purely logical considerations have given this fundamental problem adequate attention, geography nevertheless stands on no weak grounds. As we have noted in our historical survey, our field received for many years the attention of one of the great masters of logical thought. In the introduction to his lectures on physical geography, Immanuel Kant presented an outline of the division of scientific knowledge in which the position of geography is made logically clear. The point of view there developed has proved so satisfactory, to others as well as to this writer, both in leading to an understanding of the nature of geography and in providing answers to all questions that have been raised, that it seems worth while to quote at some length from Kant's original statements [40, I, §4].[51] *

"We may classify our empirical knowledge in either of two ways: either according to conceptions, or according to time and space in which they are actually found. The classification of perceptions according to concepts is the *logical* classification, that according to time and space the *physical* classification. Through the former we obtain a system of nature, such as that of Linnaeus, through the latter, a geographical description of nature.

"Thus, if I say that cattle are to be included under the class of quadrupeds, and under the group of this class that have cloven hooves, this is a classification that I make in my head—that is, a logical classification. The

---

larly, 265]; Gradmann [236, 132 f., 139–42]; Graf [156, 40–3, 90–3]; Hassinger [253]; Penck [163, 49 f.]; Granö [252, 5 f.]; Kraft [166, 20 f.]; Waibel [266, 205–7]; Bürger [11, 104–12]; Vogel [271, 7]; and Schmidt [180, 81–94]. An apparently unrelated but parallel discussion introduced in this country by Leighly [220], has been considered by Platt [221, 13 f., 33–6]; by Crowe, in England [201, 2]; and most thoroughly by Finch [223]. I do not know what basis Penck had for his prophetic remark of 1928: *"Nicht einmal das Mittelalter hat seine Geographie zu den Künsten gezählt. Aber der Amerikaner stellt sie zu den 'Arts'"* [163, 50].

[51] The following statement on terminology is also of interest: "The description of a single place on the earth is callled *topography; chorography* is the description of a region and its characteristics; the description of the whole world, *geography*." The ** entire introduction is of great interest.

* See Supplementary Note 20.

** The terms stem from Ptolemy [433].

system of nature is likewise a register of the whole: there I place each thing in its class, even though they are to be found in different, widely separated places.

"According to the physical classification, on the other hand, things are regarded according to the places that include them on the earth. . . . In contrast to the system of nature with its division by classes, the geographical description of nature shows the places in which things are actually to be found on the earth. For example, the lizard and the crocodile are fundamentally the same animal: the crocodile is merely an extremely large lizard. But they exist in different places: the crocodile in the Nile, the lizard on the land. In general, we consider here the scene (*Schauplatz*) of nature, the earth itself, and the regions where the things actually are found, in contrast with the system of nature which is concerned with the similarity of forms. . . . . .

"Description according to time is history, that according to space is geography." . . . "History differs from geography only in the consideration of time and area (*Raum*). The former is a report of phenomena that follow one another (*nacheinander*) and has reference to time. The latter is a report of phenomena beside each other (*nebeneinander*) in space. History is a narrative, geography a description. . . .

"Geography and history fill up the entire circumference of our perceptions: geography that of space, history that of time."

In much the same manner Alexander v. Humboldt described the position of geography among the sciences. As early as 1793, nearly a decade before the publication of Kant's lectures, he had distinguished clearly between the sciences that study nature according to categories of objects,[52] those that consider all the phenomena of nature according to changes in time (*Naturgeschichte*), and those that consider the objects as interrelated together in space (*Geognosie* or, later—following Kant—*Weltbeschreibung,* including descriptive astronomy and *Erdbeschreibung* or geography). There was a greater similarity between the two latter groups, he noted, than between either of them and the systematic sciences. The student of Humboldt's works finds little difficulty in correlating all of his studies to this schematic division of science, which, we may add, he restated at various times in his life—finally, and in most detail, in the first volume of the *Kosmos* [*42; 52; 60,* I, 48–73; *cf.* Döring, *22,* 46–51].   *

Ritter likewise, we saw, evidently conceived of the relation of geography ** to history and to the systematic sciences in much the same manner, though

---

[52] For all of these systematic sciences of nature, as distinct from either the history of nature or the space sciences, geography and astronomy, Humboldt in his statement of 1793 used the term "*Physiographia*"—one of the earliest appearances of this word, presumably [*42*]. He appears, however, to have abandoned it in his later writings.

* See Supplementary Note 21
** See page 77

he does not appear to have stated the comparison as clearly as either Kant or Humboldt. During the subsequent period of shifting views in German geography—the drift toward historical studies in the school of Ritter, and the counter movement that Peschel introduced in the opposite direction—the logical position of geography as defined by Kant and Humboldt was lost sight of and not restored until the early part of this century. Likewise in the spread of geography to other countries, Kant's views were of little significance, and however important may have been the influence of Humboldt —whose works were translated into many languages—his views in regard to the position of geography among the sciences appears to have been overlooked.

The methodological discussions in Germany in last quarter of the century—notably the Leipzig address of Richthofen—stimulated a number of younger German geographers to consider the logical basis of their concept of the field. In the ensuing discussions almost every German geographer of standing has taken part, either in brief statements associated with research publications or—more frequently—in separate methodological articles. In contrast with the purely personal expositions of each individual geographer's concept of the field that characterizes so many of the methodological addresses of English and American geographers, a large number of the studies of German students of this problem are based on scholarly consideration of previous treatises and of the course of historical development of geographic thought. Although this contrast may be due in part to a difference in scholarly attitudes, it is no doubt also a result of the easy access that German students have to the records of past geographic thought, since those have been written largely in their language.

There are however many indications in the more recent literature that the standard of scholarship set by the generation of German geographers now retiring from the academic stage—many of them before reaching the age of retirement—is regarded by many of their successors as out-moded. From some of these students one hears demands for "freedom" for new ideas to replace concepts of the past that are summarily rejected. On examination these demands prove to rest in large part on misunderstanding or ignorance of what earlier writers actually thought. Although this change has been commented on by many German geographers it should not be supposed that there are not numerous exceptions in both groups. The former group included some who, as Braun noted, were wont "to pass sovereign over previous works" [155, 17], and there are many among the younger writers whose methodological discussions demonstrate a high degree of scholarship.

We will not pre-judge the individual students for the reader: in the later sections of this paper ample evidence will be offered for any judgments he may care to make.

Whatever one may think of the methodological studies of the German geographers just mentioned, the intensity and thoroughness with which they have examined the problems concerned make it seem hardly conceivable that a serious student would wish to express conclusions of his own without first examining their efforts.   If he has done so and finds those to be in error, we have the right to expect him to explain his reasons for so finding.

Of all the German geographers of recent times—indeed of all geographers of any time—none has worked so long and so continuously on this problem as Alfred Hettner.   His predominant importance in establishing a major degree of unity of thought among German geographers of our time has recently been testified to by one who could not be regarded as a follower, Albrecht Penck [90].   As one of the first to have chosen geography as a life-work from the beginning, Hettner was spared the uncertainty and shifting of attitude that affected those—like Penck himself—who had transferred from one field of science to another. (Hettner, we may add from a personal communication, received his doctor's degree under Gerland but "was always in opposition to him in regard to the conception of geography, which he had taken over from his first teacher, Kirchhoff," and later modified under the influence of Richthofen, under whom he subsequently studied.)   But Hettner, Penck continues, is no mere theoretical methodologist; his work began with scientific field-work in the Andes—in the footsteps of Humboldt—and, in spite of a physical injury that appeared to incapacitate him for further field work, he has, adds Penck, "with iron strength maintained his personal touch with the objects of geography."   On the other hand Hettner's considerations of geography have not been based on a view limited to that field alone; he has the whole of science in view and, as a critic from the field of philosophy, Graf, observed, is "the only one among the methodologists of geography in recent times who had found the connection with philosophy" [156, 10 f.].   Finally, the clear logic of Hettner's discussions, the clarity of his expression, and the consistency of his thinking, have enabled him to establish a clear and unified concept of the field of geography in the long series of methodological treatises that Sölch rightly calls "classics" [237, 56].

American geographers are, of course, by no means ignorant of the importance of Hettner's position in German geography.   Unfortunately, however, the most influential interpretation of the concepts of modern German geographers in this country, by Sauer, pays little more attention than praise *

* Read, "gives little attention other than praise".

to his writings; in so far as his views are considered they are misinterpreted. In considerable part the other views that are presented are those that Hettner, and with him the majority of his colleagues, have rejected. While these statements will require demonstration in later sections, it is appropriate to note here characteristic expressions of opinion of German geographers. Sölch describes Hettner as "without doubt the leading methodologist and one of the most important leaders generally in German geographical science," whose work is "an imperishable gain for the more sure foundation and general recognition of geography" [301, review]. Similar opinions have been expressed by Penck, as already noted [see also 163, 50], Philippson [149, 14], and Tiessen [160, 1]. Krebs evidently regards the failure to consider "even" Hettner's essays in the discussion of methodological problems as damning in itself [255, 337].

The number of German geographers who base their methodological discussions on Hettner's writings is far too numerous to list in full, but in addition to his own students we may note the following: Gradmann [236, 137]; Hassinger [141, 4]; Volz [151, 241]; Banse [133, 72; 246, 43]; and * Maull [157, 44–49]. It is also significant that certain students who in recent years have been objecting to the *Weltanschauung* of the older group, direct their attack particularly at Hettner.[53] While not hesitating to come to the defence, Hettner, in proper objective fashion states: "My own importance in the construction of the methodology has been exaggerated; likewise I do not believe that my conception is 'completely peculiar.' I believe only that I have clearly expressed and methodologically established what was actually present in the development." Note also his ironic comment on another critic: *"Spricht er doch sogar einmal geschmackvoll von einer Hettnerisierung der Geographie"* [175, 382, 341].

Outside of Germany, Hettner's first major essay of 1905 made a notable

---

[53] Prominent members of the present generation of German geographers find that Hettner represents an epoch that "is already historical." In contrast with his view of science, based on the "positivistic liberalism" of the nineteenth century, they believe that geography is entering a period in which there will be more emphasis on value concepts and on the creative influence of peoples and individuals, that the *"Gleichörtliche"* will be viewed "not in terms of the sum and functional cohesion of its parts, but *ganzheitlich,* in the sense of the *Gestalt* theory," and that *"nationale Erdkunde* is the whole of geography, regarding with German eyes and from the German point of view, Germany and the world" [Schrepfer, 174, 69–71, 85]. Such views have been represented, somewhat violently, by Spethmann [251; 261], more respectfully, but nonetheless categorically, by Muris and Schrepfer. The incidental presentation, at a recent meeting of geographers, of the opposing views of the older and newer schools brought vigorous applause from different members of the audience. Because the new *Weltanschauung* involved is limited to German geographers, the writer has not considered it necessary to examine it in this study.

[138]

* Also, Philippson [417].

impression on Chisholm [*192*], and more recently Dickinson and Howarth have recognized his importance: "While Vidal de la Blache established the method of regional description on a local basis, and Herbertson, the concept of major natural regions on a world basis, the main credit for interrelating the two and analyzing exhaustively the method of treatment lies with the modern school of German geographers . . . led by the veteran pioneer Hettner . . ." [*10*, 251; see also *101*]. In Japan, Inouye and others have recognized the major importance of Hettner's work for geographic methodology [*110*]. Apparently only in France and this country is it supposed that discussions of the nature of our subject can ignore his work.

In contrast with many other students, Hettner's methodological writings were closely related to his actual geographic work. As noted above, he had made regional field studies in South America before he wrote on methodological questions. Likewise, it was not in the study, but on the shore of Lake Titicaca that he outlined his view of general, or systematic, geography. Discussed in various subsequent methodological essays this finally took actual form nearly forty years later in the four-volume set entitled *"Vergleichende Länderkunde"* [*363*]. Long before that, his concepts of regional geography had been presented not only in theory but in the two volumes of his *Länderkunde,* in which, as Sölch reports (in his review of the more recent edition) "he has applied the principles previously developed and thereby provided the test of example—indeed many of his theoretical problems had their seed in the *Länderkunde*" [*301*, review].

Our concern here is with Hettner's methodological studies which form the major foundation for the subsequent discussions throughout this paper. They were preceded by the brief informal statement of the field of geography with which Hettner introduced the *Geographische Zeitschrift,* which he founded in 1895 [*121*]. The brief survey of the development of geography, published in 1898 [*2*], provided the historical introduction for the series of methodological essays that began in 1903 and appeared intermittently throughout his forty years of editorship. The fundamental study is that of 1905 on *"Das Wesen und die Methoden der Geographie"* [*126*], the greater *part of which will still repay careful study. Most of the essays published before 1927 were collected, in part re-written, in the volume that Hettner described as his principal life-work—*Die Geographie, ihre Geschichte, ihr Wesen und ihre Methoden* [*161*]. Inouyé, in referring to this book, in an article in the *Geographische Zeitschrift,* as "the most valuable work in the history of geographical methodology," may have been expressing Japanese courtesy; nonetheless he was stating the obvious [*110*, 288]. It would be of great value to geography in English-speaking countries if Hettner's vol-

* See Supplementary Note 22

ume, together with selections from his subsequent discussions, could be published in English translation. The lengthy citations in this paper will present, it is hoped, a more adequate picture of his philosophy of geography than has hitherto been available in English. Any who wish to take issue with him, however, should study carefully his complete treatment in *Die Geographie* and his later statements; the writer has repeatedly found that ideas presumed to be original had already been discussed in those writings.[54]

In considering the logical position of geography among the sciences, Hettner, like both Kant and Humboldt before him, proceeds not from the consideration of particular branches of science but from a view of the whole field of objective knowledge [*111*]. His major division of the field of science is essentially the same as that of his illustrious predecessors, although he was not made aware of Kant's analysis until after he had written his own, and apparently did not then or later realize the close similarity to Humboldt's views [*126, 551*]. If, as is quite possible, each of these students arrived at essentially the same concept independently of each other, one may have all the more confidence in its validity. In any case, Hettner has developed the concept most fully and, since the greater part of our subsequent discussion of the nature of geography depends on a clear understanding of the comparison of our field to that of other branches of knowledge, it will be well to present his statement at length [from the most recent complete statement, that of 1927, *161*, 114–17, 123 f.].

"Reality is simultaneously a three-dimensional space, which we must examine from three different points of view in order to comprehend the whole; examination from but one of these points of view alone is one-sided and does not exhaust the whole. From one point of view we see the relations of similar things, from the second the development in time, from the third the arrangement and division in space. Reality as a whole cannot be encompassed entirely in the systematic sciences, sciences defined by the objects they study, as many students still think. Other writers have effectively based the justification for historical sciences on the necessity of a special conception of development in time. But this leaves science still two-dimensional, we do not perceive it completely unless we also consider it from the third point of view, the division and arrangement in space.

". . . The systematic sciences ignore the temporal and spatial relation-

---

[54] Fortunately Hettner's style is unusually easy for the non-German student. I regret that that is not always the case in the translations, which for the present purpose must necessarily be literal. Many students will no doubt welcome his three-page summary outline of his position, written in 1905 [*126*, 683 ff.] and his recent summary statement in the introduction to *Vergleichende Länderkunde* [*363*, I, 1–7].

* See Supplementary Note 23
** See Supplementary Note 24

ships and find their unity in the objective likeness or similarity of the subjects with which they are concerned. The common distinction between the natural and the social sciences is such a systematic distinction. Within the natural sciences we have, first, the sciences of the minerals and stones, mineralogy and petrography, of the plants, botany, of the animals, zoology, and with them, but for other reasons developed as a separate science, paleontology, the science of fossil plants and stones. Only later was the study of the earth body and its circles of phenomena taken up by separate disciplines.[55] Systematic social sciences are these of language, of religion, of the state (political science), of economics, etc.

"For the historical sciences the material relations are objects of only incidental importance. Rather, in their study they unite a number of things of entirely different systematic categories and obtain their unity through the point of view of the temporal progression of events in time. If these followed each other merely by chance, and the course of various groups of phenomena were independent of each other, science could be content with merely the systematic studies. But the connection of different times, which we express by the word development, and the connection in the same time, make necessary a separate historical study. The studies of the development of a single kind of phenomena which, for that kind alone, takes account of both the points of view named, such as the history of the animal world or the history of art or of literature or constitutional history, occupy an intermediate position between the systematic and the historical sciences.

"With the same justification as the development in time, the arrangement of things in space demands special study. Together with the systematic, or material, and the chronological or historical or time-sciences, there must develop chorological or space sciences.

"There must be two of these. The one has to do with the arrangement of things in the universe; that is astronomy . . . the other is the science of the spatial arrangement on the earth, or, since we do not know the interior of the earth, we can say on the surface of the earth."

The logically close relation of astronomy and geography, which was especially characteristic of Humboldt's cosmos, may on first thought seem confusing. That is because astronomy is often incorrectly expressed simply as an applied mechanics, or on the other hand as the science which studies the heavenly bodies as its concrete objects. On the contrary it has been a major function of astronomy to discover planets and vast numbers of stars, both by the application of mechanics to the arrangement and relations of the

---

[55] We omit here several pages in which Hettner considers the study of the earth body as a whole and finds that a "science of earth" cannot logically be developed as a single branch of knowledge. See Sec. III B.

heavenly bodies and by exhaustive search of the heavens. The investigation of the nature of the individual bodies is a part, but only a part, of its field.

The comparison between astronomy and geography, to which we will return in a later connection (Sec. XI F), is discussed in detail by Hettner, who notes in particular that a large part of so-called mathematical or astronomical geography is properly a part of astronomy [111, 274 f.]. The logical similarity of the two fields is also noted by W. M. Davis [104, 213 f.] and Vallaux [186, 66–8]. One important difference however should be noted here: in heavenly space there are clearly defined and separate bodies which can be accepted without question as unit objects of systematic study, whereas in the space formed by the earth's surface there is no such separation into parts that can be accepted as naively given objects—not even the apparent divisions of the lands by the seas correspond in total character to the divisions of the heavenly bodies by empty space.

Geography, then, as the second chorological science is the study of the spatial arrangement on the surface of the earth. "If no causal relations existed between the different places on the earth, and if the different phenomena at one and the same place on the earth were independent of each other, no special chorological conception would be needed; since, however, such relationships do exist, which, by the systematic and historical sciences, are comprehended only incidentally or not at all, we need a special chorological science of the earth or the earth surfaces.

"The historical and geographical viewpoints require that time or space be placed in the foreground and form the unifying bond of the scientific study. As history studies the character of different times, so geography studies the character of different areas and places—'the earth-spaces filled with earthly contents,' to use Ritter's expression[56]—it studies the continents, regions, districts, and localities as such" [cf. Vidal, 183, 299].

Few geographers evidently have felt themselves sufficiently grounded in philosophy to discuss Hettner's system of the sciences. The somewhat confused discussion by Schlüter will be considered in a later connection. Leutenegger, in an erudite, rather abstruse philosophical study, concludes that, though Hettner has not provided a satisfactory system of the sciences neither has anyone else; but Hettner's at least is logical and geography fits logically into it [150, 91–95]. Among others who have attempted to ex-

---

[56] The difficulty in translating Ritter's phrase—"die irdisch erfüllten Räume der Erdoberfläche"—has been noted previously. A further confusion was introduced by a number of writers in repeating it as "die dinglich erfüllten Räume"—presumably to avoid the repetition with Erdoberfläche. Ritter's concept, as his discussion made clear, included as "earthly," non-material phenomena of man that could hardly be considered as "things" [50, 156–8; cf. Kraft, 166, 5].

* See Supplementary Note 25

amine the subject from a somewhat different point of view, but arrive at conclusions that are in major part in agreement with those of Hettner, may be mentioned Michotte [*189*], Lehmann [*113*], and Schmidt [*180*].

Unfortunately, philosophers interested in the problem of the division of the sciences seldom have a knowledge of the actual field of geography sufficiently adequate to enable them to judge the system that Hettner, in agreement with Kant and Humboldt, has outlined. An objection that Wundt raised has but little importance for geography. Hettner spoke in terms of abstract (mathematics), abstract-concrete (physics and chemistry) and concrete sciences (zoology, economics, etc.), but it seems clear that this distinction can be eliminated—as I have done in quoting—without affecting Hettner's major thesis; if anything it strengthens it. Wundt's own suggestions for the position of geography among the sciences are evidently based on a very inadequate knowledge of the field; it is doubtful if any geographer would accept his conclusions [*112*, 88 f., 93 f., 277 f.; *cf.* Leutenegger, *150*, 85]. Benjamin's discussion of geography (as reported by Finch—A. C. Benjamin, An Introduction to the Philosophy of Science, New York, 1937, p. 406) appears to be based on the concept of geography as the study of relationships. If his conclusions could be justified on that basis, Finch has shown that they do not apply to geography conceived as a chorographical science [*223*, 7].

Two German students of philosophy who have also a considerable command of at least the methodology of geography have recently considered the nature of our field. Graf's treatment is based categorically on a particular philosophical theory and, though we will find much of value in his critical discussion, his attempt to fit a field of science to a particular philosophical theory does not prove satisfactory for the science. It is characteristic of his method of reasoning that he dismissed Hettner's system solely on the basis of the objection that Wundt had made—apparently assuming that the discovery of a philosophical error in a line of reasoning necessarily dismisses all subsequent conclusions, without demonstrating that the point in question is in any way necessary to the reasoning [*156*, 10; note also the subsequent discussion with Hettner].

In a brief but very instructive examination of the unity of geography as a science, Kraft does not specifically discuss Hettner's system of the sciences, but accepts at least its major conclusion for geography: "Stones, plants, animals and man, in themselves objects of their own sciences, constitute objects in the sphere of geography insofar as they are of importance for, or characteristic of, the nature of the earth surface." This gives a clear answer to the question of how far man may be included in a unified geography: "it is the reciprocal relations of man and the earth-area (*Erdraum*) that determine and limit the treatment of man in geography" [*166*, 4 f.].

In his further consideration, Kraft appears to overlook a major element in Hettner's definition of geography. It is not merely the earth surface as a whole that forms the object of geographic study, but rather—as the comparison with history emphasizes—the differences in the different areas of the earth surface. Consequently, meteorology, though a study confined to the earth surface (in the sense in which we use that term) is not a part of geography, but is a systematic science basically necessary for climatology which does form a part of geography. Likewise this viewpoint provides at least a partial answer to the question for which Kraft found no logical answer, namely, granted that geography includes the study of man, how can it justify logically the high position which it inevitably gives to human, cultural phenomena. While human phenomena contribute perhaps but a minor part in comparison with nature to the formation of the earth surface considered as a whole, in the consideration of its differential character, they constitute a factor of major importance—for example, in the comparison of Java and Borneo.

C. IMPORTANCE OF THE COMPARISON OF GEOGRAPHY AND HISTORY

The purpose of translating so much of these discussions is to emphasize the value of the comparison of geography and history. If the geographer, especially the regional geographer, finds any difficulty in developing or maintaining his concept of his field, the suggestion is that he apply this comparison as a test. If the field in which he has for years been working is suddenly and vigorously attacked, so that he fears the whole development is threatened with collapse, let him re-read the attack, substituting for the word "geography" the word "history," for the word "region" the word "period." The result will appear to be the devastation of the field of history—there cannot be, he will learn, a science of history. The history of different periods is simply a form of narrative in which incommensurable phenomena must be presented in integrated form; therefore it could be far better developed by literary writers than by professional historians. The historian then, in the sense of the general student of the history of a period, in contrast to specialists in topical history, presumably should give place to literary writers. It is hardly necessary to demonstrate that the conclusion is absurd; the necessity and validity of history as a technical subject for trained scholars is sufficiently established.

It is clear, however, that the history of a period, like the geography of a region, is a very different sort of study from the systematic sciences. The

* See Supplementary Note 26

latter, whether in the natural or the social fields, select like things wherever they find them, whenever they have been found, compare and study them in themselves and in the relationships which, in relatively simple manner, they bear to each other. Both the historian and the geographer must study a far greater range of different kinds of facts. Kroeber suggests that history is an endeavor at descriptive integration of phenomena in contrast to the method of the systematic sciences which is "essentially a procedure of analyses, of dissolving phenomena in order to convert them into process formulations" [116, 545–6]. The same might well be said of regional geography, with the difference that history is concerned with integration with reference to time, geography with reference to space [cf. also Finch, 223, 6 f.].[57]

When geographers concerned with the scientific standing of their subject think of "science" only in terms of such systematic fields as geology, zoology, or economics, they have reason to doubt whether regional geography can possibly hope to attain recognition as a "science" [cf. Leighly, 220, 131]. A "science of regions" in this sense may well be impossible, not merely as an ideal, but even as a working goal. But if geography is to look for its comparisons to history—and, essentially, to history alone—then the objections fall to the ground. Provided one has no emotional concern for the term "science," one may suppose that geographers will not need to be discouraged if their subject can occupy a place in the world of knowledge comparable with that of history.

If this comparison is so important, one may well ask why it has not been more generally recognized. Certainly the similarity has not been entirely overlooked. Barrows made use of the comparison in his presidential address, without following logically the direction in which it leads [208, 6]. Sauer also suggests the comparison in one or two places [211, 26; 84, 178], but makes relatively little use of it. Possibly he was diverted from following Hettner in this regard because he confused Hettner's concept with that which "sees in the 'Where' the distinctive quality of geography," in contrast with history, concerned with the 'When' (Sec. III D). Consequently, those who have depended solely on Sauer for their knowledge of German methodology, have not been made aware of the importance of the similarity. A few however have utilized it effectively. W. M. Davis made use of the comparison in defending geography against the charge of being merely a composite subject [104, 213 f.]; Whittlesey has made a suggestive comparison

---

[57] There are notable differences in the relative importance and character of development between systematic, or general, geography (as distinct from regional geography) and the corresponding work in the field of history. The reasons for this difference, which we will consider later (Sec. XI G) do not, as might appear, cast doubt on the logical similarity here discussed.

of the educative value of geography and history [*217,* 119] ; and the English geographer Crowe has made good use of the comparison [*201,* 2 f.], which he appears to have learned from Fairgrieve [*194,* 8, 18; Fairgrieve gives no references].

That most American geographers, when considering the scientific character of their subject, tend to look for comparisons with the systematic sciences may well be a natural result of the particular development of geography in this country. Long and close association with one of the systematic sciences, geology, tended to direct our comparative thought far from history, and indeed from the systematic social sciences, toward the systematic natural sciences. While this undoubtedly had great value in developing standards of scientific method, notably in measurement and accurate mapping, it did not contribute to a proper understanding of the essential nature of our field. Further, because of that background, it is not surprising that the greater part of the productive work of American geography has not been directed on complete geographic studies of areas, but rather has consisted of systematic studies of particular features found in different areas. Whether these are up-lifted peneplains, stream deposits, temperature efficiency, potato fields, steel mills, or political boundaries they represent areal features arbitrarily removed from their geographical settings, for purposes of systematic study. Necessary as such work in "systematic geography" ("general geography" among Europeans) may be, over-emphasis upon it tends to direct our attention away from the specific nature of the subject. Almost all geographers agree that the clearest view of geography is to be seen in regional geography (*Länderkunde*).

The comparison with history, then, emphasizes for us the special character of geography as an intergrating science cutting a cross-section through the systematic sciences. This figure of Hettner's, which could be derived also from Kant, differs significantly from Fenneman's well-known diagram of the relation of geography to other fields [*206*]. Geography does not *border* on the systematic sciences, overlapping them in common parts on a common plane, but is on a transverse plane cutting through them. A diagram in solid geometry would therefore be required to illustrate this (Fig. 1).

D. DISAGREEMENTS CONCERNING THE CHARACTER OF GEOGRAPHY AS A CHOROGRAPHICAL SCIENCE

The lengthy discussion of the comparison of geography with other sciences is not, as one might suppose, concerned merely with a question of suitable or unsuitable analogies, but rather with fundamental differences in point of view which have led to very important differences both in the con-

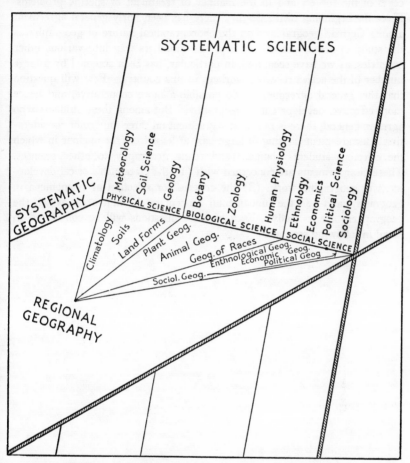

FIG. 1.—Diagram illustrating the relation of geography to the systematic sciences. The planes are not to be considered literally as plane surfaces, but as representing two opposing points of view in studying reality. The view of reality in terms of areal differentiation of the earth surface is intersected at every point by the view in which reality is considered in terms of phenomena classified by kind. The different systematic sciences that study different phenomena found within the earth surface are intersected by the corresponding branches of systematic geography. The *integration* of all the branches of systematic geography, focussed on a particular place in the earth surface, is regional geography. (See Sec. XI E.)

cepts of the subject and in the manner of treatment of specific problems. Since the beginning of the century there has been fairly general agreement among German geographers on the chorographical nature of geography as the study of areas, and this concept has found its way into various other countries, as we have seen, and, in particular, has been accepted by a large number of the active research workers in this country. Few will question that this general agreement made possible a more cooperative, and hence more effective, development of geography. But among these students who agree in general, there is radical disagreement on three important considerations, disagreements arising in large part at least from the manner in which the different students compare and relate geography to other sciences. These disagreements are concerned with (1) the place of historical development in geographic work, (2) the criteria for selection of phenomena in geography, and (3) the fundamental concept of the unit areas which the geographers are to study. Each of these questions will be considered in detail in following sections.

V. "Landschaft" and "Landscape"

A. the present confusion

In each major problem of current geographic thought which we will have to consider in following sections, the student is repeatedly confused by the many variations—some radical and clear, others more subtle but even more confusing—in the usage of a single all-important term, namely the "landscape," or *Landschaft*. The use of this word as a technical term of basic importance in geographic thought, as well as in the actual studies of regions, is of relatively recent origin in this country or England. In its German form, *Landschaft*, however, it has been important throughout the history of modern geography. Although the various possible meanings which it carries in common German speech had introduced some confusion at least a century ago, it is only with its establishment in recent decades as perhaps the single most important word in geographic language that the confusion it produces has penetrated into every field of geographic thought. It is therefore extremely unfortunate that those who have introduced the term into this country should have made no distinction between its multitudinous meanings, but, by translating it without qualification as "landscape," should have blindly introduced all the unnecessary confusion into our language.

In attempting to discuss the problems on which geographers disagree, the writer has found it necessary in each case to reexamine each of the different meanings of these terms. In itself this is not unusual. Many lengthy disagreements in methodological thought, or in philosophy in general, can be reduced to different interpretations placed on ambiguous words. If each of the protagonists involved has a different but consistent interpretation, the problem can be clarified, or even eliminated, simply by clear statements of what each means to say. But if the same writer uses the basic term now with one meaning, now with another, never letting the reader know under which shell the pill is to be found, there is no possible way of coming to agreement, excepting to stop the game, lift all shells concerned, and demand of each student that thereafter he either leave the pill under the same shell, or tell us when he wishes to move it.

It is no mere accident of translation that has brought this confusion to us. The human mind wishing to arrive at certain predetermined conclusions prefers to deal with words whose ambiguity of meaning permit it to convince itself that the conclusions desired are arrived at logically. If we may say that a particular word means the concrete unified impression that an area gives us, the objects in the area producing that impression—*i.e.*, concrete material objects—and, finally, that the word is at the same time synonymous with area itself, then we can prove with relatively little difficulty that the geographic region consists of material objects and forms a concrete unit object, and, therefore, that geography has concrete unit objects of its own to study, and so is proved to be a "science," like any other science. Indeed the fact that many German geographers, as Crowe observes, "have woven such an impenetrable web of mysticism about their '*Landschaft*' " [*202*, 15], makes it possible for them to appear to prove almost anything. Presumably geographers will wish to base their claims to scientific standing on a firmer basis. It is this assumption, and not any love for unnecessary argument over terms, that compels us to examine exactly what the "landscape," or "*Landschaft*," may mean.

The major, though by no means the only difficulty, results from the fact that the German word *Landschaft* has long been used in common speech to indicate either the appearance of a land as we perceive it, or simply a restricted piece of land. Both these concepts were introduced into German geography not later than the beginning of the last century. Hommeyer used the word to indicate an area of land intermediate in size between *Gegend* and *Land,* but at the same time appears to have retained the other concept [*11*, 8; or in *1*]. Humboldt used the term primarily in the sense of the aesthetic character of an area, though he may occasionally have used it in the other sense also. Similarly Oppel and Wimmer both regarded the *Landschaft* primarily from the aesthetic point of view, but the latter also included social features as a part of the *Landschaft* as a coherent whole [*74*, 9 f.].

Needless to say, many German students have noted this ambiguity. As early as 1896, Hözel showed that various earlier writers, including Fröbel, had arrived at erroneous conclusions just because of this confusion [*27*, 393 f.]. Banse has even insisted that there was no etymological foundation for using the word to mean other than the appearance of an area [*133*, 2]. Whatever validity that argument may have is overwhelmed by the fact of usage, both in common German speech and in German geography. Probably the majority of German geographers of our time use the word in either sense, commonly without distinguishing which one is meant. In 1923, Hettner objected to the confusion to which this practice leads in Passarge's work [*152*, 49 f.]. Since then similar objections have been raised by a number of German students, including Gradmann [*236*, 130], Sieger [*157*, review, 379], and Waibel [*266*, 207].

The corresponding French term, *paysage,* apparently also permits of this *

* This paragraph is based on an error and should be omitted.

double meaning. Vallaux has commented on the confusion that results in its undefined use in the writings of Brunhes and others [186, 93–103].

Consequently we should regard as one of the blessings of our language that common speech clearly recognizes that while "landscape" has something to do with an area of land, it is not the same as an area. No etymological demonstration of the presence of the same ambiguity in Old English would justify our re-introducing it into the scientific form of a language in which common speech has established a clear separation. On the contrary, to destroy the relative clarity of common English usage merely to follow the lead of German geographers who have not our advantage, would be little short of a scientific crime.

Unfortunately there are not only these two major concepts of the term, but almost as many gradations between them as there are geographers who use it. Even the writers who use the term as more or less synonymous with region are by no means in agreement as to what the term includes. Schlüter has repeatedly emphasized the difference between his use of the term and that of Passarge [247, 389; 169, 299]. Neither of these writers presumably would accept the varying versions of Penck, Maull, or Gradmann [163, 39 f.; 157; 236]. The differences are much more fundamental than those involved in the attempts of American geographers to define a "region"; they are not differences in determining limits, but differences in what kinds of things are included in the *Landschaft,* whatever its limits. The difficulty results from the fact that in using *Landschaft* to mean "region," all these writers wish to carry over certain aspects of its meaning as "landscape," but they are not in agreement as to which of the latter they wish to include. As James has noted in the similar situation with reference to recent use of the word "landscape" in American geography; "the geographer who uses this term has given it, perhaps unconsciously, a special redefinition" [286, 79]. Unfortunately definitions arrived at unconsciously are not likely to be sufficiently specific for scientific use, nor are the unconscious thought-processes of different geographers likely to come to the same definition. Indeed there appear to be as many definitions in this case as there are geographers using the term.

We are indebted to Waibel for a classification of the various meanings that German geographers have given this term, and its compounds [266]. Undoubtedly he would not claim that his list was complete; further, we must add others that have been suggested since he wrote, in 1933. Special attention should be called to the series of papers on this topic presented at the recent meetings of the International Geographical Congress [276–279 and 297]. Though Lautensach's discussion of the German term and Broek's discussion of its English equivalent contribute more toward clarification of terms than perhaps any previous treatments, Krebs is nevertheless justified in his conclusion that a more searching discussion of the terminology is needed [279, 209].[58]

Waibel defines the common meaning of *Landschaft* as "the section of the earth surface and the sky that lies in our field of vision as seen in perspective from a particular point," and Broek states almost exactly the same for the common meaning of "landscape." This "visible landscape" forms the core of Granö's concept which also includes, however, our sensations of sound, smells, and feelings of an area.

More specifically, Granö's concept is based on Hellpach's definition of the *Landschaft* as "the total impression aroused in us by a piece of the earth's surface and the corresponding section of sky" [139, 348]. Most students who use either of these concepts imagine an observer viewing the scene vertically, so as to eliminate the perspective.

In making the same alteration in the concept, Granö introduced a further transfer in logical thought—and unlike most other writers definitely labels the transfer [252, 23 ff.; 270, 297]. From the psychic sensation produced by an area he proceeds to the objects in the area that caused that sensation. In order to conserve the strict meaning of *Landschaft* as the sensual landscape (in the full literary sense in which sounds and smells are included) he suggests other terms, but those who follow him simply transfer the word *Landschaft* or landscape, to the objects in the area that are responsible for our sensations of landscape [cf. Broek's earlier statement, 333, 8, 136].

One might suppose that this shift, from the landscape as sensation to the objects that produce the sensation, would lead to general agreement as to what was included in the latter, but this is far from the case. Penck, for example, excludes things perceived only in immediate proximity, as by feeling. For him the *Landschaft* includes only "what is perceptible in our field of vision"; specifically, it "embraces not man himself but only his effect on the earth surface" [163, 39 f., 45]. Schlüter, on the other hand, not only includes men on the grounds that they are perceptible objects, but in order that they may appear as more than "minute grains" in the landscape, he applies the magnifying glass of our special interest to make man a major *Landschaft* element [127; 145, 26]. Passarge, in contrast, would exclude not only man, but also animal life in general, because otherwise it is too difficult to separate one *Landschaft* (region) from another. Passarge appears

---

[58] The present study was essentially completed before the publications of the International Congress came to my attention. It is all the more reassuring to me to note the large number of points on which there is essential agreement between the views expressed in various of the following sections of this study and those stated more concisely in those two papers.

also to limit his *Landschaft* to natural, non-human elements, thereby including elements like natural vegetation that may no longer be in existence. Waibel however, like many others, admits that it is difficult, if indeed possible, to understand just what Passarge means by the term, and it seems a safe assumption that the latter would disagree with any interpretations that critics may give of his statements.

The reader will readily recognize the essential cause of these disagreements. Modern German geographers are attempting to define a word which, while retaining an indefinite association with the concept of the perceptible landscape—as commonly understood—shall at the same time precisely define the objects of geographic study. Consequently, each geographer's definition of the word will vary according to what he thinks geography should study. *Landschaft* is therefore no more precise than "geographic area" but has only the appearance of being more precise. To a geographer who is primarily a geomorphologist, a term covering the visible objects of an area seems sufficient. To one primarily interested in cultural geography, something must be done to make man, the active cultural agent, assume a stature proportional to his importance. For one who feels that almost all geographic features can be interpreted ultimately in terms of natural vegetation, climate, and landforms, the *Landschaft* may be limited essentially to those features.

All of the above writers, at least, employ the word *Landschaft* as limited to physically perceptible objects, for they are willing, if not indeed anxious, to exclude immaterial phenomena as non-geographic. Those however who wish to include such phenomena as objects of direct study in a geography that studies the *Landschaft* have a greater difficulty in defining their concept. Pawlowsky, for example, defines the geographic *Landschaft* as consisting of "all those objects and phenomena that fill a certain space . . . objects that are observable with the sense organs," but his subsequent discussion of the inclusion of non-material phenomena appears to insert these into the *Landschaft,* though in just what manner is not clear [*276,* 202 f., 205–8].

Even more definitely does Lautensach appear to include non-material elements, such as "racial and linguistic conditions." He does not define *Landschaft,* but his description of what is included in it appears to represent all that any geographer would include in the study of an area. In other words, little or nothing is left of the common "landscape" connotation of *Landschaft.* Much the same is true of such writers as Maull, who consider the political state as an element of *Landschaft,* not so much on the grounds that the state affects the visible landscape, but simply because it is closely related in one way or another with other elements of *Landschaft.*

[153]

The term is therefore nothing more nor less than a synonym for "region" [*157*]. For this usage, we may add, full justification may be found in the classic authority on German speech, Grimm's *Wörterbuch*. *Landschaft* may be used to indicate a unit political area as a whole, including its population, commonly one of smaller size than a *Land*. Indeed still another of the many meanings listed for the word refers, not to the physical land itself, but to the social group inhabiting it.

Krebs, on the other hand, has urged that the distinction between *Land* and *Landschaft* is not one of size; the ending "*-land*" is not limited to such large units as *England* or *Deutschland,* but is used for such smaller areas as *Siegerland,* or *Sauerland*. The term *"Landschaft,"* he claims, refers not so much to the particular individual area, but rather to certain aspects of its character that are considered as typical of many similar areas [*279, 209* f.]. Thus, he claims, the particular character of the Alps as a mountain area may be called the alpine *Landschaft* which is repeated in other parts of the world, whereas the *Land,* or region, of the Alps, is unique. This is no fine distinction but a major one: the *Landschaft* of the Alps is limited to certain characteristics, excluding relative location and distinctive cultural characteristics, whereas the Alps, as a *Land,* includes all significant characteristics. Krebs is not alone in using the term in this way; in particular it appears to be at least one of the ways in which Passarge employs the word. But neither he nor Krebs, so far as I can find, give any justification for defining the term with this particular meaning—which is not stated by Grimm.

A further element of confusion results from the use of the same word to mean, on the one hand, a definitely restricted area and, on the other, a more or less definitely defined aspect of an unlimited extent of the earth surface. In most cases, perhaps, the context indicates which sense is intended, but this is not always the case. In this respect, therefore, the word is little if any better than our word "area," and is distinctly less clear than "region."

The majority of current German writers appear to use *"Landschaft"* more or less in the sense that we use "region," though each may define it in a different way. Waibel has suggested several reasons why the current generation feels the need of a new word to replace those formerly used for "region." Students of what was formerly called *Länderkunde* are now generally interested in areas smaller than those commonly called *Land,* and the use of *Landschaft* in this sense is justified by common German usage. We may add that Germans do not wish to use the foreign word "region" and find such German terms as *"Gegend"* and *"Gebiet"* for various reasons unsuitable. As the major reason, however, Waibel notes the general drive of our time to see things as they are together (*Zusammenschau*), to consider the totality of phenomena. Presumably the "landscape" connotation suggests that that is what is done in the study of a *Landschaft*. In addition, he notes, *Landschaft* is more impressive, more imposing, than the word *Land;* much that one today calls *Landschaftskunde* is no different from what formerly was called *Länderkunde.*

Few of the American writers who have transferred any of these various meanings of *Landschaft* into our vocabulary in the form of "landscape" have attempted to state precisely what they mean. It might be supposed that Sauer's epoch-making essay on "The Morphology of Landscape" provided a detailed exposition of the concept. The more often this essay is studied, however, the more difficult is it to perceive just what the difference is between "landscape" and "area." At times the terms are used as though interchangeable; at other times, some difference is indicated. The bare statement that a landscape is "an area made up of a distinct association of forms" may contain a definition, but it hides it from most readers [*211,* 25 f.]. A statement in a later essay tells us that "The design of a landscape includes (1) the features of the natural area and (2) the forms superimposed on the physical landscape by the activities of man, the cultural landscape" [*84,* 186]. Both from this and other statements, we conclude that "the natural area" and "the physical landscape" are synonymous; that climatic conditions, for example, are to be included with no question as to whether they are visually observable—as Schlüter argues, they are observable by the sense of feeling. Likewise we note only in passing that the combination of these two parts is not a simple sum, man not only adds to the natural area but also subtracts from, and alters, its individual features. The essential point however is that the combination of these two groups of facts does not form all that is in the area—the landscape thus defined is not synonymous with the area. Immaterial human phenomena are not facts that exist without location in the universe: they are specifically located in areas of the earth surface and are therefore included in the section of reality formed by an area (that these are excluded from the "landscape" is more definitely indicated in other statements by Sauer). Sauer's concept of "landscape" must therefore be defined either as the area minus its immaterial phenomena, or, if one prefers, as the area insofar as it is material. This is clearly the concept of "landscape" adopted by most of his followers. In addition, however, these students, as well as Sauer himself, consider that landscapes are individual units of area— "the geographic area [evidently, from the context, the landscape] is a corporeal thing, which is approached by the characterization of its forms, recognized as to structure, and understood as to origin, growth and function" [*84,* 190; *cf.* also *211,* 25].

That this writer is not the only one who has found it difficult to understand what distinction Sauer makes between the terms "landscape" and "area" is shown in the writings of those who have followed his ideas. In sum, the word "landscape," introduced to American geography in these highly stimulating and impressive essays, has been accepted by many students as the basic term in their geographic thought, in spite of the fact that no precise statement of its meaning was provided. Recognizing this deficiency, evidently, James endeavored, in 1934, to state explicitly what is in the mind of "the geographer who uses this term." Landscape refers "to a portion of territory which is found to exhibit essentially the same aspect after it has been examined from any necessary number of views" [*286,* 79]. Whether the last phrase implies that the methods of observation are literally limited to the visual views is not certain. His explanation of the other qualifying phrases need not here concern us. The essential part of this statement defines a landscape simply and unequivocally as "a portion of territory," an area, and the statement as a whole might just as commonly be used as the definition of a region. Taken as a report of current practice, James statement is no doubt essentially correct: American geographers have come to use "landscape" as though it were synonymous with "region." If they think that the geographic region should be regarded as including non-material features, they include the state as an element of "landscape," just as Maull included it as an element of the *Landschaft;* the present writer is but one of several who have used the term in this very extended sense [*216,* 946 ff.; since the term is there used consistently in the sense indicated the reasoning involved is not thereby invalidated, even though the terminology may seem absurd].

Broek therefore concludes that we now have in geography two quite distinct meanings for the same word "landscape." One meaning is derived from the common connotation of the word: the landscape consists of "the observable features of any arbitrary part of the earth's surface, but . . . in a wider sense than the common usage implies." The other meaning is "a region that has a certain homogeneity in its morphology." In the interest of clarity, Broek urges that one of these should be barred and favors dismissing the second. But we must note carefully what is involved in the "wider sense" to which he has expanded the first meaning. No longer adhering—as he did in his doctoral dissertation [*333*]—to the viewpoint of Schlüter, Granö, and Sauer, Broek now feels that the geographer may not limit himself to "the directly observable features"; he must look beneath the shaft of a coal-mine, he must see the daily and seasonal cycles of phenomena, and "more important, he must search for the invisible behind the directly discernible." In the study of a rural landscape, for example, "he must perceive the farm units, the rotation systems, the yields, the purpose behind the production." In sum, the landscape for the geographer is "an abstract landscape free from time bounds and place bounds of the observer and supplemented by invisible, but, nevertheless, significant data [297, 104]."

If we grant, as the present writer certainly would, that all of the things included in the above description should be studied by the geographer, it is still far from clear just what this "abstract landscape" actually is, unless it is simply the sum total of significant phenomena of area. In other words, hasn't the author simply followed the same trail as Lautensach, and many other German students, in adding to the common connotation of "landscape," everything that a geographer should study, so that he ends with a concept that is indistinguishable from "area?" It is significant that, like Lautensach, he did not attempt to put this concept into a precise statement. Had he done so, it would apparently read: the landscape consists of the observable features of any arbitrary part of the earth's surface plus all non-observable features of significance. (Since Broek maintains his concept of landscape consistently throughout his paper, the question raised here does not affect the logic of his subsequent conclusions. The same is true of Lautensach's paper.)

If we examine the use of "landscape" by English geographers we find a situation very similar to that in this country. Indeed many of the English students have evidently received the concept from this country. In Dickinson's recent discussion we find the same method of avoiding any precise statement of "landscape" [202]. This is all the more remarkable because Dickinson himself poses the question, "What is landscape?" His method is again that of stating one precise definition from which he immediately departs, presenting in some detail the views of various exponents—the "landscape purists"—with which he then differs more radically than he appears to recognize. Thus, if we may put several scattered statements together, we may conclude that the landscape consists of the land forms and their plant cover, together with the fixed material features, in compact areal association, which are the result of the transformation of the land forms and plant cover through human occupance. Climatic conditions, animal life and man, are apparently omitted. On the other hand many non-material factors are later added, for Dickinson is trying to fuse the concepts of Hettner and Schlüter as though they were not in opposition. But whether these are included in any sense in the "landscape" is not clear, for, as Crowe emphasizes in his comments, Dickinson has not answered the question he raised.

The fundamental requirements of logical reasoning in any science demand that the basic terms of methodological discussion must be precisely defined and the definitions adhered to in the discussions. To presume that the reader will know what the word means, is, in the case of "landscape," to presume the impossible. If a writer is unable to state precisely what he means by the term, his use of it is a confession that he does not know just what he is talking about, but is using a more or less conventional word to hide that fact, if not to permit him to perform conjurer's tricks with logic that would not be possible in straight English.

If every geographer were to adopt a different meaning for the basic concept, "landscape," clear reasoning would still be theoretically possible if each writer stated precisely what he meant, but it would be extremely difficult to avoid well-nigh endless confusion. Consequently it seems desirable for American geographers, at least, to attempt to come to some agreement of what the word "landscape," as a technical term in our field, shall denote—assuming, that is, that we are to continue to use it as a technical term. Although the present confusion would provide a strong argument for rejecting the term entirely, its widespread use would seem to indicate that it must have some advantages not possessed by terms hitherto used. In any case it is a well-nigh hopeless undertaking to weed out a term so well entrenched in our vocabulary. Practical wisdom suggests that we first see if it is not possible to arrive at least at some approximate agreement.

B. A SOLUTION FOR "LANDSCAPE"

Before attempting to find a single, precise, concept of "landscape" which would provide an acceptable technical term for geographic thought, we must recognize that perhaps no such solution is possible—that geographers may inevitably disagree in their use of the term, in major as well as minor respects. But if that proves to be the case we would be forced to recognize that the word had little or no value as a technical scientific term. Every single paper in which it was used would have to carry a re-statement of just what it meant; since few writers have found it possible to state their concept in less than a page of description, we would soon find it preferable to throw out the word entirely.

If on the other hand, geographers wish to use this term for the purpose of enlightening, rather than confusing, their readers, an approximate degree of standardization of the term is essential. Our discussion of the present confusion revealed the underlying cause: *before* deciding what "landscape" meant, geographers have announced that geography was the study of landscape; the definition of the technical term therefore is construed to fit the individual's concept of "geography." That this reduces the hypothesis itself to a mere redundancy of obscure terms has apparently escaped notice. Our first step therefore must be to reject, as a premise, the statement that geography is the study of the landscape, or of landscapes. It is not to be expected, and it is in no way necessary, that geographers should agree precisely as to what is included in geography. But in order that their disagreements on this question may be discussed intelligibly it is essential that they agree on the terms of the discussion. Only after we have decided what a landscape is, can we consider its relation to the objects of geographic study. Only in this way can we avoid begging the question, by definition, as to what should be included in and what excluded from the field of geography, or whether geography finds in "landscapes" its own concrete objects of study.

Since it is certainly not our desire to add to the present confusion, we will not attempt to construct any new concept for this much-abused word, but rather will endeavor to ascertain, if possible, the common thought that geographers have had in mind in using the word. For the moment we may ignore any question of etymological justification and simply assume that we may give the word any meaning current in geographic thought.

We have noted one very common tendency to use "landscape" as essentially synonymous with a piece of area having certain characteristics which in our minds, if not actually in reality, set it off from other pieces of area. This is a concept that can be defined, in theory, with a reasonable degree of precision and on which many geographers might agree. It represents the obvious translation from the German *Landschaft,* used by many German geographers with this simple connotation, and many American students, including the present writer, have already used it in that way. That it represents, as Broek suggests, a revival of the Old English connotation of the word appears historically inaccurate; the discovery that the word formerly had that connotation was presented long after American geographers had adopted the concept, as an etymological justification for what otherwise seemed to be but an erroneous translation from the German [James, *286,* 78 f.].

The purpose, Broek notes, of using the word "landscape" in this manner foreign to common usage—whether it be regarded as a new or as a resurrected meaning—is to provide an equivalent to the word in other Germanic languages, expressing an area that "has a certain degree of homogeneity" [*297,* 104]. We noted earlier, however, that this concept is less clearly expressed by the German word *Landschaft* than by our word "region." English-speaking geographers have long used this word to indicate an area with a certain homogeneity. Whatever difficulties have been experienced have not resulted from any confusion over the word "region," but simply from the problems involved in determining what "an area with a certain homogeneity" may be. To change the title to a term far less clear than "region" may divert our minds from the intrinsic problems involved, but will certainly not help to solve them.

To summarize, if the word "landscape" offers to geography nothing more than a synonym for "region" it has no advantages for the purpose and is therefore superfluous. In fact it has obvious disadvantages. The concept of a piece of area somehow distinct from neighboring pieces of area is much more definitely suggested by "region" than by "landscape." If it be claimed that a "region" sounds like a larger piece of area than the geographer may wish to select as a unit, "landscape" suggests a very much smaller piece. Certainly the former, in common meaning, is far more flexible in respect to size. If we need a term for subdivisions, "district" is much more clearly understood than "landscape." Finally, the additional connotations of "landscape," that cling to the term no matter how we define it, once they are openly stated are seen not to offer an advantage, as commonly supposed, but, on the contrary, to destroy the value of the word as a clear-cut technical term.

We conclude therefore that, even if English usage justified the concept of "landscape" as an areal unit, the use of the term in place of "region" is to replace an established term by one far less clear in meaning. Since, in addition, we have at the same time a second quite different connotation for the same word, we may emphatically underscore the conclusion which Broek rather mildly suggests—namely, that we do away with the resurrected connotation.

Turning to the opposite end of the list of meanings given above, we find general agreement on the elimination of the literal sense of the word "landscape" as the view of an area as seen in perspective. Significant as this concept may be in other fields, it is of little importance in geography—we need no term for it. Presumably most geographers will likewise have little or no use for any concepts based on our psychic sensations of area. Hellpach's concept is essential for the study of "geopsychic phenomena," but we may safely assume that few will regard that study as of sufficient importance in geography, as a whole, to determine our terminology [*cf.* also Lautensach, *278,* 13 ff. Hellpach, we may note, has published primarily in psychology.].

With the elimination of these two extremes (in the literal, objective sense) we still have a large number of somewhat different concepts. It will simplify our problem to note that in current usage those writers who use "landscape" to mean certain aspects of a delimited area, also speak of "the landscape," or "the natural landscape" and "the cultural landscape," without having any limited area in mind. In other words, all writers use the term on occasion to mean something about area without necessarily implying any limit to the extent of area.

We have noted that a number of writers use "landscape" to indicate all the material facts in an area. More commonly they speak of "observable" facts, but since any fact, to be a fact, must somehow be observed, that word is inadequate. Many, following Schlüter, speak of "sensually perceptible features" (*sinnlich wahrnehmbar*)—*i.e.*, all objects that are, theoretically, directly observable as sights, sounds, smells, and feelings. To many this concept is suggested by the common meaning of landscape. Granö, in particular, has attempted to develop this concept logically by working backward from Hellpach's concept of the landscape sensation as a total-impression. By this standard he selects all the features of an area that are responsible for our sensation. Working it out literally, he includes objects that affect all our senses only within the immediate vicinity in which such reactions are possible; the more distant part of the area we perceive only by sight and there he limits himself to visible objects. In both cases, however he uses other terms, in order to preserve the word *Landschaft* as a concept of sensation. Few, if any, have followed his strictly logical system, and it seems safe to assume that few will.

We may however, for the moment, accept his general line of thought, and consider the possibilities of the concept of "landscape" as the sum total of those things in an area that could produce "landscape sensations" in us if we placed ourselves in the different positions necessary to receive them. These things are of course external realities and therefore appropriate for study in an empirical science. The critical question however is, do they add up to a single something, to a unified concept? "Landscape" as a technical term in geography must mean more than a miscellaneous collection of different kinds of things, or even a collection of different things alike only in some partial respect. In other words, any statement that tells us that "landscape" includes certain, but excludes other, phenomena, is no definition of the concept, but at best an explanation of a definition. If the thing that we are defining is something, the definition must tell us not merely what it includes, but what it *is*. Granö appears to assume that the material objects responsible for our "landscape sensation" form a unit or whole (*Ganzheit*),

[161]

because our sensation of them is a whole, but the fact that the human mind has a unit impression of a collection of things does not prove for a moment that they have in themselves any relation to each other, other than juxtaposition. Lautensach therefore characterizes Granö's "unities" (or "entities") as anthropocentric—one might say, centered on the observer—and therefore essentially still in the field of psychic sensations [263, 195 f.; 278, 13 ff.]. It is not sufficient to say that these objects are in interrelation in the area and thus form a unit, for, as Granö recognizes, this interrelation includes the immaterial forces that are not productive of psychic sensations in the observer; without these, however, the unity concept is destroyed. It can hardly be claimed that these objects are unified by the fact that they can all be observed by human senses. This is a minor attribute for each object and differs radically for each, since some are seen, others heard—not to mention sounds that the human ear cannot hear, light rays that the human eye cannot see, etc. Likewise, though it is simpler to say that these things are all alike in that they are material, this is merely a similarity, it does not form them into a unit total. A collection of heterogeneous things interrelated in part through immaterial phenomena is not, with those immaterial phenomena excluded, a single thing, but only a selection of slightly similar things out of a total.

This theoretical argument may be seen more concretely if one attempts to create the concept involved into some reality that the mind can grasp. Thus, when we speak of an area as a section of the shell of the earth surface, one can imagine it as a piece of the real universe, in which there is a related multitude of material and immaterial phenomena—in short everything in the area. The religion of its inhabitants is not a phenomenon adrift in the universe, it is located in the region just as definitely as are the people. If however one says that in this area there is an X which represents all the material facts only, of the area, how can one grasp that as a unit total? It includes things seen, heard, tasted, smelled or felt—but only these; what do they add up to—other than a selection from a total?

In other words, to define any word, X, as representing the total of material objects in an area, is not to present a single unified concept, but merely a selection of things. We will later note that one effect of this concept is to focus the attention of the geographer, not on any single concept, such as the area, but on the individual things.

Despite this conclusion, we cannot escape the feeling that one common use of the word "landscape" does suggest some objective single reality outside our sensations. Perhaps we can approach this idea more readily if we limit ourselves for the moment to the visual aspect of an area. By assuming, as geographers commonly do, that we see the area from above, we eliminate the sky; the atmosphere of the earth is then simply the medium through which we see the solid and liquid forms of the earth surface. We see only, however, the external surface that underlies the atmosphere—formed by the surface of water-bodies, the uppermost foliage of trees in forests, by grass, or by top-soil in bare fields, or by the outer surfaces of buildings, etc. All of these surfaces together form a continuous surface over the area and it is this surface, and this surface alone (other than the sky) that produces the sensation of the visual landscape in our minds. But we do not need to see this surface to know it exists; a giant could feel it with its hands. The geographer, midget like, must add together a vast number of detailed measurements obtained from all kinds of observations, taken from all directions, in order to represent this surface as it is, rather than merely as it looks to him. The important thing is that he can do this, he can study it as it *is;* that is to say, it is not only in theoretical thought a unit form, it is a reality.

Can we carry this concept farther? Theoretically it should seem possible to construct a similar surface that represented the origin of our aural sensations of landscape, and so on for each of the other senses, though these are very difficult to construct mentally. In particular, the surface we *feel* consists of nothing more than those things in immediate contact with our body—a shell of air the shape of our body and the solid ground beneath our feet. Since it is obvious that any such concepts will have no utility in geography we need not pursue them further for their own sakes. We consider them only to show that they cannot be added to our concept of the "external form of the earth surface under the atmosphere" and form a unified concept—we would have merely a sum of entirely different, incommensurable things.

Further we may note that in contrast to the heated arguments that arise over what is included in the "landscape" according to other concepts, the concept described above provides a relatively clear answer in every case. Thus we cannot exclude movable objects simply because they make our concept more confusing; for we are not constructing an abstract concept but are attempting to state a reality, and that reality includes movable objects. This fact is recognized in common speech, contrary to Dickinson's implication [*202,* 5]. A view of Broadway in which no streetcar, bus, truck, automobile, or person is to be seen, is an incomplete picture of that particular landscape. No living person has ever seen the landscape of New York Harbor devoid of ships. Since the geographer cannot study or describe the constant changes of movable objects in a landscape, he is forced to generalize. This problem is not limited to the movable objects however, since many of the fixed objects of the landscape change in character in the different seasons of the year—the winter landscape may be very different from the summer landscape. In this case no generalization is possible, the geographer must recognize the significance of several seasonal landscapes.

Although movable objects in an urban landscape are too important to be ignored, in most rural landscapes they form but a very small part of the total surface concerned; though recognizing them as logically parts of the actual landscape, we make but slight error in practice in ignoring them.

Other features of areas are definitely not parts of the landscape. Mineral deposits and underground mine workings are certainly not landscape elements—we are not permitted to push the concept of landscape thousands of feet underground. An open-pit mine on the other hand is just as obviously a marked landscape feature. Further, insofar as vegetation in any place forms a complete surface above the ground, the soil beneath it is not an element in the landscape; when the grain crop is harvested and the stubble ploughed under, the top soil—literally the surface—becomes a landscape element. Finally, the amount of precipitation in any area is likewise not a landscape element but something that happens to the landscape, and may be a casual factor affecting it.

The reality that we are defining as "landscape" is essentially only a surface. The form of the surface is determined primarily by the relief of the land, but is also affected in minor degree by the height of forests and, in urban areas particularly, of man's buildings. The material character of the landscape is expressed by color and texture and may be observed by sight and feeling. To designate the material character of the landscape apart from its surface configuration, we may use the term "landscape cover." Over most of the world this consists of the uppermost surface of vegetation—whether natural, wild, or cultivated—and of the surface of water. (Whether we may use the term "landscape" in the oceans is a question that need not here be discussed.) In lands lacking in vegetation—whether permanently or seasonly—the "landscape cover" consists of bare ground, snow and ice, or of the surface of the works of man.

If it be objected that this concept excludes many elements essential to geography, it must be remembered that we are not seeking to define geography, nor to limit the objects of geographic study, but to determine a possible concept of the term "landscape" that shall represent a concrete reality. Geography existed long before it took over the term "landscape," and might get along very well without it. What shall be included as objects of geographic study, and what may properly be included in a unified concept of landscape, are two separate and independent questions. Only after each has been separately determined can we consider their relation to each other.

It should be repeated that the concept here presented is not supposed to be in any way new. On the contrary, its purpose is to express in more specific form the thought which many geographers may well have had in mind when they used the term. For example, when Finch speaks of "how an area *looks*," we assume that he is saying in common terms much the same as we have written above. (Taken literally, to be sure, his expression might infer rather "how the area looks to the observer," which could be interpreted in the subjective sense of sensation, but it is clear in the context that Finch would very definitely not be concerned with that geopsychic point of view.) Likewise the popular expression, "the face of the earth," indicates very simply the concept we have defined.

The definition that we have attempted to formulate for "landscape" is a concept that represents a certain distinct and real aspect of an area. Is this, then, an aspect of area that geography is concerned to study, and is it, perhaps, the sole aspect that geography is to study?

No doubt most readers will agree that the answer to the latter question is in the negative. If geography is greatly concerned with open-pit quarries that expose their workings as landscape scars, it is no less interested in the far more important coal mines whose underground workings may underlie a much greater area, though the representation in the actual landscape be small. If geography studies soils, it studies them whether they are continuously bare of vegetation, exposed seasonally, or perpetually covered by forests or grass. And practically all geographers agree that climatic conditions are one of the elements to be studied directly in geography. In itself the landscape is literally a superficial phenomenon and a field of science that concentrated on it alone would be superficial. If the interest in the landscape is simply in the picture that it forms, the "designs in the great carpet," to use Penck's expression, the point of view is clearly aesthetic. If, on the other hand, we are interested in the landscape as a manifestation of something else—the complex of related factors in the area—then we are merely using it as a means of studying a different object, whether defined as that total complex, or as the area itself.

If the attempt to give a specific definition to the concept of "landscape" makes it impossible to define geography as the study of landscapes, that is not because our definition has been constructed for that purpose. We have simply expressed the one possible reality that geographers can have in mind

[165]

in using the expression to mean something about an area that is not simply the area itself. On the contrary, it is the exponents of the concept of geography as the study of landscapes who repeatedly stack their cards to prove their point. While they commonly speak of the landscape as "how the area looks," the visible scene, in their methodological arguments they repeatedly add to this concept such other qualifications as may be needed in order to include within it whatever features of area they wish to have included in geography. In contrast, the definition of landscape that we have stated is independent of any particular thesis of geography; it does not therefore, by definition, beg the question of what constitutes the object of geographic study, or of what kinds of phenomena are included in geographic study, but leaves the ground clear for those questions to be decided on their own merits.

That a clear separation of the fundamental concepts of geography and of landscape is necessary is illustrated by the discussion between Dickinson and Crowe [202]. As the latter insists, the distinction that Dickinson treats as minor, between a view of landscape study as an "integral part" of the field of geography, and that which regards it as the "central objective" of the field, is the very crux of the question. In fact, Dickinson appears, without recognizing it, to have concluded that geography is not the study of the landscape: "the geographical study of area has a clear objective in landscape and society, in their association and variations *in area,* interpreted in both their genetic development and dynamic relationships" [13]. The several logical corrections that he makes of the general concept demonstrate in detail the fact so clearly expressed by this concluding statement—namely, that the study of the landscape does not offer "a clear objective" for geography. The study of "landscape and society" is not "*an* objective," but at least two, unless it be expressed as the study of the relationship between society and landscape, which is the very concept that he wishes to refute.

If, however, we return to the simple statement on which all who hold the chorographic view of geography may agree, namely, that geography studies areas of the world, and at the same time recognize that the landscape of an area is one aspect of the area that includes some, but not all of the things studied in geography, what general utility has the concept of landscape for the geographer? It is not necessary to show that the study of the landscape is an "integral part" of geography, but only that it is a part of some importance. Is this not self-evident? Presumably all will agree that one of the first steps—if not the very first step—in geographic field work is the general examination of the area as one sees it by and large, from many different points of view. Geography, Lautensach says, "starts with the *Bild* (form or picture) of the land." [278, 21]. On first entering any area, the geographer looks at the contrasts between forests, fields, and cities, without attempting to observe sub-soils or monthly temperatures. In this first step, one endeavors to get a "picture" of the area, not as seen in perspective, but as seen with an all-seeing eye; but not with a penetrating eye. It is the outer surfaces that we actually first study. We examine, that is, the great, irregular carpet of the surface form, noting the texture and designs in that surface and also the irregularities of surface, whether formed by hills or buildings [cf. Penck, 249, 8]. We study, that is, the surface which the area presents us under its atmosphere—"the face of the earth."

The very fact that we commonly start our examination of an area with what we see before us has given many the impression that that is, as Dickinson calls it, "the fundamental approach." This viewpoint is a natural product of what we may call the geomorphological point of view in geography, which Sauer in particular expresses, and which may be traced to the founder of modern geomorphology, Richthofen. But it is an undemonstrated assumption that the landscape—the visible surface—is more fundamental to the total complex of an area than, say, the invisible climate, or that houses are more fundamental than the people who build them. In fact, the amendments that Dickinson offers to the concept of the "landscape purists" are not minor amendments but major ones, made necessary by the very fact that the landscape, far from being fundamental, is only "the outward manifestation" of things that are fundamental—the interrelated factors of area. "The fundamental approach" in geography is presumably the approach to the fundamentals, whether by way of the landscape or by whatever other route.

On the other hand experience has shown us that if we start our study with the external form, the landscape, our efforts to interpret what we find in it will ultimately include the study of nearly all that is significant in the area. The landscape, therefore, is useful to geographers not merely because it comes easiest to hand, but because the study of its forms will lead us to most, at least, of the significant features of the area. The utility of the concept of the landscape—even in the realistic and necessarily limited sense in which we have stated it—is thereby amply demonstrated. But we are not to confuse a key to the problem of regional study with the essential objective of the study. Nor may we assume that the fact that the landscape is a manifestation of most of the factors significant to the geography of an area proves that any factor that has no significant manifestation in the landscape is therefore not significant to the geography of the area. That proposition cannot be proved deductively—i.e., for all possible areas—and therefore must be considered in itself in each particular area. Nor, more generally, may we assume that any feature of notable importance in the landscape is necessarily of equal significance in the geography of the area [*cf.* Broek, *297,* 104]. Just how reliable a guide the landscape is to the geography of an area will be considered later.

Finally we must return to the question which we temporarily postponed: if this concept is of high utility in geography, and puts in definite form what many geographers have meant when they have spoken of "landscape," are we justified linguistically in using this word in that sense? Since questions of terminology cannot be decided either logically or objectively, this is a question of judgment. If common speech uses "landscape" at times to mean the total sensual-perception received from an area, it also uses it in the limited sense of the visual impression, and it likewise uses it in the sense of the real superficial form that produces this sensation. If our concept is not actually simply a precise statement of one of the common concepts of "landscape," at least it stretches the concept of that word far less than is generally permitted in the transfer of words from common speech to scientific terminology. Non-geographic readers will not therefore find our usage strained—as they certainly do if one tells them that a "landscape" is an area—but may be expected to acknowledge the usage as justified.

It is even possible that some readers will find that we have gone to great effort to demonstrate the obvious. Such a feeling is possible however only for those who have not attempted to understand all of the widely varying uses to which "landscape" and *"Landschaft"* have been put in recent geographic literature. If the expenditure of ten times the effort here involved should result in clarity of use of this term in geography, the result would still be worth far more than the cost.

To prevent any possible uncertainty, therefore, we may recapitulate briefly. In an empirical science of geography there is little need for any of the concepts of "landscape" as sensations. Without disturbing the use of the term in such senses in the appropriate fields, we may define it for our purposes in terms of some external reality. It is both unnecessary and confusing to use the word as synonymous with "area" or "region" since each of those terms is far clearer. To use "landscape" as a label for the material objects of an area is to ascribe to such a selection of phenomena out of a larger total an attribute that it does not possess—namely, of constituting in itself a single reality. The same would be true if we should arbitrarily apply the word to the total of all visible objects in an area, including that is, all objects that man can see by looking under other objects. The actual single and concrete reality which, we believe, underlies the thought of many who have used this term without attempting to define it, is the external visible surface of the earth. This is the reality which produces visual landscape sensations in us. It is a continuous reality, constituting, for the whole world, a single unit whole. It is however literally a surface; it includes only that which we can see or feel from the outside. By far the greater part of it is formed by the surface of water bodies and the uppermost surface of vegetation—whether natural or cultivated—or by bare ground, or by ice and snow. Except for the water bodies all of these surfaces depart from plane surfaces, or rather the "geoid surface," primarily according to the relief of the land. Only a minor part of this actual landscape of the world is formed by the surfaces of man's buildings, concrete highways, etc.; on the other hand, a major part is formed by his cultivated fields. It hardly needs be added, however, that the relative importance of these different parts of the landscape does not necessarily represent their relative importance to the total areal differentiation of the world. Though the two concepts are closely related, they are not the same. Just what that relationship means to the geographer will be considered later.

If we may hope to have achieved a clear and useful definition of the concept of "landscape," we cannot claim to have solved the problem of how to translate the German word *Landschaft*. This is most unfortunate, in view of the predominant importance of the Germans in geographic methodology (together with such others, like the Finn, Granö, whom we know only through their publications in German) and the constant use that many of them make of this word as a fundamental concept. But we would need many terms in order to translate *Landschaft* according to the concept which each different German writer may use. Where it is clear that the writer uses *Landschaft* simply to mean a portion of area we may follow Penck's suggestion by translating it as "region" [*cf. 159,* 640]. In a relatively few cases we can correctly translate the writer's thought in terms of "landscape sensation," in others as "landscape"—as defined above. In most cases however this writer finds it impossible to translate the term within the requisite degree of certainty and therefore simply quotes the original. To know just what is meant, the reader will have to decide from the context, or by reference to the original—without any assurance that he will find the answer there. In general he may fairly safely assume that the term implies either simply "area," "region," or "district," or something very much like any one of those, plus or minus various other attributes, commonly not stated. For these reasons, no matter how many times we must repeat the word untranslated, we continue to print it as a foreign word, lest any possible encouragement be given to introducing into our language a word that is so baffling in its own.[59]

---

[59] It is hardly our place to suggest solutions for those working in foreign languages. Fortunately so, since the use of *Landschaft* to mean basically a small piece

C. NATURAL AND CULTURAL LANDSCAPES

We may complete our consideration of this term with the brief statement of the meaning of the compounds, "natural landscape" and "cultural landscape." The form in which Sauer states his concept of these two terms, in the reference cited above, as well as in his subsequent diagram, [*84*, 186, 190], would seem to indicate that they were separate components of the total "landscape," the former consisting of all the natural features of an area, the latter of all the man-made forms. Though this might be a logical division of "landscape" when understood simply as the collection of material features of an area, the suggestion that the cultural forms that man has erected could be separated from the natural base on which they are built and thus considered as any kind of a "landscape" would appear to stretch that poor word beyond recognition. Noting this difficulty, Broek suggests that we distinguish between "natural" and "cultural" elements in the landscape [*297*, 104].

It should be noted that if the cultural elements in the landscape, when considered separately, do not add up to a single unit but only a collection of parts of a unit—the total landscape—the same is inevitably true also of the sum of the natural elements in the present landscape. Considered separately from the cultural features produced by man, these do not form a "natural landscape" but only a collection of parts of the one real landscape, parts which we distinguish from the cultural parts only on an intellectual basis—in reality they are formed of the same stuff of earth, water, vegetation, etc.

Most American geographers who use the term "cultural landscape" mean simply the present landscape of any inhabited region. In this sense it is demanded only where we need to emphasize the present complete landscape in contrast to the "natural landscape" [*cf.* Lautensach, *278*].

In the German literature one finds a distinction between "*natürliche Landschaft*" and "*Naturlandschaft*." In the latter term *Natur* is used in the

---

of territory is so firmly implanted in German geography that it would appear hopeless for anyone to attempt to alter it. *Landesteil* would appear to be clearer, but lacks certain of the qualities—including not only the "mystic" qualities—of the other word. *Raum* is even simpler. Sölch's term "*chore*" has been adopted by a number, and if the multiplicity of definitions for *Landschaft* finally brings that word to its scientific death, this term created from the Greek will offer a clear substitute. For "landscape" in the sense we have defined it, *Landschaftsbild*, which various writers use, may offer the best word. But it is probably a safe prophecy that most German geographers in our time will continue to use *Landschaft* in both cases, either in no precise sense, or with some special personal definition whether stated or implied, whether maintained throughout the same study or not; and that their readers, both German and foreign, will continue to be mystified.

sense of nature excluding man, in the former it is all-inclusive, but implies "actual" or "real" in contrast to "arbitrary" or "artificial," and is used to indicate that a particular *Landschaft,* as a region, is a real unit, a *"natürliche Einheit"* [*cf.* Bürger, *11,* 29]. Waibel finds that even in German it is too difficult to avoid confusing the terms and since the literal translation into English is exactly the same in both cases, we must hope that his recommendation that the former term be abandoned will be followed [*266*].

Some German students have employed the contrast between *Naturlandschaft* and *Kulturlandschaft* on a relative basis—*i.e.,* on the *extent* of cultural development [Krebs, *234,* 84; Bürger, *11,* 63]. That this would lead to great difficulties in classification is perhaps a minor objection. More important is the fact that no matter how we define *Naturlandschaft,* or natural landscape, the connotation that it is in fact "natural," non-human, will cling to it and cause us to overlook the fact that it is only in a certain high degree, presumably natural. In other words, to call the Sudan a "natural landscape" is to beg the question whether the savanna is entirely a natural product or has not been in part produced by human action [*cf.* Waibel, *266;* and Lautensach, *278,* 20]. The contrast between the effects of man's work and those of the rest of nature is too important in geography, even as a purely intellectual concept, to allow us to undermine it in this manner.

On this basis, an actual "natural landscape" would be found, in the strictest sense, only in an area never disturbed by the hand of man. The first arrivals on any previously unvisited, but not desert, island would find * a natural landscape formed almost solely by water bodies, the contour of the land, and the natural vegetation. In the inhabited areas of the world the concept of natural landscape is purely theoretical; throughout the civilized world at least, the natural landscape has been seen by no living person. Even in primitive areas we must beware of assuming that the absence of *positive* cultural forms in the landscape proves that we are viewing the actual natural landscape, as Geisler assumes in his discussion of primitive landscapes in Australia [*277*]. "Man has influenced the living world much more anciently and much more universally than one may think," Vidal reminds us [*184,* 8]. Who will say what was the original vegetation of Illinois? Are all the patches of brush or savanna grass in the tropical rainforest areas of Africa man-made, or were there originally natural clearings in the forest? The only safe rule for scientific geographers to follow is to accept in practice what is unquestionably true in theory, namely, that the natural landscape ceased to exist when man appeared on the scene [James, *286,* 80; *cf.* Broek, *297,* 108; Lautensach, *278,* 20; Krebs, *279,* 208].

Nevertheless, terms of some kind are needed to express the difference

---

* Read: "on an island previously unvisited, but not uninhabitable,

between the primitive landscape, whose forms have been affected largely in a negative manner, if at all, by man, and those in which man has in large degree determined the character of the landscape cover. This is a contrast which the observer recognizes at a glance. Regardless of the underlying natural conditions, if the area is under the control of man, its landscape cover is ordered and arranged in definite units each of which is strikingly homogeneous and sharply separated from the others. In contrast are the areas where men live without dominating the landscape cover. However much it may be altered by man, it is not ordered; if man destroys the natural vegetation but does not replace it by cultivation, or by roads and buildings, vegetation "grows wild." Both the unaltered natural landscapes and those altered but uncontrolled by man, we may call "wild landscapes," in contrast to the "tamed" or "cultivated landscapes" of areas under man's control. Whatever objections may be made to these terms, at least they indicate clearly the concept in mind, in each case, and do not lead to an *a priori* assumption that any landscape produced by no conscious plan of man is a "natural landscape."

With the very minor exception of the glacial areas, the present "natural landscape" is only a theoretical concept. As such, it is of course limited in the same way as the landscape in general. For practical purposes it includes only the contrast between land and water, the relief, and the natural vegetation. It forms but a part of the total natural environment of an area, and those who have considered the two terms as synonymous may object on this score. But why do we need two terms for the same thing? If geographers have become bored with their predecessors' use of "natural environment" and find that "natural landscape" has the advantage of newness, do they suppose that when that wears off, their successors will not become equally bored with a phrase that in common speech certainly does not mean what it is supposed to mean in geography? The fact that "environmentalist" has been used as term of reproach in current writings is no reason why we need be afraid of a term so well established as "natural environment." The major difference between the two terms under consideration would appear to be that while the ordinary person has little idea what geographers mean by "natural landscape," and geographers do not agree among themselves as to what it includes, any one can readily understand the meaning of "natural environment," and so far as I know, there is very general agreement among geographers as to exactly what it includes. (Any argument over the use of "natural" as non-human is common to both terms, and can in any case be easily eliminated by the simple statement that in such a phrase, if not in geography in general, "natural" is used as the only convenient word for non-human. This question, however, as well as the question whether "fundament" offers us a better term, will be considered later.)

It might be questioned whether the concept of the natural landscape, thus limited and purely theoretical in character, has much value in geography. While we can construct its water and land components with fair accuracy, since in most areas man has affected these in but minor degree, the attempt to construct the natural vegetation yields results of very doubtful validity. Nevertheless one of the most interesting aspects of the study of the present landscape in any area is the attempt to determine to what extent man, rather than nature, has formed it, and to establish the successive changes that he has produced in developing it. For this purpose, the theoretical base is of course the natural landscape, even though we may in practice be forced to approach that by starting with the present actual landscape and working back through time to arrive at an approximation of the original unknown natural landscape.

This original natural landscape, however, is not the same as the present theoretical natural landscape. Nature has not been static during thousands of years that man has lived on the earth. Even if all other factors had been constant, Vidal reminds us that the progressive development of plant societies toward climax forms—actually unattained over large areas—would have produced a vegetation today different from that which was present before man, and different from that which would develop today were man to retire from the scene [184, 15]. But beyond that, all other factors have not been constant; certainly there have been important climatic changes since primitive man appeared in all the continents. Consequently we may state that the natural landscape, as of the present time, is a theoretical conception that not only does not exist in reality, but never did exist. Something more or less like it did exist in the original primeval landscape (*Urlandschaft*), which, in complete form, only the first man to arrive in any area could observe.

In conclusion, we may illustrate the use of the several terms defined with the example of an area in central Illinois. The *primeval landscape* of the area was the original natural landscape as seen by the first Indians to enter it. Whether that was much the same as, or very different from, the *wild landscape* as the first European explorers found it, we do not know. The *natural landscape* existed as a reality only in the primeval landscape; the present natural landscape is a theoretical concept which never did exist. To what extent it differs from the primeval landscape we can know only in part. The present landscape, consisting of a highly *cultivated* or *tamed landscape* over most areas, together with patches of wild landscape along some river floodplains (which is not the same in character as the wild landscape of the Indian period) has been formed by the alterations made by nature and man working simultaneously, throughout the period of human occupancy, on the primeval landscape. In this particular example the differences involved apply almost entirely to the landscape cover; the surface configuration of the landscape has changed but little. In other areas, however—for example, in Java—both nature and man may have produced significant alterations also in this aspect of the landscape, both in the change from a primeval landscape to a wild landscape, and in the change from a wild landscape to a cultivated landscape.

A. geography in history

The first of three major problems in geography today is presented by the interlacing relations of geography and history.   To what extent should these two fields include each other?   The recognition of the importance of the similarity of geography and history as integrating sciences should not lead to a confusion of their distinctive functions.   If each of them may be thought of as cutting cross-sections through the systematic sciences, these cross-sections are in separate dimensions.   The fact that an illustration of this relation requires a four-dimensional space reflects the extraordinary difficulty of producing a reasonable synthesis of the two points of view and the two bodies of materials—a difficulty which may well, as Huntington suggests, "make thinkers pause where elementary school teachers plunge without fear" [*219, 565*].   (The confusion of purposes produced by the "fusion" of geography and history in instruction was ably treated a century ago by Bucher [*51, 237* ff.].)

Nevertheless, however great the difficulties may be in relating the two fields, one can hardly question Huntington's statement that "a rational understanding of history requires a good knowledge of the changing physical background upon which the historical events occur."   Since historians are usually not equipped with such geographical knowledge their attempts at synthesis have not, he finds, been highly successful.

Of the geographers who have worked in this field, Huntington observes that most "have selected only those historical items which illustrate their special points of view" and modestly pleads guilty to his share in this offence. He evidently does not find this objection in the work he was reviewing, Griffith Taylor's *Environment and Nation* [*389*], but nevertheless concludes that its value is "that it calls our attention to the fact that . . . we are at the very beginning of our task.   We have only reached the point of accumulating materials for a foundation on which our successors may build a fair and seemly structure."   Critical reviews by other geographers would suggest that in such an extremely complicated problem it is more likely that sound foundation materials will be obtained from studies more localized in time and space than the entire history of the European continent.   Any students who have the hardihood to tackle this problem may well be urged to seek their materials not in any of the better known general studies but

[175]

in the large number of less ambitious regional studies which may be found in the German and French periodical literature [see particularly the bibliography in Vogel, *89*].

Geographers, in particular, who wish to consider this type of problem, will find a special reason for confining themselves to a limited period or area, or both. In view of much of the work that has been done by geographers in this field, one is tempted to add to Huntington's statement of what is needed for a rational understanding of history a sound knowledge of the historical events. The student trained as a geographer will do well if he can acquire this knowledge for a very limited area.

The discussion has its place in this paper because the subject considered has been studied primarily by geographers, commonly under the title of "historical geography." Huntington's statements, however, clearly indicate that the focus of attention is not on geography but on history—it is the phenomena of history which are studied in terms of their geographic aspects. Logically it appears clear that this is a problem within the field of history. One of the leading American students of the problem, the historian Turner, indicates this most clearly in describing himself "not as a student of a region but of a process" [*214*]. Probably most geographers would now agree that the geographic interpretation of history is logically a part of history, even if the contributions have hitherto been largely those of geographers [references in *216, 787–91*].

B. HISTORY IN GEOGRAPHY

Within the field of geography itself what consideration should be given to the sequence of historical development? Some geographers insist that in order to maintain the essential point of view of geography—the consideration of phenomena in their spatial relations—any consideration of time relations must be secondary and merely supplementary. Others however urge that the geographer is primarily concerned with the development of, or changes in, the phenomena which he studies; time relations therefore become of major importance.

The thesis that the geographer is primarily concerned with the changing character of an area, has been vigorously proclaimed in Germany by Spethmann, under the title of *Dynamische Erdkunde* [*251*]. In contrast with the work of other German geographers which he characterizes as "static," he proposes the "new dynamic view" of the subject. Actually, as Gradmann notes, there is nothing new in Spethmann's idea; his predecessors did not have a static concept of geography, he is merely using a *Modewort* (a word in vogue).

Spethmann's proposals aroused a great stir in German geographical circles, not only because of the vigor with which they were put forth, but also because of the unfortunate personal attacks that accompanied and followed them. While some received it with praise, others, notably Hettner [171, 5], Philippson [260], and Gradmann were highly critical [see also Bürger, 11, 93–9]. Spethmann filled a second book with replies to his critics, but few who read Gradmann's dignified review will care to wade through the former's personal complaints [261]. On the other hand, credit must be given to Spethmann for having put his ideas into a concrete example, namely in his three-volume study of the Ruhr area, in which he follows the successive changes in the development from prehistoric times down to the present. In the two volumes of this study that are required to bring it up to the present time the reader (some may prefer to rely on my review) will find ample evidence of the way in which a geographer is led out of his field when he permits "genetic explanation to become historical narrative" [Hettner, 167, 269; see also Lehmann's criticism, 181, 51–7]. *

In this country, Sauer has specifically insisted upon the use of the historical method in geography; since present "cultural landscapes" are to be studied in terms of their development from the original "natural landscapes," the study must be oriented on development [211; 85, 623]. This orientation apparently requires that we begin at the beginning. Lest this force us to start with the origin of the earth, he suggests that "geography dissociates itself from geology at the point of the introduction of man into the areal scene" [211, 37]. Starting at that point with a reconstruction of the original "natural landscape," the student of geography will work through the various different transformations in the landscape caused by man (and presumably also those caused by nature?) until he arrives at the present "cultural landscape."

Parenthetically it must be noted that geologists can hardly be expected to accept the line of demarcation suggested. The development of the earth's crust, the formation of rocks, and the deposition of fossils which the geologist studies, did not cease at that unknown time when man arrived upon the scene. On the contrary, if man's activities in any area today cause changes in the form or character of rock formations—slumping of beds of shale under heavy buildings in Cincinnati, decrease in water content of sandstone formations because of artesian wells in Minneapolis—the geologist is concerned to study those.

More important are the practical consequences of Sauer's thesis. Evidently any study of a region must include, indeed begin with, a study of its entire previous geography, chronologically arranged. In his own study of

* See Supplementary Note 28

"Site and Culture at San Fernando de Velicata" [382], the emphasis on the chronological order is so marked that it is a fair question whether the study is not to be considered as history rather than geography. In later studies Sauer, logically enough, has gone farther back into prehistoric times and it becomes difficult to dissociate the work from anthropology [383, 384]. Similarly Trewartha, following Sauer's precepts, presents as the first chapter in the geography of the Driftless Hill Land (of the inner basin of the Upper Mississippi), a chronological treatment of some nine or more French trading posts, each occupied for an average of but little over three years— a study that would be very difficult to distinguish from a study in history [393]. These comments are not intended in any sense as a criticism of the studies concerned, nor would the writer object to any who wish to make such studies. The question is whether geographers who wish to study the present geography of a region are required first to produce works for which few of us are technically prepared and which can hardly be distinguished from those of other fields.

Although no American student, so far as I can find, has presented a complete illustration of the thesis that Sauer propounded more than a decade ago, Broek's doctoral dissertation at the University of Utrecht, on the Santa Clara Valley, might possibly be considered as in some degree a product of Sauer' concepts, though it follows Granö also in considerable degree [333]. Broek's introductory statements would indicate that a regional study should consider the "landscape" of every period from the earliest to the latest, but evidently one may recognize differences in degree of significance: in any case, he omitted landscape changes before white settlement, starting with the "primitive landscape" that the first Europeans found. The major part of the work considered the changes shown in successive landscapes of six historical periods.* In each of these periods it was, of course, necessary for him to follow changes in each of various important elements in the landscape—as Creutzburg has noted, a landscape is not a whole that grows as a whole, but only as each part grows, so that "the history of a landscape becomes the histories of its parts" [248]. Consequently we do not get a full picture of the landscape in any one of these periods, but rather a genetic study of various parts. Finally, the reader must be prepared himself to fit these into the relatively limited section devoted to the present landscapes. Unquestionably, a thorough study of this work would give the reader a better understanding of the present landscape of the Santa Clara Valley than could a treatment that considered only the present without looking to the past. But, needless to say, the latter is not the only alternative, if it is even an actual alternative; the so-called "static" method of

[178]

* Not "six", but three. See Also Supplementary Note 29

regional study does not limit itself to present conditions. The question here is whether we would not have an even clearer picture if the study had been organized entirely on the present landscape, and in the analysis of each of its parts had utilized the appropriate material that was presented in the historical chapters.[60]

It follows that the student of any area must learn far more of its historical geography—and of its history in the broadest sense—than he needs to impart to the reader, in the same way—we may be permitted to suggest—that an historian of any area needs to become familiar with its geography though he may include but little of that in his historical writing. It is not always possible to know during the course of a study, which facts from its past geography will be significant for the present. The necessary selection can be made with greatest ease for reader as well as author if the facts of past periods are presented at the point in the writing where they are significant to the present.

The logical extreme of the historical point of view is to be found in Leighly's suggestion that the cultural geographer should follow the trail of the cultural historian. "This above all that one emphasizes the essential time bond of culture rather than its looser place bond" [220, 135]. Pfeifer's comment that "thereby the antithesis of environmentalism has been reached" [109, 119] no doubt correctly interprets the course of thought that led to Leighly's statement, but is it not at the same time, one must ask, the antithesis of geography?

The historical method appears particularly paradoxical in studies that claim to limit geography to visible or perceptible objects. Insofar as this concept has any merit, that would seem to be completely lost in studies in which the major part of the work consists of descriptions of things the author ascertained only from historical records, on the basis of which he ultimately emerges with the final product—namely that which he could see from the start. Indeed, as Hettner pointed out, the majority of facts that the geographer can directly observe are by nature static, the processes studied in a "dynamic" regional study must be deduced hypothetically from the static facts—or taken down from historical records [171, 5; 126, 556].

Following a similar line of reasoning, Finch recommends that the conditions of past periods should be drawn upon only where necessary to interpret present features that represent relict forms of the past [cf. the discussion of "sequent occupance" by Whittlesey, 282, and R. E. Dodge, 296, 233 ff.].

---

[60] We use this study because it is the only complete illustration found in English of the principle involved. That it is not to be taken as representative of the present views of the author may be inferred from his paper read before the recent International Congress, to which we have previously referred.

This may require the reconstruction of the geography of past periods, but such descriptions, while not extraneous, are but supplementary, "in a sense, they are asides." Far from commencing with the study of the earliest periods, Finch would commence with that aspect of the area on which the writer can be positive, the present scene [*288,* 119 f.; note also the comment of S. N. Dicken, 120 f.]. Much the same view has long been held by such students as Hettner [*126,* 556], Hassinger [*165,* 13], and Unstead [*193*], to mention but a few.

In opposition to these practical arguments is a logical argument, based on a particular concept of geography, that appears to require the geographer to study the origin and development of an area in the time sequence. If the particular objects of the science of geography are the concrete "landscapes" of the world, then geography must study these objects, not only in terms of their present form and function, but also in terms of their genesis—their origin and evolution. Though a consideration of the concept itself must be reserved for a later section, we may consider here its relation to the present problem. To be fully consistent, the study of the origin of any landscape would apparently require us to study the origin of the continents, if not of the earth itself. Sauer, as we noted, avoids this by somewhat arbitrarily cutting the sequence at the point where man entered the scene, assigning to the geologist the study of the development of the "natural landscape." (This would require more than a geologist, and the total of studies by representatives from geology, climatology, plant ecology, etc., would not provide a study of the development of the "natural landscape"—as Sauer uses the term. Someone else must put together the results of the work of these students.) The geographer is to study the development or the evolution of the "natural landscape" into the present "cultural landscape." This expression has been repeated so often, on both sides of the Atlantic, that one tends to read it without questioning its import. How does a "landscape"— by whatever definition—develop or evolve? The frequent analogy of "landscape" or region with an organism suggests, as the word "evolve" itself does, that the "landscape" is something that develops or grows as a unit from forces within itself. If, however, one clears one's mind of misleading analogies and dubious concepts and considers the question in straightforward terms one sees at once that the process of change that we may observe in either the "landscape" or the region is very different from the process of organic development or growth. This conclusion may seem so obvious to the reader as to require no demonstration, but the fact that it is so often ignored requires us to demonstrate it in some detail.

If the term "landscape" is used to mean something less than all that is in the area, in particular as exclusively the material features in the area, then it would be painfully ridiculous to say that a "landscape" sheds its forests, produces ploughed fields, or grows houses and factories. These elements of the "cultural landscape" are produced in the "natural landscape" by forces that by definition, are not part of the "landscape"—namely the forces directed by human will. In other words, though we might say that the earth surface as a whole has evolved cultural landscapes out of natural landscapes, the "landscape" as such—either the real visible landscape surface, or the collection of material objects of the earth's surface—has not *developed* but has been *produced* by factors not included in it.

There are, however, several critically important respects in which the processes of change in areas differ from those which we associate with organic development. In his discussion with Dickinson, Crowe observes that the changes that take place in cultural forms and human societies have "none of the inner necessity characteristic of organic evolution. The generation of new forms is open to human choice and initiative, it is a matter of Invention and Organization. . . . Close analogy with living matter, or even with geomorphology, is on the highway to falsehood [*202*, 15; Dickinson has indicated agreement with this conclusion]. More important, perhaps, is the objection raised by Creutzburg, that there is no unity of growth in the development of an area (or of a landscape), there is only a summation of partially related, but partially independent changes within it. The forces producing these changes—both the will and energy of the individual inhabitants and the changes in individual natural elements—do not form a unit force, but merely a summation of more or less independent, often conflicting, forces [*248*, 413]. Broek therefore rejected the term "development" and referred instead simply to "changes" in an area or landscape [*333*, 10].

The previous objections apply to the consideration of the entire area of the earth's surface. When we consider a particular areal unit, or "individual landscape," however determined, there is the further major objection that the changes within it are not necessarily produced by its own forces, but may be caused by forces entering from other areas. The landscape of the Lower Mississippi flood-lands was in part produced—and may at any time be altered by—conditions in the Northern Appalachians. Bluntschli has described how the "harmonious landscape" of the Amazon lowlands has been disrupted by forces and materials introduced from lands thousands of miles away. Indeed Schlüter has argued that most changes in areas are caused by the movement of peoples from other areas [*131*, 507 ff.]. To compare such changes with the development or growth of an organism, or even with the inanimate development of crystals or of a residual soil, is to mislead ourselves.

Likewise it would be misleading to suppose that these changes introduced from the outside were controlled by some unifying force, either within or without the area. The natural landscape is not converted into a cultural landscape as an artist applies extraneous materials to a canvas to "develop" a unit picture, for the changes are not the work of one artist, or of an organized group of artists, but merely of a collection of somewhat independent natural and human forces.

The various forces that alter the landscape of an area, whether they are internal or external, recognize no common limits to the area. It follows therefore that, in whatever manner we may consider a particular area as a definite unit, that unity can be established only as of a given time. A unit area today was probably not a unit area in an earlier period, and its present unity may be disrupted tomorrow by changes that may so alter portions of it as to include them in a neighboring areal unit. Any study of the development of the cultural landscape of an area, Broek therefore concludes, is legitimate only if we remember that the area considered through a sequence of periods is an arbitrary unit. Whatever interest there may be in studying the combination of processes of changes in an arbitrary unit of area, there can be no logical requirement that geography must make such studies. On the contrary, as Broek emphasizes, this thesis is "in principle, a survival of the idea of environmental control." Underlying it is the assumption—commonly unmentioned, and even denied—that the unit natural area (or "landscape") will remain a unit through all changes made by man and so lead to a unit "cultural landscape." In reality, however, observation shows that "cultural landscape regions are often a combination of parts of different natural regions" and we may view "cultural landscape regions as expanding and contracting areas, perhaps even as migrating areas" [*297*, 103, 107].

From these conclusions it follows that, not only is there no logical necessity for the student of a region to examine each of various stages of "development" of the area, but also that the presentation of such a sequence of historical cross-sections will not necessarily provide the best means of understanding the present situation. Since whatever degree of unity involved is to be found only in the present area, we may more readily arrive at an interpretation of the present by analyzing the features found in it, utilizing the facts of the past only to explain, for each of the present features separately, the factors that produced it.

Whereas most supporters of the "dynamic" method of regional study defend it as a means to a proper understanding of present landscapes, Stanley Dodge inverts the argument and denies scientific standing to the study of the present alone. "Much of the writing of geography treats of regions as static, as *being,* whereas it is the *becoming* which is important. What processes have shaped regions, what processes are continuing the elusive transformations, and what are the trends for the future—these are the questions which excite the field geographers" [*342,* 335]. These questions, Dodge adds in correspondence, present real problems to the geographer; merely to learn what a region is like, is no problem worthy of a scientist.

The particular problems that may excite the individual field geographer are undoubtedly of first importance to him, but that is a question of subjective reaction varying for different individuals.   If, however, one examines the question of what "is important" objectively, one must ask what importance *becoming* can have, if the state of *being* is unimportant.   In any branch of science, the study of *becoming* is significant in order to explain that which is (or was) ; it is therefore of secondary importance.   Even in that field *
which seems to be largely concerned with the processes of development, history, the purpose, as Kroeber shows, is not to be found in the sequence of changes in time; those are studied in order to explain the integration of phenomena of any particular time [*116,* 545 f.].

We may say categorically that *what is* in a region, its *being,* is important if anything is important.   It is true that this does not present the geographer with a single clear-cut "problem," such as those which the students of the systematic sciences find in examining individual types of relationships, but rather a much more complex general problem corresponding to the problem of the historian in interpreting the associated phenomena, say of English History in the reign of Elizabeth.   To comprehend fully the present geography of a region—that is to know the facts and be able to interpret them— is by no means too small a problem for any geographer, nor need we be concerned lest we shall shortly exhaust our field of problems.   "There is no region of the world," Penck notes, "whose geographical investigation can be considered as closed"; though a generation has passed since Penck made this statement, his further comments on the continuous growth of new problems indicate that he would repeat it today [*128,* 59].   Individual students will of course study whatever problems interest them, but there is no need for geography as a whole to look for new problems, as long as it has the present world before it in all its manifold and complex diversity.

We may conclude therefore that while the interpretation of individual features in the geography of a region will often require the student to reach back into the geography of past periods, it is not necessary that the geography of a region be studied in terms of historical development.   On the contrary, unless the student undertakes the ambitious task of presenting simultaneously both the history and the geography of a region he will need to distinguish

[183]

* In the parentheses, read; (or was, or is to be); "secondary" is not to be considered as implying "minor", but "subordinate". See also Supplementary Note 30

clearly between the two points of view—that which studies the associations of phenomena in terms of time, and that which studies their associations in terms of place. If the chorological manner of viewing things is the guiding principle of geography, the chronological organization of a study, even though of a "geographic region," is evidence of the essentially historical character of the work.

The relation of geography to the factor of time, Hettner concludes, is not that of a contrast with geology, according to which geography is limited to human times, historic times, or even to the present as compared with the past. Rather, geography is a field for which "time in general steps into the background." In contrast with all the historical sciences, including historical geology, "geography does not follow the course of time as such—to be sure this methodological rule is still often overlooked—but lays a limited cross-section through a particular point of time and draws on temporal development only to explain the situation in the time chosen." This one particular time must be short enough so that within it there are no major changes disrupting the cross-section. To be sure, the interpretation of the present regional geography will often require a consideration of past conditions. Geography requires the genetic concept but it may not become history [*126, 556*, or *161, 131* f.]. *

C. HISTORICAL GEOGRAPHY

A co-ordinate combination of the two different points of view under which phenomena may be integrated—the geographic and the historical—is not geography or history but both. Whether or not such a combination is humanly possible we need not further discuss, at least until someone has demonstrated its possibility in practice. On the other hand, the fact that time, in itself, is not a factor in geography, that geography studies the integration of phenomena in areas under the assumption of fixed time, does not limit geography to the present. It may strike a cross-section through reality at any point of time. Whereas we commonly assume that the term "geography," when used without qualification, refers to a cross-section through the present, we may use the term "historical geography" for an exactly similar cross-section through any previous point of time.

Historical geography, therefore, is not a branch of geography, comparable to economic or political geography. Neither is it the geography of history, nor the history of geography.[61] It is rather another geography,

---

[61] British geographers have occasionally used the term "historical geography" to refer to the history of geography as a science. Obviously every science has its own, one might say, private history; though it is most likely to be studied by workers in the

* The last sentence should be in quatation marks as a direct quotation from Hettner. See also Supplementary Note 31

complete in itself, with all its branches. Consequently, Hettner observes, "an historical geography of any region is theoretically possible for every period of its history and is to be written separately for each period; there is not one but many historical geographies" [*161*, 151].

In other words the significance of the word "historical" in this term is not to imply a direct connection with the field of history, but is used in the sense of "historical" as meaning that of the past. There is involved, to be sure, a practical relation to the field of history, in that the data for a study in historical geography must be obtained in large part from "historical" materials. Further, the historian of a past period may be interested in the historical geography of that period, insofar as knowledge of that is necessary for an understanding of its history, but this is no different from the interest which an historian of the present period may have in present geography.

We may make the parallel situation of the two fields complete by saying that, while history is concerned with the integration of phenomena in periods of time, it must recognize more or less separate histories for each major area of the world. Likewise geography, integrating phenomena in areas, recognizes separate geographies of each period of time. Geography, in the sense of present-day geography, therefore corresponds to the history of one's own country; historical (past) geographies, to the histories of other countries. A practical difference results from the fact that, while the study of foreign histories permits the use of the same methods as the history of one's own country (except for differences in language), the study of past geographies does not permit of the direct field observations of present geography. Further, there is a difference in interest: for reasons that need not here concern us, the people of any time and place have a greater intellectual interest in the history of other countries—both present and past—than in the past geography of their own country.

That historical geography is to be considered simply as the geography of past periods is a view on which there is perhaps more agreement among geographers than on almost any other question of definition in our field. It was stated early by Kant, in the distinction between "the geography of the present" and "old geography" [*40*, § 4], and later by Marthe [*70*, 453] and Wimmer [*74*, 10]. It has been most clearly expounded by Hettner, briefly in his first statement of 1895, and more fully some years later [*121*, 7; *126*, 563–4, repeated in *161*, 150–1]. Similar views have been stated since by so many geographers of different schools in various countries that we may be justified in considering it as established in geographic thought. In this

particular field, it is logically simply a specialized form of history. Of American geographers, John K. Wright has been most interested in this subject [*9*].

country, in particular, the nature of this field and the methods by which it may be studied have recently been discussed and illustrated with a detailed example, by Ralph H. Brown [334]. (For similar expressions concerning historical geography, see Hassinger [165, 14 f.], Brunhes [83, 100 f.], Unstead [193], the group of British geographers and historians reported in [197], Gilbert [198], Barrows [208, 11 f.], Sauer [84, 200], and others noted in Hartshorne [216, 790]. Vogel discusses the German literature in this field, with extensive bibliography [89].) *

Hettner observed, and Brown has demonstrated, that it is not necessary for historical geography to begin with the earliest period in the history of a region. As geography, it does not begin, proceed and end, according to a time sequence, but rather considers the geography of any period the writer desires, exactly as one would consider the present geography [Brown, 334].

Nevertheless most students in this field, under the influence of the historical point of view, have felt it necessary first to attempt to reconstruct the pre-historical landscape and then those of later periods. As Brown has pointed out, however, there are very few areas of the world for which we have anything like reliable and adequate data either for the pre-historical landscape or for those of early periods of history. The imaginative reconstructions which can be derived from a consideration of the present cultural landscape and from the meager and highly unreliable descriptions of early explorers and settlers may be very interesting, but are of dubious scientific value. Sauer's suggestion that "the major datum line from which human 'deformations' are to be measured" is to be attained from "a reconstruction of the natural 'original' vegetation," i.e., that existing "before man entered on the scene"—whenever that may have been—is simply to recommend the unattainable. Indeed more often we must arrive at the "datum line" by a consideration of the historical record of the deformations.

In order, therefore, to work with more reliable materials, Brown deliberately ignored earlier periods for which there is little reliable material, and selected the first period in the history of his area for which he found current material of sufficient geographical quality and reliability to justify a scientific reconstruction of the geography of the time.

It is true that Brown's method requires him to depend largely on historical sources, i.e., the records of the past, and to some extent therefore to utilize historical methods. This would appear to justify Hettner's conclusion that "the cultivation of historical geography should lie mostly in the hands of historians" [161, 151]. But Brown has demonstrated that to check and interpret the historical records of geographical material requires the knowledge and abilities of the trained geographer. He has illustrated Kant's

[186]

* The list might well have included E.G.R. Taylor [339, preface]; Mackinder [426]; and East [427]; but should not have included the reference to Sauer.

observation that a knowledge of the geography of the present provides the means by which historical records may be used to interpret the geography of the past [40, § 4]. In doing this he has not followed the historical development as a purpose in itself, but rather has used that for the purposes of geography [cf. Hassinger, 165, 14 f.]. Consequently his production—the ultimate test—is essentially different in character from that which any historian could or would have produced [334].

There is a further consideration that might move the regional geographer to make a detailed study of the historical geography of his area, at one or more past periods, before presenting his study of the present. The geography of certain areas is so different from what one would expect from present conditions and has perhaps been determined very largely by the geography of some particular period of the past—rather than by a consistent development of fairly permanent conditions—that the present can most readily be understood if the historical geography of that particular period is first portrayed. An urban study of Venice would offer an extreme example —a full interpretation of the geography of Venice in the fifteenth century is of first importance for an understanding of the present city. Similarly, in beginning a study of the political geography of the Mid-Danube lands in 1937, the writer found that so much of the situation of that time (the same would still apply) had been conditioned by the political geography of the nineteenth century that it was desirable to present first a fairly detailed study of the political geography of the pre-war Austro-Hungarian empire [358]. There seemed no such need, however, to present in that way the situation of earlier periods, though the detailed analysis required, at certain points, an examination of particular features of much earlier centuries.

There remains, finally, a form of study in historical geography which much more closely approaches history. If a geographer studies the historical geographies of a single region—or of a part of a region—at two or more different times, he will of course be concerned with the differences among the several pictures. This, I take it, is the type of study which particularly interests S. D. Dodge, and, though marked by a strong historical bent—the study of changes—it would appear to have its place as a form of comparative historical geography. (Compare also Volz' concept of historical rhythm in geography [243], and the discussion by Bürger [11, 103 f.].)

Since the comparison is that of the geography as found in cross-sections * through several periods of time, such a study would presumably be organized with major divisions for each of those cross-sections, but within each of these major divisions a geographical study will not follow the chronological time-table but will be organized chorologically. A specific example is offered

* See Supplementary Note 32

by Robert B. Hall's study of the Tokkaido road and region, in Japan, which presents the geography of that section of land at three different periods of time [*350*]. Sölch has commented on the contrast, between studies designed to present a temporal cross-section of geography in the past and studies focussed on the historical events and economic development that have changed areas, that is to be found in the chapters, written by different authors, in the "Historical Geography of England before 1800," recently published by Darby [*339*, review].

Theoretically one might construct an unlimited number of separate historical geographies of any region, and if these could be compared in rapid sequence one would have a motion picture of the geography of an area from the earliest times to the present. In practice however this is utterly impossible—hence indeed the separation of history and geography. We can either compare a relatively small number of cross-sections of the regional geography of past periods, or we can follow the detailed process of development of particular individual features. But the study of the development of any particular feature is a part of the systematic study of that type of phenomena; depending upon the point of view under which the process is studied, it is either a study in the systematic science concerned with that type of phenomenon, or it is a study in systematic history.

A complete combination of history and geography, in conclusion, would represent the complete integration of the areal and temporal variations of the world—it would be reality. But we have no conceivable way of constructing any form of motion picture presenting simultaneously all the changing features of even one region, even in outline. We are forced to distinguish between an historical and a geographical point of view, and in order to master the technique of either, we need to keep clearly in mind the distinction between the two.

VII. The Limitation of the Phenomena of Geography to Things Perceived by the Senses [*]

A. SIGNIFICANCE IN CURRENT THOUGHT

The question as to what phenomena in any area are proper grist for the geographer's mill has always been a difficult one. Those who have understood and accepted the concept of geography as—like history—an integrating rather than systematic field, recognize that there is no escape from the multiplicity of diverse phenomena which combine to form the character of any region. While any of these phenomena may form, in themselves, proper material for the systematic studies of any of the natural or social sciences, any of them may also be appropriately studied by the historian insofar as they show differences significant to his problem of integrating differences with reference to time. Similarly they are appropriate to the geographer, regardless of their intrinsic character, if they show differences—in themselves or in their relations to other phenomena—significant to his problem of integrating phenomena as found in space.

On the other hand the smaller group who have taken their lead from Schlüter, Brunhes, Passarge, and Sauer would limit the geographer's examination of a region to material features, both natural and cultural, excluding any non-material cultural features. Geography then includes, as it always has, man and his works, but only his material works—commonly those that are visible. Although the limitation may appear arbitrary, the motivation underlying it commands our respect. As stated by one of its first exponents, Schlüter, what is needed in geography is "limitation in the subject-matter and objectivity in the observation" [127].

In contrast with most of the methodological problems current in geography, which we have already found were repeatedly discussed in earlier periods, the present issue appears to be relatively new. Both Humboldt and Ritter, as well as their predecessors, and their followers for a generation or more, considered without hesitation many kinds of phenomena that this limitation would exclude from the field of geographic study. During the past thirty years, however, the question has been the subject of a great number of discussions in the German literature. More recently it has been introduced into this country, and, though largely confined to oral discussions and textbooks, has had, as we shall observe, significant effects on the character of geographical research. Consequently it becomes necessary to subject it to detailed examination.

It is common practice to assert this limitation without going to the trouble

* See Supplementary Note 33

of providing any justification for it. Many have been led to believe that it represented the point of view of the entire modern movement in Germany, including those representatives like Hettner, who in fact have most vigorously opposed it. Sauer, writing on geography as a social science, in the Encyclopedia of Social Sciences [85], contrasts two groups of geographers: one group is concerned with the study of relationships, which he calls "Human Geography," while "the other group directs its attention to those elements of material culture (later referred to as 'visible') which give character to area."[62]

In reality the situation in German geography is very nearly the reverse of what American geographers have been led to suppose—not with respect to the chorographical concept itself, of course, but with respect to this particular limitation of the phenomena. Though Schlüter suggested the limitation as early as 1899 [122, 65] and presented the case for it more fully in his celebrated paper of 1906 [127] he recognized in 1920 that it had "found little attention and still less approval" [148, 213] and eight years later he was no more encouraged [247, 391]. Likewise Passarge, who has followed him in certain respects, has frequently given expression to his wrath at critics who, though praising his productive work, fail to accept—worse yet, even ignore—his methodological views [Waibel, 250, 475]. The only other German writers on methodology who appear to support this limitation are Tiessen and Penck, to whom should be added Granö, Brunhes, and Michotte in other European countries. Other writers either ignore the suggestion entirely or have expressed their objections to the thesis directly. Among the latter may be listed: Hettner, Philippson, Supan, Oberhummer, Sapper, Gradmann, Hassert, Heiderich, Friedrichsen,[63] Volz, Lautensach, Hassinger,

---

[62] The student, particularly the non-geographer, who refers to the entire section on aspects of geography in the Encyclopedia [85, 86, 87] may well receive an impression of confusion even greater than actually exists in the subject. One of the principal references quoted by Sauer in support of his thesis is from Vallaux who, however, appears as the author of the following section on "Human Geography," which had been defined by Sauer as representing the opposite point of view. Further the inclusion of this reference would appear to include Vallaux among those who limit geography to material facts, which is far from the case [cf. 186, 84, not to mention Vallaux' work in political geography]. Incidentally it seems strange to use the term "Cultural Geography" for a form of geography from which a large part of the cultural aspects have been excluded.

[63] Sauer cites [84, 188] an article by Friedrichsen as illustrative of the position of Schlüter and Michotte in contrast with that of Hettner. Friedrichsen expresses himself in that article as in general agreement with Hettner in finding Schlüter's point of view "zu eng und praktisch schwer durchführbar," and elaborates the objection at some length [230, 159 f.]; another article by Friedrichsen, to which Sauer refers, could not be found in the publication to which reference is given.

Banse, Maull, Sölch, Waibel, and Bürger, to whom we may add the two philosophers who have most carefully considered our field, Graf and Kraft, and the Italian geographer, Almagià. Finally if we were to subtract from the list of supporters of this limitation those who have repeatedly ignored it in practice, none would be left except (presumably) Granö and Michotte.

In view of these facts we might be justified in discussing this thesis as of but little importance in current geographic thought. On the other hand a small number of American geographers have repeated the statement of this limitation so persistently that it appears to have become intrenched in current thought. Objections that have been raised to it in connection with political geography [cf. 216, 800–4], or references to Hettner's attack upon it, have brought no response other than mere repetition of the thesis as decreed by higher authority. Consequently it seems necessary to examine this thesis thoroughly from the point of view of the field of geography as a whole.

The restriction itself is somewhat difficult to examine because of the lack of agreement among its various proponents as to just what is to be excluded. The original conception appears to have been an outgrowth of the meaning of *Landschaft* as "landscape," but, as we saw, this term has a somewhat different meaning for almost every individual geographer. In theoretical statements most of the "landscape purists"—to use the convenient term that Dickinson suggests for those who limit the objects of geography in this way [202]—limit geography to objects that are *sinnlich wahrnehmbar*, or perceptible to the senses, whether those of sight, sound, smell, or feeling, Granö therefore protests that Hettner is not meeting the issue when he compares the concept of the "character" (*Wesen*) of an area simply with the form or picture (*Bild*) of the area [252, 177]. But if Hettner's argument does not apply to the generalized statement of the landscape purists, it does apply to the point of view that their detailed discussions present. For in these, they speak almost exclusively of the visible aspect of objects, and when they consider them together in an area, it is clearly the *Landschaftsbild,* the visible landscape, that they have in mind. To be sure, any material objects not included on this basis, such as climatic conditions, are added on the grounds of their perception by other senses, but the impossibility of combining these things under a single unit concept is reflected by the fact that none of these writers have discovered any way of expressing this total other than by the term they have created for it, or by such loose expressions as "in general, visible," or, finally, by simply saying "area" and later telling us that "area" does not include everything in the area. Even Granö himself, who has attempted to work out this concept more logically than any other student, is in practice controlled by the visibility of objects. Thus he finds the color of clothing more important than its form, and men, in general, form a part of the *Landschaft* where they appear in masses that constitute a visible form [93]. These conclusions are logical only if we are to determine what is significant in geography on the basis of how much impression they make on our senses—in which case, as Granö properly concludes, everything beyond our immediate vicinity is considered only in terms of visibility; but this is simply the *Landschaftsbild*. If the essential point is simply that the objects be somehow directly perceptible to any one of the senses, or more simply, must be material objects, how can we say that form and texture of clothing is less important than color? Are men in the mass any more material than when they are scattered?

Granö would apparently answer this question by means of an additional limitation, which was originally stated by Schlüter and has been accepted by Sauer [*85*] among others, but not by all of the landscape purists. According to Schlüter, the geographer's attention should be directed toward "those expressive features that have areal extent, as distinct from individual objects" [*148*, 148, 152]. On this basis he would include settlement forms but not house types [216], whereas the followers of Brunhes and Passarge have been stimulated to give great attention to just that feature. Schlüter recognizes that the same principle reduces men to "minute grains in the landscape," and all of these students have had difficulty in defining the proper place for man in geography. For Schlüter the problem is solved by the fact that our special interest in man acts as a magnifying glass, and thereby human geography is made of equal importance with physical geography. On the other hand, the magnification apparently must not be sufficient to permit us to observe the color of men's skin, nor may our ears observe the language he speaks [*127*, 17, 28–29, and especially 41].

Though Granö's conclusion appears more consistent, we may question whether it is actually as logical as it seems. If men in the mass form a phenomenon that is physically perceptible and covers a certain areal extent, both of these attributes are nothing more than the arithmetic sum of the perceptibility and areal extent of each of the individuals composing it. Scatter them as you will, they still cover, in total, exactly the same amount of area, and if we approach each and every one separately they are all material objects of which our total perceptions may be no less than our perception of them in mass. It is only on the basis of the part they form in the visible landscape sensation—our psychic response—that there is any difference.

For this reason, perhaps, Passarge finds no difficulty in including man in the description of a *Landschaft*. "Farmers plowing—perhaps in colorful costume—shepherds singing, merry vintagers or reapers," all these may be included in the form of the *Landschaft* [*268, 74*]. On the other hand, he bewilders his readers completely by asserting that men as such do not play any part in forming the *Landschaft* (*nicht Landschaftsbildner*) and must not be used to delimit the *Landschaft* areas [*268, 1*]. Somewhat different statements in earlier publications are even less clear, so that it is not surprising if Gradmann and Waibel should have interpreted his statements differently from the way in which they are interpreted by himself and his students [*236, 331–37; 250, 477,* reply, *166–68*].

Nevertheless, difficulties of this sort should not by themselves cause us to dispense with a concept which might otherwise be valuable. But certainly before we become involved in what would evidently be a very difficult problem of definition, we will wish to test the concept in general by three tests of more fundamental character. Is it logically founded on general principles of science or of geography as a chorographic science? Does it represent an historically consistent development in the field of geography? Does it provide a basis on which general agreement may be possible in order to restrict a field now too wide to be effectively studied by a single group of students? Let us examine the general concept in the light of each of these questions.

B. IS THE LIMITATION LOGICALLY FOUNDED?

None of the American geographers who recommend, or even insist upon, this limitation has presented a thorough consideration in which we might examine its logical foundation. Not only in presenting their own work, but in some cases also in criticising the work of others, they have contented themselves with categorical statements for which no foundation is offered other than the authority of various geographers, here and abroad. Examination of the writings of these authorities fails to reveal any logical defence of the limitation. Consequently in order to examine the reasoning underlying their thesis—for we presume that it must have some rational basis— we are limited to scattered arguments such as those brought out in discussions in the several symposia held by this Association in recent years. Though the statements were in many cases made extemporaneously, their authors had the opportunity to check them before publication. In any case, it will not be the purpose to catch them up on any unfortunate phrases, but to endeavor to understand the thought they wished to present.

In one of these symposia, both Trewartha and Finch, speaking separately, emphasized a distinction between "observable features" of area and other features that presumably are not observable. The geographer is concerned directly with the former, these are "primary geographic facts"; he is only indirectly concerned with the latter, "secondary geographic facts," to which Finch also referred as "deduced facts required for explanation of the observable features" [*287,* 111; *288,* 117]. These statements may well appear to express a fundamental distinction in the logic of scientific reasoning, but do they express it correctly?

Unquestionably we must distinguish clearly between facts as facts—*i.e.,* as observ*ed* phenomena, "the primary data of all thought . . . not to be derived from anything else"—and what we presume to be facts, on the basis of reasoning [Barry, *114,* 92]. Thus, in the example cited by Finch, if the presence of hard and soft rocks in the Jura had been observed directly by examination of the rocks themselves, it would be a primary fact; if however it was deduced from the form of the ranges, it is a fact of secondary quality, even though it is presumably "observ*able*."

It might be objected that only material facts can be directly observed and that therefore the difference in kinds of objects leads to the logical distinction in kinds of facts. But our "observation" of material facts is by no means limited to direct means; on the contrary the more exact and certain methods of determination are often indirect—notably in the most exact sciences, such as astronomy and physics. That the distinction between indirect observation and deduction from reasoning is not always a sharp one need not concern us here. It is necessary only to recognize that such facts as the following are sufficiently established by indirect methods of observation to be recognized as primary facts in any science concerned with them: that the United States has a republican form of government rather than a monarchical form, that many of the inhabitants of our South before 1861 were slaves, or that the people of Arabia are largely Mohammedan in religion.

Consequently we conclude that the sound logical distinction between primary and secondary facts is irrelevant to the difference between material and immaterial facts in geography. Our knowledge of natural vegetation in many areas of the world must be classified as of secondary quality— deduced from primary facts of soils, climate, vestigial remnants, etc.— whereas our knowledge of the distribution of peoples of different languages and customs consists of primary facts. In other words, as Hettner has observed, all facts—in the proper restricted sense of primary data—whether material or immaterial, must be perceived in one way or another by our senses to be recognized as facts [*126,* 555]. The term "observable" or the alternative, "perceptible to the senses," provides therefore no specific criteria for geography.

It is significant that while the proponents of this limitation are ultimately forced in discussion to define that which is to be studied simply as material objects, they persistently retain the concept of "observable." Thus Finch and Trewartha say "observable material features" [*322*, 5]. Apparently the redundancy is employed in order to emphasize that only those things are to be studied that are observable directly—and these are material things. That a science of geography should limit its direct consideration to these objects is defended, in the various arguments of the landscape purists, on either or both of two different assumptions, one concerned with the nature of science in general, the other concerned with the nature of geography as an "observational science."

The first of the assumptions is most clearly presented by Granö [*252*, 38 f.]. Although he recognizes that the material and immaterial phenomena of an area form a unity (*Einheit*) he insists that geography must consider directly only the material facts and leave the immaterial to sociology. By limiting ourselves in this way "we can comprehend and describe our objects naturally in the way that a normal (*gesund*) person sees them." This he claims to be an essential characteristic of science in general. In other words, geography must be made to conform to an assumption of science that excludes the social studies. In general, we may note that this argument, whether stated or not, underlies most of the arguments of those who propose this limitation. Those who insist that geography must define itself in terms of geography and leave the question of whether it is a "science" in any particular sense for secondary consideration, might dismiss these arguments as irrelevant; but it is significant to note the paradox into which these arguments lead.

This paradox appeared particularly clearly in a discussion, during the symposium previously referred to, between Finch and Colby. "How an area *is*," Finch objected, "involves subjective ideas. You *see* only things which are *objective*" [italics in the original, *288*, 122]. Actually just the *reverse is the case. Any given section of the shell of the earth's surface, containing material and immaterial facts, is a piece of objective reality. The immaterial phenomena are no less objective than the material objects. It is simply a problem of method to find objective means for observing them. When drought turned the wheat fields of the Great Plains into a "dust bowl," the widespread bankruptcy and the mental anguish of the farmers were facts no less objective than the number of inches of rainfall deficit for that summer or the decrease in humus content of the soil over many years. None of these immaterial facts could be observed directly with any degree of accuracy; each could be observed with reasonable certainty by indirect meth-

* See Supplementary Note 34

ods. On the other hand, even the sober scientist may on occasion see things that are not objective—as in the case of a mirage. In any case, even if we may assume that most of the things we see are objective, we do not see them objectively; indeed one of the most fundamental axioms of any science is that "things are not what they seem."

In passing, we may note that the assumption concerning the nature of science in general does not apply even to the natural sciences, not even to the most exact physical sciences. The latter are not concerned merely with the study of objects but also, and directly, with the study of forces which cannot be seen or observed directly by any sensual perceptions. Physicists did not delay their study of X-rays until they had convinced themselves that those were composed of particles of matter. If, according to tradition, it was the impact of the apple on his head that led Newton to the concept of the force of gravity, it was not that force itself that he observed by his sense of feeling, but simply the moving apple. Physics studies gravity, that is, only by indirect means of observation; it is not a material thing.

One may ask why similar examples are not equally obvious in physical geography. Unquestionably we would have a clear example had the earth's shape been far less spheroidal. If the force of gravity varied over different parts of the earth surface to the extent of even a third or a half, physical geography could hardly have avoided recognizing this as one of the most important of all geographic factors. It would have been reflected in major differences in most of the physical objects the geographer would study, but could never be observed directly; it would not be a material object. Any interested in following this thought might consider whether a wind is a material object, directly observable—whether it is not rather a phenomenon of motion of air particles, so that we do not actually feel the phenomenon of motion, but only the impact of the air particles.

Since these latter considerations, however sound in theory, may seem abstruse, we will attach no great importance to them. The essential conclusion is that under the broad view of science as the pursuit of knowledge of reality by objective means, science, in general, is not required to limit itself to any particular category of phenomena. The business of the scientist is to pursue knowledge wherever objective means of study permit, regardless of whether they be direct or indirect; if immaterial phenomena can be determined by indirect, but objective, means of measurement, they are phenomena for some scientists to study. There remains, however, the second question—does the particular nature of geography require it to exclude immaterial phenomena?

Sauer has insisted that geography must "claim a certain field of *observation*," within which, however, "we are to observe, describe, and explain according to the best methods at hand" [*84*, 186]. An area of the earth surface is a certain field of observation, but the study of an area would logically include the study of all significant phenomena in it, of whatever kind, observed by the best means at hand. Though the visible landscape, in the very limited sense in which that constitutes a reality—the external surface—might be considered as a field of observation, neither Sauer nor any other students wish to limit geography to that extent—for very good reasons, as we shall see later. But the total of material objects selected out of the earth surface cannot, as we found, be properly regarded as *a* field of observation. Robbed of the connecting element of immaterial phenomena, it breaks down into as many different fields of observation as there are different categories of material objects represented. To give this selection of heterogeneous objects an arbitrary single name, "landscape," does not make them into a single field of observation but merely obscures from us the fact that they are not that. This is further attested to, not only by the absence of any word—whether in English or German—that represents in common thought this total collection of material objects, but also by the inability of the students concerned to agree even approximately on what the concept they are trying to express includes.

Consequently we find that the supporters of this limitation fall back on the argument that the particular science of geography is a science in which phenomena are studied by a particular method or technique, and therefore it selects from the phenomena of area those that are encompassed by its technique of observation. Since Granö appears to be the only geographer who has ever actually used his ears and nose in the field, we may assume (for the sake of simplicity—it is irrelevant to the argument) that this technique is restricted to visual observation. But why, asks Hettner, should geographers be limited in their manner of perceiving things? What other science restricts itself in this way? [*167*, 279 f.]. If the astronomer should learn that the physicist had invented techniques for observing stellar changes without use of either the human eye or the telescope, must the astronomer decline to use them? Does a geologist trained in field work hesitate to use methods of microscopic chemistry to study rocks which he has brought in from the field, defining geology, so to speak, as the science of the hammer? (Hettner). Sauer claims that historians restrict themselves to written records [*84*, 185]. Though popular thought might accept such a statement, historians do not accept the limitation. Indeed, Trewartha has offered us a paradoxial study of early French settlements on the Upper Mississippi for which the geographer has studied chiefly the written records, whereas the historians investigated (in addition) the remains of ancient fireplaces [393, 188]. In general, historians do not hesitate to utilize whatever evidence of the past they can find, including oral traditions, maps, pictures, as well as ruins, tools and implements, indeed "all the material relics of human activity that dot the earth" [118, 310 f.; see also 117, 1–23; 119 and 120].

When we consider the actual techniques employed by geographers, including the landscape purists, this theoretical argument appears to be completely dissipated. It is not necessary, evidently, that the objects of study in geography should actually be observed by the senses, but merely that they should be, or have been in the past, "observable" (in this way). To take a specific case, the argument would appear to be that the written account of a French officer of the eighteenth century describing where on the Mississippi he located a particular fort, produces a "primary geographic fact" whereas his statement that his soldiers were Frenchmen and the surrounding natives were Indians would provide us only with "secondary geographic facts." To science as a whole the important distinction between these two sets of facts would be that the latter appears to be highly reliable whereas we may not be at all sure about the former [cf. 393].

Similarly, no one has suggested that the geographer's knowledge of climatic conditions should be based on sensual perceptions whether observed by himself or by others. On the contrary it is axiomatic that such observations are most unreliable. We depend exclusively on indirect methods of observation and obtain thereby data far more reliable than the results of our direct observations of any class of material phenomena in geography. Even more reliable however are our data concerning the areal extent of political sovereignty in most parts of the world.

In short, geographers have long recognized in practice that, for many of the features that all agree must be studied, we cannot depend upon any direct form of observation. On the contrary, as R. E. Dodge observed in the discussion of this question, our direct observations may be much less reliable and significant than those obtained indirectly [287, 110]. No one has yet suggested in what way these indirect methods of observing material objects are significantly different from the indirect methods of observing immaterial phenomena. Indeed Finch has recently recognized that the means by which we gain awareness of "the activities and forces of human dynamics" are "only slightly different" from direct observation. Further he states that these phenomena "are recognized by all regional geographers," but whether as primary or secondary geographic facts he does not specify [223, 14 f.].[64]

___

[64] Finch touches on the present question so little in this very illuminating methodological statement that it is not clear whether his view has changed from that of five

There remains one final logical argument which might appear to confine the attention of geography to material objects. Granö argues that material objects, in contrast to immaterial objects, have spatial extent, and that therefore they alone are significant in geography as the study of area [*270, 297*]. Similarly Michotte states categorically that "if geography is really a 'chorological' science, its object can necessarily be only material," because—as he adds in a footnote objecting to Hettner's view—"the concept of space is essentially tied to the idea of 'matter'" [*189,* 43].

Michotte's discussion of this question comes at the end of a long essay concerned primarily with more fundamental questions. Perhaps for this reason he did not examine this question, as he did the previous questions, in terms of specific examples of geographic work—the method of reasoning which he recommends in contrast to "a discussion of principles, which is the origin of errors committed in these matters by laymen and even a certain number of geographers" [23]. If we grant, as of course we must, that the concept of space must be tied to a material basis, does it follow that the differences in contents or characteristics of different spaces are limited exclusively to material phenomena? Whatever is produced or used by man, whether a house, a tool, a language, a custom, a political allegiance, or an idea, is specifically located on the earth surface at the point where he produces or uses it. That these things vary in their ease of transport is irrelevant—indeed, a tool, particularly a type of tool, or even a type of house, is, in fact, more often transported than is a language. The immaterial phenomena known to science are tied to specific physical objects, men, and can therefore be located specifically with reference to space—at the points on the earth surface occupied by the men with whom they are associated. This association, to be sure, is in many cases but transitory and the geographer, interested in the more enduring characteristics of regions, will ignore such cases for the same reason that he would ignore the crop production of an abnormal year. But it is an error to associate the attribute "transitory" with immaterial phenomena in contrast to material phenomena. The languages and customs of China are far older than the houses of its inhabitants; if the same crop, rice, is found in the fields decade after decade, that is primarily not because of any particular characteristic of the land, but rather because the custom of regarding rice as the staff of life endures under the impact of foreign ideas.

---

years ago, or whether we have misunderstood his statements of that time—in general, whether he does maintain the distinction involved in this limitation under discussion. In any case his earlier arguments are discussed here because they present so clearly the point of view held by those who do maintain the limitation.

"The proper object of geography," Michotte wrote, in the part of his essay dealing directly with human geography, "should consist in 'delimiting' and in 'describing' the various 'terrestrial spaces' characterized (caractérisés) by a particular form of settlement, by a particular form of house, etc. . . ." [29]. But if an area may be characterized by the form of houses that are most common within it, may it not also be characterized by the language of its inhabitants, or by any particular custom that is characteristic of that area in contrast to others? If, as Vidal says, the concept of country is inseparable from its inhabitants (Sec. IV A), then it involves not merely the physical facts of their distribution and physical characteristics, but any other characteristics that distinguish them as an areal group from the groups of other areas. The only objection to this argument that I find in Michotte's entire essay is the limitation, assumed without any explanation, of the "distinctive characteristics" of an area to those that meet the eye [30].

Unquestionably, extensity in space is an essential requirement of whatever is to be studied under a definition of geography based on the chorographic concept. But we must examine more closely the manner in which geographic phenomena have spatial extensity. As Humboldt and any number of others since have observed, geography is not concerned with individual plants and animals, but rather with "the plant and animal cover" of the earth's surface. Obviously the word "cover" cannot be taken literally: a plant society, much less an animal society, does not actually "cover" any area. We do not ignore the vegetation of the desert because it occupies less space than the intervening bare places. We do not hesitate to map types of house forms in rural areas, even though they occupy but a very slight fraction of the total area concerned: if every house in a particular rural area is of the same type and that is distinct from the houses of neighboring areas, geographers may recognize a geographic fact whose significance is independent of the slight space filled by the total of individual houses.

In other words the concept of spatial extensity does not require that the individual phenomena actually cover a particular area of the earth, but simply that they are generally characteristic of that extent of areal space. The Magyar language is an immaterial phenomena that is no less a characteristic of a particular area in the Middle Danube Basin, and is no less definitely tied to this particular space, than is the peculiar type of farm wells, or the giant agricultural villages.

Many of the proponents of this limitation appear to follow this line of reasoning, but unconsciously and therefore only part-way. Thus Schlüter does not limit man to those instances where he can be observed in the mass, but considers the density of human population, apparently because that

[200]

phenomenon is visible, in an abstract sense [*145, 26*]. If this is not plain hocus-pocus, it at least indicates that the principle of objective perception is not to be followed strictly.

Granö has, in fact, recognized by inference at least that non-material phenomena may have areal extent in the sense here defined, but rules them out of consideration because it would be too difficult to use them in determining the limits of regions—a problem which he regards as of primary importance [*cf.* Waibel's critique, *266,* 204]. The areal extent of many of the phenomena that he excludes—such as language, religion, or literacy—can often be determined with greater degree of certainty and accuracy than those that he regards as "objective." Of all maps of individual earth facts, the political map of organized states is the most reliable and exact. Language distribution may be difficult to map correctly, but it is far less difficult than, say, the distribution of landforms, not to speak of natural vegetation.

If geography could confine itself to non-human aspects of the earth, it would perhaps be free of the difficulty of studying immaterial phenomena. Once geographers are agreed—as in general they have always been—that they will study human or cultural geography, they are committed to the study of things cultural as well as natural. Culture is basically immaterial and manifests itself both in immaterial and in material results, both of which are subject to scientific observation. If culture can be geographically significant in its material manifestations, it would be most extraordinary that it should not be geographically significant in its more fundamental, immaterial, aspects. Unless that can be demonstrated it is essentially illogical for geography to confine itself to the material, to consider the subject culture only in terms of its material manifestations.

C. IS THE LIMITATION CONSISTENT WITH THE HISTORICAL DEVELOPMENT OF THE FIELD?

Far from being a consistent step in the development of the field, the general adoption of this restriction in geography would represent a radical departure in geographic work [Hettner, *167, 278* f.]. Branches of the field as old as geography itself, and which geographers have cultivated with especial interest, would be excluded, notably the geography of peoples and political geography. East is expressing a traditional concept in geography when he states: "The state structure of the world . . . no less than its physical structure, provides phenomena for geographical analysis, classification and interpretation. States, like physiographic regions, have their origin, histories, individualities and relationships" [*199, 270*]. Does anyone really think it is of no concern to the geographer studying Bohemia to know which parts are German in character and which are Czech, even if houses and farms look alike? One does not have any particularly nationalistic interest to feel there is something missing in Passarge's *Landschaftskunde* of a part of the South Tyrol in which he examines house architecture minutely but gives no indication as to whether the people are German or Italian [*268, 7–54*].

Undoubtedly geography has gained by the elimination of much miscellaneous material that has properly been taken over by economics, astronomy, and geophysics. But before we throw out wholesale time-honored sections of the field it would be well to consider certain of the consequences.

In the first place, who is ready to take over these fields which we are to disown, who will care for their proper cultivation, or are they merely to be dropped in the void between narrowly defined sciences? In the case of political geography, for example, I have elsewhere shown that there is no reason to suppose that political scientists are prepared, either by their interest or their training, to take over that part of the field that geographers have long cultivated [*216*]. On the other hand, when students like Granö inform us that economic geography as well as political geography can develop just as well outside of geography proper, one may be permitted to ask what entitles them to speak for those fields. If political geographers were demanding their independence, the question would assume a different aspect; but so far as I know, no student of political geography has suggested that his subject should be given independence. Likewise the situation would be different if political geographers were seeking admission into the field. On the contrary, political geography has been a part of geography as long as geography has been studied. As Penck at least has recognized, the new concept produces a "new geography" different from that of the classical geography [*163, 53*]. The issue, therefore, is whether the concept of geography may be changed to exclude branches of the field in spite of the objections of those most familiar with those branches.

From the point of view of those who may be interested in political geography it is no mere academic question which we are here raising. Most of its principal students have recognized that a continued sound development of political geography is dependent on the continuance of its two-thousand-year-old position as an intimate part of geography. Most of the fundamentally valuable work in the subject has been done by students who were primarily geographers, such as Ratzel, Vallaux, Sieger, Sölch, and Maull, as well as Penck, Supan, Vidal de la Blache, and Bowman [specific references in *216;* to these should be added Vidal's study of the states and nations of Europe, 79]. Furthermore, a whole host of geographers have made, and should continue to make, excellent studies in political geography as part of the general work in geography, whether these be published separately, or simply as essential parts of their regional studies. A full list of those who have made studies in regional political geography would include most of the geographers of Germany and France. That American geographers should have less reason to make such studies is natural since their special field of interest in North America presents few such obvious territorial problems. Fortunately those who specialize in Hispanic America have not hesitated to treat the geographic problems involved in such disputes.[65] All these studies provide valuable material for the student who wishes to specialize in political geography because they are based on more intimate knowledge of certain regions than he could hope to attain.

If now we consider the possibility of abandoning these special aspects of cultural geography—including the geography of peoples as well as political geography—from the point of view of the general student of geography, is it certain that our field would lose only outlying portions and not, perhaps, elements which play a direct role in its central core? Both Hassinger [264] and Whittlesey [289] have shown how a consciousness of political geography can enrich the work of the general regional geographer. For many regions of Europe, the student attempting general regional study will be seriously handicapped if unfamiliar with the principles of political geography and the geography of peoples.

Finally it is only fair to point out the contribution which political geography has made toward a clearer understanding of certain problems of geographic theory. This may seem to be a strange claim to make for that branch of the field which is most commonly considered as the "wayward child of the geographical family"—to repeat an expression of Sauer's [84, 207] which appears to have made a lasting impression. Undoubtedly the nature of the material in this field, as well as the extraneous influence of national interests, exposes its workers to the danger of loose and tendencious reasoning, but it is evident that other branches of the subject could offer keen competition in this direction. On the other hand, the obvious difficulties to clear thinking may stimulate the student to examine his assumptions and arguments more critically than in other fields. Further the fact that the conclusions may have practical importance—in the political world—and that the conclusions of students of different countries may come into conflict, may lead to a sharpening of the line of reasoning. Finally, it may be that the

---

[65] An excellent example by a student of regional geography who has familiarized himself with the concepts and techniques of political geography is Robert S. Platt's "Conflicting Territorial Claims in the Upper Amazon" [380].

oft-repeated challenge from other geographers to justify the field of political geography has required its workers to consider at greater length than other students, the theoretical problems of the entire field.

Whatever the reasons may be, some of the sharpest and soundest considerations of geographic theory have developed from the study of political geography. Particular mention may be made of Sieger's analytical study of the nature of boundaries [227], and Sölch's study dealing both with that subject and with the essential character of regions [237]—a work of the first order to which repeated reference is made in this paper.

Geographers who contemplate casting overboard whole divisions of their subject might well recall the period in which all of human geography was ignored by many geographers and consider the unbalanced situation to which geography was thus brought. Where can we find a systematic treatment of the cultural elements of geography comparable, say, to that of the natural elements which forms the great body of Finch and Trewartha's volume [322, 10–602]? The contrast results not merely, as these authors suggest, from the greater complexity of cultural features and the failure of the specialized social sciences to provide us with ready-made classifications comparable with those of specialized earth sciences [605–6]. In fact these authors have fortunately not accepted the classifications of landforms from the geologists nor of climates from "non-geographic" climatologists, but have used those of geographers (including themselves). Similarly we have no reason to expect that the agriculturists will provide us with a classification of farms suitable to geography, or that any economist will do the same for industrial landscape features. Likewise if these geographers should later realize that political features are significant in geography they would find no satisfactory classification of these had not some geographers continued to study them geographically. A concrete example is found in connection with the problem of political boundaries, which has certainly not been ignored by political scientists and historians. Nevertheless nothing approaching a satisfactory classification was developed until the problem was seriously considered by geographers, including Sieger, Penck, Maull and Sölch [see also 357].

There is a certain air of unreality in all of the discussions calling for the exclusion of these particular branches of geography. The history of science shows repeated development of new independent branches out of older stems, but it seems doubtful that this has ever come from expulsion by those less interested in the development of those branches, rather than from inner compulsion within them. Furthermore, the "landscape purists" themselves— in Europe at least—do not accept personally the limitation that they urge for geography. Thus Schlüter, in the same year in which he presented his program for limiting human geography to physical "landscape features," did not hesitate to discuss the very difficult concepts of "nation" and "nationality," in a geographical publication [385]; some years later he discussed "the influence of geographical conditions on the distribution of population and the development of culture," including the state [134, 388–417, especially 410] and has since examined "the arealy distributed societies" [145, 29] and the relations (Einklang) of state and land [154]. Penck's acceptance of the pure landscape thesis in 1928 [163] seemed to represent a renunciation of his war-time interest in political geography. (His significant rectoral address on the theory of polical boundaries [226] was followed by actual work on the German-Polish frontier.) Yet he contributed again, in the same year, to the political geography of Germany [249]. Likewise Tiessen was moved by the peace treaties not only to examine their results on the political map, but also to define the problems of political geography [390]. Passarge has repeatedly considered both theoretical and practical problems concerning the geography of peoples [375] and political geography [172; 373; 376–379]. Brunhes published two books in political geography [335; 336]. Granö and Michotte, so far as I can find, are the only exceptions.

In this country there has been relatively little incentive to develop either the geography of peoples or political geography. Nevertheless one finds that many of those whose general statements appear to limit the field to the study of material features, do not hesitate to include discussions of non-material aspects of regions. Thus, one might suppose from the introductory statements of Preston James [321]—though these are stated in no dogmatic form—that the geographer is limited to "things, organic and inorganic," but in the detailed treatment of areas, he maps the Negro residence district of Vicksburg [189], discusses the political activities of the nomads [228], the mixture of cultural ideas in the Mediterranean, "the cradle of Western civilization" [93], and bases his discussion of Manchuria on its character as a "cradle of conflict" [257 ff.]. In general, sound geographical interest in whatever phenomena are characteristic of a region will overcome any arbitrary defined limits.

Nevertheless the resultant situation is very confusing either to the specialist in non-material branches of geography or to the general student examining such aspects in his particular regional problem. Puzzled as to just what he should not do in this shadowy border area, can he look for guidance to these authorities? Since they disagree so radically, it will be necessary to examine briefly the suggestions of each.

Schlüter apparently changed his mind on this question several times. At one time he wrote definitely of the "inner" or "true geography" and of its outer branches [157, review, and in 145, 30]. More recently he appears to have welcomed *Geopolitik* as the solution of the problem, as far at least as political geography was concerned, though this still left such fields as the geography of peoples unprovided for [reference from Hettner, 167, 277]. That *Geopolitik* can provide a suitable substitute for political geography will hardly be accepted by geographers or political scientists outside of Germany, or indeed by few political geographers in Germany.[66]

Brunhes is best known in this country for his presentation of the "essential facts" of geography, from which one can readily quote statements which specifically limit the field. It is commonly overlooked that once he gets "beyond the essential facts"—and it is only then that he considers regional geography—the range of his observation is well-nigh unlimited, including epidemics, physical aptitudes, moral habits and social rules, property rights, collectivization, social organization, stock companies, and social anarchy in large cities [182, 517–68]. All these may be studied by the geographer as long as he can see any relationships between them and "the facts of the terrestrial world" [cf. Michotte's critical comment, 189, 14]. The well-nigh unlimited range which this allows him is most clearly shown in his two books in the field of political geography [335 and 336; Brunhes himself describes the former work as "a study vigorously and consistently geographic," 83, 23; cf. my discussion, 216, 803].

The concepts of Passarge appear to lead to similar conclusions. His view of geography is by no means easy to understand, because he appears to change it constantly and to publish immediately every change of thought. Anyone attempting to follow his discussions will agree with Hettner that "the route of thought from the brain to the printing press might well be somewhat longer" [152, 52]. If such students as Sapper [238, review], Gradmann [236, 335], Waibel [250, 166] and Hettner [242, 162; 167, 277] have failed to understand Passarge correctly, as he insists, it is reckless for a foreign student to attempt to analyze his theoretical concepts. But the influence of his ideas—whether because of the attractive possibilities which they appear to offer, or because they are accompanied by effective illustrations in detailed studies—has been so marked in this country, that it is necessary for us to examine exactly what they involve.[67]

---

[66] The critical discussions of *Geopolitik* by the writer [216, 960–5] and by East [199, 261–3] include references to similar discussions by French and Dutch geographers. At the time of writing my discussion I had failed to note Hettner's critique, but was all the more interested therefore to discover later that he had previously come to much the same conclusions [167, 332–6].

[67] Bürger finds that some students misunderstand Passarge, while others admit they can't understand him [11, 84]. Hettner has commented upon the difference between his methodological studies and his work as "a distinguished research worker.

Fortunately, for our purposes, experience has taught Passarge, "that it is difficult for the beginner to follow the train of thought in *Landschaftskunde*,"[68] and consequently he has gone to some pains to explain himself more fully and to illustrate his explanation with a specific detailed example [*268*]. As nearly as the writer can judge, there appear to be three grades of geography. The first, which is "pure landscape study" (*reine Landschaftskunde*), is limited strictly to those objects perceptible to the senses. Whether or not it includes animals and men is not entirely clear, but certainly they are included in various aspects of the study. A geographic description (*stadtlandschaftliche*) of Madrid includes the tortoise shell combs of the senoritas and the coquetish angle of the mantilla [*374, 693*].

Secondly, there appears to be a form of *Landschaftskunde* which we must presume is not pure, but which considers the extent to which the cultural phenomena (including perhaps man himself) of a *Landschaft* depend on its character [see also, on this point, *267* and his discussion with Gradmann, *236, 331–6*]. Thus the devout Catholicism of the Tyrolese mountaineers is considered as a result of the dangers of mountain life, just as, in an earlier study Passarge examined the characteristics of the Jewish people as "dependent on the nature of the *Land,* the *Landschaft,* and, in a wider sense, of the *Umwelt*" [*375*]. Similarly, in his brief summary description of the "major landscape belts" of the world, he notes that America "lacks the farmer class of the Old World, which because of tradition and custom clings fast to the hearths of their fathers, the dependable anchor of a people," and finds in New York "the most imposing and at the same time the most horrible of all office buildings, the sky-scrapers . . . where men are pressed together as though in an ant-heap, joyless, peaceless, breathless, caught in the chase for the dollar" [*305*, 43, 100 f.].

Finally Passarge writes of *Länderkunde* (alternately, *Landeskunde*), a term hitherto used in the German literature to refer to regional geography,

---

His field studies are excellent particularly for their exactness in detail. But when he enters upon methodological questions, this exactness and carefulness deserts him; indeed he doesn't even take the trouble first to read carefully and reflect a bit on the opinions that he is opposing, but plunges blindly at the red rag" [*242*, 162]. To this accusation, Passarge has replied that, if it means that he does not read Hettner, "stimmt"! (agreed). He explains that he cannot bear discussions of "philosophical" problems, apparently unaware of the great number that his theoretical discussions introduce [*272*]. Thereby he substantiates Braun's charge: "*Aber gerade er setzt sich souverän über Vorarbeiten hinweg*" [*155*, 17]. Both Waibel and Gradmann, moreover, have shown that this statement applies not only to methodological discussions but to the geographical literature in general.

[68] The remainder of the quotation should not be omitted: "and in relation to *Landschaftskunde,* even my university colleagues are in general but beginners, even though they may have already attained a patriarchal age" [*268*, 6].

but to which he chooses to give a quite different meaning.[69]   For Passarge, "the *Land* is an artificially bounded area which generally takes no account of the landscape-areas" but is simply a political, historical, religious, or folk area—most importantly, the state area.   His *"Landeskunde"* is concerned "primarily with the dependence of cultural conditions (both material and spiritual) and population conditions (including character and talents) on the artificially bounded land-area (commonly a state)" [*268, 79–83*]. *"Länderkunde,* carried as the crown of *Landschaftskunde,* brings a presentation not only of the present area but also of its development, the history of men, their state, social, economic, material and spiritual culture-goods. *Landschaftskunde* is thus the trunk of the tree of *Erdkunde,* which unites the roots and the crown—physical *Erdkunde* and *Länderkunde"* [*257,* 2 f.].

On this basis, Passarge elsewhere defines political geography as the study of "mutual relations between the area (the sum of *Landschaften*) and the political organizations" [*172,* 445].   He has illustrated this point of view in several articles [*373; 377; 378; 379*] and particularly in his book on the political geography of the Near East, which is, he says, "not *Geopolitik* but an attempt to give an example of what specifically should be understood by political geography" *376,* 7].

According to Passarge, therefore, all parts of geography, excepting only "pure *Landschaftskunde,"* are defined as the study of relationships, the very concept which "modern geographers" so vigorously oppose [*cf.* Lautensach, *278,* 20].   If it is scientifically dangerous to define the core of geography in terms of relationships which are to be discovered, surely this objection applies with equal, if not greater, force to those branches of the field that deal with non-material aspects of areas.   Passarge's view of geography appears as an unsatisfactory hybrid which few will accept.   It is not his concepts but only his terms which are new and the introduction of new terms, or the arbitrary use of old terms in new ways, inevitably makes for confusion.[70]

---

[69] *Cf.* Hettner, *242,* 163. It is characteristic of Passarge to use old terms in quite new meanings.   Thus, in this same connection, he defines "natural boundary" of a state "not as the boundary between two natural *Landschaften"* but as one which cuts directly across landscape-areas of different production "because every state must strive for autarchy" [*268,* 85].   Since this term has been abused before so often as to have lost all value for scientific purposes, the damage in this case is negligible [*cf. 237* and *357* or *216,* 945 f.].

[70] Gradmann suggests that the experience of Passarge, Volz, and Banse, accustomed to field work in remote and little-known tropical lands, where they had to depend solely on themselves, has led them to pay little attention to geographical literature and to think that anything they experience as new must be new to others [*236,* 139, 145, 337; *cf.* also, Sapper, *238,* review; Hettner, *152,* 46; *242,* 164; Penck, *159,* 640; Waibel, *250,* 477; and Bürger, *11,* 83 f.].

Turning to those American geographers who have spoken in favor of the limitation to material things, we find that Sauer was admittedly uncertain as to what to do about political geography [*84,* 207–210]. We are still at a loss to know what he would do about this and other non-material aspects of the field—such as the geography of races and peoples, which he does not mention—since he has not yet illustrated his methodological study by any complete regional work.

A discussion of the problem at the 1937 meetings of this Association made clear the dilemma to which Sauer's statements logically lead: we cannot exclude political geography from geography but it does not fit under our definition of geography so that it is difficult to see what to do with it. Since the dilemma was caused by the adoption of a purely arbitrary definition, in no way founded on either the logic or the history of the subject, one might suppose that the way out was obvious. Partsch has written of those who "find an attractive importance in creating out of the depth of their own judgment a limitation of the problems, and a division of the material, of geographical science, so that from a judgment seat erected on their own authority they may look down on the workers who have at any time dug their spades in the field of geography, issuing certificates to those who have, in fortunate prescience, worked according to the point of view of their successors, denying them to those who have deviated in their conceptions of their problems" [quoted by Graf, *156,* 31].

D. DOES THE LIMITATION PROVIDE A UNIFIED FIELD?

In spite of these logical and historical objections, many of those who have accepted the restricted view of geography will hesitate to abandon it, since it appears to offer a basis for restricting the multiplicity of things that the student of geography must consider. Such a need has been expressed by many geographers. Douglas Johnson, in discussing the prospects of geography as a science, spoke of "the necessity of effecting some restriction of the vast field now claimed by geography" [*103,* 221]. Granted that, from many points of view, a restriction would appear desirable, we must face the possibility that any such hope is in vain. If geographers are to study areas of the world, they must recognize that each area, like the world itself, is so full of a number of things, that these students who wish to see it in more simple terms may necessarily be as unhappy as kings. Most historians recognize that this is one of the inescapable characteristics of their subject, preventing it from becoming the sort of science which some of their number also might prefer. Yet on the whole, as Crowe notes, "historians bear their side of the burden with singular sang-froid" [*201,* 3]. Nevertheless, since geographers have been troubled for a long time by this complexity of their problem, any suggestion that appears to give hope of a solution will not be dismissed on purely logical grounds. Will the suggested limitation of the field provide a basis on which we can reduce the necessary "inner" field of geography and at the same time retain that inner core in complete unified form?

Such a basis for regional geography will hardly be found in Brunhes' system, in which regional geography is classified as "beyond the essential facts," together with considerations in political and social geography based on the principle of relationships. Helpful as his outline of the essential facts may be, it is evident that he himself does not regard it as providing a complete basis.

Likewise we will hardly look to Passarge for a more restricted and unified system for geography. He has, to be sure, discussed his concepts more frequently than any other student of this group, and he and his students have published a great number of valuable regional studies based on his ideas [he lists many of these in *268;* see however the critical discussions by Waibel, *250;* Troll, *268,* review; and Crowe, *201,* 10 ff.]. Even if we can ignore his studies in political and social geography, we found that his "pure *Landschaftskunde,*" however narrowly he may define it theoretically, was shown by his model example to contain far more than could possibly be included in any restricted concept of geography.

Sauer claims that the limitation, primarily to visible objects, will reduce the field to one in which the special experience and training of the geographer in geomorphological study "provides the necessary technique of observation and a basis for evaluation" [*85,* 623], an idea which is no doubt in the minds of those who urge a return to more thorough training of geographers in physiography. Granted that such a preparation is desirable and necessary, this should not lead to "neglect of other, even more important parts of geography," as Hettner observed in a criticism of the "Davis school" of physiography [*152,* 41–6; see also Gradmann, *251,* review, 552]. On what basis is it assumed that the special technique of the geographer is that developed in geomorphology? This is to place a very old science on the shoulders of a younger one, and one whose position with reference to another science, geology, is by no means clear. Granted that progress in geography was greatly furthered by the development of the study of landforms, it has likewise been furthered by the progress in climatology which utilizes an entirely different technique.

Parenthetically we may be permitted to suggest that the concept promulgated by the "landscape purists" may possibly represent a belated outgrowth of the period of the late nineteenth century when geographers—following

Peschel and Richthofen—were primarily concerned with landforms. The result was a greater emphasis on forms than on functions and an assumption that geography was exclusively concerned with studying the appearance of things observed with the eyes. A geomorphologist, like Penck, could accept the addition of crops and houses, since these constitute part of the landform —in a full physical sense—but the men who cultivate the crops and build the houses must be excluded. In this country a further support for the limitation of cultural phenomena studied in geography to things of material culture has apparently come from the contact of the California school with the cultural anthropologists. The limitation imposed upon the student of prehistoric, illiterate peoples, whose culture can only be studied in the remains of their material products, was apparently transferred to geographers, who are told that they may study the culture of a living area only in terms of its inanimate products—but may use those, even though they be pieces of pottery that can have but the slightest significance in the landscape. This chance combination of rather widely separated ideas has led to a peculiar paradox. It would hardly be claimed that training in geomorphology makes the student of geography better able to distinguish between a dozen different kinds of Indian potsherds than to distinguish between core areas and peripheral areas of a state [cf. 383].

If we consider the techniques used by present geographers, including the landscape purists, it is clear that they are of a wide variety of kinds. Every regional geographer must depend to a large extent on facts which he has not collected by observation; those who did observe them may have used techniques which have little to do with geomorphology. Mackinder meets the exaggerated emphasis on the study of the lithosphere, as he puts it, with the claim that geographers are really more concerned with the hydrosphere [196]. Even if this view represents exaggeration in a different direction, certainly the regional geographer must study climatic data, to evaluate which, one would hardly use the techniques of geomorphology [cf. Brunhes 182, Ch. 1].

Much the same is true of a large part of the "observable material data" used in any studies other than those of very small districts. Granted the importance of studies of small districts, such as Finch's study of Montfort [285], based on direct detailed observations, we do not add the facts observed in many different small areas in order to describe and interpret larger regions; rather we depend, and will continue to depend, for data concerning crops, animal culture, and population, on census materials which were certainly not collected by any direct observations. To be sure, they ultimately rest on somebody's observations—the farmer presumably has "observed" how many cows he has and how many children, but then all facts must be observed in some way, in order to be recognized as facts.

In Pfeifer's discussion of methods in economic geography, specifically recommended by Sauer as the proper application of the concept of landscape to that field [85], we find that we must study the course of economic events, the production and amount of transported commodities and that "the way of investigation must be fulfilled both by landscape observation and by the evaluation of statistics and the consideration of research in economics" [164, 327, 425].

Actually Sauer's outline, as well as his subsequent work, clearly requires an increase rather than a restriction in the techniques which the geographer .must utilize. He must use "the additional method . . . the specifically historical method" since the study must be oriented always on development. The geographer therefore will use all available data "in the reconstruction of former settlements, land utilization and communication, whether these records be written, archeologic, or philologic" [85, 623].

One can only wonder how these methods would be used in the development of a complete regional study since the theoretical proposals have not been accompanied, as Hettner recommended to any wishing to reform the methodology of geography, by the presentation of "a work that makes his views real and shows their superiority to the conception hitherto held" [175, 383]. In the regional studies that Sauer has subsequently published, as well as in those of various of his followers, we found earlier that the non-geographical methods, historical and anthropological, tend to predominate over any specifically geographical method (Sec. VI B). The most striking illustration of the effect of applying the historical mode of presentation to geography is to be found in Trewartha's study of the early French trading posts of the Driftless Hill Land. Whereas the author—whose field studies of Japan are outstanding—has repeatedly insisted that geographers should confine themselves to "observable features," his first chapter in a regional study is not only strictly historical in outline but is based almost entirely on the written records, since even vestigial traces of the early French occupance are scarcely to be found in the present landscape [393].

That geographic work cannot be considered in terms of one special technique, but inevitably requires the use of many different techniques, was not questioned by Schlüter. Even if we were to exclude the study of man, it "would still present a vari-colored picture in which were mixed such different colors as those of meteorology, hydrology, geology, botany, and zoology." In addition "the works of man . . . enter as ingredients into the *Landschaft* and become part of its nature," but these "cannot be understood without historical, economic, or ethnological research. . . . That geography thus stands between the natural and the social sciences, however, is not the essential point, but rather that it seeks to produce universal connections between different sciences. . . . In fact, one can now say, without great exaggeration, that, for the true geographer, the transition from the physical to the human part of his science involves no greater jump than from climate to land forms, or from that to the vegetative cover. In each of these cases it means to be transposed into a new realm of thought; in each case, however, broad bridges lead across from one to the other" [148, 145–6].

On the other hand the geographer (or at least the "pure" geographer) is not to cross any bridges that may lead from the consideration of man's settlements to that of his non-material creations—language, customs, or states. Schlüter justifies his "narrower conception" as "determined by the need for concentration" [213–14]. Unless this be done, a complete and unified study of a region is impossible. Incidental considerations of non-material features which may reflect the nature of the land destroy such unity, and a complete study of them would make the geographer's task impossibly large, indeed "would greatly disturb the cohesive unity of regional presentation." However the "narrower conception" is not so narrow as one might suppose. It includes the numbers and density of population, its composition according to sex and age groups, its growth and migrations, and the development of the entire economic structure, from production through intermediary agencies to consumption including the actual movement of goods through trade [216–217].

Beyond these, Schlüter recognizes that a complete interpretation of visible landscape features may require a consideration of non-material factors, and they may be introduced on this basis, but only on this basis. But what, as Hettner asks, is the value of excluding such features at the front door if they are to be brought in whenever significant through the back door? This is to disrupt the natural unity of a study and then to attempt to patch it up as one goes on [126, 555; 167, 280; similarly, Hassinger, 253; Almagià, 188, 16]. Similarly Penck apparently would not permit the geographer to study directly the distribution of people of German culture although he emphasizes that the Kulturlandschaft bears the particular mark of the people (Volk) who have developed it [158, 52]. In other words, as Bürger notes, the geographer may map the "cultural landscape" as the areal expression of cultural ability of its inhabitants, but may not map the direct areal expression of people of that culture [11, 64, 75].

Granö has objected vigorously to Hettner's analogy of the open back door: only those non-material factors whose differentials are interrelated to those in material phenomena are to be admitted, and only in order to explain the latter [*252, 179*]. But he does not answer the argument that on this basis he would ultimately consider all the immaterial phenomena that others, like Hettner, would have considered in the first place—with the difference that the belated consideration represents an attempt to restore unity to a study first torn apart. Granö himself starts with the assumption that all the phenomena of our environment, material and immaterial, form a unit (*Einheit*), and he recognizes that "the spiritual and social milieu . . . is in direct and causal connection, and in reciprocal relation to the physically perceptible milieu," but insists that only the facts of the latter are to be studied directly in geography, those of the former belong to sociology [*245, 5; 252, 46; 270, 297*]. This conclusion would appear to be not only the reverse of logical, but essentially irrelevant. Its frequent repetition by American geographers in one form or another is not an argument, but merely a reiteration of the point at issue. If the study of immaterial facts belongs to the various social fields, the study of material facts belongs to physics, geology, zoology, soil science, etc.—and also, in part, to such social sciences as economics. If this line of argument is followed, geography ends up with no objects whatever—excepting those hitherto overlooked by all other sciences.

Although this would appear to be the logical conclusion to be drawn from any argument that immaterial phenomena in an area belong to sociology, and therefore not to geography, fortunately none of these students in practice resign their claim to any areal phenomena. The only important consequence of their thesis is that in their procedure they resolutely limit themselves to certain kinds of things at the start—not to the primary facts of science in general, but to a collection of primary and deduced facts of a particular category—but later in their study bring in facts of another category—whether primary or deduced. Crowe therefore seems quite justified in his comment on the "vagueness of the landscape idea. As soon as a clear definition is given, it becomes apparent that the bulk of 'landscape' philosophy is a process of removing from one's hat the very things one has carefully put into it" [*202, 15*].

The point at issue may be illustrated by a consideration of some region in southeastern United States. Suppose the regional geographer limits himself at first to straight description of material things. He would describe the cotton fields, plantation houses and the huts of field-hands, carefully avoiding any mention of skin color and, particularly, of cultural heritage from slave times. These would be considered, if at all, only in order to explain the material things seen. But any reader familiar with the area will wonder why the writer thus postpones, or even omits, the consideration of what is one of the most significant *characteristics*—second in importance perhaps to no other characteristic whether natural or cultural—of our South as compared with other parts of the country, namely the fact that a large part of the population are Negroes and descendants of slaves [*cf. 359*]. In other words any feature of any area, no matter how known, which is significant to the whole complex of material features, is itself a characteristic of the area, to exclude which, even in description, will result in an incomplete picture of the region [*cf.* Gradmann, *236,* 130 f.].

One conclusion appears certain: if immaterial phenomena are going to be needed ultimately in a regional study, they had better be considered as they are encountered in the field, along with the other phenomena, and not first rigidly excluded and then, as the work in the study shows them to be needed, brought back out of the vague memories of the human observer, as distinct from the notes of the careful scientist.

There would, however, appear to be one way in which things that are strictly material form a unified object, the study of which might conceivably be regarded as the core of geography. As previously suggested, one could confine oneself to the study of the actual visible landscape, in the sense of the external surface form of the earth. Although I do not find that any student has suggested that such a study forms the core of geography—unless possibly Penck [*163,* 40]—and there is certainly no intention of proposing it here, the concept is nevertheless significant. For it appears to form the concrete base on which the vague concepts of "landscape" have permitted various students to construct different conceptions of a core of geography— or of geography as a whole.

The landscape as we defined it earlier, is an objective reality and forms a continuous object which, for the world as a whole, is a single concrete object. It therefore constitutes a unit basis for a field of study. To be sure, the moment the study passes beyond bare description the student must leave the landscape itself, must go beneath it, even to state what its form represents—to translate the outer foliage of a forest into the forest, the outer surface of buildings into different kinds of buildings, etc. If he does not do that, his study is exclusively concerned with the surface forms—meaning the plane forms in the surface. Such a study would be, not merely literally, but in any other sense, superficial, as Kraft has observed of "landscape" studies in general [*166,* 17]. Our interest in houses, factories, and forests cannot be confined to their surface form; only in the limited field of aesthetic geography could such a restriction be justified. Our very use of such words as house, barn, factory, office building, etc., indicates that we are primarily concerned with the internal functions within these structures; the external form is a secondary aspect which we use simply as a handy means to detect the internal functions—and should use only insofar as it is a reliable means for that purpose.

Further, the interpretation of the pattern and slope of the landscape forms requires us to look under the features that produce that surface. We must observe not merely the surface of the soil under the vegetation but also the sub-soil beneath that, and even the country rock still farther down—whether it be the basis for a residual soil or merely the supporter of a transported soil. The location and character of the bedrock may be essential to our understanding of such surface manifestations as slumping, springs, and artesian wells. The interpretation of a mine structure on the surface may require us to go thousands of feet beneath the surface. It is even more generally important that we leave the landscape in the opposite direction to interpret the effects of climatic conditions.

Such departures from the object of study into external objects that concern it are necessary in any science. In most sciences however they appear as incidental, and in total of perhaps less importance than that which is included in the objects of the science itself. In the case we are suggesting, it is obvious that the reverse is true. The study of the visual landscape would consist very largely in the study of things not included in the landscape itself. Further, though these things are studied in other sciences, just as each of the objects that contributes its surface to the landscape may be studied in some other science, no other science stands ready to supply the geographer with the necessary information concerning the areal differentiation of these non-landscape phenomena in their relations either to each other or to the landscape. The geographer would, therefore, be no more exclusively at work when he described the landscape than when he studied the other features necessary for its interpretation.

On the other hand, may we regard the landscape as so much more important than the non-landscape features that affect it, as to justify our considering the landscape as the core of our field? To ask this question is to answer it. Only from the point of view of aesthetics or of visual sensations can we regard the external form of a forest as more important than its contents, the surface buildings of a coal mine as more important in the area than the underground workings, or the contour of the land as more important than the precipitation that falls upon it.

We conclude therefore, that while a study which limited its purpose to the full interpretation of the actual landscape—the external surface of the earth beneath the atmosphere, which we found to be the only concept of the term that formed a concrete empirical object—while such a study would ultimately involve the consideration of most of the geography of an area, the landscape itself is not the core of the area, but merely an outward manifestation of most of the factors at work in the area. This outward manifestation is in no proper sense a center, core, or heart of the area, it is not even necessarily the most important manifestation of the area, but is only that manifestation which we can most readily observe by the easiest, but not the most reliable, of methods—*i.e.*, by simply looking at it. We may therefore recognize the value of this manifestation in reconnaissance study, where we need methods of observing large areas, even though we may depart from it as soon as we attempt more careful and detailed observations—with thermometer, rain-gauge, census data, or what-not.

Finally, insofar as the actual landscape does not reveal the presence of certain factors that we know from other means of observation are present in the area, that fact in itself is not evidence that those factors are insignificant in the area; *a priori* we can only conclude that in that case the landscape is not a complete guide to the contents of the area. Whether these non-represented features are significant or not, can be determined only by examining their relation to other features. For example, the sensible temperature of any area, at different times of the year, is a characteristic of that area and it is a characteristic of considerable importance to the human beings in that area. To question whether it is geographic or not is to beg the question. It is a significant areal characteristic that has but little if any representation in the landscape. In practice it would seem safe to say that any student who confined himself to the landscape forms and whatever factors produced them would not have occasion to examine the difference between actual and sensible temperatures. Likewise the North American student who has not happened to study across the Rio Grande or in other continents, will seldom discover any important manifestation of political geography in the landscape. In reality, however, this very lack of differentiation within the United States is in no sense to be regarded as the normal situation, in comparison with which the landscape differences between areas of English and French culture, or areas of Western European and Eastern European culture, or areas of American and Mexican culture, represent abnormalities introduced by a strange external factor. On the contrary this relative uniformity of cultural forms in the landscape of the United States might much more properly be regarded as one of the strangest of landscape phenomena, resulting only from the fact that the entire area has been developed, by civilized man, as one single political unit. But this political unit is a fact of area for which the student of the landscape itself would hardly discover any manifestation other than this apparently negative one.

[217]

We conclude therefore that the chorographical concept of geography not only does not require the exclusion of any particular category of phenomena, but that on the contrary the full study of areas of the earth surface inevitably requires the geographer to study many immaterial as well as material phenomena, and that to the student of area there is no logical basis for giving preference or higher rank to either of these groups of phenomena, or in general, to any particular category of phenomena. In the study of areal differentiation, all features that are significant to areal differentiation are on the same plane. How this significance may be determined will be considered in a later section.

E. PRACTICAL RESULTS OF THE LIMITATION

Our examination indicates that the limitation of geographic study to things observable by the senses is founded neither in the logic nor the history of the subject and does not provide a basis for restricting even a central core of the field which would be unified and complete in itself. A study limited to the actual visible landscape, the external surface of the earth, would constitute a unified subject of study significant only from the point of view of man as an observer of his surroundings. Since he observes them also in certain other ways, such a study would form but a part, though in this case certainly the major part, of the full study of the relation of an area to man as the source of his sensations. Hettner classifies such a subject as a special field of geography, under the title of "aesthetic geography." That it is closely related to psychology is obvious but there appears to be no practical need to decide whether it belongs more properly in the one field or the other.

It is true that a few geographers have been interested in this study since the very beginnings of modern geography. Humboldt endeavored to consider it objectively in terms of the external phenomena rather than the psychological sensations that they produce, but later critics have questioned whether he succeeded. In any case he regarded this aesthetic geography as but a part of geography. The scientific atmosphere of the late nineteenth century gave little encouragement to geographers interested in this field, but a few, such as Ratzel, Oppel and particularly Wimmer, endeavored to develop it. It was in the aesthetic sense, according to Friedrichsen, that Wimmer introduced the term *Landschaftskunde* into geographic terminology. "The descriptive geographer," he wrote, "is nothing other than a landscape painter and map drawer in words" [74, 9; note Wagner's critical review, 75].

Among modern workers, the English geographer, Sir Francis Younghusband, has been particularly interested in the study of the beauty of nature in geography. Two of his articles have been published in German, as *"Das Herz der Natur"* [235], and have apparently aroused more interest in Germany than did the original addresses in England. Banse and Volz in particular have followed a similar path. Banse definitely maintains the concept of *Landschaft* as the form or picture of the landscape, representing the outward manifestation of the total milieu [246, 42].

Although such students as Schlüter and Passarge might not accept the classification, various of their critics—including Friedrichsen [230] and Vogel [271, 7]—find that the point of view they present is essentially an aesthetic one. In the case of Passarge this judgment would appear to be confirmed by his notable dependence on photographs (not air photographs) rather than maps. His study of the "city landscape" of Madrid has but one very simple map of the city, showing little more than its areal growth, but uses four photographs [374]. The study which he presents as a model of his work is beautifully illustrated with 31 photographs and four panorama drawings, but includes only two maps, a section of the ordinary topographic map and a small sketch map of a minor part of his area [268].

An outgrowth of this point of view is the discussion by Banse and certain other German geographers of the importance of aesthetics in geography. This has led some to raise the question discussed in the early part of this paper, whether geography was properly a science or an art. Hettner distinguished between aesthetic geography on the one hand, and geography as art on the other [161, 151–5]. The former as a part of geography should be objective (and for that reason alone not likely to be accepted by artists even if we should wish to turn it over to them) whereas the latter is subjective, essentially artistic, so that one can hardly follow Banse in wishing to make it a part of geography.[71]

While geography is not to be restricted to any particular form of observation, this is not to say that the concept of landscape, as we have defined it, is not of value for geographers. There may be great value in focussing the attention of the regional student on the material features he sees in an area [cf. also Friedrichsen, 230, 160]. Especially on entering a region he is faced with such a host of facts that he may well be puzzled as to where he is to start; as a rule of thumb he might well start with what he sees before

---

[71] Banse's concept of the "soul" of a landscape goes considerably farther than would commonly be associated with its aesthetic aspect. The great degree of freedom that it permits him leads in some cases to descriptions of areas that should amuse, even if they do not please, those more familiar with them than he [cf. his impressions of America, 330, II, 47 ff.]. For further discussion of Banse's views and work, see the references listed at end of Sec. IV A.

him [*cf.* Finch, *288*]. If he limits himself to that, however, his study is likely to be incomplete.[72] On the other hand this is certainly not the only proper procedure. Excellent studies have been made, for example, by starting with a detailed examination of population distribution, the data for which were not obtained by direct observation. It may also be useful to assume as a rule of thumb that visible landscape facts are generally among those that are most likely to be significant to the character of an area, although this may not necessarily prove to be the case, and that non-material facts are more likely to be found to have no geographic significance, though again the reverse may be the case in any particular instance.

Whatever value this rough rule may have in field investigation, it does not provide an adequate criterion for the selection and rejection of phenomena that are to be considered. Even though we should exclude all non-material phenomena and limit the consideration to sensible, perceptible phenomena—overlooking for the moment the difficulties that the proponents have had in defining what that includes—we would still have far too many facts to consider. While the general criterion limits these to concrete facts, it provides no objective basis for selection from among them. Following the concept logically, Granö, as we found, determined the selection on the basis of the impression that the different features make upon our senses. On this basis the weather-beaten old grain elevator that yesterday was hardly noticed in the landscape may appear tomorrow as the most prominent object, thanks to its new metal covering. Though any landscape painter would accept this statement as a truism, presumably few would regard it as expressing a proper measure of geographic significance.

It may be suggested that in place of this aesthetic standard the geographer uses the standard of areal extent of landscape objects. The geographer's business is to describe and interpret the things that make up the landscape, and therefore he considers primarily those that constitute areally its most important parts. Though this would seem logical, it is significant that few students follow it in practice. Schlüter suggested this standard in his theoretical discussions, but to illustrate the systematic study of cultural landscape forms, he chose bridges, a feature that would have to be rated on this basis as one of the least important of cultural phenomena [*247*]. Further, as we have seen, he applied a special magnifying glass to human beings in order to prevent them from vanishing as cultural features of almost no areal extent.

---

[72] This conclusion is based not only on theoretical considerations but also on the result of the writer's attempt to follow the procedure in a regional presentation of a very limited area, in "The Upper Silesian Industrial District" [*356*].

In each of these cases Schlüter appears to have drawn a compromise between the logic of geography as the study of landscapes and the logic of geography as the study of areal differentiation—in which, fortunately, the latter had the preponderant influence. Most geographers make much the same decision in considering cities. Though these cover but a very minor part of the earth surface, no theoretical considerations can convince us that they are not of great importance in geography. Possibly the fact that most geographers live in cities may likewise have saved these features from any strict application of the measure of relative areal extent. On the other hand, other types of highly specialized areas are often overlooked simply because they are small in area and, though intensive in development and thoroughly distinctive in character, this distinction may not be readily apparent in the landscape.

As a specific example we may note the tendency to neglect the special characteristics of mining areas. Where the mining is carried on underground it shows but minor effects in the landscape, prominent though the structures and sidings at the mine-pits may be. But anyone familiar with an area in which mining occupies the attention of a considerable portion of the population recognizes that in countless ways the character of the area is different from, say, otherwise similar but purely agricultural areas. While this difference might be measured in terms of the quantities of minerals shipped out, and consequently in the actual movement of trains, many of the special characteristics are non-material. But they are nonetheless real and geographically significant; the social and ethnological character of the population, the labor organizations operating in what appear to be rural areas, and even the contrasted social attitudes obvious in the streets of the towns— all these are specific characteristics associated with a particular district as a result of its mining activities. The very fact that these areas possess special characters warrants greater, rather than less, consideration in proportion to any measure of their intrinsic importance, but even in terms of population they are vastly more important than the nomadic grazing areas of the world, which seldom lack of detailed consideration.[73]

In selecting out of all the multitude of various material objects in the landscape, no one appears to have attempted to apply the standard of areal extent as a basis of distinguishing the more important from the less important. All material objects occupy some area and therefore any of them apparently may be studied, though it appears to be agreed that they should be *immobilia*—things that stay in the same place. Though Schlüter, in

---

[73] This paragraph is based on an analysis of "Mining Landscape" by Lewis F. Thomas, in manuscript.

theory, ruled out the study of house types, others who accept his theory in general have fortunately disregarded this logical conclusion and have given us significant studies of rural house types, although, in terms of areal extent, these would have to be regarded as of very slight importance in the rural landscape. On the other hand, houses are obviously far more important in the urban landscape than in the rural, and the theory of the landscape purists would appear to justify studies of individual cities which include maps showing, by blocks, the percentage of colonial, mansard, or neo-Spanish styles, or even a "geographic" study of the different types of filling stations within a city.

All of these suggestions—which it is hoped no one will regard as other than absurd—would be justified by the statement with which Kniffen, in theoretical discussion, defended his study of house-types: "the primary concern of the cultural geographer is with the nature, genesis, and distribution of the observable phenomena of the landscape directly or indirectly ascribable to man" [295, 163]. In his actual study, to be sure, Kniffen has indicated that he is concerned with house types for reasons that are in no way based on the landscape or on material phenomena, but rather on the character of immaterial culture. Nevertheless we may accept his statement quoted above as the logical and necessary conclusion from the concept of geography as exclusively the study of landscape.

The landscape as an observable reality consists of individual objects (or their surfaces) which together form a whole significant only in aesthetic terms. This we found was true even if the concept be stretched to include all material objects in an area. Viewed as parts of an observable landscape their direct relation to each other has no significance except as parts of a "picture"; to study their indirect relations carries us immediately out of the visible landscape and sooner or later also to non-material factors. Although the rules of the landscape purists permit this, it is obvious that they discourage it. In any case, if the landscape features are justified solely because they are observable features in the landscape, the geographer, being neither artist nor psychologist and therefore not interested in the landscape as either a picture or a total sensation, is logically led to study each of these objects, or categories of objects, in its own right, in terms of its "nature, genesis, and distribution."

That the theoretical conclusion stated is not without significance in practice is indicated by Scofield's study of house types in Tennessee. These are classified as to types, and examined as to origin and development, with no indication of any geographic significance other than the mere fact that houses are "landscape features" [387; cf. Pfeifer's comment, 109, 120 f.].

If there is one point in methodology on which almost all modern geographers are in agreement, it is that the study of the nature, genesis, and distribution of any single type of object is not a part of geography but rather belongs to the systematic science that studies those objects. It is ironical that Sauer, who erroneously accuses Hettner of wishing to include the studies of distribution in geography (see footnote 48) should have developed a concept of geography as the study of "the landscape" which leads logically to that very conclusion. It may be claimed that no systematic science has happened to take up the study of house types, but we can be sure that any success geographers have in that study will stimulate the appropriate students to concern themselves with it and we shall have simply another case of a field opened up by geography to be taken over by another science. Insofar as our students of house types are concerned largely with morphology, genesis, and distribution only—in other words primarily with the objects themselves—they justify Leighly's logical conclusion that the professional geographer has nothing to add here to the work of the scientific historian of art and culture forms [*220*, 138].[74]

The all-important distinction that we are here emphasizing is exactly that represented by the inversion of Ritter's classic, if somewhat awkward, phrase describing geography as the study of the areas of the earth surface filled with earthly phenomena. The study of the material objects that physically fill areas is not the study of areas, it is not chorology; since it does not study all that is in areas, it logically disintegrates into the study of material things as found in areas. The study of "the phenomenology of landscape" proves to be nothing more than the study of individual phenomena found in landscape; the structure, origin, growth, and function of landscapes proves to be the structure, origin, growth and function of each material object in the landscape.

We may note in much of the work of the landscape purists an additional result in the tendency to emphasize structure or form alone. This is natural enough since our sensual perceptions can observe only forms; origin, growth and function must either be deduced from present forms—which is by no means a reliable method—or observed by indirect methods of observation involving often invisible and even immaterial factors. Furthermore, if one adheres with any consistency to the concept of "landscape," the objects in it are members of the whole only in terms of form. The landscape, as we

---

[74] It may be claimed that Leighly does not intend to draw such a conclusion, but what other conclusion can we draw from his discussion of Karling's study of Narva? His following question and answer, indicating that it "is not precisely his intention" that the geographer turn "art historian" is not reassuring since it turns out that the cultural geographer, while retaining other abilities, should also become an art historian.

have noted, does not grow as a whole; separate things within it grow separately.

The concentration on physiognomy might also be regarded as in part the result of the importance of geomorphology in the training of modern geographers. To the geomorphologist the forms of the earth surface presented significant problems because they had to be explained as the end product of a physiographic history. Much less attention was given to the functional relation of different landforms to other earth features; this contrast may be noted particularly in Penck's discussion of observation in geography, of 1906 [*128*], as well as in later work of Schlüter [*247*]. If, however, as many students now feel, the geographer is not called upon to explain the genesis of landforms, a description of forms that does not give a corresponding attention to functions appears barren. In the work of Schlüter, Passarge, and Granö, various critics have observed that the excellent analyses of the physiognomy of the landscape are not balanced by due consideration of its physiology—the interrelation of the phenomena [*cf*. Waibel, *266* and Bürger, *11*, 93]. Similarly, Finch's discussion of the presentation of the geography of regions devotes many paragraphs to form and but one line to functions [*288*, 117].

Indeed, a strict interpretation of the "pure landscape" view in geography might well rule out functions entirely, except as they express themselves in forms. But since no geographer wishes to exclude the functions of the things he describes, the tendency in many cases is to assume *a priori* that the form expresses the function—a factory looks different from an apartment house, a building with the form of a barn is not a residence. But on New England farms one may find that summer residents have converted a barn into a house without changing notably its external appearance, indeed with but minor changes in its internal form. The writer once spent several hours riding through the streets of the famous silk center of Lyons looking for silk-mills, only to discover, finally, that the silk was manufactured on small machines enclosed in buildings which could not be distinguished from the ordinary tenement buildings which they had once been and perhaps still were. Would he have been justified, on this basis, in dismissing the industry as unimportant in the geography of Lyons? As Colby writes, in answer to a recent questionnaire, "some very important human enterprises are housed in Victorian buildings. Why judge the significance of an institution by its mansard roof?" (See footnote 101.)

It is particularly in urban landscapes, as Waibel observes, that form may be an inadequate indicator of function. No doubt the well trained

[224]

urban geographer should be able to observe the difference in appearance of a city that is predominantly commercial from one that is primarily manufactural, but it would be extremely difficult actually to measure the difference in degree—since all cities are in part both. A warehouse may be changed into a factory, or vice versa, without undergoing any obvious change in external form, but the change in function may be of great significance in the life of the city. In sum, as Dickinson notes, "function and form are not necessarily in harmony" [*202,* 6].

Fortunately the importance of functions is so obvious that they appear to triumph over limitations of definition. In the treatment of cities, thanks perhaps to the influence of city planners on urban geography, even the geographers who adhere most rigidly to the limitation to material things analyze areal structure primarily in terms of function, only secondarily in terms of forms of buildings [*322,* 633; *321,* 187–9].

The landscape concept of geography leads naturally to the emphasis on form rather than function not only in the consideration of individual objects but also in that of combinations of objects in the landscape. The current enthusiasm for "patterns" has been characterized orally by some critics as simply one of the latest fads in geography. The writer is not one of those who would deny significance to pattern; on the contrary he has found that the thorough study, for example, of the street pattern of cities leads to significant conclusions as to the basic factors in their development; this was the case both in the study of a city in this country, Minneapolis [*354*], and of European cities, in Upper Silesia [*356*]. (In the latter case the results were achieved, it is only fair to add, because of the objection raised by Preston James that it was impossible for a city to have "no pattern.") Platt's series of studies of patterns of occupance in scattered districts in Hispanic America has shown the value of studying patterns particularly as an approach to differences in functions [the individual studies are listed in *221,* 13, footnote].

On the other hand, the mere presentation of patterns without further consideration of them is description in its simplest and most uncritical form. Unless the patterns presented are subjected to further study we have no assurance that they are of any important significance; the mere fact that they cover large areas on a map, and excite the interest of anyone who takes pleasure in designs, does not establish their relative importance in comparison with the host of phenomena which we are forced to ignore as unimportant. For example, the pattern of the railroads of western Ohio gives some indication of the location of the principal trade centers, but its most striking characteristic apparently is that railroads run in almost every direction, more or less indiscriminately—a reflection to be sure of the character of the surface. But if we could procure the data of freight movements over all these routes and map them by lines of proportionate widths, the resulting pattern would have far greater significance in leading to an interpretation of the relations between rural communities and urban centers within the area, and particularly of the area as a whole in relation to the Pittsburgh-Cleveland coal and steel area and the Atlantic Seaboard on the one hand, and to Lake Erie and the Northern Interior on the other.

In other words the significance of patterns depends entirely on the extent to which they depict significant relations in the location of different places in reference to each other. It is therefore most significant that many critics have found that in much of the work of the landscape purists this particular aspect is often neglected. These writers do not necessarily neglect to mention areal connections to be sure, but the constant emphasis that geographers must concentrate their attention on concrete material things naturally leads to a tendency to minimize the purely geometric factor of location—though most other geographers would agree with Bürger in regarding it as the particular element that is more distinctively geographic than any other [11, 30].

Illustrations of this effect may be found in the work of many different writers who have adopted the landscape concept. Thus Schlüter was able to state that in physical geography at least, the concept of location was of very minor importance [127, 13 f.]. Sauer's various "functional diagrams" provide no place for relative location, nor is the importance of this factor suggested in his subsequent studies [382–4], either in the text or in the small number of maps [cf. also Dickinson's criticism, 202, 7]. Likewise in Granö's work, Waibel correctly notes that "relations of location (Lagebeziehungen) are completely eliminated" [266, 204].

Much the same is true of the work of Passarge. In Die Landschaftsgürtel der Erde, the "landscape belts" of the world, including their "cultural landscapes," are not only described with great effectiveness but also explained in some detail with practically no consideration of the location of these cultural landscapes in relation to each other or to the sea [305]. Even the development of cities appears to be primarily a question simply of site—a harbor, a ford or convenient river crossing, a mineral deposit, etc. [98 ff.]. If we may judge from the model example which he presents, the same is true of the "landschaftskundliche" method in the study of a small area [268]. We have already noted the very slight use of maps in this example. From a careful examination of the text one would suppose that the cultural landscapes around Meran, in the South Tyrol, can be studied in detail without the slightest consideration of the relative location of that district either as a tributary valley of the Adige, leading to Italy, or in its more significant relations to northen Tyrol and the German area of Europe as a whole, by way of the great Brenner route or the Reschen-Scheideck [*cf.* Troll's review, and discussion, *268*].

A striking instance of the shift in point of view is offered by the use to which Mark Jefferson's widely known maps of proximity to railroads [*366*] have been put both by James, and by Finch and Trewartha. In his original study, Jefferson entitled his maps simply "Europe within 10 miles of a railroad," etc. (such areas being shown in white against the black background of areas farther from rails). His text makes clear that it is the proximity of certain areas, in contrast to the remoteness of others, with which he is concerned. James introduces these maps however as maps of *pattern*, without suggesting any significance in pattern other than that of density; but he does suggest to some degree the importance of proximity [*321, 183–5*]. Finch and Trewartha, more strictly logical, have re-named the maps "Density and patterns of railroads," because it is with these characteristics of railroads "that the geographer is chiefly concerned" [*322, 656*]. Actually Jefferson's maps show density very poorly since they allow for no distinctions in areas where railroads are well developed, as he himself clearly realized; likewise in these same areas—including most of Western Europe and eastern United States—they show nothing whatever of patterns. Pattern, at least, is shown far better on any ordinary map of rail-lines. Jefferson's maps, in fact, are not maps of railroads at all, but rather, as his own titles clearly indicated, they are maps of *areas having a certain location with reference to railroads*. Are geographers to ignore this characteristic simply because it is invisible, immaterial? Economists have long considered the density of railroads and some, like Ripley, have speculated on patterns [*381*]; only a geographer would have developed the idea of mapping proximity. The value of the idea is indicated by its wide-spread adoption by other students in relation to roads as well as to railroads, and it has already entered at least one European atlas.[75]

In general we may say that, just as the study of the individual objects in an area in terms of form, without function, prevents us from putting them together significantly as areal phenomena, so the study of the structure or physiognomy of the landscape, without consideration of the location of the differently located features in reference to each other, the factor that underlies the *physiology* of the patterns, produces a morphology that is barren.

---

[75] In his *Oberschlesien-Atlas* [*346*], *Blatt* 18, Geisler reproduces two such maps originally published by the writer [*355*], as a direct result of Jefferson's maps. See also Ralph Brown's map of "areas 10 miles beyond roads," in the Atlantic Seaboard about 1800 [*334, 228*].

Crowe has commented on the tendency in modern geography to retreat upon *things,* to become a morphological analysis of all sorts of objects in areas that nobody has previously thought worthy of study [*201,* 2].

The reader can hardly fail to have observed that underlying all the arguments of those who wish to consider only material objects—objects directly observable by the senses—is the spirit of physical science. Material things —and particularly visible objects—are the sort of phenomena that students trained in physical sciences know how to deal with. Geographers with that background—which includes most geographers today—would naturally prefer to have to study only such definitely tangible *things.* Non-material facts are grudgingly admitted of necessity where they affect the material things, but the purpose is clear: the less they are to be considered the better. In particular, if one could limit the consideration of an area to the form, or appearance, of its landscape, the decision as to what phenomena are significant appears easier. In reality, however, we found that the concept of geography as the study of the landscape, or simply of material things in areas, provides no usable standard for the selection of phenomena to be studied, and students appear to select whatever observable objects excite their interest.

On the other hand, as Bürger observes, to decide what is significant to the "character" of an area—to answer Colby's question "not how it looks, but how it is"—appears "too uncertain to the natural scientist accustomed to a sharp corporeal division of the materials of science. The critical scholar will overcome this uncertainty and in the particular case can judge what is significant and what is not. To be sure, the uncritical and superficial, who does not sufficiently engage himself in the study of his area, has not yet actually 'experienced' it, will succumb to the attempt to describe everything possible as regionally significant and geographically important. But of him, methodology can take no account" [*11,* 89].

It therefore appears paradoxical that many of the students who have expressed their theoretical adherence to the landscape concept should have endeavored to study features of the landscape whose significance in the total character of an area is important chiefly in terms of immaterial culture— namely settlement forms and house types. Whether they claim that they study these simply because they are visible features in the landscape or recognize that they are actually attempting to study the geography of culture, their readers will not be deceived, it is culture they are pursuing.

To some geographers these topics may seem of but minor importance, but others appear to regard them as of almost first importance. In a paper read at a recent "Round Table on Cultural Geography," Fred B. Kniffen included the study of individual buildings of different kinds as among the *responsibilities* of the cultural geographer [*295,* 163], and Leighly, apparently, would have the whole of cultural geography concentrate upon the study of the localization of such "cultural immobilia" [*220,* 132 ff.]. One might be justified in ignoring theoretical claims that are apparently based on no consideration of the history of geography—which indeed to many will appear to lead us completely out of geography into something like historical architecture, for which few, if any, of us are prepared. On the other hand the painstaking and illuminating field studies that have been made in these topics —particularly by European geographers, but also by Hall, Kniffen, and others in this country—require serious consideration. In particular, though such studies are clearly legitimized by the doctrine of "observable material things," it is necessary to consider whether, under that iron rule, they can become significant in geography.

F. RELATION TO THE STUDY OF SETTLEMENT FORMS

The systematic examination of the forms of houses and settlements appears in this country to be still in that elementary stage which is marked by enthusiastic observation and classification of types with little consideration of the question of the significance of these cultural features to geography. Many of the corresponding studies of European students have been motivated from the start by the desire to study the distribution of culture itself, in its differential national forms. Our students, however, have hitherto paid but little attention to the recent work on house types produced by European students [*cf.* Pfeifer's footnote, *109,* 120].

Since we are here concerned with a branch of geography that is relatively new to American geography—though we noted an early example in Ritter's studies of Asia, and Meitzen's classical study was published as early as 1895 —it is somewhat difficult to judge just what its significance may be to geography as a whole. If, as some think, the study of settlement and house forms may give us a valuable key to the study of cultural geography in general, it is appropriate to consider whether its development will be encouraged or discouraged by a particular concept of geography. For this reason, and not without hesitation, the following considerations are offered.

For some students, the study of house types is justified simply on the grounds that houses can be seen in the landscape. It is obvious, however, that the geographer cannot study everything he can see in the landscape; there must be some basis for selection. It is not surprising, therefore, that many students regard house types and settlement forms as simply the latest fads in geography—and not without reason. Leighly has suggested that a

Rhenish castle overlooking the Columbia River would be worthy of geographic study [*220,* 140] and we have actually had detailed descriptions of individual villages on the grounds that they were unique in the regions where they were found. Indeed it has been suggested that there are as many village forms in a region as there are villages and each one warrants special investigation. If one considers the total number of villages in the world and calculates the mountainous pile of manuscripts that would thus be called for, Leighly's nightmare of the surface of the earth plastered with topographic descriptions becomes by comparison a pleasant dream [*220,* 126 f.].

It would however be unjust to make too much of the natural results of enthusiasm, or to emphasize those studies of house-types that have been exclusively concerned with classification and genesis of forms. In much of the work of these students the trained reader recognizes at once geographic quality which need not—though it could—be defined. Furthermore, many readers may feel that in the wealth of detailed material presented are included new concepts, suggestions, or conclusions that should add to his general understanding of geography and to his own ability to work in it. Unfortunately this reader, at least, has found it difficult to derive such ideas or conclusions from these studies, in part, perhaps, because of the lack of any adequate statement of conclusions in many cases, but also, perhaps, because the author himself may not have thought out what conclusions, if any, could be derived from his study—*i.e.,* just what the significance of the study might be.

Of the students of these topics in this country, Kniffen appears to have most seriously considered their geographic purpose. Seeking a "logical approach to culturogeographic regions" he has classified and mapped by isoplethic methods the types of houses in Louisiana, not simply to portray the material physical character of these observable material objects, but as an "attempt to get at an areal expression of *ideas* regarding houses—a groping toward a tangible hold on the geographic expression of culture" [*368,* 179, 192; *cf.* also Leighly, *220,* 136].

It should be noted in passing that the purpose stated by Kniffen represents by no means the only possible significance which such studies may have to the field of geography. The different forms of both buildings and settlements may be significantly associated with differences in climate, relief, vegetation, or bed-rock, or with differences in any number of material cultural features, and may therefore add to the distinctive characteristics of regions. On the whole, however, the geographic significance of such features in themselves—as one group of physical features among many others—is that which they might have "in providing a tangible hold on the geographic expression of culture."

Amplifying his written statements in personal discussion with the writer, Kniffen makes clear that they mean just what they imply. Culture is made up of ideas; an idea may have geographic expression—*i.e.*, areal differentiation—but that expression is in itself intangible. We need to find a tangible representation of the idea, a representation, furthermore, whose geographic expression correctly portrays the geographic expression of the idea. Thus, we might have a general impression that in each of several different sorts of areas in Louisiana a particular idea regarding houses predominated. This is a geographic expression of cultural differences, but it is an intangible one, we can't get a hold on an idea. If the actual houses, which we can observe, classify and map proportionately, are a true expression of ideas regarding houses, we have a technique for a geographic study of an idea.

The ultimate objective, then, is a geographic study of non-material aspects of culture—indeed one can properly say, of the very culture itself, as well as the more outward and material manifestation of culture.

This detailed analysis of the fundamental point of view represented by Kniffen's study is justified, not only by the well-merited attention which geographers have given it [*cf.* Pfeifer, *109, 120*], but also by the fact that many who have discussed this study have not been fully aware of the ultimate purpose upon which it is based. Indeed, the author himself, in considering the theoretical nature of cultural geography in a later round table discussion, appears to deny completely the essential approach of his previous paper. In place of the approach from the point of view of regional differentiation of *ideas,* as in his field study, he bases his theoretical consideration of cultural geography on the familiar dictum, already quoted, that its "primary concern is with the nature, genesis, and distribution of the observable [meaning 'in general, visible'] phenomena of the landscape directly or indirectly ascribable to man." Consequently, though "the characteristic expression of a religious faith—churches, wayside shrines, and cemeteries—is of primary geographical importance," he would "consider it of little geographical concern that 90 per cent of the people of a given area belonged to a certain religious faith, if that fact had no expression in the landscape" [*295*, 162–67; final clause added by Kniffen, subsequently].

In other words, not all ideas that have areal representation are to be included in geography, but only those that are expressed physically in the landscape; the test of geographic significance is the presence of physical forms. We may agree that not all cultural ideas are geographically significant, not even all of those that we can map areally, without necessarily concluding that the sole and sufficient test is their physical landscape representation. We have seen that the character of some cultural landscape features reflects little more than the absence of cultural ideas and, on the other hand, elements of the first importance may have little landscape representation. Take for example the areas immediately north and south of the boundary between Germany and Switzerland. Though the language is the same, through the whole range of political ideas the Swiss area is notably different from that in Germany. If these differences are especially marked today, they have been well defined for centuries. Whatever representation these differences may have in the landscape, however, are sufficiently obscure as not to be perceived by a group of geographers who made a reconnaissance trip across the line. On the other hand, any discussion of political questions with the inhabitants on either side reveals immediately a wide difference, not merely in political interests, but in the fundamental political mentality of the populations.

The test of landscape representation is but one test, and by no means the most reliable test, of the differential character of culture of different areas. Furthermore, the emphasis on the physical representation in the landscape leads easily to an emphasis on the form of the representation. Indeed, one critic has argued that, though Kniffen's study of house types is primarily concerned with forms rather than with functions, he should have treated them exclusively in terms of form. The forms of wayside shrines, as seen along the trails of the Austrian Alps, for example, tell us something of the artistic ideas of the people, their ideas of what can be made of wood and stone, but surely it is far more significant to observe what they tell us of the religious ideas of the people.

It may be appropriate at this point to suggest that the assumptions of individual geographers concerning the nature of their field may be arrived at by one or more of three different methods. Some may attempt to analyze from a broad philosophical point of view the logical character of geography. This method, as we have seen, is capable of leading students completely astray, unless they test their logical conclusions against those of their numerous predecessors, and also by a thorough study of what geographers in general have actually tried to do. A much larger number of students, starting only with a very general feeling for geography—what may be called a geographic sense—develop their motivating concepts as a result of their actual research. Given the initial assumption—which is not guaranteed by the mere fact that somehow they chose this particular field—such students may arrive at a point of view which does not differ greatly from that of the most successful members of the first group. This fact is often obscured because in any direct discussion of the nature of the subject they may express apparently quite contrary views, but if one judges by their mature works, one finds that the actual differences in point of view are minor. The third group, finally, are those who have taken their concepts of geography on authority from others, usually from some member or members of the first group.

Like all simple psychological classifications, this one is unreal. Every student belongs in part to all three groups. It is to be hoped therefore that those American geographers who have made a significant start in the study of the geography of culture will not permit obedience to *ex cathedra* pronouncements to prevent the sound development of their own thought in, and as a result of, their work. On the contrary, all methodological pronouncements must, as Hettner repeatedly insists, be tested in the light of actual solutions of geographic problems. Michotte's warning merits repetition once more: "a discussion of principles," divorced from the study of specific examples of actual problems, "is the origin of the errors committed in these matters" [*189*, 23].

In the particular case under discussion, the geographer appears to be attempting to ride two horses pulling in somewhat different directions. If geography is the study of the visible landscape, the forms of rural houses constitute one element (though a minor one) in his object, but the *ideas* that people have about houses are of only secondary importance, needed only to interpret why particular forms of houses are found. To "attempt to get at an areal expression of *ideas* regarding houses" would not appear to be a proper major objective of the geographer. If, on the other hand, the geographer assumes, without conscious consideration of methodological principles, that he wishes to establish culturo-geographic regions and that for this purpose he must secure a "tangible hold on the geographic expression of culture," the attribute of visibility in the landscape is but one test, neither a necessary nor even a sufficient test, of the significance of the phenomena either to culture or to its geographic expression (its areal differentiation).

There remains the question whether in studying the areal differentiation of cultural ideas, the geographer is restricted to techniques that depend on examination of material things. Though we may find no logic in this distinction it might still represent good sense. In the discussion following Kniffen's Round Table paper to which we last referred, Stanley Dodge suggested that in cultural regional geography we are concerned with areas of uniform thinking and that we might use that as an index [*295*, 171]. But obviously we have no direct way of measuring what people think, we must arrive at that indirectly. Examination of their material products offers such a method. But is it the only possible method, or is it even certain that it is in all cases the best method? These are the critical questions, for if we can study that which we should study, by methods better than those now used, no authority in all of science, either in our own field or in other fields, shall forbid us to use the better methods.

We need not restrict ourselves to people's ideas about houses, since it is obvious that such ideas are by no means of major importance in the total complex of their ideas, nor are they necessarily the best key to that complex. For example, Kniffen found that the houses in the prairie area of Louisiana indicated that the population there had Midwestern ideas regarding houses, a fact easily explained by their Corn Belt origin [*368,* 190]. Does this indicate that these people now have Midwestern ideas in general? Possibly so, but equally possibly they may not in the slightest. They may have become entirely "Southern" in their points of view on everything else in life, but have simply continued to build houses in the inherited style (or to live in inherited houses) without even knowing that their houses were Midwestern until a geographer came to tell them. Isn't it possible that an equally reliable, if not more reliable, guide to their thoughts might be gained from a study of election returns? If it were found that year after year these areas voted Republican, one might well feel that that was a better key to their particular culture.[76] I trust that no one will suggest that geographers who can train themselves to observe and classify such complicated phenomena as house forms are not equipped to study election returns. Similarly, as Ekblaw suggested, the prevalence of certain groups of folksongs and customs in different parts of his state might provide Kniffen with another key to cultural regions [*295,* 171 f.]. If the classification and distribution of these latter phenomena have already been studied by students of other fields, so much the better. The geographer, who would not hesitate to appropriate the findings of the geologist and soil scientist in order to correlate them with his own, should rejoice whenever social scientists have provided him with similar materials ready for his use in cultural geography.

In brief, the geographic study of forms of houses and settlements can be justified as more than a fad, as a contribution to cultural regional geography, primarily if it provides a method of observing indirectly that which cannot be observed directly, namely the areal differentiation of essential cultural characteristics—the ideas, attitudes, and feelings of man. These characteristics manifest themselves in a number of different ways that are

---

[76] The suggestions here do not quite fit the particular case involved. Kniffen informs me that the houses now standing are largely those built by the first generation of immigrants from the Corn Belt, many of whom are still living. On the other hand, as a student of culture, he recognizes the significance of the fact that the area is widely known throughout the state as "the Republican section," but as a geographer he does not mention the fact. Surely the correlation of this areal fact with the fact shown on his map of house types would enhance the significance of both facts.

subject to observation, classification and proportionate mapping. Our thoughts are expressed not only in the things which we make but also in the manner and content of what we speak, sing, or dance, in what we write, vote, or tell to the census collector. Though immaterial, these phenomena are no less observable than material objects, in many cases they represent more direct manifestations of our ideas than our buildings, not to mention the settlement forms whose remote origin we may have completely forgotten. The excellent regional atlases which German geographers have recently been producing demonstrate the value of mapping such non-material culture phenomena as well as the material house types and settlement forms [371; 346; see also Schlenger, in 344]. The student of cultural geography who permits an arbitrary rule to forbid him from studying these observable expressions of culture is depriving himself of the possibility of achieving a well-rounded and, so far as possible, complete achievement of his fundamental purpose—the interpretation of cultural geographic regions.

G. SUMMARY

It may be well to summarize the conclusions which have been reached in the course of the rather lengthy discussions of this section. The doctrine that the field of geographic study must be limited to areal phenomena which are "in general, visible" admits of no precise statement other than a limitation to material features. Although enthusiasts have been preaching the doctrine for over thirty years it has been accepted by a relatively small minority of geographers; still fewer maintain it in practice. This is not surprising since it is founded neither in the logic nor in the historical development of geography, nor does it offer the basis for a more restricted though unified field. On the contrary, it would disrupt even a "central core" of "pure geography" by placing certain aspects of area outside of the geographer's field of study, except as he later finds it necessary to investigate them in order to interpret what he has studied. While it cannot exclude those fields of geography which have ever been concerned with primarily immaterial features, it would thrust them into ill-defined outer reaches of geography with no better guide to their objectives than that of the study of relationships. By reducing the element of relative location to a purely secondary position it tends to neglect the very essence of geographic thought—integration of phenomena in spatial associations. In consequence there develops an uncritical enthusiasm for patterns as such, regardless of their significance, and an over-emphasis on form in contrast to function which tends to slight those characteristics of areas whose importance is not represented proportionately by material objects.

Although the point of view of this doctrine may encourage the study of settlement forms, if strictly interpreted it would prevent these studies from achieving the significance which they may well have as part of a more rounded study of the geography of culture. The extreme results of this point of view are represented either by the tendency to look for objects not studied by other fields, which would reduce geography to the consideration of the least significant things, or, by the tendency to enter into competition with artistic workers in the subjective comprehension of landscapes.

VIII. The Logical Basis for the Selection of Data in Geography

A. derived from the fundamental concept of geography

What principle or principles should the geographer have in mind in selecting phenomena for consideration in the geography of an area? Of all the vast heterogeneity of material and immaterial phenomena that may be observed in any area, which are significant for geography? If the study is made for some special purpose, practical or otherwise, obviously that purpose will provide the measure of significance for different kinds of phenomena. Our interest here, however, is with studies whose purpose is simply to increase geographic knowledge, studies, that is, in pure geography, in the only proper sense of that expression—namely, in contrast with applied geography. We examine this question from the point of view of furthering geographic knowledge—not "for its own sake" simply, but on the assumption that the furtherance of geographic knowledge will produce values that we need not demonstrate. On this basis we may not permit our selection of data to be determined by any ulterior purpose, whether that be to fit the data to particular techniques of study or to enable geography to lay claim to the title of "science" in a particular sense that would be otherwise unattainable. Geography must aim to provide geographic knowledge, even though the nature of some of that knowledge should cause it to lose any hope of being called a "science" in the sense that certain other branches of knowledge are called "science," just as the geographer must adapt his methods of study to the material of geography rather than select the material suitable to his methods. To state the conclusion directly, the basis for selection of data must be derived logically from our fundamental concept of geography.

Among geographers who disagree in their fundamental concept of the field we can hardly expect agreement as to the basis for selecting and rejecting different phenomena to be studied. Fortunately, however, as we observed earlier, there is a marked agreement among the large number of students who accept what we have called the chorographic concept of geography. The largest number of them, including most of the geographers of Germany as well as many in this country, use a form of statement similar to that which Hettner derived from Richthofen: "Geography studies the areal differentiation of the earth's surface—the differences found in the * different continents, lands, districts, and localities" [161, 122; cf. Sauer, 211; 84].

Everyone who has traveled twenty miles from his home knows that differences exist from place to place; these differences are therefore "naively given" facts. Further, any thinking person is aware that these differences

[237]

* On the term "areal differentiation" see Supplementary Note 35

are not phenomena of independent circles of facts, but are interrelated to each other, so that the human curiosity is immediately quickened with a desire to know what these relationships are.

A person whose knowledge of the world was limited to a local area and who might assume that all parts of the world were much like his own, would, no doubt, be most impressed with differences in those elements, both material and immaterial, which vary greatly within a small area. He might observe great variation in slopes, in drainage, in certain aspects of soil, and in the productivity of farms and the relative prosperity of farmers. If he lived in the Lower Rio Grande Valley he would certainly observe the differences in the cultural character of the population on either side of the river. For many a peasant of Eastern Europe, or of many parts of China, the greatest differences in the world are those between the countryside in which he lives and the local town or city; indeed these differences between the rural and the urban environment are among the greatest differences that the earth's surface shows. They involve not only the differences in density of population and in number and character of buildings, but also in the character of the population, the contrast between "urban" and "rural" types—the "city slicker" and the "country yokel"—a contrast to which Passarge gives much attention.

Anyone who has traveled somewhat more extensively, either in fact or vicariously through reading, knows on the other hand that certain elements which may be approximately uniform over a considerable area vary greatly in different parts of the world. These might include climate, natural vegetation, major landforms and soil types, major types of agriculture, and many cultural characteristics of the population.

All of these considerations are so well known that the geographer may without hesitation start from them. Further, he would recognize in a general way that these differences are interrelated so that, where several basic features were fairly uniform over a considerable area or region, many, though by no means all, of the other features are also fairly uniform. He would therefore recognize, perhaps quite unconsciously, that each of many such areas has a distinctive character in the total ensemble of its phenomena, a character which extends over an as yet undefined but considerable stretch of territory, and which is different in some important way from that of any neighboring area.

For illustration, let us consider a German peasant youth from the neighborhood of Breslau who should happen to travel slowly to the southeast, up the plain of the Oder. For some distance he would observe characteristics much the same as those he knew at home. Fields of rye, wheat and oats, potatoes and some sugar beets on the flat plain of loess soil, would all seem familiar to him; the towns, each with its rectangular market center, its double oval of streets around the oldest part of the town, marking the inside and outside of the former city wall, and the German conversation which he hears, all this tells him that he is still in his own region, his *"Heimatgebiet,"* even though the details be unfamiliar. Farther up the plain, however, (keeping to the east of the Oder) he will note very significant changes. The farms are obviously less productive, he sees little wheat or sugar-beets, and many large stretches of pine-forest; as an intelligent peasant he immediately notes the sandy character of the soil, and needs no geographer to explain the relationships involved. But he also notes—quite possibly first of all—that the people here who look like himself and his folks at home, who live and work much as they do, are to him very different, for they are Polish-speaking, and he will certainly consider this the most remarkable observation he has yet made. Let us go with him one step farther. Wandering across the unfenced fields he may unwittingly cross a line which, however clearly marked on the political map—and in the landscape "observable" as a series of small white posts—is somewhat difficult to observe in fact in the fields. (Since he himself is an observable object this may prove dangerous for him, but we can assume that he is not observed.) It will not be long before he will realize that he has now entered a very different kind of country. For, not only do the peasants speak Polish, but so do most of the people in shops, and in nearly all of the schools and other public buildings, and if he should settle down to live here he will find that the produce of his land is not to be sold in Breslau but in Katowice, that, not only economically but also socially, he and his family are cut off from things in Germany and tied to things in Poland, that their lives will be dominated not from Berlin but from Warsaw, that his sons will see military service not in the Rhineland, but perhaps on the borders of Russia. In sum, it will not take him long to realize that the most important difference between the area to which he has come and that which he has left, is that this is Poland, that was Germany [*cf. 355* and *356*].

It is instructive to consider a similar, though imaginary, situation in the field of history. Various writers have used the fictional device of putting a person of our times temporarily back into an earlier period, though still in the same place. What are the phenomena which such a person would be able to contribute to the study of history? The historian would ask him for those aspects of the past period which could be regarded as characteristic of that period, *i.e.,* not incidental differences which had no relation to other characteristics of the time, and for those events which might have significant relations to events of previous or subsequent periods. Obviously it will often be difficult to say what things are historically *significant*, but the historians have long since recognized that there is no categorical rule for eliminating this difficulty; it must be met, as best it can, in each particular instance.

B. THE CRITERIA FOR THE SELECTION OF DATA

Hettner has stated the geographer's criteria for the selection of data in terms of two conditions that correspond logically to those governing an historical study [161, 129 f.]. His exposition, presented in translation below, was first published in 1905 [126, 561 f.], and has since been accepted by a large number, if not indeed the great majority, of German geographers. We have previously noted that his views on this question have been followed by Gradmann, Hassinger, Banse, Sölch, and Maull, among many others.

"One condition . . . is the difference from place to place together with the spatial association of things situated beside each other, the presence of geographical complexes or systems—for example, the drainage system, the system of atmospherical circulation, the trade areas and others. No phenomenon of the earth's surface is to be thought of for itself; it is understandable only through the conception of its location in relation to other places on the earth. The second condition is the causal connection between the different realms of nature and their different phenomena united at one place. Phenomena which lack such a connection with the other phenomena of the same place, or whose connection we do not recognize, do not belong in geographical study. Qualified and needed for such a study are the facts of the earth's surfaces which are locally different and whose local differences are significant for other kinds of phenomena, or, as it has been put, are geographically efficacious." (This is clearly a development from Richthofen's criterium: "insofar as they have a recognizable relation to the earth-surface" [73, 27; cf. also Penck, 163, 37].)

"The goal of the chorological conception," Hettner continues, "is the recognition of the character of the lands and localities from an understanding of the mutual existence and mutual effectiveness of the different realms of nature [reality] and their different forms of expression, and the conception of the entire earth's surfaces in its natural [real] formation in continents, lands, regions, and localities.

"Only in the application of both these points of view lies the character of geography; anyone who has not taken them into his flesh and blood has not comprehended the spirit of geography; just as an historian who does not inquire after the temporal sequence of things and the inner connections of different groups of development, has not grasped the spirit of history.

[240]

With this conception, to be sure, the choice of material presumes a previous consideration of the causal connection of the phenomena; with the development of knowledge, whole groups of facts can be won for geography or may be abandoned, and the circumference of geography will be differently conceived according to subjectively different evaluations of causal connections. But we find just such fluctuations in the historical and the systematic sciences and no objection to this principle of choice of material is to be understood as a result of them. The choice of material depends not on a single fact, but always on an entire group of facts, which one has learned to regard as causes or effects of other geographical groups of facts. Geography does not take up the particular facts only when it has recognized their geographical conditionality, but establishes their geographical circumstances descriptively, before it goes on to the causal investigation; and it can easily happen that it must introduce facts whose causal connections are not yet clear.

"The diversity of the material according to this conception is indeed large and becomes ever larger, for the progress in knowledge discovers in an ever increasing number of groups of facts a conditioning in the nature of the locality, and therefore geographical character. Present-day geography includes events as well as patterns and material conditions, facts of social life as well as of nature; but it includes all these objects only under the chorological point of view, and can therefore pass by many characteristics and peculiarities which, for the material and historical sciences, are perhaps the very most important. It can omit, not only all conditions which are everywhere on the earth the same, or whose local differences recognize no kind of rule of distribution, but also all those things whose local differences have no relation, at least so far as our knowledge goes, with those of other groups of phenomena."

We may note, in passing, a somewhat similar expression by the founder of modern geography in France, Vidal de la Blache: "Geography is the science of places and not that of men; it is interested in the events of history insofar as these bring to work and to light, in the countries where they take place, qualities and potentialities that without them would remain latent" [*183, 299*].

This distinction, or limitation of the things to be studied, unlike that of the limitation to material things, is founded on the logical position of geography as a science, as well as on its historical development. For "the necessity of a chorological science of the earth surface results from the two conditions, (1) that facts of one and the same circle of phenomena which are spatially close to each other do not lie isolated beside each other, but are mutually related, and (2) that the facts of different circles of phenomena united at one place on the earth stand in causal relationship and together

[241]

determine the character of the area." It is therefore proper for the geographer to extend his study "over all phenomena of the earth's surface so far as these two points of view are applicable to them" [126, 683].

The geographer, we may say, considers the arrangement and relationships of the phenomena within a region; to think geographically is to think of phenomena not as individual objects in themselves, but as elements determining the differential character of areas. Phenomena which exist in an area but in no relation to other phenomena in that area, such as the general phenomena of earth magnetism (where unaffected by local minerals), may be said to have no areal relationships, and are not therefore of concern to geography. *

C. APPLICATION OF THE CRITERIA

We may draw a number of conclusions of practical value in applying the criteria stated, if we first emphasize three particular concepts involved in the criteria. These are (1) the *interrelation* of different kinds of phenomena that are directly or indirectly tied to the earth; (2) the *differential* character of these phenomena and the complexes they form, in different areas of the earth; and (3) the *areal expression* of the phenomena or complexes. On the basis of our fundamental definition of geography as the study of the areal differentiation of the world, it follows logically—and has commonly been followed in practice—that only such phenomena as are significant according to all three concepts are to be described and interpreted in geographic study.

The requirement of interrelation, or *Zusammenhang,* of phenomena provides us, as Gradmann has explained, with a specific method of selecting geographic objects of study, whereby we can set aside "the dilettante and arbitrary method, formerly much in favor, of putting mere curiosities together. The individual fact enters with a degree of importance that increases with the extent to which it is interlaced, on many sides and internally, with neighboring circles of phenomena, both forwards and backwards as cause or as effect" [144, 8].

In other words, the answer to Schlüter's argument that the phenomena of non-material culture are not related in detail to the details of the natural elements, as are the material facts of the cultural landscape [145, 29 f.], is to accept the argument as a statement that is in relative degree true and to apply it to just that degree. But factory buildings, cities in general, or the distribution of population, are facts that are not related in detail to all the details of the natural elements. Insofar as the cities of our South show distinctive characteristics different from those of the North and the differences are related to other regional factors, they are significant to a regional

* See Supplementary Note 36

comparison of South and North, but they are not so important for such a comparison as the distinctive agriculture of the South, which is much more intimately related to its particular climate, soils, and population structure. Social phenomena that show relatively little relation to other regional characteristics need not be formally excluded; their relatively low degree of interconnection automatically makes them of little geographic significance, a conclusion that Richthofen clearly indicated [73, 63–5].

Consequently we may agree in part with Penck [163, 51 f.] : insofar as the state is considered as a social organization of people, it is a social phenomenon whose characteristics—whether described in terms of the form of government, the personality of its rulers, or its youthful, mature, or old-age nature—is a cultural phenomenon dependent largely on other cultural phenomena which may have but little relation to facts tied to any particular earth area, and therefore will call for little attention from the geographer. But the state (as we will later note in detail) is not only a social organization, it is at the same time the organization by man of a particular section of the earth's surface, and in this respect is inevitably closely interrelated to basic earth facts.

At this point some readers may protest that environmentalism is once again "raising its ugly head" and those in whom the traditional over-emphasis on relationships in American geography has developed a phobia against the very idea of considering phenomena in interrelation will part company with us—without noting perhaps that Schlüter and Penck depend upon the concept in their expositions. Our historical survey showed that geographers have from the beginning been concerned with the study of phenomena of different categories in interrelation, and no "landscape purist" can interpret the facts found in his landscapes without discussing "relationships." To insist that for phenomena to be geographically significant they must be causally interrelated with other regional phenomena, is not to define geography as the study of relationships; if we say that a house cannot be built of bricks without mortar, we are not saying that a house consists of mortar. (Presumably it is this misunderstanding that caused Michotte to refer to Hettner as "defending, with certain variations, the same ideas" as Brunhes—namely, the study of the interdependence of phenomena as the definition of geography [189, 12, footnote].)

The second requirement listed, the differential character of the phenomena with reference to the various parts of the earth, is based not merely on our fundamental definition as derived from the historical development of the field, but equally well, if one will, on common sense [cf. also Sauer, 209 17]. Just as the historian of the Roman era does not need to tell us that at that time people found it necessary to spend a considerable part of their lives sleeping, that they put on more clothing in the winter than in the summer, or that mothers commonly showed an affectionate interest in the welfare of their offspring, neither is it necessary for the geographer studying any particular district of the world to tell us things about it that are true for any land area of the world. If he is describing a farming area, we may assume that the population live in some sort of permanent structures built above ground in which the livestock are not housed—only in the exceptional cases where conditions are different is it necessary to describe them.

If these statements appear ridiculously obvious they are nevertheless required as contrasts to certain quite logical conclusions that have been derived from the concept of geography as the study of the visible landscape. That we are not indulging in dialectics is shown by some of the detailed work of Granö, the student who has perhaps most logically pursued the landscape concept. He has made a detailed study of a small rural district in Finland, Valosarri, to which the geographer who has never seen that country might naturally turn in order to learn the character of such a district there in contrast, say, to one in Iowa. He will find that Granö has not only mapped the individual buildings with care, but on several separate maps has shown the surfaces of the ground covered by these buildings as rainless areas, and as areas protected from wind and cold. Likewise he will learn that in this district of Finland there is less wind in the forests than in the fields, that birds can be heard more often in woods and meadows than near the houses, and that there are different kinds of odors in each of these places [252, 126–34]. This description of Granö's study is not intended to belittle it, but rather to show the difference in kind between a detailed landscape study as logically worked out, and the point of view in regional study that most students would regard as geographic.

This latter point of view, dominated by the concept of areal differentiation, we may illustrate by the contrast in geographic interest offered by a farm house and the farm fields. Though in an absolute sense the house might be regarded as of equal if not greater importance than the fields, in terms of areal differentiation of the different parts of the world, it offers much less of significance. Regardless of where the farm is located we know a great deal about the house before we see it, and to see it will tell us relatively little about the character of the area. Merely from its name we know that it is a man-made structure elevated above the ground, that includes physical facilities used by the family for sleeping, cooking, eating, relaxation, and as a retreat from unpleasant weather out of doors, and that some members of the family perform much of their productive work within it. In major degree the forms of houses are everywhere adapted to these functions, so that the differences between them are either minor, or are of minor significance in relation to other regional factors. They are minor, that is, in contrast with the notable and significant regional differences in the use of the farm fields—in crops produced, methods of cultivation and harvesting, yields attained, and ultimate manner of consumption.

It may be objected that this point of view is bound to give an unbalanced presentation of reality in any area, but this is true only if the reader of such a study lacks the universal concepts whose description is omitted because assumed. Unquestionably there is some danger here, just as for the ignorant student, history may give an impression of life in the past as made up entirely of wars and political and economic changes. To be sure, the fellahin of Egypt have sown, cultivated and reaped the land flooded by the Nile through every year of its many milleniums, but the historian is not called upon to describe the annual cycle of Egyptian life for every period of its history, any more than he needs to record the fact that in any particular generation, a million or so babies were born, although that historical fact was absolutely essential to all the following history of Egypt. In other words, history assumes countless unnamed universals and similarly in geography we may assume that a house shuts off sun, rain, wind, and cold, more or less effectively; that farmers work in the fields by daylight and spend the nights in their houses, etc.

On the other hand, the emphasis on differential factors is not to be exaggerated falsely into a concern for the rare and unusual feature, as Krebs has noted [255, 342]. An exceptional feature that attracts our attention, whether because it is seldom found anywhere, or because it is of general importance in other regions—a Swiss chalet in the Appalachians, for example—is significant to the geographer only insofar as it is significant in the area he is studying. In writing the geography of the Po Plain, one must resist the temptation to dwell on the very small districts of minor rice production, simply because they are rare in Europe, or one will lose the significance of the far more important and significant combination of corn-wheat-and-dairy agriculture with the horticulture of vine, fruits, and nuts. Similarly, Kemp has recently emphasized that geographers concerned with the Balkans should pay less attention to the scattered tobacco fields, or the rose gardens of Kazanlik, and far more attention everywhere to corn (maize).[77]

This last consideration is closely related to the third concept involved in our criteria: phenomena are significant in geography to the extent that they have areal expression. Although all might agree on this bare statement—we have previously noted its expression by Schlüter and Sauer—there is a

---

[77] In a paper read before this Association, 1937.

notable difference in its interpretation. Schlüter evidently measured this attribute in terms of the physical extent of the object concerned. Similarly Penck, in ruling studies of distribution out of geography, explains that geography is concerned not with plants but with forests, not with men themselves, but with their effects on the earth surface [163, 44]. But Krebs reminds us that, in Ritter's classic phrase, it is not the objects that fill areas that are themselves of concern to the geographer, but the areas [234, 83]. No particular category of object, not even the trees that together form a forest, nor the fields that form the major part of a rural scene, actually fill the area that we unhesitatingly map as characterized by that particular phenomenon. Furthermore, some objects that occupy but a minor fraction of the surface, may be of such great importance in their effects on other features, that their significance in areal differentiation is out of all proportion to the space they occupy physically. We have previously noted, for example, that if the types of houses in rural Louisiana furnish a reliable guide to the culture that has been developed throughout each of its different sections, Kniffen is entirely justified in giving such great attention to objects that Schlüter would exclude because they occupy so little space. Similarly, though the actual area on which the Negro population of our South stands, sits, or lies, is an insignificant fraction of the total area of the South, the presence of so many people of that particular social group is a factor of the greatest significance in the agricultural and urban character of the South in contrast with almost all other regions of the world.

Schlüter has, in fact, accepted this viewpoint, at least in part, in one of his discussions. In order to permit the geographer to study the scarcely visible facts of population density, he postulates as necessary for geographic data that the phenomena must each have "a specific location and a specific areal extension on the earth surface" [145, 27]. But it is obvious that we are not interested in the area covered by individual human beings, but rather the area of the earth in which such and such kinds of people are found in such and such numbers—it is the character of the area itself, that is given it by the presence of the human beings. If we can consider that as a real character of the area we can equally well consider the low standard of living of the population of that area, in contrast to higher standards in other areas, as a characteristic of the area.

On much the same basis we can approach the difficult problem of what size of areas may be considered. Penck has argued that one can speak of a geography of a city, but not of a market place [137, 165], but it is not clear what difference in principle is involved. If the geographer however keeps in mind the question of relative significance, he may regard size as but one of the important attributes to be considered. An understanding of the rural geography of Mexico necessitates a knowledge of the peculiar structure of Mexican rural communities. If Dicken was justified in concluding from reconnaissance that in many important respects the village of Galeana is representative of the character of many similar communities in a particular, fairly large, area of Mexico, then his detailed study of that small and unimportant village [*340*] has more significance for the geography of Mexico than would the study of such an exceptional city as the oil port of Tampico.

D. RELATION OF THE CRITERIA TO THE SPECIAL TECHNIQUE OF GEOGRAPHY

It may be instructive to apply Sauer's argument, concerning the relation of a special geographic technique to the material to be studied in geography, to that which is included by the criteria here discussed. If the geographer has any specific technique, what is it, and does it fit him to consider all the multitudinous variety of phenomena here included?

If we examine the product of geographical work, as contained in the volumes of any of the leading geographical journals, including therefore work of many different kinds, what particular technique is disclosed as common to all, and yet relatively little used in other fields of knowledge? Surely it is the technique of cartographical presentation, the one technique which geographers have developed in a great variety of rich detail. Nowhere is this better illustrated than in Bowman's unique exposition of the points of view, techniques, and kinds of data employed in current geographical work [*106*, note the large number of different kinds of maps, as well as the discussion of maps, 104 ff.].

Geologists, to be sure, are also dependent on maps, but for most parts of geology the maps are but minor means toward the ultimate purpose of understanding the history of the earth. Though the making of the standard topographic maps is entrusted in this country—in contrast to most others—to the Geological Survey, the work is placed under the direction of a "geographer."

The geographer's claim to a special technique in maps hardly seems to require demonstration. Many other disciplines use maps, just as they must also, at times, use the historical method, but no one questions that the latter is the distinctive technique of history. The field of history, to be sure, may not claim a monopoly in the historical method and neither may geography claim cartography as an integral part of its field (see Sec. XI E). Nevertheless, workers in other fields commonly concede without question that the geographer is the expert on maps, whether in making or in using them. This is the one technique on which they most often come to him for assistance, and often in such cases they come to realize that the geographer can do far more with maps than they anticipated. "The most important contributions of geography to the world's knowledge," James holds, "have come from an application of the technique of mapping distributions and of comparing and generalizing the patterns of distribution" [286, 82].

It is no accident that geographers above all others should be concerned with maps. How otherwise can the geographer assemble the facts of spatial relationships for study, how otherwise present them in their spatial relationship, except on maps? (Insofar as the photograph can do this for relatively restricted areas, it is an important supplement to the map.) Consequently we can often judge the geographic quality of a man's work by the effectiveness with which he has presented it on maps. One wishes that those editors of geographical publications who all too often reduce a valuable detailed map to the size of a half-page, or less, would post the following statement of Hettner's, written in 1905, over their editorial desks: "In consequence of the development of cartographical methods of presentation, verbal description has lost its original importance and serves now only to complete and explain the maps" [126, 685, 622–4; see also 161, 324–376].

It requires, finally, no discussion to prove that the cartographical technique of the geographer can be applied to non-material phenomena as well as to material phenomena. It has been so used hundreds of times. Indeed, the distinction in the study of a particular territorial problem, between the work of a geographer on the one hand and that of a political scientist on the other, is no more clearly indicated than by the relative degree and effectiveness in the use of maps. This distinction is not merely in the maps themselves, but rather in the use of the maps in the study—the true geographer cannot help but study the problem in terms of maps.

An excellent example of the value of this technique is offered by Bowman's *New World* [332]. While this deals with problems which many political scientists have studied, its extraordinarily effective use of maps has made it a well-nigh unique contribution.[78] Clearer examples may be found in the detailed studies of individual border areas, of which the German and French geographical literature could provide dozens of examples. Thus, the

---

[78] The writer may be permitted to correct here his description of this work in a previous publication as "consisting in large part of the materials gathered for the American Commission to the Peace Conference" [216, 785]. This statement was made on no authority other than that of a widely-held notion which Dr. Bowman assures me is erroneous. Although the stimulus to make such a study, as well as much of the general information concerned, no doubt came to Dr. Bowman as a result of his work for the American Commission at Paris, the actual materials used were all gathered after his return to this country and did not include the materials brought back by the Commission from Paris.

writer's discussion of the boundary problem of Upper Silesia depends primarily on the use of 16 maps [355].

It should not be supposed that this technique is limited, in political geography, to the study of boundary problems. The fact that it is of great importance in other parts of political geography, as well as in various other topics in what may be called "social geography," has been demonstrated many times in well known works by Ellsworth Huntington, Griffith Taylor and Mark Jefferson [cf. also Brunhes 182, Figs. 206, 211, 220–222]. More particularly, illustrations are offered by the writer's study of the problems of sectionalism in pre-war Austria-Hungary [358, based on over 20 maps], by Milojevič's study of similar problems in Yugoslavia [370], by Wright's study of regional tendencies in national elections in this country [400], and by the writer's study of the racial geography of the United States [359].

So important, indeed, is the use of maps in geographic work, that, without wishing to propose any new law, it seems fair to suggest to the geographer a ready rule of thumb to test the geographic quality of any study he is making: if his problem cannot be studied fundamentally by maps—usually by a comparison of several maps—then it is questionable whether or not it is within the field of geography.

IX. THE CONCEPT OF THE REGION AS A CONCRETE UNIT OBJECT

A. VARIOUS STATEMENTS OF THE CONCEPT

From the very beginnings of modern geography, late in the eighteenth century, geographers have been troubled by the nature of the areal units into which they divide their object of study—the world. As we observed in our historical survey, the followers of Gatterer saw a first step toward a more scientific geography in replacing the traditional division into political units by the division into "natural regions" offered by the theory of the continuous network of mountains. When this theory proved untenable, they did not abandon the concept of "natural regions" but sought to define them in less simple terms. Thus Ritter, though challenging the over-simple system of his predecessors, furthered the establishment of the concept in general as a fundamental basis for regional geography. The detailed critique of Bucher was passed over without answer and the more dramatic objections of Fröbel were lost sight of in view of the erroneous basis of the remainder of his attack. During the relative decline of regional geography in the second half of the nineteenth century, there appears to have been little concern over this concept, though Ratzel appears to have expressed it on occasion [cf. Bürger, 11, 76].

With the renewed interest in regional geography at the turn of the century, the concept of regions as definite, concrete, if not natural, units, reappeared—first, so far as I can find, in the methodology of Schlüter; whether it represented an inheritance from Ritter through Ratzel, as Hettner has suggested, is not entirely clear. But it is only in the past decade or so that we find the full flowering of this concept in the geographic literature not only of Germany, but also of various other countries, including our own. The region, or *Landschaft,* is said to constitute a definite individual unit that has form and structure, and is therefore a concrete object so related to others like it that the face of the earth may be thought of "as made up of a mosaic of individual landscapes or regions." Further, for some students, in this country as well as in Germany, the region is an organic object, comparable to biological organisms.

In contrast with the "purist" interpretation of the landscape concept, which we found to be maintained even in theory by but few geographers, the present concept is supported, in one form or another, by what appears to be the great majority of regional geographers in Germany and by many in this country. Bibliographies of the discussions centering on this question run into several hundred titles [fairly complete lists of German titles are provided by Granö, *252,* up to 1929 only, and particularly by Bürger, *11,* till 1935].

We need hardly state that this concept, in whatever form expressed, does not represent the statement of a scientific axiom so obvious that it requires no demonstration to be accepted. On the contrary, when geographers say that regions are individual objects like plants, animals, or stars, they are telling us something that is hard to believe. Though the areal differentiation of the earth surface is a "naively given fact" and the same might seem to be true of a landscape as an individual actual scene, it certainly is not true of the regional divisions of the world. Not only are laymen unaware of regions as definite objects, but in few if any cases would geographers agree on how much of the earth is included in a single region. Indeed they do not agree on how one may determine that presumed fact. Nevertheless Renner is expressing a very popular article of belief when he asserts that "regions are genuine entities." Extraordinary only is the context in which this statement is found: namely, in what purports to be a summary conclusion of views expressed by a number of "regional experts" in answer to a questionnaire, views however that showed a complete lack of agreement on the definition, delimitation, or essential character of regions [*291,* 141, 145–9]. How genuine are entities of which we know so little?

If the thesis, then, is not obvious in the nature of things, is it the product of geographical research, or has it been constructed to interpret research findings not otherwise explicable? On the contrary, no student, so far as I can find, has made any attempt to show that the thesis is a necessary conclusion from research studies, nor has anyone demonstrated that it has value in explaining facts or relationships that could not be explained just as well without it.

We are concerned, therefore, with a hypothesis that is neither self-evident nor the product of geographic research, but is constructed by what for lack of a better term we may call philosophic thinking about geography. This is in no sense a criticism—to establish concepts in any field of knowledge, philosophic thought is required at least on the part of those of its students who propose new concepts. When Passarge, who is responsible for more new and ill-defined concepts and systems than any other modern geographer, tells us that he cannot bother to read philosophical discussions of his ideas, he is merely saying that he is not willing to help sort the harvest of grain and weeds that he has sown. (He can hardly expect to be taken seriously when he tells us that "he knows only what the facts themselves tell him") [*272*].

Consequently, any pertinent discussion of this concept inevitably involves a somewhat laborious analysis and comparison of terms, which no doubt many geographers will find irksome. But the critic of current geographic thought cannot avoid the task of examining in detail the terms used to express this important concept. This task is made all the more difficult by the multiplicity of terms, borrowed from philosophy and psychology and inserted into definitive statements concerning regions, commonly with new meanings, often ill-defined, often not defined at all [cf. Wörner, *274, 340*]. It is only necessity, not a love for argumentation over terminology, that causes us to enter this jungle of terms; however, the writer can assure any who will follow with him that the path will not end in some unspecified recess of the dark forest—however tortuous the path others have laid for us to follow, we will clear our way through to daylight.

To those for whom such philosophic ventures have little interest we may indicate the path that avoids the maze entirely—though it eliminates the opportunity of discovering what element of truth the concept, or its various derivatives, may possess. Neither the major thesis itself—of the region as an object—nor any of its various forms and derivatives, has been adequately established in methodological thought or tested in research. The individual research worker is still justified in assuming the obvious—namely: that "the region is emphatically not an organism," to use Crowe's categorical statement [*201*, 10]; and that the *Landschaft* as an area is not a concrete object nor an individual unitary whole, but—to modify Leighly's statement slightly—is only a more or less "arbitrarily chosen fragment of land" [*220*, 130]. Consequently, so far as we can see at present, the face of the earth is the very antithesis of a mosaic—much closer to reality, presumably, is Huntington's expressive picture of "The Terrestrial Canvas" [*213*]. Though these statements stand just as devoid of demonstration as the assertions they deny, they have at least the validity of stating what appear to be the facts.

On the other hand, those to whom the concept of regions as objects appears to be of value in geography—in particular those who express this concept whether in methodological treatments, regional studies, or in textbooks—will presumably wish to see it subjected to critical examination in order that both its validity and its utility may if possible be established. Certainly we are not permitted to dismiss in summary fashion important concepts that may have great value for our field merely because they appear to fly in the face of the obvious. If one suspects that Crowe may be right in assuming that these ideas are possible only because "German geographers have woven such an impenetrable web of mysticism about their '*Landschaft*'" [*202*, 15], one must remember that eminent American geographers who are not given to mystic philosophy have accepted these concepts as fundamental axioms of regional geography. [*]

[252]

Before we part company from those who wish to avoid the discussion of terms in the remainder of this section, we must register a warning against one conclusion that might appear to follow from what has just been said. The fact that many geographers have stated the concept in question as a fundamental axiom of regional geography does not mean that regional geography itself necessarily stands or falls with the axiom, nor that the more general concept of geography as a chorographic subject depends upon this axiom. Such a conclusion appears to be implied in Leighly's discussions [220; 222], and also in the argument between Dickinson and Crowe [202]. But regional geography proved fruitful of results long before this concept was stated as a fundamental axiom and the chorographic view of geography as presented by Hettner depends in no way on the assumption of regions as "entities," as unitary or concrete objects.

B. THE PURPOSE OF THE CONCEPT

Since the hypothesis with which we are concerned represents an intellectual construction rather than either a statement of obvious fact or a product of research, it is appropriate first to inquire as to its purpose or value in geographic research. If we accept this concept are we thereby better able to perceive facts concerning regions? Does the concept, for example, indicate to us at once the presence of the actual regions that it postulates? Few, if any, of its supporters would claim that. Schlüter has recognized that the objects of geographic study must be constructed "by intellectual activity" [148]. Likewise Granö has repeatedly stated that areas are "not absolute Units," . . . "the determination and separation of the geographic individuals is in itself a problem of research." Furthermore these geographic individuals do not accurately represent reality: "the geographic region, representing the essential parts of reality, is brought into being by purposeful simplification of the multiplicity actually observed in reality;" in this manner, "geographic research must . . . construct the whole units that it needs" [245, 13; 270, 296–300].

In other words, regardless of what our concepts of regions may be, we still face the problem stated by Wellington Jones: "The determination of areas of homogeneity inevitably must come late in an investigation. Only after an ample and sound body of data has been gathered, and the significant relations between various categories of data have been ascertained, can the important or essential homogeneities of areas be determined. And only after such homogeneities have been established can boundaries of areas of homogeneities be drawn with any approach to precision" [287, 105 f.]. Although this statement would no doubt be granted by many of the proponents of the concept under consideration—has indeed been stated by many of them [e.g.,

[253]

Maull, *179,* 175]—they do not appear to recognize that it constitutes a challenge to the utility of the concept itself.

The proponents of the concept have, however, a definite answer to this challenge. Other sciences that have particular categories of objects to study have been able by analysis and comparison of the structure, forms, and function of their objects to classify them into generic groups on the basis of which they have developed scientific laws or principles concerning their behavior or relationships. If geography can develop such a system of generic classification of regions as objects, it may hope likewise to progress to the statement of general principles. Unquestionably this is a worthy ambition, imposed upon us, we may say, by the very spirit of science. Any concepts or hypotheses that appear to offer such possibilities deserve our most careful attention. But likewise the spirit of science imposes upon us the need to subject such concepts to the most critical examination. Since nature (reality) has been so unkind as not to present us with obviously individual concrete objects, such as those that are ready to hand for the astronomer, or zoologist, and we must construct our own, by intellectual activity, it follows that any principles we attempt to develop can have no more validity than the "objects" we have constructed as their foundation. It is no step forward merely to assert that we have objects and proceed on that basis without making positive that we really have them.

Finally, one cannot read many of the discussions of the proponents of this concept without observing, intermingled with the hope of producing intellectual tools of scientific value, the more direct ambition to elevate geography to a "higher" order among the sciences, indeed to make geography what it apparently has not been, a true science. Since each of the (systematic) sciences has its own class of objects to study, geography to be a science must, it seems, have its particular objects. It is not sufficient to say we study the areal differentiation of the world, for almost all the objects to be found in the different areas of the world are already claimed by other sciences. Unless we were content to take over objects that no other science has as yet bothered to consider, as Leighly appears to suggest, we must take the areas *per se* as our objects, and somehow, therefore, we must establish that the areas are in reality objects.

Specifically this purpose is evident in the early statement of Schlüter, although his concept of the area as a concrete object is by no means so simple and direct as that expressed by his later followers. At one point, he appears to treat the individual areas as objects, but shortly thereafter recognizes that the areas are simply parts of a total unit [*127,* 16 ff.]. But he then shifts to the concept of the entire earth's surface as the unit concrete object that

provides geography with a concrete object of study.[79]   On this ground
Schlüter endeavored to refute Hettner's contention that, like history, geography was primarily not a systematic science [127, 14–8, 52–9; similarly
Penck, 158]. As a systematic science he claimed for it the same right to
divide its object into any number of parts, as biology has to consider the
hair, skin, and other organs of its individual plant or animal. Overlooking
the question of whether the comparison of a region with an organ of an animal is sound, what kind of a systematic science would that be which was
limited to the study of one single specimen of one plant or animal, a specimen
in which no two parts were even essentially alike! Whatever name such a
science might claim, it would be in no position to develop principles on the
basis of its one object.

Schlüter, we may add in order to clarify the record, apparently arrived
at his conclusion by confusing Hettner's concept of "systematic" science with
"descriptive" science, which is to destroy the meaning [127, 56]. Insofar
as a systematic science, like zoology, is descriptive, it describes the characteristics not of a single object, but of all specimens of a kind that are essentially alike [cf. Graf, 156, 52–7]. In the case of the earth's surface, however, Schlüter wrote that that could only be studied "according to its spatial
relations" and so arrived, without realizing it, at the same basic conclusion
as Hettner [132, 631].

Schlüter's followers, however, have been more emphatic in their statements. Thus Granö writes that "we conceive the areal wholes, by and large,
just as the biological sciences do their objects" [252, 38, 47; see also Maull,
241, 12].

Likewise in the presentation of this concept to American geographers, in
Sauer's methodological treatments, one observes the repeated inference that
in the geographic area our field has objects of study comparable to those of
other sciences. "We assert . . . that area has form, structure, and functions;" to speak of "the anatomy of area" . . . "is not a gross analogy . . .
for we regard the geographic area as a corporeal thing" [211, 25 f.; 84,
189 f.]. Much more direct expression of this view is given by Finch and
Trewartha in introductory statements in their recent text, even though one
finds little or no use of the concept in the text material itself [322, 1–9,
662–6].

It is most significant that this underlying purpose of the efforts to conceive Landschaften as objects should find its clearest expression in the summary presentation that Bürger has given of the work of German proponents

---

[79] Various writers besides myself have found difficulty in following Schlüter's
reasonings; cf. Hettner [132, 627–632] and Graf [156, 142]. If Schlüter's critique of
Hettner is worthy of careful attention, as Sauer recommends [84, 187, footnote], so
also is Hettner's full and effective reply which Sauer overlooks.

of the concept. "This concept now has for us a deeper significance. The struggle over the geographical concept of area (*Erdraum*) was a fight for the validity of the geographical science in general. . . . Geography is essentially independent only if it possesses a concept of earth area of its own. The more significant this concept of area appears, the higher will be the respect given to the science of geography. . . . The areal units in question may not be artificial but must be 'given,' as it were, marked by nature [in the sense of "reality," perhaps] . . ." [*11, 27* f.].

One might well wonder whether a field that had to seek concepts on the basis of which it could lay claim to scientific standing would be likely to be granted that title. Certainly we will gain little for geography if it should be discovered that we have merely been deluding ourselves with ill-defined and confusing terms.

In particular it appears dangerous to base a fundamental statement of the definition of geography as a science on a hypothesis that is anything but obvious and for which no scientific proof has as yet been offered. This is to beg the question in the definition itself even more completely than does the concept of geography as the study of "relationships;" for no one (unless we must recognize Gerland as an exception) ever denied that relationships exist, whereas we have yet to be shown that there exists a single region that is an individual concrete unit.

If the hypothesis gave promise of being demonstrable, one might urge all geographers to concentrate their efforts to establish it. We might pick one or two large areas and employ research funds to encourage a host of students to make independent investigations to ascertain whether it does consist of unitary whole *Landschaften* or regions. Before engaging our resources in this attempt, however, it would be well to examine the logic of the hypothesis with some care; if that be at fault, no amount of field work can prove the thesis sound.

C. IS THE GEOGRAPHIC AREA AN ORGANISM?

We may conveniently begin our examination of concepts or regions with what might be regarded as the most extreme view, certainly the one that has aroused the greatest opposition, so that many who regard regions as definite concrete objects take issue with their colleagues on this point—namely, the region as an "organism" (*i.e.,* in the sense of a biological organism; the possibility of applying the term in some other sense will be considered later).

We noted in our historical survey that Ritter, in company with many of his predecessors, regarded the earth as a whole as an organism, and his description of the continents as "individuals" or as "organs" might be regarded as leading naturally to the concept of smaller areas as "organism."

[256]

Likewise Vidal de la Blache, in dependence on Ratzel, was particularly interested in the concept of the earth, or the earth's surface, as "the terrestrial organism."[80] Whatever the historical origin of the concept may be, for many geographers today the term that Ritter used in his characteristically loose fashion to describe the intricate interlacing of all earth phenomena, animate and inanimate, material and immaterial, is accepted as a precise statement of the character of a region. One of the more famous presentations of the concept is that of Bluntschli, in his study of 1921 on *"Die Amazonasniederung als harmonischer Organismus"* [231]. Two years later Krebs wrote of *Landschaften* as organisms, comparable with biological organisms [234, 81, 93] and similar views have been expressed by others; thus both Obst and Geisler speak of *"Raumorganismus"* [178, 9; 345]. Similar terms have been used in England, by Unstead [309, 176, 184 f.], and have also been introduced into this country, even in text books.

The manner in which this concept was first presented in modern American geography is paradoxical. In a discussion of Vidal's concept of the unity of "the terrestrial organism," Sauer quotes Vallaux' conclusion: though the concept had been very fruitful of results, "what strikes us above all today is the poetic and metaphoric character of the concept, the limitation of its point of view and the positive errors to which it leads" [84, 181]. The sentence immediately following in the original should however be added: "Let one conduct it [the concept of the terrestrial organism], if one will, to the scientific Pantheon with all funereal honors; but let one not neglect to seal the stone that it rise not from the tomb" [186, 49]. Only a few pages farther on in Sauer's discussion, the concept is actually resurrected, even though modified and qualified form: "we regard the geographic area as a corporeal thing" in which, following the former physician Passarge "we are to study the anatomy of area." More specifically, in his previous study, Sauer had written, "the landscape is considered in a sense as having organic quality."

Sauer, to be sure, specifically emphasizes "the fictive character of the region as an organism" [84, 189 f.; 211, 26]. Fictive analogies, however, have a way of developing into direct statements of fact. At one time Finch suggested—in a statement that offers an interesting echo to that of Butte over a century ago [cf. Sec. II A]—"a geographic region, or even an arbitrarily-chosen portion of the earth's surface, may be thought of as having some of the qualities of a human being. It is a thing . . . with physical and cultural elements so interwoven as to give individualism to the organism"

---

[80] According to Vallaux, this phrase correctly expressed Vidal's thought although he modified it on occasion [186, 41, 49]; note the use of two different forms in De Martonne's edition of Vidal's "Principles" [184, 5].

[*288*, 114]. In the work which he wrote with Trewartha, however, the concept is stated (by Trewartha, one infers) without qualification: regions are "functioning organisms" comparable with plants [*322*, 4–5, 662]. Likewise James, in changing from a symposium of scholars to a textbook, has made a similar shift from "pseudo-organic" [*286*, 79], to "organism" [*321*, 124, 155 f., 353]. We are not, therefore, dealing with a concept proposed by erratic or extreme writers.

Needless to say these suggestions have not passed uncriticized. To say that a region, or *Landschaft,* is an organism, is to imply more than that it is in some particular unstated manner like an organism; the statement asserts that the region has the qualities that are inherent in organisms. Thus it is not sufficient to establish that a region includes animate and inanimate things in close interrelation—one could say the same for a spade-full of soil. Likewise, as Finch notes, there is no difference in this respect between a "geographic region" and any other section of the earth surface, however chosen. At the most, therefore, one would have, not a closed, individual organism, but simply something organic, a part of an organism.

An organic combination of animate and inanimate matter, however, involves something more than a close interrelation of those elements; it involves, according to Wörner, connections of the Whole as the superstructure to which the physical elements are subordinate. We find this, for example, in the human organism, whereas in any section of land, or even for the whole earth surface, that form of organic, final superstructure cannot be recognized [*274*, 346].

We may make clearer the difference involved in the previous statement by listing, with Vallaux, some of the characteristics of organisms: powers of adaptation, of cohesion, of reaction, and of re-creation. In his lengthy critique, Vallaux observes in some detail that neither the earth surface nor any areal portion of it can possess these powers. A region is acted upon, of itself it never reacts [*186*, 50 ff.].

Bürger has drawn sharply the distinction between a region—which he regards as a Whole or *Gestalt*—and an organism, at just the point where they might most easily be confused. In both cases he notes a multiplicity of different parts combined to form a whole. In the organism however this whole is capable of life because there is a differentiation between the parts such that each part has a particular function prescribed for it by its place in the whole organism. There exists, therefore, a general functional harmony that subjugates the individual part (organ) to the laws of the whole, and limits its separate existence; it cannot be understood by a consideration that is separated from the entire organism [*11*, 45, 47 f.].

It follows from this description of an organism that its parts cannot be considered in themselves as organisms, but only as organs, members, or organic parts of an organism. As Penck stated more briefly, an organism is essentially indivisible, whereas any regional unit of the earth surface can be divided into smaller units and these into still smaller units [*249*, 8].

In view of the persistence of this concept in geographic thought it is evidently not sufficient to demonstrate, as did Vallaux, that neither the earth surface nor any section of it is an organism. We evidently need to pile still more stones on the tomb to which he consigned this concept, and we may properly do that by concluding that a region is not even like an organism—not in the sense that there are no resemblances, for one can always find resemblances of some sort, but in the sense that the degree of resemblance is far less than the differences. Even as an analogy—if the concept could be kept within the bounds of analogy—it is dangerously misleading, as Vogel, Schmidt, Broek, Lautensach, and Crowe all emphasize [*244*, 197; *180*, 51–4; *333*, 10; *278*, 16; *202*, 15].

The analogy with an organism is apt to be particularly misleading at just the point where it might be supposed to have value, namely when we consider what is frequently called the "growth" of regions, or—to use one of the most widely favored phrases—"the development of the cultural landscape out of the natural landscape." In organic growth, all the individual parts develop from a common origin (the fertilized seed), are nourished from a common food supply, and are controlled in their growth by some common directive agency. External elements introduced into a single part of the organism are either converted into materials that are spread through the whole, or are expelled, or, in the abnormal case, are immediately recognized as "foreign bodies" and isolated, as in a cyst. What do we find comparable to this in the alteration of an area of the earth? The soil erosion of any single slope may be entirely independent of all conditions in other parts of the area; the growth of a single tree is dependent only on the immediately surrounding conditions; what takes place in all the rest of the area may be of no importance to it whatever. The rainfall conditions are largely the result of external forces quite independent of changes in the area itself. Finally the cultural landscape developed by man cannot be understood either as a growth within the area nor as a process of digestion of external materials by the area as an organism: cultivated plants are introduced not into the area as a whole, nor into any common digestive organ, but first into some particular field. Foreign capitalists and engineers may insert factories in a region of primitive subsistence economy, as though a surgeon were to put a backbone in a star-fish.

As Creutzburg concludes, the region as a whole does not undergo changes, but only the complex of different regional elements changes with changes in its elements. The development of a *Kulturlandschaft* is therefore nothing but the development of the complex of its cultural elements [*248*, 413]. To this, Broek adds that the changes may be largely the result of external influences, as he demonstrated in the case of the Santa Clara Valley. Consequently he speaks not of "development" but more properly of "changes" [*333*, 10]. Again, we may note, that insofar as the analogy of an organism is at all applicable, it can be applied only to the whole earth surface: the interpretation of changes in any single area of the world must be extended into neighboring and even remote areas. It is as though one man's lungs were connected with those of some of his neighbors, his heart with those of others, and his lymph derived from people living many miles away.

Finally, we will have occasion later to discuss a further confusion that results from the attempt to consider types of areas as analogous to species of organisms. In the organic world the relation of individual members of a species, and of different species to each other, is the result of a past organic relation—their evolution from common origins. That any attempt to carry this principle over into the relation of different, widely separated areas of the world can only lead to confusion, hardly requires demonstration.

D. HARMONY AND RHYTHM IN THE GEOGRAPHIC AREA

If the concept of the region as an organism is an expression in scientific terms readily subjected to critical analysis, the increasing use by many German geographers of the concept of "harmony," and Volz's suggestion of "rhythm," in the *Landschaft*—whether as landscape or as region—introduce concepts from non-scientific realms of knowledge with which few of us are familiar.

Presumably one might trace the concept of harmony in the interrelation of regional phenomena to Humboldt, though I do not find that he ever used the term in other than a generally descriptive fashion. Modern German geographers who have discussed it in detail commonly refer to Bluntschli's description of the Amazon floodplain as *"ein harmonischer Organismus"* [*231*]. Many who reject the comparison with an organism accept the concept of harmony, but here again the reader must take care, for each student interprets the concept in his own way. For some, it implies, as it did presumably for Humboldt, simply the concatenation of all the phenomena of a region. In this sense there is harmony in every part of the earth surface; changes in any category of phenomena will produce greater or less effects in others and the new total situation is simply another harmony [*cf.* Gradmann's interesting and suggestive discussion, *236*]. Vogel however finds that, even in this "dynamic" sense—as distinct from an "aesthetic" sense, "harmonious" should be used only in those cases where there actually exists "a mutual accommodation of the forces to each other, a complete interrelation"; this, he concludes, is not the case in most areas, since certain factors, such as climate, or the geological formation of the land, are unilaterally determined; other features adapt themselves to those. "There is a concurrence of forces that leads to a particular *resultant* that characterizes the region, but the forces are in large part causally independent factors" [*244*, 196 f.].

Bluntschli's original conception was evidently of a different order. The changed situation brought about by the introduction of foreign commercial features had produced a new Amazonia which was not a harmonious organism. Since the essence of his discussion of the new form is that he does not like it, we must assume that he intends the word "harmonious" to be understood in the aesthetic sense. If we may assume that geographers are not to be prohibited from considering the aesthetic character of landscapes and of regions, there would appear to be no objection to the use of "harmonious" in this sense; it is only necessary that the reader understand that the description is not intended to be purely objective.

Other German writers have consciously followed the use of the term "harmony" in the field of music. Thus Creutzburg speaks of primary and secondary harmonies in the *Landschaft* [*248*]. Even more striking is the suggestion that Penck has derived from Partsch [*249*, 4–8]. Not only is there "a certain harmony in each chore (the geographic region very specifically defined by Sölch [*237*]), but the chores may be found to form a larger unit area, a *Landesgestalt,* characterized by a particular "symphony." Thus, only in the region of German land (*deutscher Boden*) does one find the *Dreiklang* (triple tone) of Alps, *Mittelgebirge,* and plain; Poland is a *Zweiklang* of plain and fairly high mountains; the Mid-Danube area presents a double *Zweiklang* of mountains and plain, forests and steppe, arranged not zonally but concentrically; farther east in Russia the tones descend to a lighter *Zweiklang* of forests and steppe, passing off into faint notes of the treeless tundra and barren steppe. Suggestive as these descriptions are, one wonders if geographers can safely embark on analyses of the symphony of harmonic chores. [For other interpretations of the concept of "harmony," see Krebs, *234*, 81–90; Granö, *252*, 27 f.; and Bürger, *11*, 99–102.]

Another variation on the theme of harmony is presented by Volz' concept of "rhythm" which he describes as "the harmony of change" [*243*]. Under the same term Volz considers three different classes of change in the

Landschaft: (1) the rhythm that may be observed as one changes the view over the landscape at any one time, (2) the rhythm of the seasons, and (3) the wave-like changes in the whole area over longer periods of time. It is difficult to see what is new in this suggestion, other than the confusion of combining three very different things under a single term which, generally speaking, may only safely be applied to one of them—the seasonal changes. If in any area the arrangement of different forms—forested hills and cultivated valleys, for example—constitutes a rhythmic pattern the observer will no doubt note the fact, but there is no apparent reason for assuming that such patterns will generally be found. Neither will the scientific mind assume a rhythm in the highly irregular and often inharmonious course of history [*cf.* also Gradmann, *236;* and Bürger, *11,* 193 f.].

E. IS THE GEOGRAPHIC AREA A CONCRETE, UNITARY OBJECT?

Our previous discussion of various concepts that have been applied to geographic regions hardly touches the essential question, which we must now examine. May the geographer consider areal sections of the earth surface— whether established by non-human nature, by the totality of all real phenomena, or by himself—as concrete objects that are individual, unitary wholes, which "have form, structure, and function, and hence position in a system" [*211,* 25 f.]. As already noted, this proposition has been stated in the positive by a large number of geographers, particularly in Germany and in this country, including many who have been mentioned as objecting to the concept of regions as organisms. Other students, apparently less in number, have more or less definitely opposed the concept. Since many on both sides of the question regard it as of fundamental importance in the whole field of regional geography, it commands a more thorough attention than perhaps any other single question raised in this paper.

Although the origins of this concept are to be found, as already noted, in the writings of the pre-classical geographers, from whom it may be traced through Ritter and possibly Ratzel, its great development in our time appears to have been furthered first by the work of Schlüter and Passarge, both of whom have presented it in opposition to the methodology of Hettner. Passarge in particular emphasizes the purpose of presenting the total form of a *Landschaft* in opposition to "the merely analytic method of Hettner" [*cf.* Bürger, *11,* 85 f.].

To clarify the issue we may note one point, already touched upon, on which there may be general agreement. Geography has at least one individual, unitary, concrete object of study, namely the whole world. If we add to this world—the earth surface—certain external factors that influence it, not as a whole, but differentially in different parts, namely, the earth's interior, and the sun and moon, we may consider the whole thus formed as a sort of unit mechanism that has an "organic" arrangement of its parts—in the non-biological sense of "organic" [cf. Hettner, 269, 142; Vallaux, 186, 38 ff.]. This conclusion however is essentially irrelevant to the question as to whether parts of the whole world are themselves individual, concrete objects.

In the discussion of various concepts of *Landschaft* or landscape (Sec. V A), we noted the danger of drawing apparently logical conclusions by the use of the term in different meanings. By defining "landscape" as more or less synonymous with "area" and, at the same time, retaining the connotations of its other meaning, as a visible scene, one appears to have proved, without argument, that an area is an objective unit. Thus Sauer, after making "landscape" appear identical with "geographic area," refers to either of these apparently interchangeably as "a reality as a whole," "a corporeal thing" that has "form, structure, and functions" and he contrasts this "concrete landscape" with "De Geer's 'abstract' areal relation" [211, 25, 47; 84, 190]. Other statements, to be sure, appear to give a different view: the task of geography is "to discover this areal connection of the phenomena and their order;" it is "the phenomenology of landscape" that is to be studied [211, 22, 25]. But where then is the contrast between this view and that expressed by De Geer's "abstract areal relation?" *

The validity of an hypothesis cannot be made to depend on the use of any particular word; on the contrary, if it appears to depend upon a term that is never precisely defined, it is for that very reason suspect. If an areal section of the earth surface, delimited in some particular way to constitute a "region," can be demonstrated logically to be a corporeal thing, a concrete unit whole, it must be that by whatever name we call it. If we wish our logical demonstration to be clear, we will avoid words whose lack of clarity is likely to lead to dubious conclusions. If area and region are regarded as inadequate by some because geography is not to include everything in an area but only its material features, and there exists no term that expresses the area minus its immaterial features, it would be better to invent a term, such as "geographic region," that could be precisely defined to mean that, since it has no established meaning in common thought. The student who has had experience in examining the logic of recent methodological studies in geography will look with suspicion on any hypothesis whose logical exposition depends on the use of "landscape" or *"Landschaft."*

The particular problem before us is still more complicated as a result of the terms that various German geographers have introduced from philosophy and psychology. In many cases these have been given new meanings, frequently without any statement of what that meaning precisely is. In his

[263]

masterly examination of the use of these terms in our field, the psychologist Wörner[81] rightly insists that "it is urgently necessary that these concepts be unambiguously defined" [274, 340]. Our present consideration will be simplified if we enumerate briefly at the start the various meanings that have been given to the terms used.

Confusion has resulted in some cases from the use of the word "individual" (the German and French forms of the word are essentially the same as ours). Presumably all would agree that any area, whether a region or not, has "individuality" in the sense that its particular combination of interrelated features makes it different from any other area. In this sense, Hettner speaks of the "individuality" of any area, or rather of any point on the earth [142, 21; 161, 217; and 269, 143 f.]. It is only through misunderstanding of his statements that some have supposed that he regarded the area in question as an "individual" in the additional sense of a definitely limited object or entity. It is quite possible that French geographers likewise are concerned simply with the unique character of areas when they refer to them as "individuals," though their interest, as reported by Musset, in the "personality" of each area would seem to imply a more concrete concept [93, 275]. To make the point clear, we may consider the contrast between an ordinary painting and a mosaic. In the former any square inch of the painting may be unique in its particular combination of color and line, any appropriately selected portion might appear to have "individuality" of its own, but actually no part is a distinct unit individual. (Again we call attention to Huntington's very effective use of this comparison in "The Terrestrial Canvas" [213].) A mosaic on the other hand is formed of individual unit pieces, any one of which taken alone however does not necessarily have "individuality" in the sense of uniqueness, since it may be identical with others, in form as well as color. We may avoid this confusion by using the term "unique character" for the former sense of "individuality" and confining "individual" to definitely limited objects.

More common is the confusion resulting from the several meanings of the word "unit," or the corresponding German *"Einheit"* and, particularly, its adjective form *"einheitlich."* The latter may be translated either as "unitary" or as "uniform" or homogeneous. Granö uses *Einheit* in the latter sense in contrast with "geographic individuals," which are definitely limited areas that may or may not be homogeneous [252, 33 f.]. Although Granö

---

[81] Wörner studied geography, also—under Lautensach. The latter has indicated that in this relationship between the two fields, the teacher learned also from his student. In part for this reason, Lautensach's recent study is perhaps the clearest presentation of the problem of the geographic area in current German literature [278].

defines his terms clearly, it does not seem likely that this usage will end the confusion. Whatever may be the best solution in German, we will use "unitary" in the sense of an individual unit, which may or may not be uniform or homogeneous.

In the discussion of the historical development of modern geography we noted a distinction between two concepts of unity which we described as "the vertical" and "horizontal" aspects of unity. Humboldt repeatedly emphasized the concept of the *Naturganzen* of phenomena of many different categories, and such later students as Richthofen and Hettner refer to the unity of phenomena of any given section of the earth surface. In this vertical sense the concept is applicable to any area—for example, to a county of one of our central States. Whether it be homogeneous or heterogeneous, and where its boundaries may be, are irrelevant questions; each of its parts is related in some way to the others. Nor is the relationship of contrasted parts necessarily less than that of similar ones; a single farm in the Salt Lake Oasis has vital relationships with the rain-swept slopes of the Wasatch Mountains.

Many German writers have called this vertical form of unity a *Ganzheit* or Whole [Volz, *262;* Lautensach, *173,* 30; Granö, *270, 296*]. Both Hettner and Wörner, however, remind us that it is in reality nothing more than the sum of interrelated parts, and is not to be confused with what has variously been called a primary Whole, or an "organic Whole" (not necessarily biological). In such a Whole (*Ganzheit*) the parts are dependent not merely on each other, but on the Whole, and cannot be properly interpreted in terms of each other independent of the Whole—in other words the Whole is something more than the sum of its parts, and, further, is relatively independent of outside elements [*269; 274;* see also Webster's Unab. Intern. Dict., 1935]. In the example suggested, we do not need to know the Totality of the Salt Lake Oasis to understand its rainfall conditions, but only a few of its other elements. Wörner reports that most students concerned with these terms in general would call such a loose unity simply a Sum (*Summe* or *Undverbindung*) [*274,* 341]. Driesch, he notes, would use the word *Einheit* for a "relatively closed system with strong reciprocal dependence," but it remains to be established that the vertical combination of phenomena in any area constitutes a "relatively closed system." Our "vertical form of unity" is therefore but a total sum complex. Though names in themselves are not important, we must beware of calling a total sum complex a Whole, and then assuming that since a Whole has certain properties we have proved that whatever we started with must have those properties.

To this wealth, if not welter, of terminology, Penck has contributed a new term *Gestalt,* borrowed from the psychologists [specifically from

[265]

Köhler, *249, 2*]. He apparently uses it to express the unified form of larger areas, each consisting of a particular arrangement of regions, but others, like Bürger, have adapted it for the geographic *Landschaft.* According to Bürger a *Gestalt* (our psychologists, I believe, do not attempt to translate the term) is a dynamic structure in which the parts reach into each other functionally and can be understood only in view of the whole [*11,* 44–46]. Wörner reports, for the psychologists, that a *Gestalt* is generally the same as a *Ganzheit* (Whole), but that Krüger uses the term only for a Whole that is markedly differentiated into members [*274,* 342]. We may therefore define the two terms simultaneously (following Wörner and Ehrenfels). (1) A Whole is more than the sum of the parts, in that, not only does the interrelation of parts bring out latent characteristics in each, as in any complex, but the complex as a whole takes on a new character not explainable out of the parts—for example, in a melody. (2) The parts of a Whole are *members* that cannot be characterized outside of the specific Whole of which they are a part, whereas the *elements* of a complex sum, or even of a unity, if transferred from one total to another have each the same amount of matter. (3) Wholes are transformable without loss of identity: we may change all the notes of a melody and have the same melody; one may disturb the arrangements of cells during egg development (by centrifugal action) and still get a normally developed individual; or we may add, the members of the government of an independent state may all be changed and yet we have the same government as a Whole.

As examples of Wholes (or *Gestalten*), Wörner lists: all living organisms; certain kinds of groupings of living things, as in a family and, to a certain degree, in a people (*Volk*); and certain phenomena of our experience. His discussion of the latter type is particularly significant in relation to Granö's concepts which appear to be based on landscapes as we experience them. Thus, our experience of a circle is a Whole, but geometrically a circle is merely a sum; that these do not come to the same thing is proved by the contrast between geometric facts and the experience in optical illusions of them. Even though we may grant that the optical illusion is a fact observed in much the same way by different observers, one sees that the concept of landscape unities as phenomena of experience would convert geography into a psycho-geographic subject.

To what extent may these various concepts be applied to particular sections of the earth surface somehow defined as regions? Any area we select will have structure and will include forms that are in functional relation to each other. But not all the forms present are functionally interrelated. That is shown most clearly by the extreme, but by no means uncommon case, of relict forms; as Lautensach notes, these may be very important in the physiognomy of the area but of little importance in its physiology [278, 18 f.]. In general the interpretation of any form may require the consideration of other present forms—as well as of past forms—but it does not necessarily require the consideration of all other present forms, and, fortunately, it does not require consideration of the whole. I say "fortunately," because otherwise we would have to conclude that geography has never interpreted a single regional element, *in itself,* whether of climate or of country rock or of landforms, since we have not yet had an interpretation of a regional Whole.

Even if we are not to consider the region as a (primary) Whole, something may still be gained if we can consider it as at least a loose unity, in the sense of a complex of related elements forming a relatively closed system—"the unitary self-enclosed *Landschaft,*" as Maull puts it [157, 36]. Only if this be true can we properly say that the region—as distinct from things within it—"has structure, form and function and hence position in a system." To have these attributes the region must obviously have fairly definite limits and, since the region can only be defined in terms of area, that means we must have fairly definite areal limits. Granted that we are somehow able to arrive at these in any particular case, as the result of research in the area in question, can we then say that the region thus determined is so constituted in the structure and function of its parts as to represent a relatively closed unit, in contrast to adjacent, or distant, similar units? Though some might wish to answer this question at once in the negative, so many geographers have assumed the opposite that we must examine the possibilities in greater detail.

The problem of drawing the limits of our regions, which we have just seen to be critical to the whole theory, is not to be solved simply by substituting boundary zones in place of lines, as Volz, Unstead or Pawlowsky would have us believe [262, 104; 309, 185; 276, 205]. Even when we limit our consideration to a single factor we do not necessarily have a situation analogous to the colors in the rainbow, which one might recognize as forming distinct bands even though we cannot distinguish exactly where one merges into another. Bucher showed, a century ago, that this analogy of Wilhelmi's does not fit certain elements of our problem. The Bohemian Basin cannot be considered without including part of the surrounding mountain walls, since without these it would not be a basin [51, 88 f.]. As Lehmann notes in general, "there are relief forms that on the same surface are at the same time valleys and elevations. The foot of the ridge may lie in the middle of the valley, the valley may begin at the crest of the water divide" [113, 226]. The two concepts depend upon each other and necessarily overlap; we are

[267]

not to seek a boundary between them, for as Philippson had previously noted "the slope of a mountain is at the same time the side of the neighboring valley, there is no boundary between mountain and valley" [*143,* 12 f.]. In quoting this statement Graf [*156,* 83] appears to overlook its essential significance: we are not concerned with a difficult transition zone, but with a piece of land that is essential to two different units. We are not dealing with bodies that can be delimited from each other, but, as Philippson concludes, "with mutually interpenetrating parts of a single, great, uneven surface, the earth-surface."

Even if it were possible to solve this problem of delimination of areas in respect to a single element, the problem takes on a very different aspect when we consider the full association of regional features—even if we limit these to the natural (non-human) elements. Once again, we find arguments presented by Bucher to which I find no satisfactory answer in the writings of the succeeding century. The upper portion of the Rhine forms a part of the Swiss mountain area; the middle and lower portions, parts of other relief areas; but from a different point of view one may just as properly consider the whole Rhine basin as an areal unit [*51,* 88 f.]. How can we decide which of various elements are to be considered as decisive?

In the systems of *Landschaftskunde* that Passarge has constructed, one might believe that the answer to this question had been found. To be sure, Passarge has recognized, in theory at least, that *"Landschaft* areas are not closed unit forms like an animal or plant" [*229,* 56], but his system of *Landschaft* types has led others, at least, to the concept of regions as specific concrete objects. He greatly simplifies his problem by considering essentially but two elements; vegetation (or climate) and land forms, all others being placed in definitely subordinate position [*268,* 6, 92–8]. Indeed in most of his work he depends almost entirely on vegetation, though often this is confused with climate directly. The implication is that unit areas of (natural) vegetation are necessarily unit areas of climate. Where this is not the case, for example in the humid prairie of Illinois and Iowa, the problem is solved simply by establishing an area of "Forest-steppe land" which includes all of Minnesota and Wisconsin as well, *i.e.,* a purely climatic province ignoring vegetation differences [*305,* 8 f.].

Nevertheless Passarge recognizes that even with but two factors it may be necessary to resort to arbitrary selection of lines of convenience. Thus, if forests overlap from mountains into the margin of a grass plain, the forested border may, he suggests, be included in the region of grass plain. If the mountains grade through foothills into the plain, but the forest-grass boundary is sharp, then this line may conveniently be used as the *Landschaft* boundary. If, however, the transition area is of considerable extent it may be recognized as a separate *Landschaft;* but Passarge appears not to realize that this simply introduces the double problem of delimiting the transition zone on either side [*229,* 14; *268, 62*].

Maull has suggested a method for solving this which in itself is undoubtedly valuable, whether it actually provides the necessary solution or not. We may draw a series of borders, one for each significant element (overlooking now the problem of finding such a border, in certain cases, even for single factors) ; the bundle of lines thus formed he calls a "boundary girdle" [*157,* 601–8; Maull's concept includes cultural as well as natural features]. To one who has not attempted it, it might seem probable that the borders of the separate but interrelated features would fall relatively close together so that the boundary girdle of any region would be sufficiently restricted to constitute a limited border zone. Unquestionably examples of such situations are to be found. Since drainage, soils, vegetation, and animal life are all related to climate, the boundaries of each of these elements may correspond closely to that of climate, and where mountains extend in approximately the same direction, the boundary of land forms may also correspond. To a student who, like Graf, was more familiar with the theory of geography than its practice, it therefore seemed possible to think of the natural surface of the earth as divided into *"Landschaft* units, cells, of which the earth-organism is composed" [*156,* 96; note also Hettner's discussion].

It might seem superfluous to remind geographers that in innumerable cases such close correspondence is entirely lacking. And yet, well-known though the difficulty is in the practice of regional geography, it is frequently lost sight of in discussions of theory. To say that "the elements of a region are geared together so that they are interdependent, each element reacting upon all the others and in turn being reacted upon by them" [*322,* 663], is to give the impression that if any one element varies in major degree the others will likewise vary in major degree and hence the complex of elements as a whole would change in marked degree. Certainly the analogy with an organism, however fictive, strengthens this implication. In reality, however, we know that this is often far from the case. Evidently the deductive theory is incomplete. It has been more clearly developed by Hettner: "the different realms of nature and circles of phenomena are closely knit together and consequently differences present in any one realm of nature must extend into the others. . . . Regions which one sets up in one realm of nature and one circle of phenomena will correspond to a certain degree with those of the others. But the correspondence is seldom complete, because each realm of nature and each circle of phenomena has its own particular law." This leads to a lack of correspondence in several different ways [*161*, 291 f.; originally in *123*, 211–3].

In the first place it is obvious that there is but little relation between the climatic divisions of the earth and the major divisions of land forms; either climate or land forms may vary greatly with but minor differences in the other.

Secondly, insofar as surface forms, and particularly soils, are affected by climatic conditions, they have been developed less by present than by past climatic conditions which differ in notable, even though minor, degree from the present. Indeed many hold that this applies also to the natural vegetation in certain areas, such as the humid grasslands.

A third and most important factor which prevents any neat correspondence between the regions of different elements is formed by the entire group of transport or "circulatory" phenomena—in the air, the waters, and even on dry land (mass movements of soil) [*cf.* Whitaker's excellent brief statement, *284*]. The extreme of such conditions is found of course in the seas. Schott's attempts to establish the regions of the oceans have indicated extraordinary difficulties which both Vallaux and James have discussed [*186*, 165; *275*]. The confusion developed on the lands is similar, even though not so great. Thus the ancient Egyptians could measure, but could not understand the functioning of their life source, the Nile, for the very reason that its headwaters were not included within their region as they knew it, or as any geographers today, presumably, would delimit it. Similarly the "unity" of the North German Plain, as formed by the major relief features and the sand marine deposits and the climate, is notably disrupted by the deposits of silt soil that the rivers have brought down from the central uplands. More common is the occurrence of the same phenomenon in mountain valleys, where we may have but minor differences even in local relief, and essentially no differences in climate, but radical differences in soil.

Fourth, the regions of vegetation and animals, which might appear to be most completely dependent on such basic factors as climate, landforms, and soil, may be notably different simply because of separation, by seas or areas of other life conditions. No matter how identical the inorganic conditions might be in, say, a district in the Amazon Basin and one in the Congo Basin, the two regions are in many other important respects very different [*cf.* Maull, *179*, 184–6]. When we map the mid-latitude rain forest area of Chile as of the same landscape type as that of British Columbia, we are creating an illusion which will not deceive the practical lumberman. On the other hand we can have the greatest differences possible in both vegetation and animal life which produce but the slightest differences in any of the inorganic elements, other than soils.

[270]

In sum, then, the interrelations of the various natural elements are not to be thought of as simple formulae, or as equations of the first power, so that variations in one produce equivalent variations in all the others, but rather of such a kind that major changes in one may be associated with but minor changes in others. Consequently the combinations which we find in reality vary from place to place in such manifold and complicated ways that it is impossible to say that one specific combination extends from A to B, where another begins—unless A and B are immediately adjacent points.

Indeed we may conclude that the thesis of "natural regions"—individual and distinct unit areas determined by the non-human elements—whether in its original teleological form or in its more modern presentations, is unconsciously based on the assumption of a single sequence of control in the interrelations of natural phenomena. In contrast with this concept of the earth as a (logically considered) simple edifice, we may present Hettner's picture of it as "perhaps the most complicated building that we know."

"It is almost as though different architects with entirely different ideas had worked on its construction, so that the internal arrangement is not in harmony with the design but originates in entirely different considerations, and it is as though both architects had changed their views many times during the construction. The earth surface owes its nature not to a single cause but to a great number of causes which have nothing to do with each other. On the one hand, because of its separation from the surrounding cosmic nebula, it has its fixed tellurgic character; on the other hand, it is constantly under the influence of other heavenly bodies, particularly the sun, both under the influence of their force of gravity and under the influence of the sun rays. Of tellurgic origin are particularly the forces of the earth's interior, on which the construction of the earth's crust is based, and which thereby give the occasion for movements based on the laws of gravity. The sun's rays cause the differences in climates and thereby lead to movements of the air. The distribution of climates is primarily dependent on the latitude, which has nothing to do with the interior construction; only in second and third degree are the climates dependent on the inner construction. The tectonic and climatic phenomena create, therefore, two separate though coexistent groups of causes. Most of the other geographical facts are conditioned by these in one manner or another, not, however, certain ones by one of these, others by the other, but most of them by both together. Not only simple dependency is to be considered, but also the movements and transfers of characteristics caused by the differences in climate and relief. Both the telluric and the cosmic factors have changed in the course of time; the effects of the past remain, but in part are affected by, and combine themselves with, the present influences so that many important geographical conditions are not based on present causes but on those of the past" [*161,* 308 f.; originally in *300, 96*].

How much more complicated is the situation if we add to the complexity formed by the natural elements, the alterations that man has made in producing the present face of the earth! The difficulty would not be so great if only these alterations were such that there was "necessary and inevitable correspondence between the natural regions and the regions of cultural or human geography," to quote Vallaux' logical challenge [*186,* 165 f.]. No one, we presume, would claim that that was the case. On the contrary, one might say that it was not in the nature of man to conform to nature outside of himself. Regardless of the configuration of the land, he may, and often does, lay out his fields as rectangles, and plant the same crop on flat portions and slopes, on silt soil and on sand. Though his farm may have originally consisted of patches of forest and patches of prairie, his present use of the land may show no distinction whatever between those different parts. The plant cultivated by the Amerinds of the tropical highlands has become a staple crop of tropical rainforest areas, the mountain farms of the Southern Appalachians, or the flat plains of humid microthermal climate and glaciated soils farther north (whether originally grassland or forest who will say?).

Recognizing the difficulties which these facts present, Finch and Trewartha apparently abandon the former's previous characterization of a geographic region as an organism which is "both physical and cultural" [*288,* 114] and confine the concept of organic unity of regions to the natural elements [*322,* 663; in his more recent presidential address, Finch appears to have dropped the concept of the region as an organism entirely, *223*]. Even if it were possible to establish the theoretical existence of unit regions on this basis, which, as we have seen, is highly dubious, the return to the traditional "dualistic" point of view in geography does not solve, but merely avoids, the real problem of considering a region as we actually know it. If a (populated) region has unity (even if only in the vertical sense) this unity must either include man and his works or it must exclude them, for otherwise what unity have we? But our picture of the region without man—its natural environment, "natural landscape," fundament, or what you will—is, as James notes, an intellectual concept without real existence. "There is after all only one landscape" [*286,* 80]. Even if it were possible to reconstruct with certainty the fundament which existed before man, which is far from the case, that would not represent correctly the present natural environment, since natural elements have changed in some degree since the time of early man.

[272]

In other words, a unity based on the natural environment or fundament of today can have no reality since its basis is not reality. If before man there was a natural unity in the fundament of a region, once man enters, he and all his works, whether or not they become a part of a new unity—which without them is not a unity—inevitably destroy the previous unity. This primeval unity which we could never study directly, has in any case long since ceased to exist.

Most of the authors who uphold the concept of regions as actual units attempt to include at least the material features of culture as well as natural features. Though we find this the more logical basis, it necessarily encounters the greater difficulties because of the lack of correspondence between cultural and natural features [cf. James, *321, 353* f.]. None of these students, of course, would claim that there is a complete correspondence. Nevertheless, as Hall observed, although they "have openly expressed resentment against environmental determinism and the teleological approach," they "have betrayed a naive faith that natural regions would be found to be also human regions" [*290,* 134 f.].

In the discussions presented by a number of these students, one has the impression that they have assumed general validity for conclusions that may be drawn only from certain special areas of the world where there does appear to be a close correspondence in character of all the significant elements, both natural and cultural. In areas where perhaps two natural elements—particularly climate and relief—are fairly uniform throughout and one is of such an extreme character as to dominate most of the other elements, the area may appear to have such a degree of interlocking and indivisible harmony that we are led to think of it as a unit, indeed as an organic unit. Bluntschli arrived at his concept in the single study of the Amazon lowlands, and Gradmann illustrates his conception of the *"harmonische Landschaftsbild"* as an organic unit with examples exclusively drawn from the deserts and tropical rain forest areas [*236,* 130–7]. But neither of them has put the theory to the test of drawing definite boundaries—even boundary zones—to show where one "unity" ends and another begins. Even if these could be established with some degree of agreement in these particular cases, we would still not know whether the thesis was valid for more ordinary areas.

In his keen and thorough treatment of the problem of the limits of regions, Sölch has considered the results obtained when one studies areas of no exceptional character. "The boundaries of the areas of the individual factors are frequently such broad zones and at the same time so far distant from each other that . . . the transition areas are often much larger than the actual cores [*237,* 43 f.; see also Sieger's emphatic agreement, in his review, and also Lautensach's repetition of the same conclusion, *278, 22* f.]. Vogel would solve this problem by recognizing the wide areas between regions of definite character—not transition zones in the ordinary sense, but areas in which some, but not others, of the elements differ—as "characterless *Landschaften*" [*271, 3*]. Though this device has long been honored in practice, geographers cannot be satisfied with a classification of regions that assigns perhaps the greater part of the world to the "miscellaneous column."[82]

Granö finds it possible to reduce greatly the extent of these intervening areas that belong in some respect to regions on one side of them, in other respects to those on another, by recognizing as distinct wholes, areas "that are homogeneous in respect to more than one of their characteristics" [*270, 299*]. Although he has worked out his system in great detail, illustrated by regional divisions of Finland and Esthonia, these are based on maps for each particular factor that are themselves considerably generalized, apparently not on any objective basis [*252, 143–62*]. In his far more detailed study of a single district, the transition zones are larger than the core areas [*169–71*]. Valuable as the result produced by this system may be, regions* determined by what may be but the lesser part of their total features are hardly to be called objects [Lautensach expresses the same conclusion, *263, 196*]. To call these more or less satisfactory but essentially arbitrary divisions, Wholes (*Ganzheiten*), is to pour still more water into a term that, as has been noted, Granö had already reduced to nothing more than a sum of related elements. Indeed Granö recognizes that his "Wholes" are but intellectual constructions, since the elements do not correspond areally, so that what he calls the "geographic individuals," in the world of reality "pass over into each other and consequently the face of the earth is like a confused mosaic" [*270, 297*]. One may therefore accept Granö's concepts only if one translates his terms according to his particular definitions; in the ordinary sense of the words, individuals do not pass over into each other, and a confused mosaic is not a mosaic.

Many of the other students interested in this problem have recognized the undoubted impossibility of establishing unit regions on an objective basis delimiting all the factors concerned [see Hassinger, *225, 471*; Philippson, *143, 13*; Passarge, *258*; Lautensach, *263, 195–7*; Bürger, *11, 53* ff.; Obst,

---

[82] In a later section (X F), we discuss two striking examples. The so-called "Corn and winter wheat belt" in the United States is little more than a catch-all for districts along the southern margin of the Corn Belt that differ notably in their agricultural types, as is clearly shown in Baker's description of the different parts of that "belt" [*312*]. Similarly, in the Bureau of Census study of farm types in the United States, many parts of the country are classified as "General Farming," simply because they are not sufficiently specialized to fit into other categories [*320*].

[274]

* See also Granö's more complete study of 1931 [*441*].

178]. Nevertheless some of them believe that with increasing study by many students, the subjective element will be reduced to a minimum. As Obst puts it, "the more often and more conscientiously a *Raumorganismus* is studied, the greater, *without question,* will be the agreement in the judgment of what factors are dominant" [*178,* 12; italics added]. One wonders if there is a single case where various geographers who have studied the same area are ready to agree on its limits in any approximate degree. Bürger's statement, on the other hand, we may accept, though we draw a different conclusion from it than he does: the many differences of opinion over the boundaries of regions are "no worse than the varying views of the limits of historical periods" [*11,* 55].

Consequently, when Finch recognizes that regions are "in a sense, mental constructions rather than clearly given entities" he is pointing at the necessary conclusion without accepting it [*223,* 12]. To be sure, there are entities in reality that are not "naïvely given," but must be discovered by research. But if they can be discovered, they can be established definitely, within a reasonable margin of certainty. We have not achieved that in one single case. The problem of establishing the boundaries of a geographic region— which Finch supposes "to be no longer a problem"—presents a problem for * which we have no reason to even hope for an objective solution. Consequently we not only have not yet discovered and established regions as real entities, but we have no reason ever to expect to do so. The most that we can say is that any particular unit of land has significant relations with all the neighboring units and that in certain respects it may be more closely related with a particular group of units than with others, but not necessarily in all respects. The regional entities which we construct on this basis are therefore in the full sense mental constructions; they are entities only in our thoughts, even though we find them to be constructions that provide some sort of intelligent basis for organizing our knowledge of reality.

We may summarize our discussion of the concept of regions as concrete, individual objects with the following quotations from Vallaux and Hettner, both of whom have subjected it to critical analysis.

"It is essential [to the thesis] that the demarcation of the regions can be applied to the earth's surface without effort and in an incontestable manner; it is necessary to demonstrate the existence of regions without mutual trespassing; and it is necessary that all of the facts of geography, physical and human, can be grouped in concordant regional definitions. We have not arrived at that point by far" [*186,* 164; Vallaux's further conclusion on this point is discussed later].

"The differentiation is not the same in each circle of phenomena and they

* The parenthetical reference to Finch is based on misunderstanding of what he wrote and should be omitted.

are stratified in the most manifold ways. Any particular large portion of the earth surface is therefore homogeneous only in one relation, in the others heterogeneous; and in the strictest sense only a particular spot on the earth has complete individuality or character of its own" [Hettner, 269, 143 f.; cf. Banse, 330, I, 41].

F. APPARENT AND PARTIAL FORMS OF AREAL UNITS

The reader who has followed our reasoning in the previous pages may agree that the conclusions drawn are logical and yet still feel that "there is something in the idea" of regional unity, even of "organic unity." We have already noted a special type of area in which there is marked appearance of unity over a wide extent of territory—namely areas of extreme conditions such as deserts or tropical rain forests—but concluded that this was merely because the dominance of one or two factors causes a relatively high degree of homogeneity and of simple co-ordination of other elements. In a somewhat similar manner, all our work in regional geography appears to produce unit areas, each organized largely within itself and sharply bounded from the others, and these units placed upon a map form a mosaic of the earth surface. In this case the word "mosaic" is well chosen, since the mosaic is one of the more highly conventionalized, unrealistic, forms of artistic presentation. As an expression of reality, however, it may only be seen from a great distance: the more closely one examines any part, the greater is the falsification of reality. It may be that we must produce such mosaics, since reality is too complex for us to present in all its details, but we are only deluding ourselves if we ascribe to our arbitrarily determined regions an actual character as unit-whole areas.

There remain however a number of special respects in which the concepts of Unity and Wholeness are of value in regional geography. After examining this field critically from the point of view of the science that has been particularly concerned with these concepts, psychology, Wörner suggests a number of cases in which the concepts may appropriately be utilized [274, 343–5].

A very special form of "holistic" thinking that Wörner recognizes as justified is the intuition that a student may have concerning an area that he has studied—or, as others have put it, in which he has worked so intensively that he has "experienced" (erlebt) the area. Although this is an experience or intuition concerning what exists in reality as only a sum total of inter-related phenomena, in itself the experience is a Ganzheit—as a psychological phenomenon. This is the concept which Gradmann, if I understand him correctly, expresses in somewhat different form [236], and perhaps the

French geographers who speak of the "personality" of an area have much the same thought in mind [93, 275]. No doubt some students would immediately brand such a concept as pure mysticism which science cannot tolerate. We can agree that intuition is not a scientific form of thinking, but such students of the theory of knowledge as Cohen have recognized that such non-scientific forms of thought may be necessary, and therefore justified, aids in guiding us to scientific analysis that would otherwise be overlooked—in somewhat the same sense as the scientific experimenter may follow a "hunch." But just as the hunch must be recognized as non-scientific until demonstrated —when it ceases to be a hunch but becomes a scientific conclusion—so any holistic intuitions of an area, no matter how strongly felt, are not to be considered as scientific until they have been broken down by analysis to demonstrate the valid correlations previously "felt," and this process will inevitably demonstrate also that the area is not in reality the Whole that intuition pictured. In other words, the psychological impression of the Whole, having served its purpose, disappears in the process of analysis.

There are, however, several other respects in which the relation of the earth to man appears in the form of unit Wholes, at least from the point of view of the psychologist. The most obvious is the appearance of the landscape to the observer. This landscape sensation is not an objective fact outside the observer, but is a fact only in his consciousness and, though there is similarity in the impressions received by different observers in the same area, it is difficult, Wörner concludes, to minimize the influence of the individual constitution and disposition of the student. As previously noted, this is the form of Landschaft study that Hellpach has developed as a geopsychical field. Granö, if I understand his view correctly, endeavors to translate this concept, by reversion, one might say. Starting from the sensory landscape, a psychological unit phenomenon, he attempts to analyze and synthesize the objective earth phenomena that have produced that result in the observer. In doing so however, he apparently transfers the concept of the Whole back to the earth phenomena, overlooking the fact that these represent a Whole only in the psychological reaction in the observer, not in the reality outside him—and they do not produce the same Wholes in different observers.

Man not only perceives an area through the effects of its physical features directly upon his senses, but he also experiences the area, as his milieu, both directly and indirectly. This Landschaftserlebnis, Wörner finds, is a whole, like any experience. Obviously however this phenomenon depends not only on the area but also on the character of the inhabitants and their inherited talents—the inhabitants considered both collectively and individually. It is obvious that any geographers who undertake to study such experience phenomena of the earth will need to be thoroughly competent in psychology. Whether such a border field does not belong more to psychology than to geography is a question which we need hardly decide.

This experience of earth areas as Wholes may be translated into reality in terms of the *Lebensraum* of man. In the general, and clearest, case, this is represented by the entire earth surface, plus the sun and moon, and also the starry heavens as a part of the visible scene. For any single man, or group of men, Wörner defines the *Lebensraum* as that sector of surroundings that has any form of psychic connection with them—reciprocal connections between the milieu and man. In the eyes of the man concerned, this *Lebensraum* is a Whole and is therefore to be regarded holistically (*ganzheitlich*). Whatever value this concept may have from various other points of view, it obviously does not provide geography with distinct concrete objects, since the areas forming the *Lebensräume* of people living in quite different places will overlap inextricably. Only in most primitive areas could we establish relatively restricted, but still overlapping life-areas; that of a modern commercial city would include almost the whole world.

For geography—as commonly understood by geographers at least—this is a meagre harvest from all this controversy, and Wörner's major conclusion is that geography has gotten along fairly well in the past without considering the concept of Wholes, because in reality it is but little concerned with Wholes, other than those obvious units, such as plants and animals, whose examination as individuals is performed for geography by the systematic sciences. On the other hand, he does suggest certain forms of areal combinations, for the most part very small in areal extent, that may be regarded in particular aspects as forming unit Wholes. So long as we consider them only in those aspects we may, and should, study them as Wholes; any Whole, he has noted, is, from certain points of view, nothing but an arithmetical sum—*e.g.*, a man is physically a simple sum in terms of gravity [*274, 346*].

The existence of these concrete areal wholes is dependent on the one factor in the real world that is capable, within limits, of producing distinct areal units and of organizing them in terms of structure and function into *Wholes*—namely man. A farmer's fields are distinct units, in each of which the crop grown and the method of cultivation may ignore the variations in soil—though the yield will reflect it. These units, however, in a more important sense, are not wholes, but merely parts of the farm which is organized in many respects as a primary Whole. Its parts depend on the whole—including the barns, animals, farmhouse, and farm family; the use of the fields cannot be understood except in terms of the farm as a whole. In general,

Wörner recognizes that any single work of man may be considered as a Whole: the parts can be understood only in terms of the whole, which requires, he adds—and we may emphasize the point—reference to the manner of thought of the builder.

In urban areas the concept may be extended somewhat further, though with increasing reservations. City blocks are distinct and separate areal units within what we may consider, in certain respects at least, as an organic whole, the city. The various parts of the city—residential districts, factory districts, and central commercial districts—cannot be understood merely in terms of their mutual relations, but only in terms of the whole; they are not merely elements existing in interrelation, but are members of an organized Whole. Further, the relations of the city to the surrounding country—for example, in the movement of food supplies, water, etc.—are in large degree not relations directly of one element in the city to those in the country, but of the organized Whole to the country. Finally, the city area can be delimited within a relatively narrow boundary zone—a zone that the city may, if it desires, reduce to a sharp line.

In general, we observe that because man cannot conveniently adjust his use of the land to the multitudinous and gradual variations in nature, a major aspect of his conversion of the natural landscape into the cultural landscape is the creation of sharply distinguished land units within each of which he seeks for homogeneity. In this way he tends to create a mosaic, a scene, that is, in which the parts are each homogeneous but sharply distinct from each other. Nature, we may say, produces no mosaic; such a landscape is the surest sign of the work of man. The more thoroughly man has converted a land to his uses, the more he has "tamed" the natural "wild" landscape, the more nearly perfect does the mosaic appear. One cannot examine the face of the earth from the air—whether from an aeroplane or from air photographs—without being impressed by this contrast. In a landscape which is largely tamed and reconstructed by man, any piece of wild land stands out like an incompleted section—or perchance a damaged section—of a mosaic. This is not because it may be wooded in contrast to fields, for orchards fit into the mosaic, but because of its lack of homogeneity in itself and, frequently, its irregular, uncontrolled edges, merging indistinctly into pastureland. Similarly, air views of typical farm landscapes in different parts of the world immediately suggest to us the relative degree of perfection to which man has converted the natural landscape into a cultural landscape. Significant comparisons could thus be made between areas in southern New England, in the Corn Belt, in the Paris Basin, and in the Yangtse Delta, while the antithesis, in an area that cannot be called unpopulated, is to be found in Central Africa.

Though man has thus produced landscapes which approach mosaics, and has created organic space units of farms and towns, and could, no doubt, produce larger units in regions if he chose, this latter he has not in any real sense attempted to accomplish. Except for the limited degree represented by his planned and organized routes of transportation, man does not, either consciously or unconsciously, organize a region as a unit. Recent discussions of regional planning, or of dividing a country into more or less fixed regions for unitary administration, have illustrated the difficulties in organizing such larger regions because of the lack of even the crude form of organization which, without conscious planning, evolves in the growth of cities [cf. Report of National Resources Committee, 291]. The absence of any such organization of regions is likewise reflected in the difficulty of finding "scientific" boundaries for political states. If the geographer could show that regions were functioning unit organisms, the problem of finding suitable boundaries would be reduced to that of deciding simply to which state each such region was to be granted. Those who use these terms might well study the *logical* conclusions drawn from them by the German students of *Geopolitik*—an over-simplification of the problem which I have elsewhere discussed [216, 960-5].

On the other hand man has organized certain large areas to some degree as functioning units, namely the independent states which he has arbitrarily created and which function not merely as political wholes but, to a considerable extent, as economic and social wholes. These are far larger than what is commonly called a region, and may, indeed, totally disregard any sort of "natural regions" which geographers might recognize. Nevertheless, since they are areally defined with exactness, and function in many respects as organized units, one might call them "areal-organisms," provided we recognize clearly the respects in which they are not like biological organisms [from Maull, 157, 78]. But to apply this term to a part of a state, as Geisler does in his study of Silesia, is to destroy its valid meaning [345].

This concept of the state as an areal-organism requires considerable reservation, however. As Hettner has noted, in criticizing Braun's concept of the state as *"organisches Lebewesen,"* the state is only one of the many factors in human life [167, 340]. For example, the relations of the Ruhr area of Germany to the Lorraine iron-ore area of France are not the relations of two interdependent organisms, Germany and France, to each other, but are relations basically independent of those organisms as a whole, between a part of one and a part of another. Many similar examples could be found in the relations of the United States and Canada. In other words, in economic geography we have amorphous and undefined forms of "areal organisms" that include parts of two or more state "areal organisms."

G. CONCLUSION: PRACTICAL RESULTS OF THE CONCEPT

From the point of view of the areal divisions with which we are commonly concerned in regional geography, we conclude that it is not possible to define sections of the earth surface as regions that form units in reality, that we cannot correctly consider them as concrete individual objects. As Pfeifer emphasizes in a quotation from Hettner, "the ideas and conceptions of geography are of a complex nature rather than of the nature of systematic objects" [164, 425].

If one inquires as to what effect the introduction of this concept has had upon regional geography—other than the great amount of energy expended in the innumerable methodological discussions concerning it—the answer is difficult to find because of the relatively limited use to which the concept has actually been put. Possibly we might regard the systems of type areas of Passarge, to be discussed later, as an outgrowth of this concept, but no doubt he would have constructed those in any case. Vallaux urges that the effort to fit the facts of surface into definite individual areas "leads us often to seek limits and to invent forms for that which has and can have neither form nor limits" [186, 50]. This in itself, we may add, is no scientific crime, provided only that we do not suppose that the limits we have set and the forms that result, represent anything other than more or less arbitrary compromises with reality.

A more serious result may be observed in much of the work of the proponents of this concept. In the attempt to elevate geography to a position of equality with sciences that have each a particular class of objects to study, they have apparently lost sight of the one essential element that distinguishes geographic thought from that of all other sciences, namely the constant consideration of the relative location of things on the earth surface with reference to each other. Though we have previously suggested that the tendency to overlook this factor was a result of the refusal to consider non-material elements as "primary geographic facts," and it might also be considered as a result of the emphasis on systematic geomorphology in the training of geographers, it appears to result in addition, and particularly, from the consideration of a region as a relatively closed unit in itself. In attempting to interpret the phenomena of a region in terms only of the facts within the region, the element of location is reduced to the internal facts of location. Theoretically one might talk of the relative location of the region as a unit in reference to other such units, but any attempt to analyze a concrete situation in such terms reveals the fundamental fallacy: regions do not have relations to each other as units—only particular elements and complexes of elements within regions are related to those in others.

We do not hesitate to repeat our reference to the slight consideration of this factor in the work of such students as Schlüter, Passarge, Granö, and Sauer, because we are concerned with the factor that Bürger rightly describes "as the unquestionable possession of our science, our very own, that can be called 'geographic' with higher right than any other elements of the *Landschaft*, since these belong, from the point of view of their substance, to other sciences and to geography only insofar as they underly the geographical-areal consideration" [*11*, 30].

The manner in which this essential element in geographic thought may be lost sight of in the effort to establish regions as relatively closed objects may be seen most clearly in regional world surveys. We have previously noted this with regard to Passarge's *Landschaftsgürtel der Erde* and may examine the results here in a work in our own literature that is markedly influenced by his ideas—a work, moreover, that has with good reason been praised as the single major attempt of an American geographer to present a world view of geography, physical and cultural together, on chorographic principles—namely, James's *Outline of Geography* [*321*].[83]

In the introduction to this work—intended for elementary as well as advanced students—the factor of relative location is omitted entirely. More important, it is difficult to find consideration of its importance in much of the detailed regional treatment. Tropical plantations, fruit areas in California, and the wheat regions of the grasslands are all discussed with little or no reference to the all-important factor of distance. Cities are mapped and discussed in terms of their size and growth in the past century, and their external form and internal patterns, but so meagre, by comparison, is the discussion of even general principles of the location of cities, that the reader

---

[83] Particularly because the writer of this paper has made no attempt to produce such studies while freely subjecting major works by others to criticism, a word of explanation, if not of apology, is required. Our concern here is not with the wealth of subject matter nor the high quality of these volumes, nor with their value as texts, but only with their implications on the nature of geography, whether in the statements of theory or, by inference, in the arrangement of the material. Because the authors have chosen to present their views, other than by oral expressions, in text-books rather than in methodologic articles, one has no choice except to discuss these or ignore their views entirely. On the other hand, the very fact that these books have been highly praised by many, including this writer, may well cause them to have marked influence on geographic thought in this country, which makes it all the more desirable that their underlying philosophy be subjected to critical examination.

will gain little understanding of why they have developed in the particular places where they are found, possessing the particular, differing functions that they have.

No doubt the difficulty here is one of organization. If a world survey is organized by regions—which this writer would heartily support—it is not merely inconvenient but extremely difficult, repeatedly to leave the region under consideration in order to follow the connections of certain of its features to those in neighboring areas—far worse where the connections lead to remote parts of the world. Those who have not made the attempt to organize such a survey, therefore, are not in a position to cricitize the author for failing to solve the problem involved; we can only register the fact that he has not solved it. On the other hand, it would appear possible as well as necessary, in a single regional study of the Brazilian coffee area to consider the significance of the location of the whole area with reference to consuming areas overseas, and particularly the differential location of different parts of the area with reference to the ports [365].

In the final summary of his world survey, James appears to recognize that the "organic unity" of a city "extends to the most remote parts of the earth" [321, 353], but he does not draw the logical conclusion from this, namely, that for man today there is no complete unit less than the entire earth surface. Once this is realized, Hettner's conclusion becomes clear: "no phenomenon of the earth surface may be considered for itself; it is understandable only through the apprehension of its location with reference to other places on the earth."

It is interesting to note that the followers of the doctrine of relationships, however misleading that may be in many ways, have had a clear understanding of the importance of relative location for the proper understanding of the features of any area. One thinks of Whitbeck's illuminating study of New Jersey [397] and, particularly, of the exhaustive analysis of Semple [204, Chap. 5]—even though this does tend to confuse climatic location, which is simply climate, with relative location properly speaking.

The significance of the concept of relative location can be made clearer if we consider again the character of geography as an *integrating* science akin to history. In a statement previously referred to, Kroeber suggested that the distinctive feature of the historical approach was not the dealing in time sequences, but the attempt at descriptive integration. We might say that both history and geography are attempts at descriptive integration, but that in history the integrating factor is time—the association of phenomena taking place at approximately the same place but related to each other in the sense of time—whereas in geography the integrating factor is space—the association of phenomena at approximately the same time but related to each other in spatial terms, *i.e.,* in terms of relative location. Just as the historian is concerned with the historical association of phenomena not only more or less contemporaneous, but also those rather widely separated in time—for example, the relation of the Napoleonic Wars to the World War—so the geographer is concerned with the spatial connections of things not only close together in a single area, but things as far apart as a dairy farm in New Zealand and a grocer's shop in London.

Since the concept of geography as a "science of relationships" approximates the concept of integration, it is not surprising that its followers have constantly recognized the importance of spatial relations. However, it should not be supposed that the two concepts are the same. To say that we are concerned with "descriptive (and interpretative) integration of phenomena" is not the same as to say that our purpose is to find integrating connections between phenomena.

If one wishes to distinguish between the natural and the cultural elements in geography it may be noted that the factor of relative location or *locus*—to suggest a shorter term—is not in its essence a cultural element, as such phrases as "transportation facilities" or "accessibility to markets" would suggest, but is inherent in the natural environment. The locus of Egypt in reference to the tropical highlands to its south was a major factor in the development of the soils and natural vegetation of the Nile Valley long before the first pre-historic men appeared on the scene. Possibly we must recognize the factor as a *geometric* factor belonging neither to the natural nor the cultural features of an area, but ever essential for a geographic interpretation of any of them individually, as well as of their combinations that form the total complex of a region.

The region itself, we find, is not determined in nature or in reality. We cannot hope to "discover" it by research, we can only seek the most intelligent basis or bases for determining its limits—in general, for dividing the entire world into regions. How this may be done will be considered in the following section.

X. Methods of Organizing the World into Regions

A. theoretical principles of regional division

Many of the critics of the concept of regions as units existing in reality *
have concluded that in a scientific geography there is place for no concept
of regions whatever. Over a century ago Bucher's most careful critique
of the regional concept brought him to this negative conclusion—though he
did leave the door open for consideration of any particular area for certain
special purposes [51, 94]. Similarly, a few years later, Fröbel ruled the
regional concept out of scientific geography, but he gave it a much more
important place than did Bucher, in his non-scientific half of geography [56].
Likewise, in our time, Vallaux concludes his critique of the regional concept
with the following statement: "We are not even sure that for many parts of
the world, the regional synthesis is anything more than a logical artifice and
a method of instruction, that is to say, after all, an actual deformation of
truth, excused but not legitimized by the arguments which one could with
validity present in its favor" [186, 164].

Our previous discussion undoubtedly justifies the major part of Vallaux'
statement. Any regional division of the world which takes into account all
the significant elements is not a true picture of reality, but is an arbitrary
device of the student, more or less convenient for his purpose—and for that
reason differing from student to student, depending on what elements appear
to him as most significant. Even the apparently obvious division into continents does violence to regional continuity in climate and many aspects of
human geography. Banse, among others, has emphasized the false character of this conventional division, but apparently fails to see that any other
division of the world into major realms will also necessarily do violence to
regional continuity in one way or another [Hettner, 152, 54; cf. Graf, 156,
113–8]. In view of what has previously been said, it is not necessary to
demonstrate that sub-division into smaller units will not solve the problem
until infinite sub-division brings us to the individual spot on the earth's surface, and this spot is a pure abstraction.

Fortunately, however, the long developed system of regional geography
is too sturdy an infant to be thrown out of the window with the bath water
simply because some theoreticians have credited it with mystic attributes
which are found to be erroneous. Likewise Leighly's threat that regional
geography may be read out of the field of science [220; 222] has been properly answered by Platt [221] and Finch [223]. Whether or not regions are
genuine unit realities, no one can question that the features of the earth sur-

* See Supplementary Note 39

face form interrelated complexes that differ in its different areal parts, or regions, and that the phenomena of any one region are in large part determined by its particular complex. In this qualified sense we may say with Finch, "here *are* the realities of regions everywhere about us;" he who is interested in gaining an objective knowledge of the world as it is, will not shun the difficult task of attempting to study the totality of phenomena in regions for "fear of losing scholarly caste" [*223, 9, 6*].

In passing, we may question the unstated assumptions concerning the nature of science that are involved in Vallaux' statements, no less than in the arguments of both Bucher and Fröbel of a century ago. Any science concerned with the study of reality—as distinct from pure mathematics—must use concepts that represent actual deformations of truth, however slight. The necessity of reducing the incomprehensible complexity of reality to comprehensible systems necessitates, and thereby not merely excuses but legitimizes, these deformations of truth. All that science requires is that the scientist recognize always that his concepts are but approximate and arbitrary alterations of reality.

Bucher arrived at his conclusion in part by utilizing the analogy of geography with anatomy and physiology [*51*, 91 f.]. The biologist in studying the human body does not, he claimed, study a particular part, in terms of its portion of the skin, veins, nerves, muscles, etc., but studies these systematically over the whole body. Though this is undoubtedly true, surely the student of human physiology must, later, seek to know and understand the particular character, in terms of forms and functions, in which these elements are combined in, say, the human fore-arm in contrast to the thigh. In other words he must have more than a systematic knowledge of all the different kinds of things in the body, he must have an accurate knowledge of the formation of these things together in the different parts. However that may be, Bucher, with characteristic scientific caution, noted that the analogy was not to be driven too far, but only so far as the two things compared—the earth and the human body—were actually analogous. We may note two major differences. One is that in geography we have only one body to study, the physiologist has millions. Far more important however is the very fact that the earth is not an organism, that, though its surface is notably differentiated, it does not form an integrated mechanism of highly differentiated areal parts comparable to the human body. In other words, in spite of our insistence in the previous discussion upon the importance of the relations of elements of one region to those of another as a phenomenon that prevents us from considering any area alone as a closed unit, nevertheless it is obvious that these relations are far less than those of the corresponding relations of the human body—as represented by the nerve system, the blood circulation, the digestive system, etc. Though no part of the earth can be completely understood by itself alone, it is to a large degree, nevertheless, the product of its internal conditions and therefore justifies our concentration of study upon it in far higher degree than upon, for example, the human hand.

In any case, we are not to be diverted by any argument claiming that a "science of regional geography" is an impossibility. Such a conclusion, based as it must be on a particular meaning of the word "science," would not eliminate the need for a subject of regional geography, by whatever title one wished to call it. Geography exists, whether or not it is to be called a science, and the same may be said of regional geography. If, to qualify for any arbitrary requirements of "science," the very nature of geography must be changed, we can more easily abandon any claims to high titles than we can change the essential nature of our subject. As we saw at the start, the existence of geography as a field of study rests on the general human interest in the different character of different parts of the world. Men may be interested in a systematic classification of each particular element, or of a specific group of elements, as it varies in different parts of the world. It may be desirable to study the whole group of natural, non-human, elements in their natural arrangements in different parts of the world, and likewise to study the whole group of human or cultural elements in their associations in different areas. But this dualistic arrangement, made necessary, according to Vallaux, because of the lack of concordance between the two groups—the obstinate refusal of man to conform to his natural environment—this dualistic system, even though tied together by remote telegraph connections, to use W. M. Davis' earlier simile [203], will not satisfy the fundamental interest upon which the field of geography depends. This asks us to describe and explain, not two separate sets of elements over the world, but *all* the features of a particular part of the world which are distinctive of that part in contrast with others. Geography has, therefore, imposed upon it from the start, the difficult—in a complete sense one may say impossible—task of integrating many kinds of phenomena found in areas, in relation to which the differences in these phenomena are not concordant. If we accept the responsibility of geography no difficulties or objections may divert us from this task [cf. Lautensach, *278,* 12].

Likewise, we may add, those who accept the task of developing regional geography as a branch of learning, whether of science or not, will take that task seriously and will therefore pay little or no attention to advisors who indicate that they are unable to take it seriously. Negative criticisms must be given consideration from whatever source they come, but a suggestion that it doesn't matter how regions are determinated, or that political divisions will prove "generally satisfactory," merely indicates that the advisor has little concern with the subject and is convinced, rather, that it will be generally unsatisfactory no matter what is done.

We will assume that it is the business of regional geography to seek to understand the differences—as resulting from the total complexes of phenomena in different places—in the different parts of the world. We must therefore study these total complexes in particular areas; in order to provide a systematized organization of the total earth surface in regional studies we must have some basis for dividing the world into regions. Of course one could study divisions determined on any possible basis—such as political—and we have already noted that there are no actual unit areas in reality for us to study, but it does not follow that there is nothing between these extremes. On the contrary, experience has shown—and theory has demonstrated—that certain bases of division make the problem of regional study exceedingly difficult, unduly repetitious, and unproductive of new concepts. That this is the case in respect to the use of minor political units was not only abundantly proved in the conventional treatises of the writers before Gatterer and Ritter (of which Fröbel's "hack-job" on Peru, etc., is a late example [53]), but unfortunately is being redemonstrated today in some of the state "guide-books" produced under "W.P.A. projects" in various States.

We therefore can agree with Maull that "in the very foreground of research in regional geography is the attempt to divide the world, or any parts of it, into regions of various orders of size" (*Länder, Landschaften, Örtlichkeiten,* etc.) [*179, 173*]. Though we do not seek to discover regions that actually exist, we do seek to find the most intelligent and useful method of dividing the world into regions. We must therefore concern ourselves, as Hettner did as early as 1908, "with the principles of a division of the earth's surface" [*300; cf.* also Granö, *252,* 48, 139 ff., and Almagià, *188,* 19–21].

How can we hope to solve this problem? Ideally, one might say, the geographer is concerned with the admitted individuality of each particular spot on the earth. We could therefore justify, theoretically, the study of any number of minute places, but since the number possible is infinite, it is clear that there must be some test of significance. Some places may be so significant in their relations to large parts of the world as to warrant individual attention—*e.g.,* the Rock of Gibraltar, the mouth of the Chicago River, or London Bridge. Others, in themselves of little importance, may be more or less representative of similar places in their vicinity and so justify study as typical, in limited degree always, of much larger areas—*e.g.,* the valley of the Isère in the French Alps [*331*], a single hacienda in the District of Antón, Panamá [*221*] or the Princeton community in the Corn Belt [*341*]. We need not at this point concern ourselves with the question of whether these particular places are in fact typical of their regions. The obvious similarities in character among many French Alpine valleys—in contrast, say, with the Cevennes or the Tirol—justify the assumption that they can be roughly grouped in one or more types, the determination of which is a problem in research (see Sec. XI H).

Where we find that a particular, larger area of the earth consists in considerable part of smaller areas of the same type or types, whereas adjacent areas are different in this regard, we can comprehend the character of the larger area, or region, by studying it as a whole. (Actually a degree of synthesis has already been made in the consideration even of such a small unit as a single valley.) Since the types are not clearly distinct in actuality, even less so than are different species of plants, and the different parts are at the same time not as similar as are plants of the same species, the regions delimited on this basis will likewise be neither unified within themselves nor distinct from each other, but will grade into one another in respect to many different elements, and will overlap in respect to some elements. Because the actual situation is much too complex for our minds to comprehend directly—quite aside from the problem of teaching it to others—we are forced to simplify actuality by arbitrarily assigning somewhat definite limits to the regions studied. But we should not lose sight of the fact that these limits, indeed the regions themselves, are essentially arbitrary simplifications of the actual complexity which we are attempting to comprehend.

Such a simplification, and therefore distortion, of reality is by no means peculiar to geography; arbitrary divisions are common in other sciences. Even the division into species becomes arbitrary in some parts of botany, and is certainly so in paleontology.[84] Even more arbitrary are the geologist's divisions of rock formations, and even of types of rocks. The clearest analogy, of course, is in history. The historian finds it necessary to recognize a particular, but somewhat indefinitely limited, extent of time in the history of Europe as the Reformation Period, and no one questions his use of this device so long as he does not think of that period as a distinct and separate unit in history, related as a complete unit to preceding and subsequent periods.

The comparison with history would further emphasize what has already

---

[84] It is, however, an error in logic to use this comparison, as Graf and others do, to claim that geography has in its regions objects of study no less defined than those of the zoological sciences [*156*, 83 f.]. The concrete objects of the zoological sciences are not the species but the individual plants or animals.

been suggested by our study of geographical regions, namely, that these divisions of the world are not clearly indicated in the nature of things; we cannot start with them as given facts. Neither are they facts which we may hope to discover simply by exhaustive observation. We cannot even hope that by examining the world through a reversed microscope we may ultimately find individual cells which together form the whole; it is all too clear that we are examining a single cell, a cell in which there is not even the relative degree of separation of parts characteristic of a biological cell.

Upon what basis, then, can we divide the intrinsically complex and indivisible world? One thing is clear: we can distrust from the start any simple solution. We are not looking for the one true method of division, since there can be none; we are looking only for a more or less suitable method. Under any possible system the divisions will be, in one sense, only "arbitrarily selected fragments of the land"; in another sense, however, our division will not be arbitrary, but will conform to the general requirements of science if it is based on appropriate principles consistently carried out. Clearly, what principles are appropriate depends on the ultimate purpose for which the division is to serve, namely, the understanding of the character of the different areas of the world. Since these are so complex in their nature that no single survey of the world can provide an adequate place for all aspects, even in outline, it follows that there may be many different bases for regional division, each more or less suitable for different geographical purposes. "Agreement and uniformity," as Bowman says, "are not desirable or even attainable goals in a classification that depends upon such variables as time, space, and function" [106, 144; cf. also Finch, 223].

The question as to which criteria shall be chosen for determining regions likewise finds no answer in nature, says Hettner. The choice must be made by "the geographer, according to his subjective judgment of their importance. Consequently one cannot speak of true and false regional divisions, but only of purposeful or non-purposeful. There is no universally valid division, which does justice to all phenomena; one can only endeavor to secure a division with the greatest possible advantages and the least possible disadvantages" [161, 316].[85]

On one point there appears to be complete agreement. We are concerned not merely with a method suitable for delimiting any particular region

---

[85] This point of view, which Hettner first stated in 1908 [300, 107], has been reflected by many others, including Unstead [309, 176]. A contrary view, which Graf expounds, is based on a philosophic concept of an "idealism which presumes the possibility of absolute truth," so that in the division of the earth surface there must be "only one geographic truth" [156, 114].

that a geographer may wish to study, but also, since the world, in countless respects, is a unit, with a method by which we may divide the world into regions, or, if one will, by which we may fit regions together into a world system. The first question then is, which of these two procedures should be followed?

Should we commence from a consideration of the world as a whole—as we know it from all the branches of systematic (general) geography—and divide it into realms, regions, districts, and localities? This might appear to be the method of the biological sciences—the subdivision of the tree of life into orders, genera, and species. Hettner notes, however, that the method in those fields has actually been the reverse—namely, to study the characteristics of individual specimens, group these into species, and the species into genera, etc., and only then to test the conclusion in the whole system of classification. Since in geography the relation of different areas to each other and to the world as a whole forms no such simple pattern as that of species to the tree of life, it is all the more necessary that geographers start with the thorough exploration of the characteristics of particular areas and only on the basis of the similarities and differences discovered establish their regional division. Humboldt did not arrive at the geographical concept of the Llanos by subdividing South America, but by the comprehension of the homogeneity of that region [*161, 307*].

The inductive method alone, however, is not sufficient to provide a satisfactory division of the world into regions which can be in any way compared. There must be some sort of a general system of division into which studies of individual regions may, in some way, be related [*cf.* Hall, *290, 127*]. "The theoretical consideration must make comparative examination of the regions discovered and perceive the foundations which have been instrumental in their development; only on this basis can theory proceed to its positive problem and undertake a division of the earth's surface on the basis of the nature of the earth as a whole." The two methods must constantly supplement each other [*161, 307* f.].

Similarly, in determining regional divisions at any level, are we to proceed by dividing larger units into smaller, or by combining smaller to form larger? An example of the former is Maull's method of the girdle of boundaries, mentioned above. As early as 1915, Maull utilized this method in an effort to divide a portion of the Balkans into regions, and he has illustrated it since with other examples [*157; 179*]. Needless to say, the method is not strictly objective, for it not only involves the subjective decision of reducing a wide—often extremely wide—zone of different boundaries to a single boundary, but in the process forces one to distinguish between major and minor factors; one may not, as Lautensach notes, merely take the geometric average of the lines [263]. Granö, working independently, has utilized a very similar system, in which he attempts to state fixed, though arbitrary, rules of procedure [252]. In a somewhat different manner Passarge formerly worked in the same direction: he superimposed transparent maps of different features on each other and endeavored to draw his divisions on this basis.

More recently Passarge has given up this simple method and uses a method of "inspection" of the individual *Landschaftsteil* (areal part of a region); regarding the complex *Landschaftsteil* as the real building stone of the Landschaft, or as the cell of the regional organism, he adds these up to form his region [258; cf. Bürger, 11, 81]. Passarge apparently regards this as a major contribution, giving validity to his claim that his *Landschaftskunde* represents something new in geography. If, however, we ask how the individual regional-part is determined, the answer must be that it is determined in no other way than that by which he formerly determined the region—except that, instead of using the partially objective method of map comparison, which could of course be applied to the smaller unit, he proceeds by a method of "inspection" which would appear to be far more subjective.

We arrive therefore at the answer to the question stated above: regardless of the level of regional division, we proceed in both directions, by division of larger units into smaller, and by building from smaller into larger; in every case, though the two methods partially check each other, in part we are forced to make compromises between them.

B. TYPES OF SYSTEMS THAT HAVE BEEN CONSTRUCTED

Efforts to find a system of regional division of the world are no doubt as old as geography itself; indeed the very nature of the subject requires some basis of division of the earth surface. Although most early workers were content to accept the division already provided by the political map, the efforts of early Greek geographers to divide the world into climatic belts represented an early attempt in a different direction. The first major effort in modern times was that of Ritter, which however was based on a teleologic philosophy which modern science denies. Nevertheless, "to Ritter remains the great merit that he recognized the necessity of a division of the earth spaces based on a universal consideration of all features and made the attempt to put it into effect" [Hettner, 161, 306]. Freed from the teleological concepts which dominated the thought of Ritter's time, and aided by the progress in geographic knowledge, Hettner followed Ritter insofar as the latter utilized not one element, but the combination of various elements and, further, has developed his system in part inductively from the character of individual small areas.[86]

A different direction was given to the problem by the epoch-making effort of Herbertson in 1905 [307]. Whereas Ritter, as well as Hettner, and other German authors, including Philippson, had associated individual regions together as they are in fact, into larger realms and continents, Herbertson considered his regions as specimens of types, thus associating widely separated areas in the same type. His system was therefore a classification of regions regardless of location and association in space, in contrast with an areal division of the world into major parts, these each sub-divided into sub-divisions which are contiguous and together form an associated whole. One system is a classification according to internal character, the other an outline of the actual areal relations. In the first case the regions might be called *generic,* in the latter *specific* (to follow the suggestion of the committee of the British Geographical Association [310, 254]). Whereas most works dealing with the regional geography of particular continents—including many that cover the world—employ a division into specific regions, most of the recent attempts to map the world into regions follow the generic or comparative system.

It seems probable that geography can benefit from both methods. The generic method reveals the inherent similarity of remotely separated regions; this similarity, however, is fundamentally incomplete, since the location of a region is one of its important inherent characteristics which may have marked effect on other inherent characteristics [cf. Penck, 159, 640]. Even though we find "Mediterranean" climate, vegetation, and crops, combined with landforms similar to those of the Mediterranean Region itself, in such areas as Southern California and Central Chile, these areas, as Krebs observes, are not and cannot be "Mediterranean" in any full sense [234, 94]. On the other hand, if we merely associate each of these areas with their neighboring areas in a system of specific regions, we lose the advantage provided by the comparative (generic) method.[87]

---

[86] Hettner's system first appeared in part in his text for Spamer's Atlas, 1897, was in course of preparation in 1905 [126, 674], and was published in complete outline in 1908 [300]; the complete regional treatment now appears in two volumes, *Grundzüge der Länderkunde* [301].

[87] For what we are here calling comparative systems of regional geography the term "systematic regional geography" has been suggested. Though this is briefer and in some respects expresses the nature of such studies, it appears almost certain to lead to misunderstanding. The contrast between "systematic" and "regional" geography is fairly well established in American thought and might easily be confused by a term involving both words; furthermore, the distinction between "systematic geography"

The significant difference between the two methods is often obscured by a rather loose use of terms. In the system of specific regions, which we might call the *actual* system of regions, sub-regions added together form a true single region, whereas in the comparative method they commonly do not. It is only by a misuse of words that we can speak of a single region consisting of eastern England, southwestern and northern France, and the North Sea and West Baltic Lowlands [Unstead, *309*, 179], or of that geographical monstrosity, "the Anglo-Flemish Basin." One can speak, in a certain sense, of a region which includes eastern England and at the same time the opposite lowlands from Cherbourg to the Elbe, but this is not a region simply of "Mid-Temperate Lowlands," but rather a region of lowlands and narrow seas, in which the seas form a major part.

Although Herbertson's attempt at a division of the world into "natural regions" met with little favor for many years, ultimately his idea was taken up by his own students in England, and by various followers in this country, * as well as by Gradmann and Passarge in Germany (though Passarge appears to have been ignorant of any previous systems at the time he first worked out his own [*236*, 336 f.]). The interested student will find discussions of the older systems in Hettner [*161*, 294–306] and of the more recent systems in Hall [*290*], Bowman [*106*, 154–163], and in the committee report to the Geographical Association, which includes the world maps of Herbertson, Stamp, Passarge, Unstead, and Van Valkenburg [*310*].

Most of the earlier systems are called "artificial" by Hettner on the grounds that they proceed purely from description and are based solely on a single, readily identified characteristic. Such systems of division, common to every science in its early stages, have their utility in the development of the science. Since they take no account of the causal relation of the phenomena, however, Hettner feels that they do not touch the real nature (character) of the things concerned. "It is a peculiarity of the artificial classifications that they are more concerned with clear demarcations of things than with a complete comprehension of their character" [*161*, 294].

The clearest demarcation of areas is that provided by the political divisions. The disadvantages of such a system of division for geography are so well known as to require no discussion. On the other hand, it represents a swing to the opposite extreme when geographers pretend that regional study can ignore the importance of this division. In areas where the political boundaries are antecedent to the development of present cultural landscapes the latter may be clearly divided in their associations, both exter-

---

and "systematic regional geography" would be very difficult to maintain in common usage.

* See Supplementary Note 40

nal and internal, and in many respects they are sharply in contrast to each other on either side of the political frontier. Such a situation is described in the writer's "Geographic and Political Boundaries in Upper Silesia" [*355*]. The spring wheat region of North America, on the other hand, offers an example of marked separation by an antecedent political boundary without major differences in character [*367*].

Furthermore the fact that the state does tend to organize the different parts of its total area into an economic, social, and political unit justifies the student in giving some consideration to state-areas. Few, however, would support the view that the political division provides a suitable basis for major world division in geography (except, of course, in political geography itself). In particular, the general geographer will pay little attention to political sub-divisions of states, which in this country include the so-called States themselves.

Likewise little need be said of divisions of the world based on the races of man or the peoples. Undoubtedly these are important as geographical factors (*i.e.*, factors leading to areal differentiation), so that world maps on this basis are of value to the geographer. But few would consider such factors as of such great importance as to make them the fundamental criteria for regional world division [Hettner, *161*, 296 f.].

In the greater part of southern and northwestern Europe the marked degree of separation of different parts of the land by the sea provided an apparently obvious basis of division which has been accepted in daily use as well as in our science. Italy was a "geographical expression" centuries before it was a state; indeed the geographical expression has more than justified itself. If all the world were as clearly broken up into islands and peninsulas, this system might well have sufficed for many purposes. Since, however, the seas actually perform the function here assigned them—of bounding limited regions of the land—most inadequately, the idea of the waters as a dividing factor was made to include the rivers. Absurd though it appears to us, this idea had far-reaching consequences in the political field —the thesis that "the Rhine is the natural boundary of France" was maintained at least as recently as the Peace Conference of 1919.

On the other hand, other students emphasized the unifying effect of rivers and so divided the lands on the basis of their drainage basins. When it was realized that in many cases the valleys of captured streams were more intimately related to other drainage basins than their own, the attempt was made to substitute mountain ranges for the drainage divides. This however presumed that mountains formed a continuous system of sharp divides over all the lands, but "now that such mountains have disappeared from our maps,"

[295]

mountain crests can no longer fulfill the most important requirement of all artificial divisions, that of easy and universal applicability [161, 297 f.].

C. SYSTEMS OF "NATURAL REGIONS"

The reader may wonder why we have given consideration to outmoded systems of division which are not only artificial but, in our eyes, extremely naive. This brief survey of past systems, however, may enable us to look as with the eyes of future geographers at those we are now using. It may be that we are deceiving ourselves in somewhat similar manner.

A common form of self-deception consists in the loose application of terms which appear to have a precise meaning. Though the author may be conscious, at first, that the term is used loosely, in time the technical term itself will lead his readers, and himself, to believe that what he has produced is what he says it is. If we believe that earlier writers were deceiving themselves, for example, in calling drainage basins the natural divisions of the land, are we on safer ground in our present use of such terms as "natural regions?"

Unfortunately common usage permits the term "natural regions" to carry any of several different meanings (the difficulty is much the same in French and German—perhaps in all the Indo-European languages). In its broadest sense, "nature" obviously includes man, together with all his works; in this sense it is nearly synonymous with the universe, or better, with the reality of the universe as distinct from our thoughts about it. Thus Kant, Humboldt and other writers have used "nature" to mean the objective world outside of the observer's mind.

On this basis some students use the word in the term "natural regions" to indicate "something inherent and not arbitrarily imposed" [310, 253]. This appears to have been Herbertson's intention in his original use of the term [307] and it is perhaps significant that in his later publication he should have dropped the word "natural" and written simply of "major regions" of the world [308]. The word "region" itself, as used by geographers at least, carries the connotation of "inherent character"—in contrast to the word "area." If "natural" is added in order to emphasize the concept of "inherent regional character" as opposed to arbitrarily imposed divisions of area, the addition seems not only redundant (one anticipates the appearance of a system of "Truly Natural Regions") but in addition somewhat pretentious: it implies that there is in nature, *i.e.,* in the real world, an unambiguous division of the earth surface and the problem is simply to recognize it correctly. No such division exists in reality, as we have seen; any attempt to divide the world involves subjective judgment, not in the determination of the limits *

[296]

* Insert "merely", following "not".

of individual factors, but in deciding which of several factors is to be regarded as most important. The "inherent character" of an area is composed of a multitude of incommensurable elements which are interrelated in no complete sense: temperature, rainfall, relief, slope of land, physical texture of soil, underground minerals, and relative location are all in large degree independent of each other—not to mention the degree of independence which human elements exhibit. In a world in which these factors all vary from place to place, the variations being in large degree independent, it is, as any number of writers have observed, impossible to draw the divisions on the basis of all the factors simultaneously. One must, therefore, determine which variations are of greatest importance. By what measuring rod can we determine this? Which is bigger, a mountain or a lack of rain? The decision can only be subjective; the regions so constructed are in this sense arbitrarily imposed on reality.

In some cases authors who have used the term "natural regions" do not, perhaps, intend the title to be taken literally but use it merely to indicate that the basis of the regional division is to be found in nature as a whole, including man, in contrast to a division based on a single element, as in the case of "climatic regions," "agricultural regions," etc. (In most cases, to be sure, this usage also appears pretentious since the division is actually based almost entirely on one factor, namely climate.) On first thought "natural regions" may seem a legitimate abbreviation of "regions on the basis of nature," but when we reflect on the full connotation of "natural" it is clear that the shorter phrase claims much more than the longer one; as though we shortened "regions on a basis of real conditions" to "real regions." The conditions are reality, the regions are intellectual conceptions.

The difficulty could easily be avoided by using the term "regions" without any qualifying word, as Herbertson did in his later publication, or, if one will, by use of the term "geographic regions." It is true that "geographic" has sometimes been used in the sense of natural, exclusive of human, but in a science of geography which includes human aspects this misunderstanding should not be possible [Dickinson objects to this usage by British geographers, *101*, 258, 268]. The adjective of "geography" includes all that "geography" includes [James, *286*, 80].

The distinction between "regions on a natural basis" and "natural regions" is strictly maintained in Hettner's *"natürliche Einteilung der Erdoberfläche"* [*300*]. He recognizes that, whatever the basis for division, its application requires so many subjective decisions of judgment [107] that the resultant regions cannot, in any proper sense of the word, be called "natural." On the other hand, his phrase—whether or not it is entirely clear in German—if translated as "natural division of the earth's surface," is far from precise. In his discussion, Hettner contrasts his system of division with those which he terms artificial (*künstlich*); his system, then, is *"natürlich"* in the sense of realistic; more fully, he seeks a logical basis founded solely on reality and on the whole of reality (*Natur*).

Each of the interpretations discussed in the preceding paragraphs depends on the use of the word "natural" in its broadest sense of including all the phenomena observed outside the observer's mind—objective reality, one might say. Unquestionably this usage is well founded in geographic literature as well as in other fields of thought; for Humboldt, as well as his predecessors, it was the common understanding of the word.[88] During the nineteenth century, however, scientists in general came to distinguish more and more sharply the study of non-human phenomena as "natural science," and in geography, as well as in other fields of thought, "nature" was used as a word to express the contrast between that part of reality that was independent of man and that part that was "human." Although in a sense one may agree with James that it is "supreme arrogance" to contrast "Man and Nature" [*286, 79*], the fact that man alone in nature is capable of contemplating such a contrast is sufficient justification for terms with which to express it. All the fields of science and philosophy that are concerned with "the proper study of mankind" find it necessary to distinguish between that part of the universe that is of man and that which is not, if only in order to state intelligibly the intimate relations that exist between them. James, indeed, recognizes this need by recommending the use of the word "fundament," "in thinking about the relations between man and the earth" [80]— as though man were not of the earth, earthy. While the term fundament has certain advantages, it likewise has obvious limitations—in particular the fact that in the adjective form it becomes an entirely different word. Illogical and confusing though it may be, scientists wishing to distinguish non-human elements from human elements have become accustomed to use the word "natural" and will presumably continue to do so. It is only necessary that we make it clear that this is what we mean by the term.

Whereas in most sciences the point under discussion is of little impor-

---

[88] Granö uses a different version of this meaning of "nature," adopted from the philosopher Rickert: *"die Natur ist das körperliche Sein,"* in other words, material things—whether human or non-human—in contrast to immaterial [*252; 270, 297*]. Rickert apparently defines "natural science" in this way. Since there is no lack of terms for distinguishing between material, corporeal, physical things and the opposite, there does not seem to be any reason for introducing this further element of confusion, unless it be to prove some particular thesis as to what a "natural science" should be.

tance, in geography it is obviously of greatest importance since geography intersects both the "natural sciences" and the human or "social sciences." To be sure, many modern geographers would not regard "the connection between the natural and the social sciences" as a proper description of geography; like Schlüter, they would emphasize the fact that geography intersects all these individual sciences, forming connections between all of them, irrespective of the conventional division into two major groups. On the same basis some writers consider "human geography" as but a part of "biological geography" in contrast to "physical geography" (meaning, in this case, the study of inanimate geographic features), these forming the two divisions of systematic (general) geography. If this appears logical, it is nonetheless unrealistic. As Richthofen observed, man in his relation to non-human factors reacts in an entirely different manner from all other living things; his reactions are "cultural" reactions, involving such items as clothing, houses, tools, etc.—even on the lowest step of his existence [73, 56 f.]. Furthermore, since geography is a science developed by and for men—as distinct from all other biological beings—the relations between the world of man and the non-human world are of greatest concern in geography. In this sense, the geography of wolves, for example, is as remote from the geography of men as is the geography of landforms. We are therefore justified in accepting the popular conception that in the universe the things that are cultural,—*i.e.*, of man—are fundamentally different from everything else— which for want of a better term, we call "natural" [*cf.* Vogel, *244*, 192].   *

In this limited technical sense of "natural," the term "natural regions" would presumably indicate regions considered in terms of their non-human elements. We have noted that the concept of the earth surface exclusive of human elements is an intellectual concept not represented in reality (Sec. VC), so that any such system of "natural regions" is purely theoretical. Further, we found that the world, even in theory, does not consist of a simple system of natural regions; even though we exclude the complicating factor of man, the complex combination of natural elements, many of them essentially independent of each other, does not admit of any single regional division except by arbitrary decision.

It is also possible, as was the case in the use of "natural regions" in the broader sense, that "natural regions," in the limited sense to which we adhere, might be used merely to indicate that the world has been divided on the basis of the natural, non-human elements. In addition to the objection raised in the previous connection, this usage is misleading for two reasons. Nature, as distinct from man, provides no basis for a division which will be significant or suitable at the same time for all natural features, not even for two or three natural features of great importance. Neither does nature indicate

* See Supplementary Note 41

which of its elements is of more significance than others. We have, for example, no grounds for presuming that the natural elements significant to a mosquito, whether in New Jersey or in Alaska, are less important to nature than those significant to a sequoia. In other words a map of "natural regions," or of "regions based solely on natural elements" with reference to mosquitoes, would be entirely different from one made with reference to sequioas, but any such division must be made with reference to some such ulterior concern. Needless to say, all such divisions by geographers have been made with reference to man's point of view—nature as man is concerned with it. If we are to think accurately we must call these "regions based on natural conditions as significant to man," as Herbertson, at least, has recognized [*310* and *308*].

The reader may feel that many words have been used to bring us to a conclusion that was obvious from the start, but it is evident from many studies that others have not found it so obvious. The importance of the conclusion is seen when one considers the disadvantages of a regional system in which we must always take account of the human significance of natural elements instead of considering them in their own right, "in their inherent characteristics," as some have suggested. There is introduced a factor which affects every one of the elements and which varies notably both from place to place and from time to time. A regional division of North America, even though confined to natural factors, is one thing in terms of the pre-Columbian period, another in the period of settlement, and indeed has been changing constantly, though less rapidly, ever since [*cf.* Broek, *297, 107*]. Likewise for any single period, the relative importance ascribed to the different natural factors in the United States cannot be assumed to be the same in a regional division of China—unless one wants simply a regional division of the natural environment of China as it would affect Americans if they happened to inhabit it.

Is there then no possible way by which we may divide the world into regions that will give due account to the whole complex of natural elements involved and yet be independent of the human point of view? Such a "natural division of the world into regions" would, if it were even theoretically possible to construct it, provide a framework independent of man against which we could study the development of cultural regions. The great value which a framework of this sort would have for regional geography requires us, in spite of all the logical arguments which make it seem impossible, to consider any serious suggestion that appears to provide it.

A great part of the enthusiasm with which various students have seized upon the concept of the "natural landscape," as Passarge and others have expounded it, is perhaps to be ascribed to a feeling that the answer to our

[300]

question is to be found in this concept. By observation of the "natural land-scape" we will find in nature itself the basis for the regional division of the earth surface. In one of his more recent statements Passarge asserts that "the individual spaces, *Räume,* [each more or less homogeneous in a par-ticular category] of the surface forms—of the rocks, of the plant-cover, and of the water—are combined according to rules (*regelmässig*) into areas of certain landscape character" which he calls "landscape spaces" [*268,* 6].

We found that the landscape that we can observe as a reality consists in its land portions largely of the relief of the surface and the vegetative cover. But how can we coordinate even these two largely independent elements? What amount of difference in one is to be regarded as more important than what amount of difference in the other? The landscape itself gives no ob-jective answer. We can determine it only on the subjective basis of our sensations—or in terms of the relation of the landscape to something else, such as man's life.

Passarge, without stating the basis for his decision, answers the question in favor of vegetation. That determines his major division of the world; landforms are considered only in subdivisions [*268,* 92]. Obviously this is a purely arbitrary measure of importance; why is the difference between the rugged slopes of a mountain area and the wide expanse of an adjacent plain less significant in the landscape than the difference between forests and grass on the plain? In actual practice, on his map of landscape belts of the world, Passarge found no room for these subdivisions; the Rocky Mountains are simply included in "dry landscapes."

A second and even greater objection to this system is its essential unre-ality. It is not, in fact, based on the real landscape as we see it, but on a hypothetical landscape which no living person has ever seen, namely the natural landscape. This is not entirely clear in Passarge's presentation; in places where "cultural landscapes" are but little developed one looks appar-ently at the present landscapes, which are loosely called "natural landscapes," but in other places one looks at presumed primeval landscape, unaltered by man.

Unless we are to be completely befuddled we must clearly understand which stage of the landscape of the world we are to view and divide into spaces. Theoretically there are three distinct possibilities and they cannot be combined without destroying whatever validity the system may have. The present, cultural, landscape is the only landscape, as James has recog-nized, that actually exists. The present natural landscape, we found, is a theoretical concept which not only does not exist but never did exist. The natural landscape that once existed, the primeval landscape present before the arrival of man on the scene, would not have remained the same had man not arrived. The first of these must be ruled out in any attempt to study the natural landscape of the highly cultivated parts of Europe, Asia, and America. The second could provide no basis for observation since it is purely theoretical. Most students therefore rely on a reconstruction of the primeval landscape, the original landscape, but commonly without recognizing that it is not the same as the theoretical natural landscape of to-day.

What value, as an observational basis, has the original natural landscape which has been seen in few areas of the world by historic man? Is it possible that in the present landscapes we can make observations of the original natural landscape, including particularly the major landscape factor of vegetation, that will provide an approximately correct framework for regional division? Obviously this is impossible in areas that have been thoroughly cleared and cultivated for centuries, as in Europe and southern and eastern Asia. We might have supposed that the same was true of the agricultural areas of the New World, but at least one member of the Passarge school appears to hold a different view. In reply to Waibel's critique, Ahrens protests: "I do not believe that he seriously means to suppose that the landscapes of the continents outside of Europe appeared essentially different a hundred years ago, yes even four hundred years ago, than they appear today" [250, reply, 166]. In counter-reply, Waibel, passing over the revolutionary change in the landscapes of eastern North America, refers to Schmieder's demonstration of the major changes that white settlement has caused in the vegetation of the Pampas.

In all of these cultivated lands, however, we have patches of unused land which appears to offer us specimens of the original vegetation, on the basis of which we deduce the original vegetation for the whole area. But we know, from the work of Gradmann and others in northern Europe, that this is not reliable; northern Europe was not all forests in pre-historic times. Further, we cannot assume that the patches of wild vegetation in populated areas represent accurately the character of the primeval vegetation. When Penck reports that on the abandoned farms of northeastern United States "a new primeval landscape has grown where formerly were fields" [quoted in 11, 47], he must be using the term, *Urlandschaft,* in some different sense from the usual. The differences in the humus content of the soil, in the stream gulleys, and in the character of the forest itself, are by no means insignificant.

That these facts are the case in the parts of the world in which we live is well known. Why should it be supposed that they were not the case in remote areas of which we know less? Alteration of the original vegetation and of the soils that supported that vegetation is not limited to peoples of a higher stage of agriculture. When Volz tells us that a generation after the

Malays in Sumatra have abandoned a clearing, the site can "scarcely be recognized by anyone ignorant of the facts who passes through the resultant forest," he is tacitly admitting that some change, however slight, is left as a result of this single cycle in a series that has been going on for untold generations [*262*, 101]. We cannot say with certainty that there is a square mile of lowland in Central Africa in which the character of the vegetation and soil has not been significantly modified by centuries of shifting hoe culture of the natives. According to Lautensach, we cannot even be sure that the tropical savannas, in general, are "natural" [*278*, 20].

We conclude therefore that the "natural landscape," whether it be the theoretical natural landscape of the present, or the actual landscape of prehuman times, is something that the geographer of today cannot expect to see in any but very exceptional parts of the world. Whatever utility there is in judging landscapes as we see them, however, must presumably be lost in a system based on landscapes never seen. Or, is it possible that we can achieve much the same utility from the reconstruction in our minds of the original natural landscape? If we have reliable knowledge of the original natural vegetation, can we not substitute that for the present cover? Only if we have such reliable knowledge.

Certainly our literature includes maps of natural vegetation of the world —any number of them. Which are we to take as reliable for our basic purposes? Unfortunately the Passarge school, including particularly Passarge himself, have given but scanty attention to this very difficult problem. In adopting Passarge's system, James wisely ignored the former's map [*305*], not merely, we may presume, because he questioned the suitability of its classification, but also because of the extraordinary extent to which it appears to ignore the facts that are known about natural vegetation.[89] James himself, however, would not ask us to place too great reliance on his own maps. He recognized that we have no certain knowledge as to the original vegetation of the Prairies of this country and the Pampas of South America [*321*, 211], and on his map he uses the device of an honest scholar —a blank area with a question mark—over almost the whole of one of the most important regions of the world, China. If his world map of vegetation were on a larger scale he would likewise have to pepper the map of northern

---

[89] To take but one example of many, an area of "humid steppe lands" is found to consist of the double line of states on either side of the Mississippi (including Wisconsin, Iowa and Mississippi) in contrast to forest areas on the east, whereas the Pampas of both Uruguay and Argentina are in "dry steppe lands." Waibel likewise finds the facts "grossly generalized and in many cases incorrect," whereas Ahrens defends the "*Landschaftsgürtel*" as "established with astonishing accuracy" [*250*, 168, 170].

Europe with question marks, since European geographers and allied students are still uncertain as to extent of natural grass clearings before the coming of man.

Passarge clearly recognized the difficulty involved, though he does not appear to have seen its undermining effect on his system: "Man in his work practices such a determining influence on the landscape that, not only is its appearance completely changed, but it becomes actually an important and difficult problem to reconstruct the original landscape" [268, 71]. The statement implies nevertheless that such a reconstruction is a possibility. Likewise Schlüter, though recognizing that in cultivated areas it may be almost impossible to reconstruct the primeval landscape, nevertheless regards the problem as not merely important and attractive but necessary; for the original landscape is the starting point for the consideration of the *Kultur-landschaft* [145, 19 f.; *cf.* also Sauer, *84*, 202]. Granted that the reconstruction of the original landscape offers an attractive possibility as an end-goal for research, does it provide a useable basic framework for regional geography? Since we have no historic record of the "original landscape," *i.e.*, before the arrival of *prehistoric* man, we are faced with the well-nigh insuperable difficulty of discovering, and correctly interpreting, all the alterations that have been made since, both by man and by nature.

In view of the impossibility of reconstructing the natural vegetation of the landscape as it would have been if man had not entered the scene, and of the extreme difficulty of reconstructing the original natural vegetation before man, it is not surprising that even those who have been most enthusiastic for a regional system based on vegetation have not developed it in a manner consistent even in major outline. Their systems are in reality based on a combination of a (presumably) visible vegetation and an invisible climate. Over and above the vegetation divisions of Passarge, other than those of the "dry areas," is the familiar division into a "hot belt," a "middle belt" and the "polar caps."[90]

Similarly the terminology of James's regional classification reveals the predominance of climate behind his thought. "The Mid-Latitude Mixed Forest" may differ, in appearance and in types of trees included, from the "Boreal Forest" on the one hand, and the "Tropical Forest" on the other, but few will suppose that those differences are of the order of "major lineaments of the earth." The very arrangement of the sections, in which the "Grasslands" are placed between the Mid-Latitude Mixed Forest Lands and

---

[90] See also his theoretical outline [268, 92 ff.]. The writer should admit that, like Gradmann, he may well be lost in the several very complicated systems of Passarge, each with its very similar but complicated terminology. Even advanced students must find this very difficult, not to mention a beginner, as Passarge suggests.

the Boreal Forest Lands, reveals the underlying dependence on differences in climate.

We end, therefore, where we started, with the problem of measuring the relative importance of differences in climate as compared with those in vegetation—hot deserts *versus* cold deserts, mid-latitude prairie *versus* mid-latitude forests, etc. No matter how we term our system, whether "natural regions" or "regions of the natural landscape," we are fundamentally measuring the different natural criteria in terms of their importance to man. Since the relative importance of the different natural elements to man is certainly not determined by nature as distinct from man, it follows that our systems inevitably have a human basis; in this sense all might be called "artificial."

D. SYSTEMS OF "SPECIFIC REGIONS": HETTNER'S GENETIC SYSTEM

We have repeatedly stressed, in many previous connections, the essential importance of the factor of relative location in geography. Unquestionably any system of regional division that leaves this out of account is notably incomplete [*cf.* Hassinger, *225,* 474 ff.; Krebs, *234,* 94; Lautensach, *263,* 195 ff.; Waibel, *266,* 204]. If it is considered, then the location of the regions with reference to each other becomes of major importance and we have a system in which areas are divided as they are found on the map, rather than by types based on internal characteristics. We may call the system a "realistic" division in contrast to a "comparative system" and the areas, as the British geographers have suggested, are "specific" regions, rather than "generic" types [*310,* 254].

Almost every professional geographer, no doubt, has attempted to divide some larger area, such as a continent, into specific regions. What principle or principles are to be considered as offering sound guidance for dividing, not merely one particular area, one continent perhaps, but any large areas, or the whole world?

To divide any area into regions involves necessarily more than the recognition of the distinctive character of certain parts of it, ignoring those less distinctive in character. An incomplete study may limit itself to the "cores" or "hearts" of regions, although this involves the questionable assumption that such areas are geographically more important than the areas of less distinctive character. A complete study must give attention to every portion of the area, so that the student sooner or later must face the problem, as Wellington Jones has noted, of drawing definite, even though arbitrary, boundaries between the regions [*287,* 106]. The device of "transition zones" does not eliminate the problem, since these zones.are themselves areas which must be delimited. (For an example of the correct use of transition zones with specific boundaries see Jones's map of agricultural regions in central northwest United States [283].) *

The most thorough study of the problem of dividing the world into specific regions is that made by Hettner, to which we have often referred [300]. Recognizing the manifold character of the world, he realizes that a division based on all of reality is not possible on one principle, but can only be built up on a combination of several grounds. The problem, therefore, is to compare the different bases and measure their relative importance [106]. The degree of importance is, one infers, to be measured in terms of the effect on organic life, notably that of man.

What is the order of importance of the multitude of factors involved? On one point there may perhaps be complete agreement. The presence of land in contrast to sea is of first importance not only for man but for most organic life. Indeed most systems of division confine their attention to the lands. In any realistic system of specific regions however it is a debatable question whether the regions of the first order must be based solely on this fundamental distinction, *i.e.,* be represented by the continental land-masses. In view of the similarity of certain land areas separated by smaller seas— the Mediterranean Region, the Caribbean Region—and in view of the extent to which these seas not only separate but also in part connect, some students, like Banse, would dispense with the conventional continental division in favor of one involving a combination of factors. Whichever method one follows the result cannot be completely satisfactory.

If we now confine our attention to the lands, we find no fixed order of importance of the different elements that is applicable to all parts of the world: in one area, climatic differences may be of major importance, in another, differences in relief, in still others, soil, mineral deposits, or simply relative location [*cf.* also Geissler, *277,* 6 f.; Lautensach, *278, 19*]. Consequently it is necessary to use different grounds for division in different parts of the world, although within a single division, Hettner argues, logical reasons require one to maintain the same ground. On the other hand different grounds may be combined in recognizing a single region. Thus the Spanish peninsula is distinguished as a subdivision, of major size, because of the separating effect of the sea on three sides and of the mountains on the fourth, even though these factors, sea and a type of land, are actually opposing elements.

Furthermore, the same factor may work simultaneously in both ways: the Pyrenees separate major divisions, but form in themselves a minor

[306]

subdivision. This difficulty has caused "many a student to rack his brains in the vain attempt to find a solution, not realizing that it is an attempt to square the circle" [106]. The only way out is to divide the major subdivision by a line through the mountains, but in the consideration of smaller divisions to consider the mountain area as a whole. If the reader objects that such a method does not fit into a logical system of division and subdivision, one can only say that it is the earth which is a fault. While the Pyrenees do, in many important respects, form a homogeneous region which must be recognized in any realistic study, it is also true that their southern slopes are a part of the Spanish peninsula as a major division.

A system of division based on a large number of elements whose relative importance varies greatly from place to place may appear hopelessly complicated. These elements, however, are not completely independent of each other. Will not the study of their interrelations lead us to a somewhat simpler basis? Hettner argues that "every realistic (*natürlich*) system of division must be genetic, *i.e.*, it must show the causal relations present in reality. It must search for the creative forces of the earth, it must seek to know the manner in which the phenomena of the earth surface result from the composite influences of those forces, and it must likewise learn to reconstruct in mind the edifice of the earth and thereby learn to understand the individual parts and spaces in their character and their significance" [*161, 308*]. Philippson likewise concludes that order in a regional system can be attained only through a genetic system [*143, 13*].

To what extent can we consider the various geographic features as functions of basic causal factors? How many such causal factors must we recognize as independent? If we assume, for the moment, that all the variations of human elements are to be explained in terms of natural elements, we might theoretically consider that the total differentiation of the earth-surface—the whole of geography—rests upon two variables: the angle of the sun's rays in different parts of the world, and the tectonic forces that have given the earth its highly differentiated crust. Even if it were held to be the function of geography to study these fundamental forces in the formation of the earth surface, it is obvious that the manner in which they have produced the specific combinations of features found in different areas is so inextricably complicated that we could never hope to attain from such a consideration a workable basis for organizing the earth areas in our thought. In other words, although we recognize that the existence of land areas and sea areas, the form of the land, the country rock and mineral deposits, the soil and vegetation, etc., may all be considered as varying manifestations of the two basic forces named, geographers do not consider them as products, but rather start with them as elementary factors. To say that this is incomplete is merely to remind ourselves that no science is complete in itself, but is merely a branch of one single universal science. We cannot, in every branch, analyze our problems down to the fundamental electrons and protons. "What constituent is to be considered as an 'element' is a relative matter," Hellpach concludes, "each scientific branch of study decides about that in each case according to its practical, scientific needs" [*139, 351*].

Hettner would explain most geographic differences genetically in terms of three basic realms: the relation of land to sea, the inner construction and form of the land, and the differences in climates; no one of these, however, is to be explained, in any major degree, in terms of the other two.

The resulting difficulties may be illustrated by a single example. In the division of Europe into major regions, Hettner decides that the differences in landform are of greater importance than differences in climate, whereas in Africa and Australia he finds the climatic differences of greater importance. These conclusions would presumably meet with general approval, but many American geographers would object vigorously to his conclusion that, in North America, landforms were of greater importance than climate, thus placing the Mississippi Basin as a major region in contrast to the Atlantic Seaboard [*300;* also in his *Länderkunde, 301*].[91] On the other hand, there is no logical ground on which we can say that this conclusion is false, the contrary conclusion the only correct one. In other words, as Hettner clearly states, the genetic principle does not lead to a system free of subjective decisions.

In fact, however, the situation seems to this writer more complicated than Hettner himself admits. A close analysis of his discussion indicates that the number of features which we are forced to consider as essentially independent factors is much greater than three. "The inner construction and form of the land" is palpably not a statement of a single factor. The character of the country rock—necessary for an understanding of soils—is not to be combined as a single factor with the character of the land forms, and the nature of subsurface mineral deposits is, for our purposes, likewise independent of both of these. Similarly we conveniently use the word climate to cover several elementary factors, notably temperature and precipitation,

---

[91] Note also Waibel's more general statement: "It must be admitted that relief is the most important phenomenon of the earth surface and the foundation of all life" [*266*, 200]. Such a statement is understandable in the background of Central Europe, and indeed the great concentration of German geographers on geomorphology might be regarded as another illustration of a theme Whittlesey once discussed, the influence of the natural environment on the development of the science of geography in different countries [*399*].

* See Supplementary Note 43

which in fact, however, we must consider separately—at least until we have a satisfactory single genetic basis for the classification of climates. In addition, the differences in geographical location (the importance of which Hettner also recognizes) may cause otherwise identical combinations of conditions to have quite different expressions in the resultant features—*e.g.,* the Arabian desert produces no cactus. Finally, if we are not to beg the question of the determination of human elements by the natural environment in the very foundation of our study, we must accept certain human elements—such as differences in customs, stage of development, etc.—as primary factors leading to areal differentiation. *

It is clear, therefore, that a system of regional division cannot be completely genetic—we cannot, in the effort to construct a basic framework for regional geography, pursue all of our elements back to their origins, and if we could, we would find many independent origins. One can accept this conclusion only with regret. For if it were possible to trace all elements in a particular region back to a common genesis of the area, and similarly those of another area to a separate common genesis—in the sense that this is possible for a biological organism—we could hope that the genetic principle would ultimately lead us to a classification system comparable to that of the biological sciences. But we know from the start that this is definitely not the case; we found this to be one of the most important respects in which an area differs from an organism. Though the use of the genetic method would reduce the number of independent elements with which we have to deal, we would nevertheless still have such a large number of separate elements that it is not clear, to this writer at least, why the genetic principle must be regarded as essential.

On the contrary, there is reason to believe that the insistence on the genetic principle leads us into serious difficulties. These become clearest when we consider the relation of human, or cultural phenomena to the system of regional division. Though the works of man may be largely regarded as resultants in a genetic explanation, at the same time human factors are causal factors in the development of many features, and though these causes —for example a particular custom—may conceivably be explained as results of non-human causes, we are seldom in the position of being able to demonstrate that. Indeed the question could properly be raised whether Hettner's system is not based on a philosophic assumption of the theoretical possibility of explaining human elements as products of natural elements. Hettner has insisted that "science as such must be deterministic" [*130,* 411 f.; *176*], to which Maull quite properly asks, "Why?" [*179,* 182 f.]. The business of science is to study reality as it finds it; although we cannot deny the possi-

[309]

bility of truth to the deterministic hypothesis, recent studies in philosophy cast great doubt on its reliability as a basic assumption.[92]

In any case that philosophical question would appear to have but academic interest for geography. Whatever may be the fundamental character of the relations between human phenomena and non-human, as they are presented to us in geography, they are often of such an involved character, the products of historical evolution of centuries or millenniums of unrecorded time, that we cannot hope to explain them, and therefore for all practical purposes in geography must accept certain major human phenomena as independent elements. On the other hand there are other human phenomena—of land use, for example—that we can study in relation to natural phenomena, as well as to these elemental unexplained human phenomena. For this purpose, can we accept Hettner's principle that "one will be able to grasp the geographic phenomena of man in their natural conditioning more clearly if one regards them without prejudice in the framework of a "natural division" (a logical genetic division based ultimately on non-human factors), than if one makes them the basis of the division beforehand?" [*300,* 106; repeated in *161,* 314].

Unquestionably a framework for regional geography based on "the geographic phenomena of man in their natural conditioning" would beg the question from the start; it would not represent a framework, but an ultimate conclusion. But is it possible to regard the phenomena *"unbefangen im Rahmen einer natürlichen Einteilung?"* Doesn't the very act of putting the phenomena of man in a particular framework determined by natural conditions prejudice the consideration of "their natural conditioning"? In contrast, if our framework is simply based on the geographic phenomena of man—regardless of their natural conditioning—then we could without prejudice consider the relation of those phenomena to natural conditions.

On this particlar point, the greater number of Hettner's colleagues who have attempted to divide larger areas into specific regions take issue with him. On principle they exclude no factor, but endeavor to base their regional divisions on all pertinent factors, whether natural or human, including in most cases such non-material factors as language. Obviously, however, such systems cannot maintain the genetic principle, and the problem of deciding subjectively the relative importance to be assigned to each particular factor becomes vastly greater. Further, they recognize that regions defined on such bases are not stable but change in course of time. (Among many

---

[92] Particular mention should be made here of Lehmann's recent study of *"Der Zerfall der Kausalität und die Geographie"* [*181*], a work called to my attention too late to enable me adequately to utilize its pertinent suggestions on this and other current problems in geography.

others who have discussed this question, with specific examples, we may list: Hassinger [225; 141]; Gradmann [303; 144; 236]; Maull [228]; Sölch [237, 25 ff.]; Lautensach [263; 278]; and Bürger [11, 51–3], who lists in addition, Friedrichsen, Hassert, Machatscheck, A. Schultz, and Troll.)

Though Hettner's system leads ultimately to a relatively permanent division of the earth surface, it is clear that its establishment requires far more accurate knowledge of the elements employed, and their relationships, than we as yet possess. It represents therefore a goal for geographic work rather than a base; no doubt this is his intention. That we may wish to reject his division of our own continent is of no importance; on the basis of his principles we would be justified in reconstructing it to attain a more "purposeful" result. In any case, his is the only system that, in execution, as well as in theory, attempts to give due account not merely to one or two major elements, but to all natural elements of significance to man. When we have examined other systems suggested or constructed we may conclude that Hettner alone has presented a single system that can be utilized as a framework for all the significant features of areas. *

E. COMPARATIVE SYSTEMS OF GENERIC REGIONS BASED ON ELEMENTS OF THE NATURAL ENVIRONMENT

In a "realistic system" of world division, such as that which we have been discussing, every region is recognized as unique and the regions are associated in their actual relationships. With such a system it is of course perfectly possible to take widely separated regions which show marked similarities in certain elements and study both their similarities and their differences. In this sense, comparative regional geography is nothing new, as Hettner and others have pointed out [167, 284–6]. On the other hand, a system of regional division which is based on certain major similarities of regions, regardless of where they are located, a system, that is, which distinguishes regions not so much in terms of their unique reality but in terms of types into which they can be classified, obviously facilitates the study of comparisons. I have therefore called such a system a "comparative system" of division of the world into "generic" regions.

In the comparative systems which we are to consider, one major division of the world is assumed at the start, namely the division between lands and seas. Only the lands are considered for division into types.

In many of these systems, the subdivisions are called "natural regions." We have already noted the confusion which results from the various meanings and connotations of the word "natural" in this term wherever it is used. Entirely aside from those objections, the use of the term in various compara-

[311]

* This concluding sentence is neither clear nor necessary and may well be omitted.

tive systems appears to claim far more than is actually represented by the subdivisions produced; the same is true of the clearer term "geographic regions."

It is perhaps a minor objection that the subdivisions in these types are not "regions," in the sense in which most geographers use that word, but merely areas of a certain type. For this reason perhaps, Finch and Trewartha use the word "realm," while James simply calls his subdivisions, "lands" of a certain type. Krebs has urged that, in German, the word *"Landschaft"* should be used only in this limited sense [*279*, 210]. Even if his suggestion were followed by his colleagues, which seems very unlikely —the word could not be translated in this sense into any single English word, but only into "kind of land" or, possibly, "land-type." Not finding any suitable word in English, the writer follows the current custom among English-speaking geographers of using "regions" in the loose sense of "areas of a certain type."

The major objection, in the present connection, to the use of either "natural" or "geographic" regions, is the fact that in almost every case the comprehensive term, "natural," or the even more comprehensive term "geographic," is found, on examination, to be reduced to a single natural factor. Most of them, like Herbertson's, are nothing more than maps of major climatic areas of the world. Herbertson appears to have recognized the overly simple character of his major division, even from the start. Particularly in the last discussion of his system, in 1913, he wishes to include not only other natural elements, but men as well; nevertheless the criteria which he adopted for their demarcation were almost purely climatic [*308; cf. 310, 262–6*]. The whole question is indicated nicely in the title which Stamp gives to his modification of Herbertson's system, "The Major Climatic Regions of the World or the Major Geographical Regions" [*310, 266*]. Another map which appears to be almost purely on a climatic basis, in spite of its title, "Natural Regions Based Mainly on Climate and Human Use," is that of Van Valkenburg [*314*].

Unstead recognizes clearly that the term "Geographical Regions" properly includes all the significant features, both natural and human, but finds, in the determination of major regions of the world, that climate is the main criterion, so that his world map is admittedly climatic, save only that high-land areas (*high* mountains and *high* plateaus) are classed separately, in four, climatic, divisions. With this exception, the factor of landforms is considered only in the subdivisions, where—particularly in the divisions of much lower order—cultural factors may also be recognized [*309*].

A different direction, as we have seen, is that taken by Passarge and followed by James [*321, ix*]. In spite of the underlying consideration of climate of which we spoke, in the systems of both these authors the individual regions are defined specifically in terms of natural vegetation. The actual classification of vegetation regions however is very different in the two cases; in addition, James puts the high mountain areas into a separate category.

We do not mean to suggest that these students have naively identified all the natural elements, or all the geographical elements, with the single element upon which they base their systems—at least in the first major division. Their procedure might be justified on the basis of either or both of two possible claims: (1) that the element used as criterion was universally of such great importance (to man), that all others could be ignored, or (2) that this element was so intimately related to other important elements that a classification on the basis of one element represented in fact a classification of the greater part of all the elements. If the basic consideration included cultural as well as natural factors, the term "geographic" would be appropriate; if limited to natural factors, the system could properly be described as one based on the natural environment.

It is necessary therefore to consider to what extent either or both of these claims can be substantiated with reference to each of the two groups of systems—that based largely on climate, and that based largely on natural vegetation. Of the four possible theses involved, none can be accepted as logically obvious, none has been previously established by geographical research. The question, therefore, with reference to each of these groups of systems, is whether its proponents have presented sufficient evidence to establish either or both of these claims, or whether their systems rest simply on naive assumptions.

The majority of these systems, as we saw, are based on a climatic classification. Few will question that climatic differences in themselves are of very great importance in determining the character of areas from the point of view of man. Likewise, however, there is a general agreement that climatic differences are not universally of first importance. Most authors recognize the high mountain areas, at least, as exceptions.

There are other exceptional areas however, not always recognized, in which climatic elements are not clearly of first importance. From Long Island to the Florida peninsula, including a large part of the latter, is a belt of land in which the character of the soil and the drainage are far more important than the fact that the belt extends across at least two major types of climate. Though most climatic maps indicate but minor differences in the area from Nova Scotia to the Dakotas, what extraordinary contrasts in character of landscape are revealed in the flat pine-barrens of northern Maine, the fertile plain of the St. Lawrence, the rocky barrens of the southern margin of the Laurentian Upland, the rough moraine of brown forest soils in central Minnesota, and the chernozem soils of the Dakota plains!

These are not highly exceptional instances; on the contrary if one strikes traverse lines at random across the major climatic areas, in perhaps the majority of cases one's traverse will reveal areas in which the particular climatic characteristics are of less importance than some other natural factor. In mountain areas not included in the special category of high mountains—in such areas as the Southern Appalachians, the Vosges, or even the relatively low Pennines—it is at least questionable whether the slope of the land is not more significant than the character of the climate.

If we consider the relative importance of climate to other factors in areas of primitive occupance, we learn that our basic standard of importance, the significance to man, is not itself a constant factor, either with reference to space or time. In general the great importance which we attach to climate, aside from the debatable question of its direct physiological and psychological effects (which would likewise vary in reference to culture), is based on the importance of heat and water for cultivated plants. For primitive peoples, that may be of minor importance in comparison with the wild vegetation, which, as we have noted, does not necessarily correspond with climate. For the aboriginal inhabitants of North America, for example, the difference between prairie and forest, whether or not based on a minor difference in climate, was a difference of the greatest magnitude, in comparison with which the differences in climate within the forest lands, from the Great Lakes to the Gulf of Mexico, were of minor importance. Very similar is the situation in the areas of tropical rainforest climate: large parts of them do not have the tropical rain forest.

In discussing the first claim, that of the superior importance of climatic differences, we have indicated also the limitations of the second claim, namely, that a classification based on climate will at the same time classify such factors as soils, vegetation and cultural development. No one will question that there are important relationships connecting climatic differences with differences in these other features, but presumably we have passed the naive stage of assuming this to be single-determinant relationship. As Finch and Trewartha state, all we have is a "tendency for many natural features, each adjusted to and in balance with all the others, to develop a considerable degree of similarity throughout." And this is but a tendency toward a "mature landscape," and completely mature [natural] landscapes are rarely found [322, 665].

That a climatic classification of regions will not correspond to a classification of cultural landscapes is clearly recognized and well expressed by the authors last quoted. Though a limited degree of correspondence might reasonably be expected, and is often the case, "it is unreasonable to expect similar environments to produce similar types of land use." For this reason their classification is limited to natural elements, though the title, "The Geographic Realms," would seem more inclusive.

The climatic basis of regional classification, to summarize, does not classify cultural regions; as a classification of natural regions (as significant to man) it is incorrect for a large number of areas of widely varying size; finally, even in those areas where climatic differences are of first importance and where they are correlated fairly completely with vegetation and soils, these together cannot be credited with a monopolistic determination of regional character. A map of climatic areas is a map of climatic areas—in scientific terms, nothing more; that it is an indicator of variable reliability for many other elements, including cultural, cannot be denied, but to call it a map of "natural" or "geographic regions" is, in either case, to invite misunderstanding.

We have already rejected the special claim made for a regional classification based on natural vegetation, namely that it was a classification of the visible natural landscape and so independent of man's concern (Sec. X C). The most that could be said is that in areas not notably altered by man such a system may to some degree classify one major element in the appearance of the natural landscape. In fact, those who use the systems developed on this basis classify areas of very similar appearance of vegetation as areas differing in major degree and, on the other hand—at least in the case of Passarge's system—classify in the same type, vegetation areas as different in appearance as forest land and prairie. The value of a system based on natural vegetation cannot be based on its direct visibility, but must stand the same tests as any of the other systems, namely the test of the significance of its criteria in indicating the character of regions in their relation to man.

Since the significance of the criteria from the point of view of man is fundamental to a system of regional division, it may seem strange that there should be so much serious discussion of systems based largely on vegetation. For most of the human race it has been a long time since vegetation (meaning always natural vegetation) was of major importance—indeed few of us have ever observed natural vegetation. In a world geography produced by natives of Surinam it might seem the logical basis, and it is perhaps not without significance that Passarge first developed his concepts of *Land-*

schaftskunde in tropical forest areas. In such areas the actual vegetation *
does dominate the visible landscape and, together with the animal life associated with it, is of very great importance to primitive man. Furthermore,
the actual vegetation—which may be approximately similar to the original
vegetation—does dominate the visible landscape and is of itself of great importance to primitive man. More than that, the great forest, comprising not
only vegetation but wild animals as well, ready at any time to choke out the
small clearings made for cultivation, may seem a living embodiment of the
greater part of the natural environment—vegetation, climate, soils, and animal life all included in one expression.

Undoubtedly this association of the other natural elements with vegetation holds true also in non-tropical areas. In short, it is suggested that, as
a key to natural environment as a whole, vegetation is a better guide even
than the primary element, climate; while that tells only what the other elements might be, the vegetation represents, as it were, a natural synthesis of
climate, soils, animal life, and perhaps even, to some extent, landforms, ready-
made by nature as an indicator for the geographer.

Likewise we are informed, whether directly or by inference, that the vegetation is a guide, perhaps a better guide than climate, to the manner in which
man will use the land. It is, therefore, not only the key to many natural
elements, but also in large degree to the cultural landscape which emerges
under the hand of man.

These are no small claims. It is not surprising that Passarge's system
has created such a stir in geographic thought; if he has discovered the touchstone by which, on a single basis, we may classify truly geographic regions,
one of the fundamental problems of regional geography will have been solved.
Although the thesis has not made great headway among German geographers, it has made a definite impression on recent geographic thought in this
country, and therefore calls for serious consideration of its claims.

We have already indicated that any claims made for a system based upon
vegetation are valid only so far as we know what the natural vegetation is,
or rather—since it does not exist and since we cannot base our work on a
purely theoretical natural vegetation that would have been here if man had
not come—we must know what the natural vegetation originally was. For
large and important areas of the world, not only did we find that our knowledge is as yet very uncertain in detail, but that, quite possibly, we may never
have the necessary knowledge. Indeed, to secure that knowledge we will
have to study first that which the unknown is supposed to reveal to us, the
features built by man's culture. Thus, we might ultimately learn what was
the original vegetation in different parts of China by careful analysis of soils

[316]

* Delete, on the first line, the words "the actual vegetation", and all
of the next two lines.

which, after forty centuries of intensive cultivation and fertilization, are perhaps more largely cultural features than natural features. Though everyone recognizes the difficulty in the case of such areas, it is only by contrast that the problem appears easy in newer areas, or in areas of primitive occupance. How can we map the "original vegetation" of Central Africa; can we assume that all open clearings or areas of brush scattered through the rain forest are the result of man's activities, that none of them were originally there, perhaps because of particular soil conditions, no longer distinguishable as a result of innumerable burnings by the natives?

Finally, insofar as we can reconstruct the original natural vegetation of the time when man appeared upon the scene, we might be forced to decide that, because of climatic changes which have taken place since then, it was no longer of significance for the regional framework of the present world.

Because of the difficulty of reconstructing the original natural vegetation it is common practice to deduce its character on the basis of the climatic and soil conditions; in other words, to construct the synthesis which the natural vegetation itself is supposed to provide. In Passarge's division of the world the vegetation areas might just as well be called climatic areas [cf. Waibel, 250, 476].

So far as we do know, or may hope ultimately to discover the character of the original vegetation, and so far as that would represent the theoretical natural vegetation of the present, would it provide us with a scientifically reliable key to such natural elements as animal life, soils, and climate? Certainly not for animal life, since that may be profoundly different merely because of lack of connection of areas of similar vegetation. As for soils, until our knowledge of this field is more complete, it may be necessary for us to continue to infer the character of the soil from other factors which we do know, which would certainly include climate as well as vegetation. Further, we trust that we have passed the first stage of reaction against the former classification of soils according to bedrock. James recognized that the same general type of vegetation may cover great differences of soil, differences important to man [321, 140 f.].

No doubt there is a general correlation between vegetations and climates. "Vegetation types are the visible reflection of the climates" as James says, "they are the great climatic emblems" [321, 6]. These expressions are well put; they indicate accurately the degree of reliability of vegetation as a key to climate. But why should scientific workers wish to use reflections and emblems—which are not to be seen and whose character cannot be accurately determined—instead of facts which we are able to establish with high degree of accuracy? On few of the natural elements have we such a large and relaively reliable body of data for many parts of the world as for climate; on few so uncertain as for natural vegetation.

In spite of all these objections, we still might justify a regional division based on natural vegetation if it appeared that that element, combining as it does effects of climate, soils, and to some extent relief, could be shown to be of greater significance in relation to things of human interest than any one of those factors alone. It might then be considered as a natural summation of these various factors applicable to the problem, say, of man's use of the land. That this is, in fact, the case is the assumption, stated or implied, of many who use this basis.

In his critical discussion of the work of the Passarge school, to which we have frequently referred, Waibel finds that this concept, the relation of natural vegetation to the use of the land by man, is the core of the method of thought—that Passarge's *Landschaftskunde* might with some exaggeration be called *"angewandte Vegetationskunde,"* or applied plant geography [*250,* 476]. No doubt the majority of modern geographers of all schools would agree with Waibel that it is dangerous to argue, as Passarge implicitly does, from his types of *Naturlandschaften,* to resultant phenomena concerning animals and men [*266,* 202 f.]. As an example, from the school of Passarge, he quotes Ahrens' statement: "man in the same *'Landschaftsgürtel'* arrives at the same subdivision of economy (*Wirtschaft*)" [*250,* 170].

The thesis may be found clearly stated in at least one recent American text: "The plant cover of the earth is a key to human occupance, for the plants that have grown naturally in a region . . . roughly indicate the way in which the land will be used agriculturally." This is illustrated by the example of eastern United States in which the original vegetation is described as "trees and native tall grasses" and the agricultural development is described as the cultivation of the grasses, "wheat, oats, rye, barley, and corn, together with root crops, like potatoes and turnips" [*323,* 6, 107]. From the omission of such crops as cotton and tobacco, not to mention the domestic animals of the Eurasian steppes, one infers that the thesis is not to be taken too literally; furthermore the text in which this appears does not use a regional system based on vegetation.

Nowhere in his book does James express any such definite thesis. Indeed, in his introduction and elsewhere, he specifically denies any such principle of environmental determinism. Furthermore, he distinguishes clearly between the character of the occupance by peoples of different major cultures. In the detailed treatment, however, the reader can hardly fail to perceive the assumption that within areas of any particular type of natural vegetation the peoples of one major cultural group will tend to develop the same form of agriculture.

[318]

It is clear that we are concerned here with a thesis of the first importance in geography. If the character of the natural vegetation is indeed the key to human occupance—a better key than the character of the climate—all geographers should promptly recognize it and proceed accordingly. Needless to say this thesis cannot be established by mere assertion, nor can it be proved merely by deductive reasoning. On the contrary it calls for laborious and painstaking search for evidence. As Waibel has emphasized, such research has not been undertaken by those who assert or assume the thesis [250].

A primary step in an effort to demonstrate such a thesis, one may presume, would be to compare the maps of different types of natural vegetation with those of different types of agriculture. Although a close correlation between the two would not in itself establish the thesis, some degree of correlation is surely a necessary condition. What do we find if we compare the natural vegetation maps which James presents with the maps of agricultural regions in North America and Europe—whether those which he uses, by O. E. Baker and Jonasson respectively, or the maps of Hartshorne and Dicken [324] or of Whittlesey [319]? In each case it is the lack of correspondence that impresses one. The comparison cannot be made with any certainty in the Orient, since the natural vegetation of much of China is unknown, but even there it is clear that the areas of paddy field agriculture do not remotely correspond with what is known of natural vegetation.

There is no intention of denying the possibility of significant relationships between the areas of distribution of natural plants and that of cultivated plants as man has spread them over the world. But a proposition for which the readily available evidence is negative will require a great deal of careful research. One positive contribution has been made by Troll, based on Engelbrecht's work on the relative importance of different cultivated plants in Europe [394]. He has found some very marked correlations, though the student who is familiar with small details of the agricultural geography of Europe will regret that Troll's map is not sufficiently detailed to provide an adequate test of his findings. Assuming, however, that these will be found to be valid, it is significant to note that he has not correlated the plant associations of agriculture with those of the natural vegetation but rather individual crops with individual wild plants, of which some, at least, are of very slight importance—*e.g., Schmerwurz* or *Tamus communis* (black briony).

Finally, it would be well to repeat Troll's observation that our knowledge of the production of cultivated plants, for large parts of the world, is far more accurate than our knowledge of vegetation—whether the original natural vegetation or the present wild vegetation. Consequently a system based on natural or wild plants places us in the unsound position of arguing from the unknown to the known.

In conclusion, therefore, a regional system based on natural vegetation—even to the limited extent in which it is possible to construct one—does not justify the claims expressed by such titles as "natural" or "geographic" divisions. A division of the world based on vegetation is simply a division of world vegetation (so far as we can infer it) and nothing more. Like a map of climatic division it *suggests* other things, but *tells* us nothing more upon which we can rely.

Should we then look for a system of world division based not on one or two natural elements alone, but on all the natural elements significant to man? On first thought one might wonder why this has not been done. If one considers the synthesis of major natural elements—climate, relief, soil, and natural vegetation—which is to be found in, say, a part of Mongolia, is this not also to be found in parts of Patagonia and our Great Plains? If we can overlook the difference in native animal life as no longer important—though once of greatest importance—and disregard the difference in relative location, we can presumably agree that here is a type of natural environment which is repeated in various areas. No doubt many other examples of such types of areas can be offered, but anyone who attempts to divide the world on this basis soon realizes that examples do not make a map. Where will we find the counterpart of the North German Plain—in terms of climate, relief, and mantle rock? If we speak of the somewhat different combination of these same factors in the Carolina Piedmont as the "Carolina Piedmont type" of natural environment, can any other region in the world be included in this type? To put the matter directly, the synthesis of natural elements in areas represents a combination of several factors of importance to man (ignoring for the present the variations in the importance of the same factors to different groups of men) and these several important factors are sufficiently independent so that they can be associated together in a great number of different combinations. In our actual world, on whatever scale of division we use, the number of different combinations found is very nearly as great as the total number of areal divisions, *i.e.*, nearly every area is a type in itself.

It may be suggested that this difficulty is to be avoided by making our types more generalized, so that each separate region could be considered as a subtype of the more general type. But the combinations of several factors, each of which varies independently, can be reduced to common types only if one assumes that some of the factors are of minor importance, no matter how much they vary.

[320]

This proposition is so important to the whole problem of comparative regional division, on whatever basis, that it needs to be clearly understood. It might be thought, for example, that classifications of climate that are based on actual climatic conditions—rather than on genesis—represent the organization of many semi-independent factors, rainfall, temperature, winds, humidity, etc. But the systems which we have—such as those of Koeppen, Jones and Whittlesey, or Thornthwaite—depend primarily on but two of these factors, precipitation and temperature, and largely ignore the others. Taking a limited number of gradations of each of these factors, x gradations of precipitation and y gradations of temperature, the number of possible combinations of precipitation is, of course, $x \times y$. If we took a half dozen gradations of each of the two factors we would have, theoretically, 36 climatic types. Some of these however do not occur on our planet, and others we may arbitrarily ignore on the ground that, if either factor represents an extreme handicap to organic life, all the variations of the other factor are of little importance. Consequently the number of actual combinations that must be recognized need not be too large and each type may be represented by several areas.

If, however, we try to combine, say, six gradations of temperature and six of rainfall with four of relief, the number of theoretical types becomes 144. To complete the formula for all the independent elements of the natural environment we would have to add at least several gradations of bedrock (as being a determining factor in most actual soils) plus additional gradations for other elements such as drainage, surface waters, and even vegetation, insofar as these elements cannot be fully expressed as functions of the previous elements.

A specific example is provided us by Granö's system of regional formulae, in which he designates a single type of area by means of seven independent symbols (six symbols for natural elements, *not including climate,* and one for human elements). If one recognized but four gradations under each of these symbols, the total number of possible types exceeds sixteen thousand. To be sure, many of the elements can be regarded as factors of minor importance in any particular area, but none can be eliminated from the entire consideration, since any one may be, in some particular area, the most important factor. Consequently the number of actual combinations which we would have to recognize as types would be little if any less than the total number of regions that would be recognized; we would have a system of regions but not of types. Needless to say, this is no detraction from the value of such formulae as providing a shorthand method of describing regions for comparative purposes. This writer has used a similar method in describing divides between areas, but found that it did not provide a single system of classification, but as many systems as there were factors considered [357].

It might, however, be supposed that we have merely confused the situation by arithmetic, that no matter how many types are recognized, we can always reduce the number *ad lib.* by generalizing, that is, by combining those of lesser differences to form a smaller number of types. This we can do only if we have some standard by means of which we can arrange the elements in order of importance. Since the elements in themselves are mutually incommensurable—we cannot compare degrees of slope and degrees of temperature—we can determine the order of their importance only in terms of their effects on something else. It is absolutely essential that this external feature, which is to serve as the barometer, must reveal the same order of importance of the elements in any two places where the elements are each the same; it is essential to a logical system of division and subdivision that this order be the same even though the elements vary greatly. Needless to say, no such barometer exists or can exist, no matter what feature is chosen— inorganic, plant, animal, or human—elements which commonly affect it in but minor degree may, in their more extreme forms, affect it to major degree.

The validity of these statements can be seen if we examine for a moment the difficulties involved in classifying climates. Any system of types which is to permit logical reduction of the number of types by generalization must be arranged in terms of major divisions, each subdivided into types. Such an arrangement is, of course, provided in most climatic systems, such as those of Koeppen and Thornthwaite. The distinction between major and minor divisions, however, is intelligent only if the criteria used for the major division are, in fact, more important than those used for minor division. But which is more important, rainfall or temperature? Obviously the question has no answer except in terms of some external feature affected by both of these factors.

Natural vegetation can be considered as a synthetic measure of differences in both precipitation and temperature; on that basis there is ground for deciding, as Thornthwaite does, that precipitation is more important than temperature, even though the conclusion is certainly not true in polar areas. In other words, at some unknown point between the tropics and the pole, the relative importance of the two factors is changed and the system becomes illogical.

On the other hand, one may use cultivated crops as the standard against which the differences are measured, and decide, with Koeppen, that differences in temperature are more important than differences in precipitation, except in the dry areas where the reverse is the case.

We have stated only the first difficulty. "Cultivated crops" obviously does not constitute a single standard, since the effects of temperature and rainfall are different for different crops, but we can, of course, select a limited number as standards. Many systems, in fact, depend both on certain cultivated plants and on natural vegetation. Consequently the decision, at any point in the system, as to which of the two factors, rain or heat, is more important, is a decision which may be significant for one group of plants but not for another, is, in other words, purely arbitrary. The result, therefore, is not a significant system of division and subdivision, but simply a convenient system of just so many separate type-combinations of two factors. For example, in Koeppen's system, there is no more reason for relating the types **Cfa** and **Csa** as subdivisions of **C**, than there would be to relate **Cfa** and **Dfa** (in the form **Fca** and **Fda**) as subdivisions of a major group **F** (ample rain all year).

Any student of the climatic problem will recognize that, far from overstating the complications involved, I have simplified them. Either precipitation or temperature, alone, represents no simple factor to be measured by a single figure, but a highly complicated factor whose variations in different parts of a year must be indicated in any proper classification. This would lead to such unanswerable questions as: what amount of seasonal variation in precipitation is more important (in its effects on what?) than what amount of seasonal variation in temperature?

It is little wonder, therefore, that there is so little agreement between different systems of climatic classification. Indeed one might question the logic of any attempt to combine the two separate factors. It may be that we have simply been misled by the fact that our language uses one word, climate, to include factors which, whatever association they may have genetically, are largely independent in their actual variations and in those effects of the variations with which we are primarily concerned. (The relation of temperature to the effectiveness of precipitation for plant growth is provided for by Thornthwaite in his formula for precipitation efficiency, which therefore stands independent of temperature efficiency [391].) From the point of view of plants there would appear to be no more reason for combining water and warmth than for combining water and soil—perhaps less. In other words, a classification that combined differences in precipitation and in soils might be just as reasonable as that which we call a climatic classification. Obviously a proper classification, from the point of view of plants, would combine all three as coordinate factors. If it is objected that soils are in considerable part products of the conditions of precipitation, temperature, and the previous plant life, we are only led into greater difficulties: we would have to combine conditions of precipitation, temperature, slope, parent rock, and drainage, not to mention the variations of these factors in the past. (We may merely point to the well known difficulties in the classification of soils, dependent as they are on such a large number of relatively independent factors forcing us to decide more or less arbitrarily as to which are more, which less, important.)

In brief, the attempt to synthesize in one system of classification all the elements whose variations are important for plant growth cannot produce a system which is either logical or feasible. Nature, to be sure, did produce an actual synthesis of these factors in the original natural vegetation, but we have seen that this is a synthesis which we cannot observe and therefore cannot analyze and that, considered as a whole, it reveals facts of but minor importance in the most important parts of the world.

In view of the difficulties encountered in the attempt to consider at one time the factors significant to plant growth, how much more difficult is the task which Passarge undertakes in attempting to develop types based on all the elements that give character to *"Landschaft"* types. (These appear to include all the common natural elements and also many elements of the *"Kulturlandschaft,"* but not men or animals. Gradmann, among others, finds this anything but clear [*236*, 333–35].) It is not surprising that Sauer should find "more success in the statement of problems and methods than in the content" [*84*, 191].

It might appear that Passarge had eliminated many of the difficulties which we have noted in other systems by establishing theoretical rather than real types. His typical *"Landschaft"* is "an ideal *Landschaft* which does not exist at all, but has the advantage of presenting an object of comparison for many real *Landschaften*" [*236*, 333]. Granö says much the same of his types [*252*, 34]. Since such an abstract system might well prove valuable, it merits our further attention.

The system which Passarge presents has a logical structure of major and minor divisions, leading to individual types and subtypes. It is intended to provide us with a classification of ideal types similar to the Linnean classification of organic life into orders, genera, and species [*268*, 91–8]. Passarge did not seem to realize that his final subdivisions, in which all the factors were represented, would be "species" of which the earth can show not more than one specimen each. That, however, is a minor matter.

A serious difficulty results from the fact that, in contrast to the organic world, the different elements of areas are in large degree independent. To be sure we do not find all the conceivable combinations on the earth; we do not find forests in desert climates. But we do find all the major types of relief in almost all types of climate; we not only have radically different soils in the same type of climate but also very similar soils in areas of notably different climate. It is as though there were animals with mammary glands among the invertebrate as well as the vertebrate, or as though some quadrupeds had wings.

The situation is still more complicated if we include cultural forms. In areas of very similar climate, relief, and soil, man may develop quite different cultural forms, but he may, on the other hand, extend any one of these cultural forms, within broad limits, over areas of different climate, relief, and soil.

We have suggested (Sec. IX C) that the possibility of classifying organic life into a logical system, such as that of Linnaeus, is a result of the common genetic origin of all specimens of a single species and the common origin of species of the same genus, and so on back through the system of evolution. We know that no such system of evolution of individual areas is conceivable. Individual features of all areas have the same planetary and solar origins, but the combinations of these which give character to different areas have not evolved through generic types of areas to specific types and individual specimens. The climate of any area is genetically a part of the world's climate, not a part of the genetic development of that area, certainly not a part of the genetic development of all areas of that type as distinct from those of other types.

Various students, therefore, have urged that the attempt to compare types of regions to species could not be successful [cf. Penck, 163, 41]; Philippson, indeed, had noted the fallacy long before Passarge presented his system [143, 13]. The individuality of each region, they insisted, was of a much more special character than the individuality of specimens of a species [cf. also Creutzburg, 248, 413]. Passarge has since come to the conclusion that the comparison with a Linnean system was unfortunate [272, 57]. In agreement with Creutzburg, Philippson, and Penck, however, he believes that types of regions, rather than species, can somehow be established.

These types, it is clear, are not ready in reality for us to discover. We could however construct a logical system of classification of types if we could determine the order of importance of the different elements concerned. Passarge has, perhaps unconsciously, attempted to do this. Since his actual studies, as distinct from his theoretical outline, have progressed little beyond the first major division, which is based presumably on vegetation, his system appears to be little more than a system that determines types chiefly in terms of vegetation, as Hettner notes [242, 163 f.]—though Waibel finds that the manner in which vegetation is determined results simply in a system based on climate [*250, 476*]. Indeed, Passarge says as much, in at least one of his many and varied treatments of the problem. His "normal types," he writes, are based on climatic differences, other factors being assumed to be "average." The normal types may be influenced by *"Modifikatoren"*: extreme conditions of any natural or human factors. By arranging these under such group heads as "dry," "wet," "fertility," "destructive," and "cultural," Passarge asserts that "one may speak of a logical system of *Landschaft* types" [*259, 704*]. But this can be done only by assuming a fixed order of importance of the different factors, and, as previously noted, such assumptions are not only arbitrary but in a great number of cases can be demonstrated to be contrary to fact; a factor of minor importance in many areas may be of first importance in others.

We must conclude, therefore, that even a system of theoretical types cannot be constructed on the basis of a large number of more or less independent factors of varying importance. A comparative division of the world into generic type areas arranged in a logically sound system can be based on any one single natural element or element-complex (*e.g.*, vegetation) but only on one. Undoubtedly such single-feature systems are of great value in geography; the question here is whether any one of them alone can provide a suitable background for regional geography in general. We cannot expect a perfect system; in spite of all the objections which we have raised, practice might show that either a climatic or a natural vegetation system provided a fairly satisfactory background. Though we found that a system based on climate could not be logically divided and subdivided, it is nevertheless a feasible system, since only two major factors need to be combined, and we can secure the data concerning those factors. The data for a system based on natural vegetation is not, and can never be, complete, but possibly we have enough for a fairly adequate system. Finally, though neither system provides a sure key to other natural features or to cultural features, there may, in either case, be a sufficient degree of correspondence to make the system useful as an outline for regional study.

Our judgment of the utility of either system must be based on the results and we must examine the results where we find them. It is unfortunate that American geographers, in contrast to those in Germany and England, have presented systems of regional division almost exclusively in textbooks. (If any criticism is implied in this statement this writer is included among those to whom it applies.) For teaching purposes it may be necessary to subordinate reality to organization. But the problem of organization is not confined to textbooks; it is, in fact, the problem of organizing reality with which we are here concerned. It may be that inherent difficulties which can be obscured in a detailed research study are brought into clearer view when presented for instruction. However that may be, it is to be understood that nothing in the following discussion is intended as a reflection on the value of these works as texts, nor is it our concern to examine the detailed work included in them, in itself. We are concerned solely with the results which apparently follow from the use of a regional system based on a single natural element or element-complex.

The very fact that the study of the world is organized on the basis of one particular feature inevitably leads the reader—if not the author—to think of that feature as the most significant feature in every area studied. If the discussion of a certain type of cultural development is largely confined to a chapter named in terms of a certain type of natural vegetation or climate, a relationship is thereby suggested, whether or not it has validity.

Further, if the feature used as criterion is not itself important, there appears to be an unconscious desire on the part of the author to justify the system by emphasizing the indirect significance of that factor to other natural features and to cultural features—an emphasis which, in many cases, is out of proportion to the actual significance involved.

These disadvantages, inherent in any system based on a single natural feature, are most apparent in one based on natural vegetation because of the marked discrepancy between the agricultural regions and the areas of natural vegetation used as the outline. Of the many examples which might be selected we will limit ourselves to one.

Geographers in common with less technical writers recognize the marked similarity in characteristics in that part of the United States which extends from central Ohio to eastern Nebraska, south of the Great Lakes—the Corn Belt. Few areas of its size are so well established as specific "geographic regions." In the high percentage of cultivated land, the specific crop and livestock associations, the high yields, the character of farm equipment, and even in the social characteristics of the rural population, this is one of the more clearly defined major regions of the world. Nevertheless James, who in a previous publication recognized this unity by twice using the Corn Belt as an example of a major region [286, 79, 86], is forced by his system to divide it between two of the eight major world types of lands—the Mid-Latitude Mixed Forest Lands and the Grasslands [321, 239–45]. (In reality, of course, there is but little forest land or grassland in either part—both are primarily farmland.) By way of justification he presents sections of two maps showing the contrast between the woodlots of farms in the eastern part and the strings of less controlled woodland along stream valleys in the western part. Apparently because the most productive part is in the former grasslands, the area as a whole is discussed under Grasslands, and the general impression given is that this type of agriculture is a phenomenon developed from the natural characteristics of the prairie, though overlapping somewhat into the forest area. Actually, of course, the historical development was just the reverse.

The situation may perhaps be more clearly understood if we roll out the Appalachians in our minds so that the interior plain continues to the seaboard with the same soils and vegetation as in Indiana. Who will question that the Corn Belt would not, in that case, extend equally far east, terminating only in the sandy soils of the coastal plain? In fact, the agriculture of southeastern Pennsylvania is in general the same as that of the Corn Belt, and its unusual characteristics may be found in various districts within the Corn Belt [324, 107]. In other words, given that change in the relief of eastern United States, one would tend to consider the Corn Belt as a phenomenon of the Mid-Latitude Forests overlapping into the Grasslands.

The similar, but even greater, difficulties in the Cotton Belt may account for the fact that that well-recognized region is barely mentioned in the work referred to.

Whichever basis is used, whether climate or vegetation, one is in the difficult position of starting from a single element to explain features which are the result of a complex of factors (see the later discussion on this point, Sec. X G). Waibel has commented critically on the errors to which this has led several of the followers of Passarge in Germany. The form in which he states his criticism is of particular interest to us: "the manner in which causal connections between the *Landschaft* and the phenomena of organic life are constructed by largely deductive processes, reminds one of the most primitive 'in response to environment' processes of many American geographers" [250, 478 f.]. Perhaps one difference should be recognized: in place of individual responses to each of the environmental factors considered in turn, we now have a complex of regional correlations with one single environmental factor.

The difficulty, not to say impossibility, of reasoning from a single element to a feature dependent on many elements apparently leads the more careful of these students to limit themselves to suggestive observations of apparent correlations of the areal extent of a cultural feature and a type of vegetation. When one is informed that the crop system studied in the Corn Belt, under the heading of Grasslands, is also found in the grassland areas of the Danube Basins, but not in the neighboring forest areas (we overlook any question of fact here), but no analysis of the factors involved is attempted—no mention of relief as a possible common factor—the innocent reader is led to assume a direct correlation [*321*, 245 f.].

The problems involved in interpreting areal correspondence of phenomena have been well discussed by James, in another connection [*286*, 82–4]. He draws attention there to the possibility that the phenomena found in correlation may both depend on a third factor as common cause. He does not mention, however, two other possible interpretations which geographers in particular need to have constantly in mind.

One is that the correlation, even when very high, as Cohen has emphasized, may be "entirely accidental—*i.e.,* we can find no reason why the two factors should be correlated at all" [*115*, 92].[93] In a field like geography in which the amount of data is commonly meager the possibility of purely chance correlation is relatively great. Correlations that involve very few cases, and for which no logical demonstration is to be given, hardly seem worth stating—for example, the fact that three major trade centers of the grasslands in this country, Chicago, St. Louis, and Minneapolis-St. Paul, are located on the edge of the forest.

The other possible interpretation is that the author has unconsciously forced his areal mapping to fit the conclusion. In the determination of our boundary lines in mapping any phenomena, there may be a wide range within which any of various lines are equally justified; if one of these corresponds closely with the boundary of a natural element which we have taken as a fundamental basis it would not be surprising if we should place our other line there also. I am not of course referring to such gross misrepresentations of fact as are to be found on Passarge's map, but to very much slighter alterations which can only be checked by very detailed study. Neither have I noted such purposeful inaccuracies in any of the works under consideration (James, in fact, uses many maps which do not fit into his system). Indeed, in many cases we could not speak of an "inaccuracy" or an "alteration," but rather of an unconsciously purposeful choice of one particular boundary line where any of a number of others are equally justified by the primary facts, but would not produce the correspondence aimed at. We may illustrate this point with a potential case not found in any of the works to which we have been referring.

In determining the boundaries of the manufacturing belt of North America on an objective, statistical basis, one finds relatively little room for

---

[93] Cohen cites an example demonstrated by other students: over a considerable period of years, membership in the International Association of Machinists correlates to 86% with the death rate in the state of Hyderabad.

disagreement on three sides, but great uncertainty on the western boundary, from the Rock River district in northern Illinois to the Ohio at Cincinnati [326]. One could find some justification for almost any boundary drawn between the straight line connecting these two districts on the one hand and a westward sweeping curve that included Davenport and St. Louis on the other. On the basis of the primary facts, the boundary was actually drawn so that it happens to lie not far from the margin of the forest and prairie areas, but this fact did not occur to the author at the time. Had he been making the study as a part of a regional consideration of the "Mid-Latitude Forest Lands," however, it might well have seemed just as legitimate to draw the boundary a little more directly southward, in closer correspondence with the vegetation boundary. He could thereby have suggested a new explanation for the concentration of manufacturing in northeastern North America—comparable with De Geer's correlation of the limits of this area with those of the Ben Davis apple.

In sum, no matter how close the correspondence between the distribution of two geographic phenomena may be, we must be suspicious of any assumption of either a direct or an indirect relation, unless such an assumption can be shown to be theoretically reasonable. In the example cited, that is, we must be able to deduce some logical grounds to explain the difference in industrial development between, say, Des Moines and Indianapolis, that is related to the difference between the prairies of Iowa and the forests of Indiana.

The conclusion to be drawn from our consideration of regional systems based on natural elements is in agreement with that of Finch and Trewartha: "it is deemed unwise to try to fit world culture patterns into the framework of the physical regions" [322, 663]. In addition, however, we conclude that it is hardly possible to fit the summations of natural elements which they call "physical regions" into a framework based on one or two natural elements.

F. COMPARATIVE SYSTEMS OF GENERIC REGIONS BASED ON CULTURAL ELEMENT-COMPLEXES

We turn to the possibility of constructing a regional division based on cultural elements. Some European geographers, who have apparently also come to negative conclusions in regard to so-called "natural" divisions, have retreated to such simple divisions as those based on political boundaries, states and provinces, or on racial or folk areas. Few will wish to follow such suggestions, other than in the specialized fields of political geography and the geography of peoples. To use such systems for geography in general is to abandon the hope of finding any significant regional division; we might as well divide the areas into square sections of equal size.

If the attempts to provide a regional framework based on all the significant natural elements have failed because of the complexity of natural environment, many might regard the possibility of synthesizing the cultural aspects of regions as even more remote. Furthermore, geographers have, until recently, given far less careful attention to the analysis and classification of cultural phenomena. Consequently it is not surprising that so few attempts have been made to approach the problem from this side. Indeed most of the examples which we will consider have not indicated clearly that they were intended to lead to a system for world division on a general cultural basis. A theoretical approach to the problem is therefore needed.

What are the major aspects of culture which are of greatest importance to regional geography? If we refer this question to our fundamental concept of the nature of geography, it reads: what major aspects of man and his work are relatively homogeneous in limited areas, but in different areas differ notably in ways that are significantly related to the total character of the area. If areas could be found for which these conditions were true of all aspects of man and his work, these would be perfect cultural regions. Since we know that the areal differentiation of the various cultural aspects will not coincide, we must limit ourselves to major aspects.

The distinction between major and minor, among phenomena that are mutually incommensurable, is not a question permitting objective answer, but on a subjective basis it is possible that there might be fairly general agreement.

Presumably all will agree that the areal differentiation in numbers of people, population density, is of major importance. Indeed it might be suggested that it is so all-important that we need go no further. Our problem then would be relatively simple, since population density is easily measured as a single mathematical factor and we actually possess the necessary data for large parts of the world. The very simplicity of this factor however suggests its inadequacy; areas of similar population density and distribution may be very different in many other geographical features. *

Much more could be learned, no doubt, if we had a more detailed picture of the way in which population is locally distributed in different areas of the world—actually a picture of the variations in local density within all the small areas. Hall, among others, has frequently emphasized the need for such a study of the whole world [295, 167 f.]; he has also illustrated the suggestion with a detailed map of Japan [351]. How effective such a basis would be for world regional division can be judged better when we have

[331]

such detailed maps on a common basis for at least some areas of the world of notably different culture; it is to be hoped that the "experimental maps" which Hall has made will be completed for publication. From what we already know, however, it seems probable that the same form of distribution may be found in areas of different culture and may represent in many ways entirely different regional character. The *Strassendorf* villages which Hall found in Japan [*348*, 112] are functionally different from those in Germany even though the form be the same.[94]

We may conclude therefore that population density, even in its most detailed and complicated form, though providing us with one major key which is not to be overlooked, is not alone sufficient.

Aside from their numbers, what are the ways in which people differ, in different parts of the world? We can eliminate, for geographic purposes, all those differences which, however important as between members of any local group, are not significant as between peoples of different areas. Further, in comparing regions we can disregard certain differences that are of major importance within a region—including the whole complex of differences called to mind by the terms "urban" and "rural," though we will need to distinguish between areas primarily rural and those primarily urban. On this basis we can eliminate a whole host of cultural phenomena of man, since in so many ways peoples the world over are alike, do the same things, and do them in more or less the same way.

However much we eliminate, there remains certainly a great multitude of cultural phenomena which differ significantly from region to region. For this very reason, it might be well for geographers in general not to plunge headlong into intensive studies of any one or two phenomena without first considering their importance, both in themselves and in relation to the other cultural phenomena that are geographically significant. If one were to attempt a full list of such phenomena it would include a large number of material objects which men make, and make differently in different parts of the world, and it might be that small artifacts were more significant geographically than large ones. But culture is represented not only by what we make but in a thousand other ways. These would include: the physical characteristics of the population; the manner and substance of thought, speech, and writing; the way in which people eat, dance, walk, or ride; the character of their clothing, shelter—both for man and beast—and the grouping of these shelters in settlements; the way in which the people work and play and the tools and implements used in doing each; the domestic animals which

---

[94] Hall himself has come to this conclusion, as I learn from personal conversation since writing this, and has therefore decided not to use the term "strassendorf" for villages of this form found in the Orient.

they use in different ways; the various materials which they use for all these and other purposes, including notably those for food and drink, clothing, tools and implements, shelter, fuel and power; and, finally, the various alterations to which they subject the earth surface, including alterations of vegetation, soils, land-forms, bed-rock, and even underlying formations.

Any one of all these phenomena may have its own geographic regional expression and might furnish something of a key to the geography of cultural phenomena—the difference between a set of Mah Jong and a roulette wheel might be considered to express the difference between the culture of the Orient and that of the Occident as specifically as the difference between the Ming Tombs at Peking and Westminster Abbey. This frivolous comparison is deliberately selected in order to suggest the principles that should govern our choice of criteria from this list. Serious attention should be given to those criteria which are in themselves of greatest importance, or which, because of their intimate relationships with a large number of other cultural phenomena, provide a key to a larger complex of phenomena which, as a whole, is of major importance in determining regional character. Finally, if theoretical considerations indicate that there are such criteria, we can use only those which we are able to observe, classify and to some degree measure quantitatively. This condition, however, is secondary to the first; we will not select things easy to observe regardless of their significance.

On the basis of both these conditions we can largely eliminate the physical characteristics of the population [cf. Hettner, 161, 248 f., 289]. It is instructive, however, to reflect how different our conclusion would be if races of men differed as much as those of dogs; in that case the racial map of the world might be of even greater importance than that of mere numbers of people. On the other hand, the geographer is indirectly concerned with such a minor physical difference as skin-color, in those areas where that characteristic is regarded as the outward material sign of a cultural distinction of enormous importance to almost all other cultural phenomena (the social distinction involved, however, may be based not on actual differences in color, nor on differences in "race," in the proper biological sense, but merely on the presumption of color; cf. the writer's study of the "racial geography"—"racial" in the social sense—of the United States [359, 277]).

Since most, if not all, the other cultural phenomena listed are in part products of the way in which people think, Stanley Dodge suggested that we might determine cultural regions by finding the areas in which people think alike [295, 171]. Overlooking for the moment the fact that this ignores the extent to which human thought is independent of regional environment—in Minnesota we think a great deal about the desirability of mining coal but we have no coal mines—it is obvious that we have no direct way of observing what people think. What indirect criteria can we use? Because of the general human tendency to think more or less like other people with whom we talk or whose writings we read, there is some degree of uniformity in thinking by people who speak the same language in contrast with those who speak a different language. To the extent to which this is true, linguistic regions may be expected to show similarities in a large number of the cultural phenomena listed. But this is no sure guide. Centuries of German control in Upper Silesia have stamped many characteristics of German thought and culture on people who nevertheless continue to speak Polish. In its material culture, furthermore, this area is much more like that of Germany, and all western Europe, than it is like Poland, and eastern Europe [355, 201-9]. Important as linguistic regions are in the interpretation of the cultural geography of Europe, many of the human phenomena listed above are but little affected by differences in language, possibly but little affected by what people think in general.

We see that we are faced at once with the problem of determining the comparative geographic importance of different individual cultural phenomena. Have we any yardstick, even in theory, for comparing these apparent incommensurables? If we remember that we are concerned with areas and with man, I think we have. If we combine the areal spread of a phenomenon with its relative importance to man—in terms of number of people concerned and importance to each—we have the major factors for a theoretical basis of measurement. Without attempting a complete analysis, we might measure the importance of a phenomenon to each individual in terms of the part it bears in his total activity. On this basis the chopsticks of the Chinese peasant are less important than his plough, and the plough, both on this basis and in terms of areal extent, is less important than his paddy field.

In an absolute sense, the house on a Chinese farm might be measured as no less important than the fields, since the women and children may carry on a part of their activity within it. But to compare the relative importance of these cultural features in areal differentiation we must cancel out all that is not areally differentiated. In other words, insofar as the house in question, and the activities within it, are essentially the same as those carried on in farmhouses over much of the world, we eliminate those factors, and what is left—the particular distinctive characteristics of architecture, for example —is clearly minor.

The reader will no doubt have perceived the conclusion to which this reasoning leads. For most of the people of the world, a major part of their activity is concerned with ways and means of keeping body and soul together, or more correctly, with making a living—*i.e.,* economic activities. Likewise the greater part of the land area of the world used by man is used for economic activities.

These economic activities, particularly those involving the use of the greatest amount of land, show marked areal differentiation in different parts of the world. These facts, we may note, find definite expression in the landscape. As Broek concludes, "economic forces are by far the most influential agents in transforming the landscape" [*297*, 107 f.]. Krebs's comment, that this view is too exclusive, that such features as settlement forms and land division may be determined by cultural elements that are economically not rational [*279*, 211], is less a correction than an addition to Broek's statement. Land is divided and used, and buildings are constructed, primarily for economic purposes, even though the manner in which these things are done and their resultant character may be influenced by cultural factors other than economic.

Furthermore, the importance of economic activities to the areal differentiation of the world is not limited to their direct importance. Differences in economic forms are closely connected, both directly and indirectly, with a large number of the other cultural differences listed above.

If we can agree that in the complex of cultural phenomena associated with economic activities we have the single group of phenomena of greatest importance to cultural geography, the next question is, which particular phenomena will be of greatest utility as criteria for regional differentiation.

We will simplify this problem if we first eliminate a considerable group of economic features which, important though they may be in the economic structure, are of little geographical significance because they are essentially similar, both in form and function, wherever they are found. This will include a large proportion of urban features—shops, offices, etc. We need, therefore, pay little attention to them, except in terms of their relative numbers, which we can assume to be measured by the general extent of urban development in a region. Not included in these, however, are special manufacturing and commercial features which give distinctive character to one city as compared with others, or to the cities of one urbanized region in comparison with others. On the same grounds we can eliminate, for the most part, the economic activities of women in homes—cooking, sewing, etc.

Of the remaining economic activities there can be no question as to which is of greatest significance for our purpose. By far the greater part of the earth surface that is used by man is used for agriculture, grazing, and the securing of forest products. Since only one of these is commonly found at the same time in the same place, we can combine them as one—the use of the land to secure plant and animal products. In terms of intensity of use—number of people concerned—it is true that this group of activities may be less important locally than mining or urban activities, but, particularly since we have eliminated a large number of the urban activities including the more "universal" types of manufacturing, there is no question that by far the largest number of people are engaged in land use for plant and animal produce, primarily for agriculture. Not only is this true for the world as a whole, but likewise for almost any large portion of it. Except for smaller areas, therefore, we can consider primarily the differentiation in these forms of land use, and, in certain areas, the extent of urban activities as a whole.

If, therefore, we can select observable criteria by means of which we can classify and map the intensity and manner in which the land is used in these activities, we will have a basis for regional division of the world that will be significant for the largest possible number of cultural phenomena important to man and to the earth surface as well. In terms of differences in the intensity and manner of agricultural land use, the land areas of the world can be divided into three major classes: (1) those which are not used at all—the uninhabited areas; (2) those in which man utilizes the wild vegetative and animal life as he finds it; and (3) the areas where the surface is dominated by man's cultivation, that is, where he has removed the wild vegetative cover and replaced it with plants of his own choosing. While the largest part of the land area of the world is included in the second class, the most important areas are, of course, those of the third.

Within the cleared and cultivated lands, the differences in manner of use which are geographically most significant are the differences in the (domestic) plants and animals produced. Possible additional factors are certain major differences in the manner of cultivation—whether with plough or a mere hoe, but in most of the important areas of this type, the plough is fairly universal. All other differences in technique, patterns of the fields, etc., are minor in comparison with those first stated—either in terms of their actual importance or in terms of their significance in areal differentiation.

The intensity of agricultural land use in these cultivated areas is measured, in the first instance, in the extent to which they are cleared and cultivated rather than left in forest or wild grass. Beyond that, the particular association of crops and animals provides commonly a rough measure of intensity, but not one that is adequate to distinguish, for example, between land use in the United States and that in western Europe, not to mention North China. These differences must therefore somehow be added to our system of criteria.

The first class of lands, the uninhabited areas, are undifferentiated in

[336]

* For "universal", read "ubiquitous", as defined by Garver [347].

terms of cultural geography—they are all simply zero lands. Theoretically they could be distinguished in terms of the factor which prevents their being used, whether permanent frost, permanent drought, barren soil, or whatever. Since these are areas of extreme conditions in one or more natural elements the preventive factor commonly dominates the actual landscape—in these cases, the natural landscape.

The basis for subdivision is less clear in respect to the intermediate, and largest, class of lands, those sparsely populated areas where man is a part of the organic world but does not dominate it. In large degree the manner and intensity of his use of the land is dominated by the wild vegetative and animal life (whether this be approximately the same as the original natural vegetation and animal life does not here concern us). Consequently it might be logical to subdivide these lands in terms of the actual vegetative cover (since that largely, though by no means completely, determines also the wild animal life).

In sum, therefore, we could establish a common basis for dividing each of the three major classes of lands—namely, in terms of the actual vegetation (and animals). These are real, rather than theoretical, features which we can observe and measure by various methods, and can therefore classify and map. That such a system would provide at least a major part of a valid basis for establishing culturo-geographic regions is, we may repeat, no *a priori* assumption but has been derived from our preceding consideration of all the cultural features significant in geography. That a world division on this basis will show significant associations with the areal differentiation of natural features—though by no means a simple or exact correspondence —is a proposition whose general truth the informed reader will presumably accept without demonstration. (Particular problems which that proposition raises will be discussed later.)

On the other hand, it is clear that the system so far outlined is much too simple to provide a final solution for our complicated problem. It is obvious that it provides no method for measuring the relative importance of fundamental cultural features not associated with production from the surface of the land—such as mining and manufacturing. Nor does it provide a sufficiently inclusive classification of characteristics of rural areas. Thus different areas may have nearly the same association of crops and animals, but the methods, tools, and equipment of production, and particularly the use made of the products—upon which a large group of important cultural characteristics depend—may be sufficiently different in the several areas as to require recognition in the regional classification. Finally, a closer examination of those areas which we may loosely describe as populated but not culti-

[337]

vated, those still dominated by the wild vegetation, reveals an error of generalization which leads to considerable difficulty. In any particular region of this class—whether in semiarid steppes or tropical rainforest—the greater part of the area may be covered with wild vegetation, and the people may depend, directly or indirectly, on this vegetation for a large part of their sustenance; but at the same time, in all but very few such areas, they also carry on some cultivation in small patches of land. Though the patches of cultivated land are small and used commonly for but a few years, their importance to the population of the total area is far out of proportion to their size. In other words, this intermediate class of lands represents, in part, a transition to that of the thoroughly cultivated lands, to some extent we must apply to them the criteria used in the cultivated lands, that of domestic crops and animals. But the transition is not represented by differences in crops, since they may grow the same crops. It might be measured indirectly in terms of intensity, *i.e.*, yields per acre (if the data were available), but in reality it is more significantly and directly represented by the methods and tools of cultivation—notably the shifting character of the cultivation and the use of the hoe or a simpler tool in place of the plough.

The difficulties which have just been enumerated present themselves when one attempts to apply the system theoretically developed, to the world as it is. They reveal, of course, the over-simplification of our theoretical consideration, but before attempting to correct and complete that, it will be well to examine the actual systems of world division which illustrate—in part, at least, demonstrate—the general theory. The theoretical consideration has not, as one might suppose from the manner of our presentation, been developed independently of practice; far more, it has been evolved from the study of actual systems.

Attempts to construct a regional system based on cultural features are unfortunately few, and most of them have been confined to a few major aspects of cultural geography, with perhaps no consideration of their relation to the general problem of culturo-geographic division. Nevertheless sufficient progress has been made to indicate the direction and goal of possible development. These may be seen more clearly if we summarize briefly the development to date.

In many studies in cultural geography the attempt is made to proceed directly from individual elements—once they are classified and their distribution determined—to regional determination. This appears in its crudest form in the attempt to divide an area into "agricultural regions" on the basis of single crops. The device of over-shading—not to mention the addition of letters and other special symbols—is simply a confession of the inadequacy of the method for regional division. Furthermore, the method is fundamentally unsound since it assumes the farmer's fields as the fundamental units rather than the farm as a whole. A Corn Belt farm is more than a corn farm, it is commonly an organized unit producing corn, wheat, oats, hay, and various livestock, not to mention other elements.

While Engelbrecht and others in Germany, as well as many in this country, attempted to determine agricultural regions on the basis of single crops, Hahn struck at a more basic element in considering the manner of cultivation—in particular the contrast between hoe culture and plough culture [298; 299]. Schlüter has likewise emphasized the importance of the "*Wirtschaftsform*" in agricultural land use [145, 22 f.]. Working further in this direction, Waibel has studied the distribution of forms of land use by the fallow system, the three-field system, and various forms of rotation [395].

In this country there has been, during the past ten or twenty years, an increasing interest in attempts to study agricultural geography in terms of particular associations of crops and livestock. Without developing, in print at least, any theoretical foundation various students perceived that the interpretation of the distribution of farm crops could not be explained in terms of each individual crop but required the study of the association of crops and livestock actually found on single farms as units. More recently, Kniffen has emphasized that, in the study of cultural geography in general, a necessary step is the recognition of areal associations of elements—"element-complexes" as he calls them [295, 163 f.].

An early attempt toward agricultural "element-complexes" is represented by the map of Agricultural Regions of the United States prepared by a group of workers, including O. E. Baker, and published by the Department of Agriculture in 1915 [311]. Over a decade later, Baker's very detailed study of this topic for the whole of North America constituted one of the first of the comprehensive series of similar studies covering the world that appeared serially in *Economic Geography*. (Beginning in 1926, these extended over a decade, major world areas being treated by Olaf Jonasson [313], O. E. Baker [312], Samuel Van Valkenburg, George B. Cressey, Robert B. Hall, C. F. Jones, Griffith Taylor and H. L. Shantz.)

Valuable as these studies are, they cannot be combined into a single uniform system, in part because each of the authors constructed an independent system. Furthermore, in some of these, the regional determination is based directly on the individual elements, perhaps because the study of element complexes had not, at the time of writing, been sufficiently developed. Finally, no clear distinction is maintained between "specific" regions, each unique in itself, and the "generic" regions of a comparative system of division.

Examples of this confusion of both types of systems may be found on Baker's map of North America, "The Columbia Basin Wheat Region," the "Spring Wheat Region" and the "Hard Winter Wheat Region" are each indicated and described as specific regions, whereas the "Hay and Dairying Region"—from Minnesota to New England—is not a specific region but an area of a certain type of agriculture. More important is the difficulty created by recognizing the Corn Belt as a specific region. In terms of crop and livestock association the Corn Belt is simply a part of a larger area which extends south to the Cotton Belt. In order to distinguish it, it was necessary to introduce the criterion of intensity of production, so that the so-called "Corn and winter wheat belt" is a collection of regions differing notably among themselves, but in general having much the same crop and livestock association as in the Corn Belt, though the production is in general distinctly less [312 (1927), 309–27, 447–66; cf. 324, 105–7]. Had this criterion been applied elsewhere it would have shown even greater distinctions, notably in the Cotton Belt.

By far the most detailed study of farm types is the United States government publication, *Types of Farming in the United States,* published in 1933, with text by Foster F. Elliott [320]. The system of classification, into which the more than 800 agricultural districts are classified, recognizes primarily the more specialized types. Districts less specialized are thrown into a large miscellaneous type called "General Farming"; since this includes areas whose principal product may be cotton, tobacco, or wheat, range cattle, hogs, dairy products, or poultry, or forest products, it is of course, no type at all. Nevertheless, the material in this work is of great value for any study of agricultural regions of the United States.

So far as is known to the writer, the first attempt to classify types of agricultural land use the world over, in terms of "element complexes," and to divide the world map on that basis, was the system developed over a long period by Wellington Jones and Derwent Whittlesey. This appeared in several photostat and mimeograph forms, the world map in printed form in 1932 [315], and finally, in 1936, Whittlesey revised the treatment and published it with a new map, as "Major Agricultural Regions of the Earth" [319; note also 316–318].[95]

During the latter part of the period in which Jones and Whittlesey were working upon this system, the writer, in collaboration with Samuel

---

[95] Pfeifer is, of course, misinformed in classifying this work "in this new direction in cultural geography that grew out of the criticism of the regional school" [109, 120].

Dicken, arrived at a very similar classification [*328,* lithoprint; a major section was published in these *Annals* in 1935, *324,* and the whole system is presented in very generalized form in *327*].

The striking similarity of these two systems of classification is not to be ascribed simply to the influence on the writer of the point of view of his former teachers. Both the determination of the particular types and the examination and detailed mapping of the basic data were carried on entirely independently, except that in both cases the technique of measuring the relative importance of different crops and livestock products proceeds along lines laid down by Wellington D. Jones, in his "Ratios and Isopleth Maps in Regional Investigation of Agricultural Land Occupance" [*283; cf. 319,* 209; *324,* 101].

Each of these systems seeks for a comparative regional division of the world which will be significant for the greatest possible number of features associated with land use. In view of the large number of different characteristics which are considered, the fact that two pairs of students working largely independently arrive at fairly similar results at least suggests that the general approach is valid.

It might be claimed, however, that in both systems the problem has been over-simplified. Although a great number of characteristics are discussed, most of the individual types are determined primarily, if not exclusively, in terms of the crop and livestock associations.

Much more detailed than either of these two studies, is a recent study of agricultural zones in Germany, by Busch, a student of agriculture who depends in part on the (German) geographic literature [*306*]. Busch likewise determines his agricultural types and the areas in which they are found, primarily on particular combinations of crops and livestock. His study is worked out far more thoroughly than either of those discussed at length here and demonstrates that the methods used for world studies can be employed equally well, if not better, for lesser areas. So far as I can find, its use of the method of crop-animal complexes is unique in the German literature. Though we cite it here as the best example of what the method may accomplish for a limited detailed study, it otherwise throws no light on our particular problem of organizing the world into regions.

Unquestionably the criteria of crop-animal complexes is more easily applied, and gives better results, in areas limited to a single major cultural type. Hartshorne and Dicken found it most nearly successful as applied to areas of European culture—*i.e.,* in both Europe and North America [*324*].[96]

[96] I refer in this manner to the studies of which I was co-author, not merely for

They have likewise attempted to set up similar criteria for all the world types [specifically stated in *328*]. In certain cases, however, the statement of criteria may be misleading. Thus, areas of "paddy field rice" are determined on the basis of the proportion of land in rice, but the thought behind this—as shown by the fact that the rice districts in the United States are not included—is evidently the peculiar form of the Oriental paddy field and the special methods of cultivation associated with it. Even more clearly, the type of "Oriental small-grain, soy-bean, agriculture"—in which there may, in fact, be no soy beans—is based not on the crop association, nor primarily on the small number of livestock, but rather on the intensive methods of cultivation.

Whittlesey, in his introductory discussion, presents five sets of criteria, but, unfortunately, does not state specifically how they were used in determining the individual types. From careful examination of his system, however, it appears a fair inference that the majority of his types—including particularly those within the well cultivated areas, the lands of permanent agriculture—are likewise limited chiefly in terms of crop and animal associations.

This simplification, however, can be justified to the extent that differences in the other criteria can be directly related to differences in the crop and animal associations. Though this is true, to large degree, of the other four sets of criteria which Whittlesey lists, in no case is it completely true. The methods of production, the intensity of production, the degree to which the production is commercial, and the character of farm buildings, commonly differ with different crop-animal complexes but are not necessarily the same in areas where that complex is the same. In particular, Pfeifer, in his necessarily very brief comment, objects to the inadequate consideration of "important primary differences," such as for example, the *Wirtschaftsform* (form of economy) as studied by Hahn and Waibel [*109*, 120 f.; *cf.* with Waibel, *395*]. In particular, we may note, little or no consideration is taken —in either of the two systems under consideration here—of the manner in which the land and workers are organized in relation to each other—*i.e.*, the differences between one-family independent farms consisting of a single unit of land, similar farms formed of scattered parcels out of a larger unit of land, large estates operated by a differentiated labor system, etc. [*cf.* Schlüter, *145*, 22]. The consideration of such elements in connection with the plantation areas, or areas of extensive grain farms, are exceptions that emphasize

convenience but also because I have endeavored here to consider these studies in the same way in which I examine that of Whittlesey, as though they were publications in geography in which I had no part. To what degree this is accomplished only the reader can judge.

their omission in other areas. On the other hand, how these features can be included in combination with the others considered is an extremely difficult, if not impossible, problem.

It should not be supposed that either of these systems is based solely on the crop-animal complexes, together with such of the other criteria as are automatically classified by those. In each of the two systems, one or two major types are distinguished on the basis of intensity of production in addition to the crop-animal complex. But neither system recognizes the differences, in all of these sets of criteria, between farmlands of northern United States and northwestern Europe. Similarly, each system, in classifying one or two types, utilizes the difference in degree of commercial, as distinct from subsistence, production. This single division, however, is hardly adequate to distinguish differences along the whole range, from farms that are nearly 100 percent subsistence to those nearly 100 percent commercial. Indeed, in many areas where the difference in degree is certainly very great, no distinction is made. To be sure, this is a criterion for which we have few statistical data, and it is certainly not clearly reflected in the crop-animal complex.

It should not be supposed that the limitation to one set of criteria would necessarily produce much the same system of types and regions. In view of the large number of factors involved in the crop-animal complexes and the wide room for disagreement as to which most fully indicate the sum total of cultural characteristics involved, as well as the different numerical limits that may be used for the individual criteria, one might well expect far greater differences than are found.

In both cases the specific limits for the ratios regarded as significant were determined by the technique illustrated by Wellington Jones's division of the Central Northwest of the United States. By making rough isopleth maps of many significant ratios of land use, one finds that certain areas stand out as distinctive core areas, each having a fairly definite character, represented on most, or all, of the maps. Further, the comparison of the different maps suggests which criteria appear to be more fundamental. On this basis the essential criteria of the particular agricultural type may be roughly determined, and the area extended as far as those criteria are found. If the isopleths show no sharp gradients, one must either select a limiting line, on a very arbitrary basis, or, recognizing a transition type, select two limiting lines, likewise on an arbitrary basis. In more cases than one might expect, however, the gradients are fairly sharp. In many cases therefore—as along the northern boundary of the farming-plantation area of Southern United States (chiefly the Cotton Belt) or the eastern boundary of the commercial grain area of the Central Northwest—it makes relatively little difference which numerical figures are taken as the limit for the critical criteria. In other words, man tends to develop sharper limits than those set by nature. Where an important crop can successfully be grown, it is commonly grown as a major crop; where it cannot be grown successfully, it may not be grown at all. Other borders, however, notably those between the cultivated lands and the grazing lands, are in reality wide transition zones of mixed cultures in which any line on the map is largely arbitrary. (This method of starting with the criteria determined in the core areas and working outward, appears to be similar to that used by Lautensach in determining specific regions—a method derived, he tells us, from Hözel, Hettner, and Gradmann [263; 278, 23].)

While the two systems here under discussion show similar results in many details, certain of the differences illustrate the problems involved in selecting criteria for world regional division.

A number of areas that Whittlesey classifies as "Commercial Dairy Farming" are classified by the other authors as "Hay, pasture and livestock farming." These titles represent no mere difference in terminology, but a difference in the criteria selected. For the former, the criterion is, I understand, the amount of milk produced in proportion to the crop land (as developed by Jones [283]), for the latter, the relative acreage of pasture, hay, and tilled crops. For the most part the areas are the same in each case, but in some areas of "Hay, pasture, livestock farming"—notably in parts of Great Britain—dairy animals are less important than meat animals, and on the other hand some areas that are unquestionably "commercial dairy areas" have too high a proportion of land in tilled crops to be classified as "Hay, pasture, livestock farming."

Significant examples of this last-mentioned difference may be noted in southeastern Pennsylvania and in the strip of land extending west of Chicago to include the northeastern corner of Iowa. According to Whittlesey's criteria, these are certainly areas of "commercial dairy farming" [as shown by Jones, 283, and following him, with more detail, by Hartshorne, 325]. But according to the criteria of Hartshorne and Dicken, these areas must be included in the areas of "corn, wheat, and livestock farming." The latter system is evidently based primarily on the appearance of the landscape; the difference between meadow and pasture on the one hand and tilled crops on the other is held to be greater than the differences in the farmstead resulting from the production of dairy rather than beef cattle. More is involved, however, than the contrasts in landscapes in each case. In the methods and intensity of production, the differences associated with tilled crops in contrast to grass may be greater than the differences that result from the contrast between dairy and meat production.

In other words, the problem of determining which factors in the crop and animal associations are most effective in determining types that will reflect the largest number of significant farm features is not one that admits of clear and unquestioned answers. Nevertheless the criteria, which have been shown to be significant, are sufficiently interdependent so that even when different ones are chosen the final results are in fair agreement.

The comparison of these two systems of world division shows greatest differences in the methods of handling the problems presented by areas of primitive occupance, notably in the tropical forests and savannas. Here, Whittlesey depends primarily on the methods of production. The areas of "Shifting Cultivation" include the greater part of both tropical forest and savanna lands in Africa, whereas the savannas of South America are classified under "Livestock ranching"—but with no distinction, either in map or text, between the livestock ranching of the Campos and that of the Pampas or Western United States. In contrast, Hartshorne and Dicken, though they do not indicate it in their terminology, based their division on the criterion of the landscape as seen to-day. The areas which they call "hoe culture" are made coextensive with the tropical forests (overlooking the simplification of boundaries for student use), and they recognize a special subtype of grazing, "savanna grazing" in the savannas. Though this may be satisfactory for South America, in Africa it leads to difficulties which they frankly dodged by adding question marks to their map.

In a later, simplified, presentation of the system, Hartshorne resolved the dilemma by frankly accepting the present landscape as the criterion [327]. (Though no attempt was made at that time to determine precisely what this concept involved, it appears to have been considered in approximately the sense defined earlier in this paper.) On the map of major landscapes of the world the areas previously entitled according to the form of culture are classified simply as "Tropical forests" and "Tropical savannas." (For the sake of consistency, the title "Grazing Areas" should be changed to "Grasslands," *i.e.*, present grasslands.)

It seems to the writer now that this solution places too much emphasis on the areal extent of what one sees in the landscape. If people practice much the same sort of shifting hoe cultivation in small clearings in either forest or grass, it is of lesser importance for cultural geography to consider the vegetative character of the surrounding unused landscape; we might say that the important landscapes are the clearings. On the other hand the work of making new clearings every few years is presumably very different where the wild vegetation is forest rather than grass. Further, insofar as the areas surrounding the clearings are used, the manner of use differs enormously in the two cases.

Probably neither system has adequately classified these areas, but it is possible that the essential data are inadequate. We can, of course, recognize shifting cultivation in forests and shifting cultivation in savannas as separate types, but in doing so we are simply assuming important differences in cultural features—methods of production in this case—which we have not demonstrated and which, if demonstrated, might better be stated directly.

G. THE PRACTICABILITY AND VALIDITY OF COMPARATIVE
SYSTEMS OF CULTURAL REGIONS

The comparison of the two alternate systems for dividing the world into type areas of land use has demonstrated a number of the difficulties that we listed as inherent in the problem. Though neither system has solved all these difficulties, sufficient progress has been made to indicate how they may perhaps be solved. We may therefore use these systems further—in spite of the fact that neither one would claim to represent more than a stage in the development of a finished system—to examine theoretically how certain of these problems may be solved and to consider finally, the validity of systems of this general sort for regional geography.

We have already concluded that the problem of world regional division into cultural regions cannot be solved on the basis of a single set of criteria. Whittlesey definitely bases his system on five sets of criteria; the other authors appear to base their system primarly on one general set—the present vegetative cover—but we found that was not adhered to consistently and does not prove adequate. What sorts of criteria may be logically brought together and on what basis can they be combined?

Whittlesey alone appears to have considered this question, but answers only its first part; he does not tell how he determined their combinations. From the results, one infers that this was done by no systematic method, but by subjective decisions in each particular case.

To establish a sound system of division it is necessary to know exactly how the division is arrived at. If several sets of equally important criteria were used, each of which varied independently, no solution would be practicable. With five criteria, each with but four gradations, the number of possible types would be over a thousand. Even if we were to use but two independent and equally important criteria, the system, though possible, would lack essential unity. It could lead to a list of types, but not to a genuine classification of types.

A proper classification of types must produce something more than a list of 50 to 100 independent types which cannot be logically arranged in groups. Since the world does not, in fact, consist of so many individual parts which we are to recognize and enumerate, but is a whole which we are to divide into more or less similar parts, a logical division will provide for recognition of different degrees of similarity in different stages of division.

When one has wrestled with this problem repeatedly in the actual attempt to produce a system of world division which one can then defend as sound, one is forced to accept the logical requirement previously discussed (Sec. X E) : only those criteria can be used which can be compared with each other in terms of their importance to some single standard, and whose order of importance on that basis is more or less constant.

The two systems which we have been discussing suggest two different possibilities : one is based primarily on the present landscape cover of the world—*i.e.,* the texture of the present landscape, ignoring relief—the other is based on the synthesis of features involved in land use. Let us examine each of these in turn.

The present landscape cover consists of vegetation, or in its absence, whatever forms replace it—roads, fallow fields, quarry-pits, buildings, etc. Whether the vegetation be natural, wild, or cultural is for the moment immaterial. In contrast to the concept of the natural landscape, or natural vegetation, this concept of the present landscape cover is a reality; it can be observed and analyzed [*cf.* Unstead, *309,* 185]. Further, its various elements can be arranged in an order of importance which does not vary. For example, if we decide that the difference between a cultivated field and a forest is greater than the difference between a forest and a stretch of wild grass, the statement will hold true in any part of the world, no matter what differences in climate, or soil may be found.

Furthermore, the landscape cover of the world is not a hopeless kaleidoscope of varying features, but in major aspects shows marked similarity within limited areas with greater contrasts between those areas. This results from two separate forces both of which produce real syntheses in the landscape. Nature produces the synthesis of the plant associations, theoretically, the "natural vegetation"; where man has affected these without destroying them, we find the associations of the wild vegetation. Where man has more or less completely wiped out the associations of this character, he has substituted new plant associations of his own choosing and, commonly, maintains these over fairly wide areas. Unquestionably, however, we will find areas where this process is incomplete, areas of disrupted associations, of combinations of wild and cultivated associations, which will be difficult to classify.

[347]

We have therefore the basis for a classification of types of areas which can be arrived at by major and minor divisions [this is developed to some extent in *327*]. Following the primary division of the world into lands and seas, the first major division of the lands is that which has already been suggested and which might be briefly characterized as the division between "wild lands" and "domesticated lands." In the cultivated lands, as Julian Huxley has commented, "man has done more in five thousand years to alter the biologic aspect of the planet than has nature in five million years" [quoted in *321*]. This conversion of the "natural landscape" into the "cultural landscape"—in the sense of a landscape dominated by man—Penck describes as "the great, perhaps the greatest, act of man, a geographical and historical event of the highest importance" [*158, 52*]. From this point of view the distinction of importance in present landscapes is not that between those areas where "natural landscapes" may actually be found—*e.g.,* the glacial regions—and the inhabited lands, all more or less affected, positively or negatively, by human occupance. The contrast which both Huxley and Penck have in mind is that, expressed in a multitude of different aspects, between areas whose landscape cover is largely under the control of man, and those where any effect of man's occupance is minor in comparison to the forces of nature. Within these wild landscapes one could distinguish theoretically between those that were exclusively natural and those that have been negatively affected by man, but in practice the distinction is difficult to determine and in many areas is of but minor importance.

Within the areas of wild landscapes we may subdivide on the obvious basis of appearance, distinguishing areas of ice, bare earth (deserts), tundra, grass, savanna, and forest. (The differences among these might be regarded as no less than the differences between any one of them and the cultivated lands. On that basis they could all be included in the previous major division; there is no necessity for agreement on that point.) Within any one of these major subdivisions one could distinguish between areas of continuous wild landscape and areas of wild landscape with cultivated clearings. Possibly, however, one might recognize an intermediate major division between the wild lands and the cultivated lands to include all areas of predominantly wild landscapes broken by numerous cultivated clearings. This is not a transition phase, either in time or place, since vast areas are permanently characterized by this combination of features. The problem presents difficulties which may force a compromise with a logical system, but the areas are fortunately limited largely to tropical forests and savannas.

In the cultivated landscapes the landscape cover is represented not merely by the cleared and cultivated fields whose character changes so markedly through the seasons, but also by fences, farm-buildings, villages and roads, and even by railroads, towns and cities. Insofar as these latter features are functionally associated with the farm features they may be considered as parts of a complete areal complex. Any division of the cultivated landscapes consistent with the general system, however, would presumably be based on the major features of the landscape cover, namely the cultivated fields. Presumably it would be consistent to distinguish between fields ploughed annually and those in more or less permanent grass cover, but it would not appear consistent—in the present system, that is—to divide the crops on the basis of their use, or to consider livestock at all. Likewise one might recognize important differences in intensity of crop growth but not the differences between Oriental and Occidental methods of producing the same crop. The difference between commercial and subsistence agriculture is in itself extraneous, but one might recognize a distinction in the surface landscape between farm areas where there is little development of roads, railroads, and towns and those in which the crop cultivation was directly associated with a marked development of those features.

It is unfortunate that the system described has not been consistently developed so that we could judge its efficacy; it has had but a partial demonstration. Nevertheless, if we have in mind those cultural features which we listed earlier as of greatest geographical significance, it is apparent that most of them are represented, directly or indirectly, in the present landscape; a regional division based on that should therefore be significant for a great many features.

Would it be possible to extend this system so as to include all surface features of the landscape, urban as well as rural? The fact that these features are not superimposed on each other in reality might lead us to suppose that we have only a technical problem of generalizing from large-scale maps to small-scale maps. [For a very generalized world map of such a type, see Hassinger, 360.]

We have already suggested that certain aspects of urban development may be included as associated parts of a largely rural area, but only insofar as the urban development is in reality associated with, and not merely adjacent to, the rural scene. Towns and cities form a characteristic of all cultivated lands in contrast to most of the non-cultivated lands. Furthermore the particular characteristics of the urban development of any area may be largely determined by the character of its rural development. The urban and rural features in such a case form a real association which we can consider as a unit.

In many areas, however, important features will be found which are not associated, except in position, with the dominant rural element-complex of the area. While the "agricultural towns" of central Illinois can be considered as a characteristic part of a dominantly rural scene, the coal-mines and mining towns form an essentially separate form of element-complex. The addition of an agricultural town to a surrounding farmland forms an agricultural area, the addition of farms to mines forms nothing except "farms and mines." Needless to say, a full regional study of such an area will require the consideration of all element-complexes within its limits, no matter how loosely associated. Our present concern, however, is not with individual regional studies, but is with a comparative study of type areas.

Though one might waive the problem raised by mining features as of small importance, that raised by the specialized manufacturing cities of such areas as northeastern United States and northwestern Europe is not to be dismissed in this fashion. The writer attempted to solve the problem by the crude method of superimposition [327, 339–373]. On the world map of regions determined in terms of rural landscapes, there are superposed the areas that have been determined, by statistical methods, to have a relatively high development of non-local manufacturing [326]. The result, however, is not one system, but two systems on one map. In order to form a single system it would be necessary to add a new criterion to the determination of all our divisions. While this would, in fact, affect only the cultivated lands, it would affect many of the subdivisions of them. We would not have simply a "hay-pasture landscape" but rather two types, a "hay-pasture and industrial urban landscape" and a "hay-pasture landscape without industrial urban features." In reality this would be insufficient, since the landscape of cities of heavy manufacturing is notably different from that of cities of light manufacturing, not to consider more detailed classification. To attempt to combine different types of industrial cities would destroy the logic of the system. There is no method of determining whether differences in city landscapes are more important than differences in rural landscapes.

A comparative system of world division on the basis of the total actual surface landscape, we conclude, is neither logical nor feasible. We can have one system based on the dominant element-complex of the rural scene, possibly we can construct a separate system based on the urban scene—though there is reason to think this would require more than one, and, to be complete, we would need also separate systems for mining features, fishing features, etc.

The alternative to a system based on the present landscape cover is one based on the synthesis of all features involved in the productive use of the land surface. Since man can produce from the land surface only in terms of vegetable products—whether from wild or cultivated plants, whether for direct use or indirectly through animals—the two bases are closely related. They are not however the same. The former is limited definitely to the appearance of the land used; the latter includes all cultural features associated with a particular form of using the land, whether those features are obvious in the landscape or not, whether they are material or immaterial—if characteristic of a particular form of land use, they are to be considered. This concept is therefore much broader than the other. It may appear to some to be less distinctly "geographic" because it is not based directly on the land, but we are not to be misled by that idea. In itself, as a classification of a certain wide association of cultural phenomena it is not geographic in any case, even though the cultural phenomena have to do with land use. It is the consideration of the areas in which the cultural types are found that is of geographic significance.

Are the various cultural features associated with land use in reality combined in such a way that genuine types can be recognized? The individual cultural features are not independent in their variations, they are in fact synthesized by man in his individual organized units of land use. The basic units are represented by the individual farm, plantation, or ranch. The farm, as an organized unit, includes not merely the land and the plants and buildings on it, but also the livestock, tools, methods and intensity of production, and the use of the products. In other words, the farm represents not merely an element-complex—such as is found in areas of wild vegetation—but is a primary Whole—according to Wörner, a *Gestalt* [*274*, 343–5]. Each and all of the elements listed, whether material fields, buildings or tools, or immaterial methods of production, can be understood in form and function only in terms of the whole farm unit.

Particular types of land-use units have areal expression. The actual extent of an individual farm, to be sure, is too slight to concern us, but over considerable areas we find most farms, whether considered as a whole, or measured in terms of their individual cultural features, are very much alike and differ notably from units in other areas.

We may note further that land-use units are associated in areas to form larger element-complexes, of various orders of magnitude and—inversely— of coherence. In many European areas, for example, the farms of a particular community in which all the families live in a compact village form an element-complex that may well be considered a unit, though hardly a Whole. The association of a much larger area of farm-land with a neighboring agricultural town or city, likewise, as noted earlier, forms an element-complex of a relatively loose form, in which the highways and railroads constitute connecting elements. Though this larger areal complex is by no means a Whole, it may show a definite character that can be considered in the recognition of generic types of regions.

That the areal distribution of different types of land use is significantly related to other geographic features, both cultural and natural, requires here no demonstration. We have already indicated that the land-use complex is, in fact, one of the most important, if not the most important, of all element-complexes in human geography.

The difficult problem is to determine how the elements in a land-use complex can be combined in a system of major and minor divisions. It is necessary to determine the order of importance of the elements to the complex as a whole. Major stages of division must be based on differences in factors of major importance, subdivisions on those of lesser importance. In any single process of subdividing a single division we must employ the same criterion throughout, though different divisions of the same order may be subdivided by different criteria.

Whittlesey groups his criteria according to their character, rather than their importance. In addition to the crop and livestock associations, he recognizes the intensity of production, the methods of production, the degree to which the production is commercial rather than subsistence, and the farm buildings. Each of these sets, we may note, includes criteria of major and minor importance. Thus, the difference between cultivation with the plough and that with only the hoe is of very great importance, but whether farm machinery is pulled by horses or by tractors is minor.

Obviously the problem of determining major and minor criteria is more difficult than in the system based directly on the landscape cover. Without attempting to list specific criteria, we may suggest the possible bases for dividing at each level.

The first division of all types of land use can be based very generally on major differences in methods, as follows: permanent cultivation, shifting cultivation, grazing (on wild vegetation), production of forest products, no production.

We need not here follow the divisions of each of these (areas of no production can, of course, have no subtypes in terms of land use). In the grazing areas the greatest differences are associated with the degree of commercial development; the grazing of nomadic tribes, producing few permanent buildings, roads or cities, is in sharpest contrast to commercial ranching with its permanent ranch-houses, barbed wire fences, roads, railroads and many towns and cities. The semicommercial grazing of the American savannas, associated with some rudimental agriculture, is presumably a third type.

Our chief concern is in the lands of permanent agriculture. In both of the actual systems which we have discussed the authors have evidently used, without clearly indicating it, a primary subdivision based on important differences in methods of production—differences which are, to be sure, reflected in intensity of production, but are clearly determined on the basis of methods. These differences are of lesser order of importance than those used in determining the first world division, but are of a greater order of importance than, for example, the difference in methods used in eastern United States and Western Europe. They are roughly indicated by the division into: rudimental agriculture (to use Whittlesey's term), plantation agriculture, Occidental agriculture, and Oriental agriculture.

Within each of these divisions further subdivision is again independent of that in the others. Within Oriental agriculture there is general agreement that the greatest differences in land use are found between areas where the cultivation of paddy fields for rice is of major importance and that where it is not found.

In Occidental agriculture the situation is more complicated (or perhaps only appears so because we are closer to it). Major differences can be found within this division in three somewhat independent groups of characteristics, each of unquestioned importance. In addition to the differences in crop and animal associations are differences in the degree of commercial production—as between Western and Eastern Europe, roughly speaking—and differences in methods and equipment of production—as between northern United States and Western Europe. As the effort to combine all of these would lead to an extreme multiplicity of types, we are forced to take separate steps of division and must therefore decide which differences are of greater, which of lesser importance.

It is not necessary for us here to decide this difficult question. In the actual construction of a system a decision must be made and followed consistently. Consistency however does not mean that somewhat similar criteria may not be used at different stages. Thus, one might decide that the first division of Occidental agriculture is to be made on the basis of the *major* characteristics of crop-animal complexes, *e.g.,* grain and fruit culture (Mediterranean) ; grain and livestock culture; hay pasture, livestock farming; and specialized grain farming. Each of these might then be subdivided on the basis of degree of commercialization, and these subdivisions then later divided on the basis of particular crops—*e.g.,* a commercial subdivision of the grain and livestock agriculture could be divided into a commercial corn-wheat-livestock farming and a commercial small-grain-livestock farming.

[353]

It is clear that at some stage of the subdivision one ultimately arrives, in every case, at a subtype which is represented by only one area. But the system has grouped the subtypes into types on a significant basis and therefore provides a proper basis for comparative regional geography.

From what has been previously said, it is clear that this system cannot be extended to include any cultural features of an area whose relation to the features of land-use in that area is one merely of juxtaposition and not of real association. The cultural features associated with mines, or with specialized manufacturing cities will require separate systems of world division.

Neither of the two systems here suggested in outline, therefore, will provide a regional basis suitable for all important cultural features. Nevertheless, either one of them offers a basis for regional division significant to the largest number of cultural features, concerned with the greatest part of the world.

All the students who have been concerned with the construction of these, or similar, systems of world regions based on cultural elements presumably regard their published productions as having only tentative value. This is true not merely because of the inadequacy of data available for many parts of the world but for a much more fundamental reason. Even if all the necessary data for each cultural element concerned, for every part of the world, were immediately at hand for the individual student preparing such a regional division, he would still not be in a position to prepare more than a tentative outline. Far more work remains to be done on the essential step of establishing the element-complexes present in reality. For this purpose the methods of statistical ratios and isopleth maps are inadequate; they indicate the general associations concerned and may even suggest the particular element-complexes involved, but cannot alone establish them. That can only be done by field studies. (See also R. E. Dodge's "plea for systematic approach to the study of modes of occupance" [296].)

In particular, if we are to base our culturo-geographic regions on the areal distributions of land-use units considered as primary Wholes, then those units should, as Wörner indirectly suggests, be studied in fact as actual Wholes. We cannot be satisfied with the abstract "average farms" that Hartshorne and Dicken have constructed from county statistics and used as illustrations [324]. Possibly it is a fair assumption that in any fairly homogeneous county the normal farm (one whose element-complex is that found on the largest number of farms concerned) will correspond approximately to the average farm—but they have made no effort to demonstrate that assumption. To a certain degree we know it is not the fact. Various specialized crops do not appear in small amount on the normal farm, as the diagrams of "average farm" types suggest, but are, in many cases, major crops on the exceptional farms in the county and completely absent from the normal farm. Similarly, the size of the average farm, and its division into woodland, pasture, and cropland, as indicated in Dicken's diagrams, does not necessarily represent the correct picture of the normal farm.

In order to comprehend either element-complexes of the more highly organized and closed units that we call Wholes, we must study individual cases—the element-complexes as individual complexes, the Wholes, each as a whole. Lest any one misunderstand us, as Fröbel misunderstood Ritter, we hasten to add that the study of a unit as a whole requires, of course, analysis as well as synthesis. But the elements analyzed—the individual crops, animals, methods of production, and consumption, etc.—are to be studied, not over the whole world, nor over a whole region, nor even a county, but within the unit farm; it is essential first to establish the relation of each of these elements to the total farm (land-use) unit as a unit whole. There is of course nothing mystical about this; we will merely follow in scientific form, the thought of the individual farmer who every year must organize these elements into a whole. This point of view and method, we may repeat, is not only necessary in order to understand the unit as a whole, but also in order to understand the significance of each of its members or elements. As our critic from psychology, Wörner, has concluded in his most instructive examination of the use and abuse of the concept of the Whole in geography, though there are relatively few cases in which geography is concerned with actual Wholes, they are found in the unit works of man, such as a farm. In such cases, the individual factors in the complex whole can be correctly evaluated only in view of their position in the complex [274, 346 f.].

This essential step in the interpretation of culturo-geographic phenomena has not been taken by the authors of these world systems; obviously if they sought first to do that for all the types necessary for a world study they would never in a lifetime be able to even start on their world maps. But their results indicate, it seems to me, not only the need for such studies but also the valuable results that we may expect to follow from them. Likewise their further development of the statistical-geographic methods of Wellington Jones offers at least a partial answer to the difficult question, where and how is one to look for the normal or typical land-use unit. Isoplethic maps of various kinds, drawn as detailed as the basic statistics permit, reveal clearly where we may expect to find counties of relatively homogeneous character of farms and where we would expect to find exceptional or mixed conditions. Finally, though average figures for a county do not represent correctly the normal farm, nevertheless they indicate approximately what character of farms may possibly be normal, and exclude definitely the exceptional farm as abnormal.

No doubt there will be loud protests at such a suggestion of "ultra-microscopic" studies in a field that should maintain a "macroscopic" point of view, but we are not to be frightened by belittling words that any may hurl at us. The geologist concerned with the movement of a glacier is not criticized for attempting to study the movement of one particle of ice with relation to another. The historian concerned with the evolution of parliamentary government may devote a major work to the debates in the House of Commons during a period of less than a decade. All that need be required—and it may well be insisted upon—is that the student making a microscopic, or ultramicroscopic, study should at the same time continue to maintain the macroscopic point of view. That is to say, if a geographer makes a detailed study of a single cotton farm in the Yazoo delta, he is not to forget that the geographic world is interested in the farm only in its significance to an understanding of the Yazoo delta, or even the Cotton Belt in general. From this point of view he may well ignore many details about the farm, as non-representative, or non-significant. But to understand land-use in the Yazoo delta, we must know more than the facts about cotton, corn, alluvial soils, etc., now readily available, we must know the relation of all these factors to each other and to the unit cell of land-use, the single "normal" farm.

Throughout our discussion of regional systems of world division based upon cultural element-complexes we have frequently raised a question that must now be answered. Can such a system—whether based exclusively on cultural features, or on the actual features of the landscape cover, both cultural and natural—provide a sound framework for a field that is concerned with all features, natural as well as cultural? Will we not encounter insuperable difficulties when we compare such a system of essentially cultural complexes with the complex of natural elements—whether the present, theoretical natural environment, or the actual original fundament? Are we not, in fact—as some have suggested—attempting the reverse of a logical procedure in making the comparison in this direction?

We may note first that this comparison is not to be stated as "the relation of cultural regions to natural regions." The determination of regions in either case requires the measuring rod of significance to man; though this presents no difficulty in the recognition of "cultural regions" it eliminates entirely the concept of "natural regions" in any literal sense of the term. Consequently we do not run the risk of dropping back into the environmentalist position of which Sauer warned in the case of the opposite procedure:

that it might lead—as we found does lead—to the attempt "to identify natural regions with cultural regions" [84, 191].

What is the situation when we compare the culturo-geographic regions with the regions of the individual natural elements? Either system of culturo-geographic regions is based on a large number of cultural features no one of which is dependent on a single natural element; rather each is dependent, in part, on a number of natural elements, and the manner and degree of dependence is different for each cultural feature. In consequence, the sum total of cultural features as represented by the cultural regions will differ so markedly from the regional classification of any one of the natural elements, or of any combination of a few of them, that there can be little temptation to try to presume any complete relationships. That is to say, if one takes the maps of cultural regions, either of Whittlesey or of Hartshorne and Dicken, and compares them with world maps of climate, relief, soil, etc., it is obvious that the explanation of any single type of cultural area in the cultivated lands will require consideration of all the natural elements, and, further, that it cannot be completely explained by that consideration alone. The radical difference between the regions in China and those in North America will at once demand consideration of cultural factors.

It has been claimed, however, that this procedure is the reverse of a logical one. As Sauer puts it, in a statement abstracted from Vallaux, "to describe the human landscape without knowing how these landscapes are constructed is to put the cart before the horse. The first solid basis to establish is therefore the physical geography which supplies it. There are to be reconstituted therefore in general the natural landscapes, in which the activity of the living world is comprised, such as nature made them, as though there had not been a living person on the earth" [84, 202, the specific reference to Vallaux is not given].   *

Undoubtedly the method which Vallaux and Sauer recommend, and which many geographers have followed, is in accordance with the logic of nature (in the broadest sense of the universe), assuming that we can impute a logic to the universe. But the logic of science is not the logic of nature. Molecules were not discovered from the study of atoms, nor atoms from the examination of the electrons and protons of which they are composed. The movements of the heavenly bodies were not, in the first instance, determined by the laws of celestial mechanics, rather the reverse was the case. Once scientific laws have been established the scientist may reverse them, and thereby discover hitherto unknown planets; but few would suggest that geography had arrived at that point of development, if indeed it ever can.

In fact there is only one basis on which geographers, from the analogy

[357]

* The statment is not, as it appears, a paraphrase from Vallaux, but a direct translation [186, 200].

with the more exact sciences, could argue that logic required us to proceed from the "natural landscape" to the "cultural landscape," namely that the character of the former determines the character of the latter. This procedure is therefore, as Broek remarks, "in principle a survival of the idea of environmental control." Broek has noted the logical flaw in the argument: "the natural elements are not the dynamic elements in determining the character of regions" and are therefore not suitable criteria for a regional classification [*297*, 103, 107].

There is, however, another sense in which the procedure from the natural environment to the cultural features might appear logical, namely, that it corresponds to the actual sequence of development in time. But science does not recognize that logic as essential. Biologists do not find it necessary to make a complete study of all the steps in the evolution of man before they study his present physiology. Even historians, who must deal with the sequence of phenomena in time, realize that it may be studied in either direction. Modern history cannot be understood unless one reaches back repeatedly to events of earlier periods, but no historian supposes that we cannot study modern history without first establishing *completely* the beginnings of history.

The one essential logic of science is, of course, to start with what is known and to proceed through intermediate relationships that can be demonstrated with certainty, to conclusions concerning what was not known.

What do we know in geography, what are the fundamental facts with which we have to work, and what relationships of those facts do we know with certainty? We know, to be sure, quite a few things about some of the natural elements of a region, about some we admittedly know little, perhaps never can have adequate knowledge. Of the relationships between the different natural elements we have some fairly certain knowledge, but much is still unknown. Of the relationships between these elements, on the one hand —whether taken individually or collectively—and individual cultural elements, on the other, our knowledge is far from sure. Is it then logical to attempt to put together the known and the unknown—in the form of the "natural environment," the "natural landscape," or what you will—then to derive conclusions from this on the basis of relationships of which we have very limited knowledge, in order finally to arrive at something which is already known, namely, the cultural features of an area? The primary facts of geography, the facts upon which our reasoning must be based, are the present features in an area; these include only some of the natural elements —since some no longer exist, or exist in changed form—but all of the present cultural elements.

[358]

Either of the systems outlined starts with what is known. They put first the observation and classification of elements and the recognition and measurement of element-complexes of phenomena for which we have the facts. That, I take it, forms the "solid basis" of any scientific work. We must free ourselves of the notion inherited from our past that facts concerning rocks or soils or the atmosphere are any more "solid" for scientific work than facts concerning acres of corn or numbers of people or even the language of peoples. Presumably the preference which many geographers show for natural features as compared with cultural features is based on the idea that they are of more permanent character. We are, however, prone to exaggerate the unchanging character of natural elements. It is particularly surprising to find in a work based on natural vegetation, a feature that in many areas no longer exists, the description of man's settlements as "engravings on the older and relatively more enduring background of the physical earth" [321, 4]. Since "physical" is another word used differently by different geographers it is possible that in this case it does not include vegetation. In that sense the statement might be accepted, though we should not forget that the street-pattern of a European city center may have changed less in centuries than the landform, soil, and water of a neighboring flood-plain—e.g., Vienna.

In any case, an argument over the durability of our facts is irrelevant to the question as to which are known with sufficient certainty to form the basis of scientific work. In that category belong exclusively the present features of areas.

Finch has therefore recommended the procedure of starting with the present features of the "landscape." Although he makes this recommendation specifically for the study of an urban area, his statement can be applied to the study of any cultural feature. This procedure, he concluded, "puts to the fore that aspect of the subject upon which the writer speaks with authority: the present city. It puts in its proper place that body of deductions about the past which, however well they may have been developed by a rigorous historical method, still are deductions. A century hence the recorded observations of landscape may be worth much, the deductions little" [288, 118]. By way of illustration we may alter Sauer's statement concerning the value of Humboldt's work in Mexico to say that had he used his time there in an attempt first to reconstruct the original landscape—say, that existing before any form of Indian civilization had developed—his results would probably have little value today, whereas his detailed description of what he actually found is of unique and imperishable value [84, 185]. Ralph Brown calls our attention to a similar example in the work of Arnold Guyot. In connection with studies made in order to demonstrate certain physiographic theories that are of no more value to us today than his teleological theories of the relation of the earth to man, he recorded incidental descriptions of cultural features of the Southern Appalachians at that time which have permanent value for the student of historical geography [65].

Even were it possible to know as much about the natural elements of an area as we can learn from observation about the cultural elements, it would still not be a safe procedure to work from natural causes to cultural results. Any specific cultural feature represents a development in the past from a host of factors, including not only a large number of natural elements but also many human elements some of which we cannot hope to know. In this complex "the nature of the land," as Schlüter says, "is of secondary importance, the determinant and creative factor is man" [148, 214; cf. also Waibel, 266, 203 f.; Broek, 297, 107 f.]. To interpret these highly complex phenomena we have as yet no adequate body of scientific principles, we are in no position to start with the host of factors involved, natural and human, and attempt to put them together in the way in which they operated in reality to form a single cultural feature. In other words, we are still in the stage where we must analyze the process of development of a cultural feature by starting with the present known product and working back as well as we can toward its original causes. Whether we shall ever be in the position of reversing the method, as many of the natural sciences have been able to do, need not be discussed at this point. In any case, the mere desire to emulate their success will not give us the necessary knowledge to follow their methods.

This point of view, as Hettner noted in 1907, is regarded as self-evident in the other sciences that study human problems [130, 414]. Only geographers have attempted to put the watch together without knowing all its parts and without having first studied its mechanism while taking it apart. (See also the quotations from Hettner and Barrows in Sec. III C.)[97]

The insistence that the study of the cultural features of an area must start with the cultural features does not mean that in the general study of the geography of any region one must start with cultural features. Assuming that one may beg the question of delimiting the region, the only scientific requirement of course is that one start with available facts. The available facts, as Finch, among others, has vigorously stated, are the present, actual features of the area, regardless of whether they are cultural or natural [288,

---

[97] Although there can be no uncertainty as to Hettner's view on the question of how cultural features are to be studied in research, his insistence on the genetic method leads to the opposite procedure in presentation, and in the establishment of a regional system (see Sec. X D). His arguments in the latter case, to be sure, refer to a realistic system of specific regions and do not apply to our present consideration of a comparative system of type areas.

115 ff.]. Although he presents four different structures for different types of "landscape," these all fit into the general rule that one starts with the "present landscape" [121]. The interpretation of the natural features will, by definition, depend exclusively on non-human factors, so that the study of an area but little affected by man might quite reasonably begin with the consideration of natural features. But the consideration of the cultural features, even within such an area, must, to be sound, proceed from the observations of the cultural features.

We are here concerned, however, not with single studies of areas which may be quite arbitrarily delimited, but with a system of world regional division that is to serve as a primary base for geographic work. We have seen that both logical and practical considerations require such a system to be based on a single element-complex or synthesis whose varying forms have areal expression. Since neither nature nor man synthesizes all geographic features in this way, a single system of regional division including all geographic features is not possible. We can establish as many regional systems as seem desirable, based on individual elements or on individual element-complexes, and can compare these with each other. If we could superimpose all of these on one map, the total would represent the actual geography of the world as it is, but it would not establish definite regions nor would it establish definite types of areas, but rather would show all the differences actually existing in specific areas.

For some purposes the world division into type areas based on actual landscape cover may prove the most useful. The largest number of different features, however, is represented by a world division into type areas based on the great number of cultural features synthesized by man in his productive use of the land.

H. SUMMARY

Although the earth surface—the world—in terms of all its geographic features, is not divided into distinct areal parts, the fundamental function of geography—the understanding of the differences between different areas—requires the geographer to divide the world arbitrarily into areal parts. A general principle of any science requires that its material is not to be left in patches, but is to be organized into a logical system of knowledge.

The areas of the world which the geographer studies may be organized into logical systems of division in one of two ways, either in terms of a single system in which specific regions are recognized as they are actually arranged on the face of the earth, or in a combination of several systems in each of which the areas of the world are classified under *types* as determined by individual features or element-complexes, but regardless of their actual arrangement on the planet.

The realistic system which recognizes specific regions in their actual spatial arrangement is the only one by which all regional knowledge may be organized into a single system. It alone provides a system for a complete regional geography. Such a system of specific regions is not inherent in the world which the geographer studies—neither in the world of nature nor in the actual world which nature and man together have made. Neither does our geographic knowledge of the present tell us with any degree of certainty even the major lineaments of such a system. Further, because the different areas of the world have not been separated, either as individuals or as types, in their development, but rather the development of each particular charactristic of any area has been a part of the development of that element elsewhere, we know that this process has produced no simple system of classification of areas, whose general outline can be recognized on the basis of our present knowledge of the field. On the contrary, to construct a sound system of specific regions will require the greatest amount of research, and additional knowledge will repeatedly lead to modifications, even serious modifications, of the system. In other words, such a system represents the ultimate organization of regional knowledge, rather than a basic framework in which to seek knowledge. Nevertheless, it is at any time the logical system in which to present the regional knowledge that has been acquired to date.

A logical system of specific regions cannot be organized on the basis of any one principle. It must consider all geographic features of significance and, in determining any particular level of subdivision, must select those features which are found to be of most importance in determining the character of the area—as measured in terms of human interests. Any element whose variations leads to variations in other features has, therefore, an importance based not only on itself, but also on those other features. It is not possible however, even in theory, to reduce the number of independent factors to be considered to less than three, perhaps not to less than six. In actual practice it will be necessary to consider many more as independent, since we cannot hope to trace their causal relations. Consequently it is not clear that the genetic principle is essential.

Although any number of geographers have developed systems of division into specific regions—at least for particular continents—very few have attempted to study the principles necessary for a sound regional division, and there has been little attempt to examine the various regional systems developed in the light of any body of principles. The most thorough investigation of the theoretical requirements of such a system has been made by Hettner, who has also demonstrated it in the actual development of a system over all the lands of the world.

SUMMARY

A fundamentally different method of organizing our areal knowledge of the world is represented by systems which classify areas according to their character as determined by particular aspects. Areas cannot be classified logically according to their total character; that consists of many aspects which vary independently so that an area determined by one does not correspond to an area determined by another.

A sound logical system of classifying areas by types must be based on but one element or (actual) element-complex. We may therefore classify the areas of the world into major and minor types according to any one of the natural or cultural elements significant in geography. Of this kind are maps of the world showing types of temperature conditions, rainfall, landforms, soil, houses, religions, or languages. Classifications based on element-complexes are represented by world maps of types of natural vegetation, or crop associations. All of these systems of classification are useful helps in regional study. None of them covers a sufficient number of features to offer, alone, an adequate background for regional study, or to provide even a tentative system for organizing all our regional knowledge of the world.

The most comprehensive synthesis of features into element-complexes which have distinct areal expression is found in the present landscape cover of rural areas. In large parts of the world this consists of vegetation which is largely a natural synthesis, though partly modified by man. In less extensive, but for us more important, areas it consists of a variety of features, chiefly but not exclusively cultivated plants, which have been synthesized areally by man in his use of the land.

The fact that such syntheses exist in reality offers us two possibilities for *
classifying areas. One is confined strictly to the landscape cover itself and provides us therefore with a background based, not on the total present landscape, but on the present landscape cover. Since, however, the synthesis of the landscape cover is actually formed by two independent forces, namely nature (the sum total of all factors which produce the natural vegetation) and man, it is not strictly logical. It is, however, practical, to the extent that the actual landscape cover may be regarded as either overwhelmingly natural —in the wild landscapes—or overwhelmingly cultural—in the domesticated landscapes—or as a genuine synthesis with significant areal expression, rather than merely the result of chance, produced both by man and nature— *e.g.*, in the areas of shifting cultivation in tropical forests.

The alternative method is to base the system of classification not on the landscape cover but on all the cultural features that are areally synthesized in the use of the land. This system is perhaps more significant than the other in the areas of important human development. It is based on element

[363]

* I.e., two alternative methods, for which the systems discussed above are examples.

complexes organized by man into unit wholes. Further it is more strictly logical. It is, of course, based on the assumption that in any area one single type of land use, one single form of land-use unit, may be regarded as predominant and therefore characteristic. It is valid therefore to the extent that that assumption correctly represents the facts. Undoubtedly there are areas in which that is not true, *e.g.,* portions of the Great Plains where both livestock ranching and grain farming are important but separate features. Presumably such areas are confined to transition border areas. They cannot be classified in the system logically, but where they are not too wide, they may be divided, by some arbitrary limit, between two types.

Either of these systems classifies areas of the world in terms of a comprehensive group of individual factors. Either one, therefore, offers a more comprehensive background for the study of regional geography than any system based on one or two elements. Since each is based on primary facts, facts of observation, it provides a more solid base for scientific work than a system based partially on deduced facts. Further, since both are based in large part—the second system completely—on cultural features, they encourage us to study the relation of known complex features to all possible individual elements, in contrast to a system based on a single causal feature which encourages the student to explain cultural features in terms of that cause.

Neither of these systems, however, provides a complete background. They must be supplemented by systems classifying those features which they do not include, notably the features associated with mining and manufacturing, and also, of course, by all the systems of classifications of the individual natural elements. The sum total of all these systems of classification does not add up to a classification of type areas of the world, but merely shows, for every spot in the world (say each unit of a square mile), the various types of areas, according to the different criteria, to which it belongs. In the nature of things, there can be no classification of areas logically based on all these characteristics of areas. When one superimposes all the maps of world division, the boundaries of the different categories cannot be reduced to a common boundary except by extremely arbitrary decisions that will result in a false picture for a greater number of places than those for which it will be true.

Finally, there is one factor of great importance in regional geography which we cannot classify, namely, the factor of relative location of one place with reference to another. An essential feature of any area is its location with reference to the other areas of the world, both near and remote, both land and sea. Clearly this cannot be classified in a finite number of systems. In terms of its relative location, or *locus,* each area is unique; the facts cannot even be adequately expressed in words but can only be shown on a map—or rather, on a globe. Once this factor is introduced, therefore, we must shift from systems that classify areas to a system that recognizes specific areas as they are actually located.

Systems of classification of types of areas therefore may, at best, provide us with a method of approaching the ultimate problem of recognizing specific regions. If used as a background for regional work the limitations, even of the most valuable ones, must be realized; any one of them can only tell us something about an area—some more, some less. All taken together—insofar as that is at all practical—cannot tell us *all* about the area even in outline, since they cannot include one of its most important fundamental characteristics. The pursuit of regional geography must, therefore, work toward an actual division of the world into specific regions.

XI. What Kind of a Science is Geography?

A. "what's in a name?"

The question that is more commonly placed at the beginning of methodological treatments we have intentionally reserved for the end. We wished to avoid any attempt to determine the character of geographic science by logical deductions from assumptions. For that form of reasoning can have no greater validity than the basic assumptions, and in this case we could have but little reliance in the assumptions. We therefore proceeded by examining directly the subject whose character we wish to determine—namely the field of geography. We traced the development of the nature of the field, as understood by its significant students, since its beginning as a modern field of study in the late eighteenth century. We examined the logical foundation that such students have given to geography, as the study of the areal differentiation of the earth. In the light of both these considerations, the historical development and the logic of its concept, we studied how geography selects the phenomena that it regards as significant in areal differentiation, and how it may conceive of the areal divisions that it is to study. This survey should now enable us to determine inductively what kind of a science this subject is. It should enable us to answer the questions that geographers have often discussed and which Douglas Johnson [103], and later Colby, have raised more specifically, as to whether geography is, in this or that particular sense, a science, entitled "to take its place in learned circles and in public esteem with other basic sciences" [107, 2].

The reader will observe from the heading of this section, however, that we still persist in begging the question whether geography is a science of any kind. Finch has considered this question directly and, to my mind, effectively [223]. If an issue is made of this point, it commonly resolves itself into a debate over the meaning of a particular word, a word that is not essential to our further thought. In particular we feel justified in ignoring the various undefined concepts of the term "science" upon which many recent methodological discussions have been based. It is a common, but nonetheless naive and erroneous, assumption of many students of the natural sciences that the nature of "science" is sufficiently well known as to require no statement. When they do attempt to state it, their definition commonly excludes the social studies without indicating in what other kind of knowledge these are to be included. But many physicists and chemists define "science" in terms that could not include zoology or geology; I have even heard such a definition stated by an eminent geologist who did not appear to realize what a small part of his field could be included under it.

[366]

Whatever may be gained from the discussion of such a question, there is no need for it here. One would gladly avoid the problem by using some other term possessing less emotional connotation than "science" seems to have acquired for some of our colleagues in other fields, if another suitable word were at hand. In a recent public address, a renowned natural scientist, whose position as university president requires him to consider the problems of the social fields, declared that the "scientific approach" was not applicable in the social studies and regretted that we had no word like the German word *Wissenschaft* to apply to such studies; but he failed to note that German students seem to get along fairly satisfactorily without any word corresponding to our word "science"—that is, other than the word *"Wissenschaft."*

If we were to use the word in a definitive sense, it would be necessary to attempt to determine its meaning specifically. But our need here is simply for a convenient handle to apply to that general form of knowledge that is distinct from either common sense knowledge or from artistic and intuitive knowledge. For that form of knowledge our language provides only the word "science" and we will therefore use it in that sense without, for the moment, wishing to claim that geography is entitled to any of the distinctions granted to "science" as used in some other, undefined sense. At most, one may add, any such claims influence only those who make them; no matter how logically we might demonstrate in theory that geography should be granted a title, the esteem that presumably goes with that title will be granted only in recognition of more solid contributions.

Though we consider, then, any question of titles as of little importance, it is of very great importance for us as geographers to know what sort of a study geography is. We have noted many disagreements among geographers that have been produced by the desire among some to make geography into a certain kind of science—the only kind, perhaps, that they would call by that name. But any efforts that require geography to change its essential character must be in vain; we cannot make over geography in any fundamental way, we can only fulfill that which it has been and is. Ignoring, therefore, any question as to what geography should be, let us consider what kind of a study it is.

B. THE CHARACTER OF GEOGRAPHY AS DETERMINED BY ITS POSITION AMONG THE SCIENCES

If the classification of the sciences were in fact, as is frequently supposed, analogous to that of the species of organic life, we could expect to derive the character of geography in major part from a consideration of the generic character of the order and genus of sciences to which it belonged, and simply add to that the specific differences between geography and other sciences of the same genus. But Hettner reminds us that no branch of science is in reality a separate and distinct science [*161*, 110 ff.]. There is only one science, which human limitations require us to divide more or less arbitrarily. The classification of these parts of science involves, therefore, difficulties similar to those which we found in classifying the areas of the world that are simply parts of a single whole.

Consequently, it represents a distortion of science to attempt to arrange its parts in any simple system of classification, such as that which recognizes the natural and social sciences as quite separate groups within each of which various classes of individual sciences are distinguished. "All knowledge of the inorganic, organic, and human world is one interlaced whole," as Heiderich has emphasized [*153, 212*]. Only the fact that this whole of science is far too much for any one person requires that it be divided into more or less conventional branches, and the necessities of academic organization may require that these be grouped in major orders. This conventional grouping, however, proves in many cases to be anything but convenient. Since geography, in particular, must examine phenomena in the actual complexes in which they are found, it is impossible for it, in practice, to separate natural and human phenomena.

When we consider geography, in this particular aspect, in comparison with the single unity of all science, rather than in comparison with any other particular branch of science, the charge that geography is dualistic because it includes both human and non-human phenomena has no weight. As Penck comments, "a dualism is felt only by a person who sees boundaries rather than zones of contact between the sciences, who emphasizes the differences between the social and the natural sciences more than the interconnection of all sciences, their belonging together in one great unit science. The divisions of that unit science do not lie beside each other like the lands on a map. They stand in manifold relations with each other" [*162*, 41].

Almost all modern geographers are agreed that geography cannot adapt itself to the conventional division between natural and social studies; not only does geography as a whole fit into neither group, but neither can it be divided into two halves, natural and human. It is not the position of geography, however, that is illogical: the separation of things natural from things human is possible only in theory, in reality they are interwoven. Geography, like psychology, is evidence of the arbitrary character of the conventional division of science.

To be sure, there are geographers who assert that they are interested primarily in "the physical aspects of geography," but one will look hard and long to find any of them who do not contribute published studies involving human aspects of the subject. Fortunately, when such students become concerned with a particular area, they quite forget that they have labelled themselves "physical geographers" and proceed to study all features interconnected in the area.

Indeed, it is somewhat misleading to over-emphasize the position of geography as "a bridge between the natural and the social sciences." Though Penck has used this analogy a number of times he would be one of the first to insist that, insofar as there is a gulf between the two groups, the gulf is of man's making, it is not present in the reality that science is to study. We cannot, however, accept his further inference that the concept of scientific laws has been developed only on the one side of this artificial gulf and the bridge of geography is needed to carry it across to the social sciences on the other [158, 54; 163]. A concept of this kind requires no bridges. On the other hand, Penck may mean that scientific laws in the social sciences can be developed on a sound basis only if they are connected, through geography, to the natural sciences. Even in this sense we would be claiming too much, for the social sciences have other connections with the facts and relationships of the non-human world, notably through human physiology and psychology [cf. Kraft, 166, 12].

Whatever conclusions may be drawn with respect to that question, geography is not to be thought of as a connecting link between two groups of sciences, but rather as a continuous field intersecting all the systematic sciences concerned with the world. It therefore has not two, but many facets, as Schlüter observed; the difference in methods between studies of climate and of landforms is in many respects greater than the difference between the study of natural vegetation and of cultivated crops [148, 145 f.].

The most that we can learn about the nature of geography from the conventional classification is that geography necessarily shares in whatever difficulties or limitations the social sciences are heir to, and that, on the other hand, it shares in part in the greater ease with which facts and relationships can be determined if the human element is not involved. Since the developments of the last generation have destroyed the faith in absolutes of the nineteenth century physicists, we know that there is here no difference in kind, but only in degree, between the two groups and among the different sciences in each group. Furthermore, this is a difference which applies only in general, not necessarily in the particular instance. Failure to recognize this fact has led many geographers to presume that geographic work had a major degree of soundness if its feet were established in the natural sciences, regardless of how wildly it might leap from there to conclusions in the uncertain atmosphere of the social sciences. In reality, few facts of the natural environment can be established with such a degree of certainty as the rate of population growth in the United States, or the areas included within the dominions of the political states of the world.

We can secure much more insight into the character of geography if we consider it in terms of the classification which we discussed in the fourth section of this paper. According to Kant, Humboldt, and Hettner, it is necessary to look at science as a whole from different points of view. From one point of view, all reality may be regarded as a collection of many different kinds of phenomena which can be sorted into groups according to the kinds of objects with which they are concerned. The student who approaches science from this point of view endeavors to learn everything he can about the phenomena of one particular group of objects regardless of where and when they may be found. Since it is possible to classify all objects, roughly, as animate and inanimate, of non-human (natural) or human origin, this "systematic" point of view permits a fairly clear subdivision into different "systematic sciences."

In the reality which science is to study, however, the phenomena are not arranged according to the classification which the systematic point of view constructs. Consequently this point of view gives an incomplete view of reality. If phenomena were simply piled and mixed together in reality without meaning, it would perhaps be sufficient simply to state that fact. We know, however, that there are significant relations between the different kinds of phenomena that are found together in any particular section of reality, and also between phenomena in different sections of reality. That is, there is some degree of system or order in the actual arrangement of phenomena in reality. To comprehend reality more fully, therefore, we must not only study phenomena, but must also study the different sections of reality in order to understand the character of each section in comparison with the character of other sections. To understand the character of any section of reality we must attempt to comprehend the integration of phenomena of different kinds that are actually integrated in it.

Although this integration can be stated theoretically in the singular, the nature of reality forces us to take two separate points of view. The whole of reality may be divided into sections in terms of either space or time. Though a single section combines these—here and now is one point in reality —it becomes practically, if not theoretically, impossible to consider simultaneously differences in time and differences in space. Only if the phenomena are relatively simple, as in astronomy, or the data relatively meager, as in paleogeography, have efforts to combine the two met with success (see Sec. VI A). The consideration of sections of reality in terms of time is the historical point of view, represented by historical geology, prehistory, and history in the narrower sense. The consideration of sections of reality in terms of space is the chorological point of view, represented by astronomy and geography.

Every one of these historical[98] and chorological sciences must study all the kinds of phenomena that are found in its particular sections of reality. Theoretically these could include phenomena of all the systematic fields, whether physical, biological, or social. Only special circumstances limit the range within certain of these fields. W. M. Davis recognized this common characteristic of geography, history, and astronomy. "Dealing with things or events of many kinds in definite relations to time or place, they cannot have the singleness of content which subjects like mathematics and physics and chemistry possess." Astronomy, he continues, is essentially the mathematics, physics, and chemistry of the universe, and only the fact that evidence of organic life has not been found in the heavens has prevented the astronomer from overlapping into biology, or even one might add, the social sciences [*104*, 213 f.]. Likewise, it is only the circumstance that natural conditions on the earth have changed but little in historical (not *human*) times that largely limits history—as distinct from "prehistory"—to human phenomena. Nevertheless the eruption of Vesuvius is a phenomenon of concern not only to the geologist but perhaps even more to the historian—as is indicated by the fact that the reader knows at once to which eruption we refer. Likewise anyone studying the history of Holland in the Middle Ages must consider the changes consequent upon the formation of the Zuider Zee.

It may be particularly instructive to glance at that special division of the historical view of science known as historical geology. The innocent layman might suppose that one could study the inanimate rocks of the earth's crust without overlapping into the fields that study the phenomena of life. But since the historical geologist is the only scientist who is presented with material for studying the history of the world in remote times, he finds that he must include historical botany, zoology, and human anatomy, and even to some extent historical social anthropology.

This consideration of the nature of different kinds of science should enable us to meet "the oft-discussed assertion" of which Colby speaks in his

---

[98] The use of the term "historical sciences" to refer to the sciences that study man—whether because they grew out of history or because they find much of their material in history—appears illogical and is misleading. The science which studies the history of the earth is "historical" both in name and in character.

presidential address, namely, "that geography has no distinctive phenomena[99] at the center of its interest, as have, for example, soil science, botany, and chemistry" [107, 2]. The geographer need not hesitate to acknowledge the truth of that assertion, even though it establishes an essential difference in character between his field of study and the systematic sciences like chemistry, botany, or political science. The group of sciences among which geography is thereby classified should not, one would suppose, prove humiliating to the geographer.

Geography does not claim any particular phenomena as distinctly its own, but rather studies all phenomena that are significantly integrated in the areas which it studies, regardless of the fact that those phenomena may be of concern to other students from a different point of view. The astronomer has no monopoly of the study of the stars, he is not disturbed if physicists and chemists study the elements of the stars. Similarly, geography need not look for any concrete objects as its own. The rocks which the historical geologist uses for his data are equally the concern of the dynamic geologist and his pet fossils are proper objects of study for the botanist, zoologist, or anthropologist. Likewise the historian is not disturbed if told that his field is an aggregate of economics, political science, and sociology.

Finally, geography does not distinguish any particular kind of facts as "geographic facts." As Barrows has often insisted, any particular fact— meaning a primary fact, not a relationship loosely considered as fact, nor a deduction from relationships—is not a "chemical fact," a "geological fact," or an "economic fact"; it is simply a fact, and any branch of science may use it. It is only because various kinds of facts are more commonly studied in certain sciences than in others that these conventional, but misleading expressions are in common use. Thus the facts concerning the price of wheat in different places and different times may be considered most frequently in economics, and therefore are called "economic facts," but they could equally well be called "historical facts" or "geographic facts." Geography in particular cannot accept either the popular misconception which classifies under "geographic facts" only the facts of location, or, the misconception common in scientific circles which considers this term as including, in addition to the facts of location, only the facts of natural phenomena. In the broadest sense, just as all facts of past time are historical facts, so all facts of the earth surface are geographical facts. And just as history does not use all facts, but only those—of whatever kind—that are "historically significant," so

---

[99] It may be that Colby uses the word "phenomena" in a different sense from that used here and so only appears to have come to an opposite conclusion. One cannot be sure, since he intentionally does not discuss either his question or the answer which he only suggests.

geography will determine which facts it will utilize, not according to their substance, but according to their geographical significance, *i.e.*, their relation to the areal differentiation of the world [*cf*. Sec. VIII].

To state, for example, that Vesuvius is (and was) a volcano located at 40° 49′ N., 14° 46′ E., is to state a fact which is no more geographic than geologic or historical—it is of course simply a fact. In the systematic geography of volcanoes, we are concerned with this fact in its relation on the one hand to the zone of diastrophic action that runs through the Mediterranean region, and, on the other hand, to the fertile ash soils of the neighboring Campagna, to the ruins of Pompeii and buried Herculaneum, the hazards of life of the population of the area, and the landscape effect of the volcanic mountain in the level plain.

In sum, then, geography, like history, is to be distinguished from other branches of science not in terms of objects or phenomena studied, but rather in terms of fundamental functions. If the fundamental functions of the systematic sciences can be described as the analysis and synthesis of particular kinds of phenomena, that of the chorological and historical sciences might be described as the analysis and synthesis of the actual integration of phenomena in sections of space and time.

Both history and geography might be described as naive sciences, examining reality from a naive point of view, looking at things as they are actually arranged and related, in contrast to the more sophisticated but artificial procedure of the systematic sciences which take phenomena of particular kinds out of their real settings.

It is not surprising, therefore, to find both history and geography developed as fields of study in the earliest period of scientific thought. Furthermore, it was natural enough that each of these should have become a "mother of sciences." The attempt to integrate all kinds of phenomena in space or time leads to the discovery of many kinds of phenomena, any of which may then appear to be worthy of study in themselves; indeed the attempt to understand their significance in a total integration requires that they be studied in themselves. Consequently we may expect this evolutionary process to continue indefinitely, so long as the new kinds of phenomena discovered are deemed worthy of study in their own right. Thus, if geographers have discovered the phenomena of house types and can demonstrate that they are sufficiently significant, we may expect some branch of systematic science to make these objects a subject of special study.

On the other hand it should not be supposed, as has often been done, that the recognition of the independence of daughter fields thereby reduces the extent of the field which geography or history is to study. On the contrary the mother field remains exactly what it was before. Furthermore, as Richthofen observed [*73, 27* f.] and as Hettner has repeatedly emphasized, the progress in these related fields enriches the materials to be studied in geography. Just as the development of economics and political science has greatly increased the ability of historians to interpret history, so modern geography has benefited enormously from the development of systematic physiography, climatology, soil science, etc., and should benefit from the findings of economics and other social sciences. What contribution geography can make in return will be considered later.

The failure to understand that geography is to be defined essentially as a point of view, a method of study—just as all science is a method of study—has caused many to suppose that the growth of the daughter sciences had left nothing for the parent science to do. Attempts have been made to save the day by claiming for geography a particular type of phenomena, such as relationships between man and nature, or by searching for new objects of study which no one else has previously considered worth studying [Crowe, *201, 2*], or by attempting to metamorphose abstract concepts of area into concrete objects. Each of these efforts, to a greater or less extent, has caused geography to depart temporarily from its path of development in directions which have proved, or will prove, to lead either into fields that other sciences will not cede to geography, or into the bog of mystical thinking.

One answer to the question at the head of this section, therefore, is that geography is a study which looks at all of reality found within the earth surface from a particular point of view, namely that of areal differentiation. This might be called the *position* of geography as a field of knowledge. More significant to the general question is the *character* of geography as a field in which knowledge is acquired.

C. THE CHARACTER OF GEOGRAPHY IN RELATION TO THE GENERAL NATURE OF SCIENCE

To understand the character of geography as a field in which knowledge is acquired, it is necessary to understand the essential character of the whole field of knowledge of which it is a part. We are concerned here, not with all knowledge, but with that sort of knowledge—by whatever name one chooses to call it—that is distinguished from either commonsense knowledge or from artistic perception "by the rigour with which it subordinates all other considerations to the pursuit of the ideals of certainty, exactness, universality and system" [Cohen, *115, 83*; the discussion of these principles in the following pages is based largely on Cohen, pages 83–114; see also Barry, *114, 3–88*].

Geography attempts to acquire knowledge of the world in which we live, both facts and relationships, which shall be as objective and accurate as possible. It seeks to present that knowledge in the form of concepts, relationships, and principles that shall, as far as possible, apply to all parts of the world. Finally, it seeks to organize the dependable knowledge so obtained in logical systems, reduced by mutual connections into as small a number of independent systems as possible [cf. 115, 106-14]. It is in terms of the manner in which geography pursues these ideals that we will attempt to describe its character as a field of study.

It should be noted, in general, that our definition of the kind of knowledge of which geography is a part is based not on what is known, that is, on what has been learned, but rather on the pursuit of knowledge—i.e., the fundamental principles governing the manner in which the unknown is to be learned. Different branches of this "kind of knowledge" differ in the degree to which they have been able to approach the ideals stated; no branch of science can claim to have attained perfect certainty or exactness, actual universality, or complete organization of all its knowledge in a single system.

If one compares these ideals, as the fundamental requirements for that form of knowledge, which hereafter for convenience we will call "science," with the ideals of artistic perception (in whatever form one finds them stated by students of art), it is clear that there can be no logical combination of the two and no transition from one to the other. The artist certainly does not subordinate all other considerations to an ideal of exactness, nor to one of certainty. He may require that his work express a fundamental universality but that concept does not control the details of his work. Likewise he may require that an individual work of art should be organized, but does not seek a common organization of all similar works of art. In other words, we are dealing here with two essentially different approaches; both artists and geographers may attempt to acquire and present knowledge of an area of the earth, but neither can adopt the ideals of the other without sacrificing his own [cf. Kraft, 166, 20 f.].

That geography seeks to make its knowledge of the world as accurate and certain as possible is an assumption which presumably would not be questioned by professional geographers. It is not, however, a correct corollary of this assumption that geography should arbitrarily limit itself to particular kinds of facts because they appear to be subject to more accurate and certain means of measurement than others. Even if it were possible to demonstrate —as would not be the case—that the facts and relationships of immaterial phenomena could never be determined as accurately or certainly as those of either visible or material phenomena, no principle of science would require us to limit ourselves to those phenomena that could be studied somewhat more accurately and certainly, excluding those that could be studied somewhat less accurately and certainly.   On the contrary, in order that our knowledge of, say, the cultural character of an area may be made as certain as possible, we are required to consider all the facts that bear upon that knowledge, whether the material products of cultural ideas or the immaterial manifestations of those ideas.   Both sets of facts must, of course, be observed as accurately and certainly as possible [115, 83–99].

Likewise these ideals admit of no limitation in the specific methods of observation in geography, but rather require that we utilize all methods that will lead to more accurate and certain knowledge.   Thus, because statistical data are neither sufficiently accurate nor sufficiently detailed, we must employ the technique of field work—both direct observation and personal interview [cf. Jones, 287]—to check and supplement the knowledge gained from statistical data.   On the other hand, even if it were physically possible to make a complete field observation of every part of a region, the observations obtained reflect temporal conditions, commonly of a single season, which, as R. E. Dodge noted, may lead to erroneous generalizations concerning the continuous use of the land [287, 110].   Consequently the findings of field observations must be checked with statistical data, even though these were very indirectly based on observation.

The scientific ideal of certainty commands that the terms and concepts of description and relationships be made both as specific and as certain as possible—we cannot develop a sound structure on a marsh foundation in which ambiguous concepts shift their meaning whenever pressure is applied to them, and concepts apparently specific prove to be but dubious analogies.

The ideals of accuracy and certainty apply not only to the manner in which primary facts are established and to the formulation of fundamental concepts and technical terms, but also to the processes of mathematical and logical reasoning by means of which we induce relationships of observed facts, and thereby deduce further conclusions as to facts.   When a single student studies a particular scientific problem, no matter how hard he strives to live up to these ideals, no matter how critically he examines his own work, there remains the possibility of error, whether through carelessness or through subjective influences affecting his observations and reasonings.   Since this is a generally accepted axiom in those sciences whose facts can be measured with the highest degree of accuracy and whose relatively simple phenomena make logical reasoning most certain, how much more uncertain are the findings of one student in geography !   In order to make possible a higher degree of accuracy and certainty, therefore, it is a recognized principle of all science that studies should be carried on, organized, and presented in such a manner as to provide an accumulation of evidence of different students on the same problem. Every scientific study should, that is, be scholarly, by which we mean both that it should make use of all previous scholarly studies bearing on the problem and that it should be presented in a form usable by subsequent students.

It is one of the handicaps of the social sciences, and in part also of geography, that in much of their work it is not possible for later students to have access to the primary data used in any study; much must be taken on faith in the professional ability and reliability of the individual student. Consequently it is all the more essential in these fields that similar studies of other students should be utilized and referred to, not as a matter of professional courtesy, but for the sake of accumulating evidence in order that results may be more certain.

Science will accept on faith from any scientist no more than is absolutely necessary. In order that subsequent students may reexamine the findings of a particular study—whether in order to utilize them further or simply to check or correct them—it is essential that there be a clear understanding of the generic concepts employed, of the methods of observation and reasoning by which the results were obtained, and, finally, of the relation of the study to the field as a whole. Obviously such a clear understanding will be easiest if the field has developed standard techniques and organization; if it has not, they must be specifically indicated in each study.

It is not the function of this paper to pass judgment on the scientific quality of work actually produced by geographers—it is their ideas about the nature of geography, and the consequences of those ideas that concern us. On the latter basis, however, it is a fair question to raise, whether the research publications in geography indicate that either geographers in general, or the editors who, to some extent, control their publications, have accepted these standards as essential to geography—standards dictated by the scientific ideals of accuracy and certainty.

With respect to these ideals as essential principles in the pursuit of knowledge, there can be no differences among the different branches of science, but only differences in degree of attainment. Undoubtedly the attainments of geography in this respect are not of such a high degree as to tempt us to compare it with other sciences in such terms. But neither need we feel humiliated if those who choose to make such comparisons should assign our field to the "lower classes." In any competition the understanding observer will judge attainments not only in terms of degree of success but also in terms of the relative difficulties of the tasks undertaken.

The consideration of the manner in which geography may pursue the other two scientific ideals, universality and system, is of the greatest importance in understanding the character of geography as a field of study. Since these are not so readily understood as the two ideals which we have just considered, each will require detailed consideration.

D. GENERIC CONCEPTS AND PRINCIPLES IN GEOGRAPHY

If Galileo's most famous experiment had merely demonstrated that when he, Galileo, dropped two specific objects of different weight from the Leaning Tower of Pisa they fell together at the same rate of speed, that fact would have found but a small place in scientific knowledge. Its great importance, of course, was that subsequent experiments showed that he had illustrated a universal, a relation that was true regardless of where, when, or by whom the weights were dropped. Few will question that it is an essential function of science to seek for such universals. On the other hand, if later experiments had shown that the same results were not obtained elsewhere or at other times, the fact that they did take place on that one occasion, if substantiated as a fact, even though never explained, would represent a bit of scientific knowledge. "The business of science," as Barry puts it, "is to learn as much as it can" [*114, 122*]. Though it seeks for universals, it cannot ignore certain and accurate knowledge which it is unable to express— which perhaps never can be expressed—in terms of universals. One might say that it is an axiomatic ideal of science to attain complete knowledge of reality—expressed as completely as possible in terms of universals, but in any case expressed in some way.

While everyone recognizes the importance of universals in science, it is a common error to overlook that part of our scientific knowledge which cannot, as yet at least, be expressed in universals. Many have assumed that science was concerned exclusively in the development of laws and principles. This concept, according to Hettner, represents an outgrowth of the great development of laws and principles in astronomy, physics, and chemistry in the last century [*161, 221–4*]. Indeed, it came to be assumed that physics and chemistry were exclusively abstract sciences, concerned only with laws and principles. More recently, however, many scientists and philosophers of science have recognized that no branch of science concerned with reality— as distinct from theoretical mathematics—can limit itself to laws and principles. Though science strives for universals, these do not exhaust the study of reality, there is always an individual remainder that is not described or explained. If this be ignored our knowledge is less complete than it might be, a *quod est absurdum* which no science can accept. "Research," as

Lehmann says, "is a seeking for the most complete presentation possible" [*254, 299*].

The contrast between that aspect of science which can be expressed in universals and that which is concerned with the individual object or phenomenon as worthy of study in itself, received the particular attention, at the turn of the century, of two German philosophers, Windelband and Rickert [see discussions by Hettner, *111*, 254–9, or *161*, 221–4, and by Graf, *156*, and the discussion between them]. They appear to have classified the different branches of science on this basis, distinguishing between "nomothetic" or law-making sciences, and "idiographic" sciences, those concerned with the *einmalige,* the unique. In a rough way this seemed to conform to the * conventional division between the natural sciences and the social sciences.

It seems clear, however, that these two aspects of scientific knowledge are present in all branches of science [*cf.* Schlüter, *131*, 510 f.]. The common idea that generalizations and laws themselves are the purpose of science is characterized by Hettner as an extraordinary adherence to medieval scholastic realism. On the contrary, they are "merely the means to the ultimate purpose, which is the knowledge of actual reality, the individual facts, either conditions or events" [*cf.* also Kraft, *166*, 11–13]. The astronomer devel- ** ops laws of celestial mechanics not in order to prove that the universe is governed by law—which is a philosophical rather than a scientific thesis— but in order that he can rightly understand the motions of the heavenly bodies. He does not forget his interest in the latter as individuals. The students who have mapped the surface of the visible side of the moon or who study the rings of Saturn are no less astronomers if they have not thereby been able to establish laws.

Nevertheless it is clear that in some sciences the nature of the phenomena studied permits of a much greater development of nomothetic knowledge, whereas in others the greater differences between the phenomena studied, and also their greater importance individually to man, requires the students in those fields—whether they will or no—to concern themselves in large degree with the unique. This is, then, a significant respect in which there are important and permanent differences of degree among the different branches of science.

A comparison of various branches of science from this point of view may aid us in understanding the character of geography, may help us to perceive what kind of science, in this particular respect, geography is.

Although we have noted that all branches of science are concerned to some extent with the study of the unique as well as of universals, there are some sciences in which the actual work of many students may be largely, if

[379]

not completely, absorbed in the search for universals, the study of the unique being left to other students or to other divisions of science. I presume that this may be the case in physics and chemistry, particularly in their theoretical aspects. To a lesser extent the same is true of the biological sciences and psychology, and also of certain branches of economics and, perhaps, sociology. (It is both impolitic and dangerous to discuss the nature of a field in which one has not been trained. I might say, however, that the statements in this and the following paragraphs are based on discussions with colleagues in the appropriate fields.)

On the other hand, there are certain branches of most of the sciences in which the students are very largely concerned with the study of the individual object, the unique. We have already noted the importance of this form of study in astronomy. Idiographic studies are certainly of great, if not major, importance in geology. This is obviously the case in paleontology. It is true of mineralogy in so far as the student concentrates on certain mineral deposits for the purpose of securing complete knowledge of those deposits rather than of developing new principles of mineral deposition. Even more does the statement apply to the great and certainly valuable work of the geological survey in mapping the areal geology of particular areas (though this might logically be considered a branch of geography carried on by geologists).

Among the social sciences, economics has developed its nomothetic knowledge possibly to a greater extent than some of the natural sciences listed above. In political science, probably only those branches commonly referred to as "political theory" are primarily nomothetic. Most of the students of comparative government, for example, are concerned not merely with general principles of government, but—perhaps even more—are concerned to know and understand the differences between individual governments. In their persons, these students combine the nomothetic and idiographic interests, whereas in physics or biology or perhaps economics, these may be separated between different students. Since the conditions which have resulted in this situation in political science are very similar to conditions in geography it may be instructive to analyze them.

The first condition is that for most of the phenomena of political science a generic description provides only a bare outline of common features, the individual specimens differing notably with respect to other features. The phenomena are much more complex in structure than those of physics, or even of economics, and, though less complex than the specimens of organic life, do not have the high degree of similarity that results from the common origin of specimens, species, genera, and orders in the biological sciences.

Even though it be true that, to the research physician—in contrast perhaps with the student of human physiology—"every human body is different and unique, every spine to some extent different from every other one," these differences are very much smaller than those found among, say, the individual governments of the same type.   In other words, a generic description of human spines gives a much more nearly complete picture of any one spine than is possible in a similar approach to governments.   If the political scientist is to learn as much about governments as the physiologist learns about human spines he will have to give a much greater share of his attention to the unique characteristics of each specimen.

In addition, more attention will be paid to an individual government, because of its greater importance, in comparison with an individual human spine.   This second condition holds generally: the phenomena of political science are, as single cases, so important (to man) that it is necessary that each be understood as completely as possible.   Whatever the personal preferences of a student may be, mankind will not be satisfied with a science that studies dictatorships, for example, only to the extent that generic concepts and social laws can be developed, and declines to consider the differences between the present government of Italy and that of Germany on the grounds that these are unique cases.

A third consideration is of practical import.   With millions of human spines in the world, a single scientist, even a whole group of scientists, may be kept fully occupied in considering only those aspects of the spine that are relatively universal, and may be able to derive sufficient universal knowledge thereby to justify his self-limitation, leaving to others the consideration of individual characteristics.   But the political scientist cannot find, even in the great storehouse of history, more than a relatively small number of dictatorships of all kinds to study, hardly more than a handful of cases of a particular type of dictatorship.   Under these conditions, to limit himself to the study of generic features and the development of principles, and to decline to study individual specimens as unique objects would be absurd.   The reverse consideration is perhaps even more significant: human physiology cannot find time to study the individual characteristics of all human spines, whereas individual studies of all the governments in the world would not be impossible for the field of political science.

It is obvious that the same considerations apply in large part, and with even greater force, in history.   To be sure, an individual historian, or particular group of historians, might limit themselves to considering the history of constitutions, of wars, or even of battles, and from these studies derive laws or principles of value to the general historian interested in the study of periods of history and the sequence of events. Whether such historians would not thereby become students of military science or of constitutional government is a question which is not for us to consider; certainly such work represents but a small part of history.

If, now, we consider our own field there can be little question that geog- * raphy is very much concerned with the study of individual phenomena. Indeed some critics might feel that geographers tend to lose themselves entirely in the consideration of the unique, regardless of its importance. If we were to study each and every feature of the earth surface that could be shown to have geographic significance, without considering their relative importance, it is clear that geography would become hopelessly lost in an insurmountable mass of detail. Not only every small district might be studied, but every single village, every mountain, every little rill on the face of the earth.

On the other hand, a geographic description that is confined to generic terms is inadequate. Even a brief and superficial description of a major area must name and describe a great number of individual features. The Alps and the Himalayas, Mount Rainier or Vesuvius, the Amazon, the Mississippi, or the Niagara, states like Germany or the United States, cities like London, Paris or New York—if included within the area described— these will be considered not merely as examples of types or as illustrations of principles, but for their own individual significance [cf. Hettner, 161, 221–4].

One reason for this, as we have already seen, is that any generic description, however detailed, is completely inadequate to depict these individual features. When a biologist has described the nucleus of a certain type of cell he has provided a fairly complete picture of the nucleus of any individual cell of that type, whereas a full knowledge of the character of the commercial core of Chicago would give one but little understanding of the corresponding district in Minneapolis. It is only a very general similarity which justifies the citizens of the latter city in calling their commercial core, in imitation of the larger city, "the loop"; actually the street car lines do not loop in Minneapolis, as do the street and elevated car lines in Chicago.

If the previous example appears to depend on details of secondary importance, consider such geographic features as the Great Lakes System or New York City. There are, to be sure, a number of great lakes in Africa, but, as a system of connected fresh-water seas, the Great Lakes of North America have no counterpart on earth—or, so far as we know, in the universe. When one considers New York in its fundamental aspects as the

[382]

* See Supplementary Note 47

focal point of trade of the greater part of a continent, and then looks for its counterpart simply in this one respect. ignoring entirely its local character, one finds no other city of the type; Shanghai, Buenos Aires, and Hamburg are all centers of trade for extensive areas but no one of them functions for its particular continent as New York functions for North America.

The informed reader will recognize that while these are conspicuous cases, they represent a rule that applies to most geographic features, even to those of slight importance. Because these features are not adequately described in terms of types, geographers, as well as laymen, fall back on analogies. Possibly these have their value; but if taken too seriously they are deceptive. To call Hankow, "the Chicago of China," or the Pittsburgh-Cleveland area, "the Ruhr of America," is to give a false as well as true impression.

Our examples have likewise illustrated the other consideration which requires geography to examine many phenomena as individuals—namely, their great importance as individual phenomena. By this is meant not merely their utilitarian importance in the world, but also their importance to the earth surface on a purely academic basis. Even if Niagara Falls had no material value, whether actually or potentially, a geographic science would not be satisfied with a description which simply classified it as of a certain type of falls along with hundreds of others. In itself, in all its individual characteristics, it is a phenomenon of scientific concern to man.

The concern for the unique in geography is not limited to phenomena but applies also to the relationships between phenomena. Thus, if the relationship of Winnepeg to the lakes and barren land on the north and the international boundary on the south is essentially unique, we do not disregard it on that account.

It might appear as though geography were as largely limited to the study of the unique as is history, and, until the last century, that was no doubt the case. As Hettner notes, "geography was confined largely to such idiographic description . . . general concepts were to be found only in the crude form expressed by such common typewords as mountain, valley, city, etc." [*161, 222* f.]. Insofar as the situation was unavoidable, such a statement would represent no criticism of geography as a science, any more than it does of history. Barry's statement may also be expressed in negative form: it is not the business of science to learn what cannot be learned.

On the other hand, a geography which was content with studying only the individual characteristics of its phenomena and their relationships and did not utilize every opportunity to develop generic concepts and universal principles would be failing in one of the main standards of science. We can therefore agree with Hettner that "the greatest scientific advance of geography has been the fact that, by taking over and developing the results of the systematic sciences—earlier in one branch of the field, later in others— geography has gone over to general observations depending on generic concepts" [see also Sauer's quotations from P. Barth, *211, 27*].

Undoubtedly the development of generic concepts has progressed much further in respect to the natural phenomena of concern in geography than in respect to the cultural phenomena, partly because of the close relations which geography has had in most countries with geology and other natural sciences, and also because of the greater development during the past century of the systematic sciences concerned with natural phenomena as compared with those concerned with cultural phenomena. Nevertheless, examples of generic concepts could be cited from every section of systematic geography.

Thanks to such generic concepts it is possible to express many characteristics of a particular feature in one word or phrase—or in a symbol, such as **Cfb**—so that one can give a relatively brief description which can easily be kept in mind. Furthermore, the use of such generic concepts has made it possible to develop principles of relationships between different factors which are found repeated in different parts of the world. These principles can then be used in studying the integration of phenomena in specific areas.

While general principles have been most fully developed in connection with the relationships between various natural features, progress has also been made toward general principles concerning the relations between any particular cultural item, such as a crop, a particular kind of factory, or a city, and all the other features that are significantly related to it. Such principles have been developed in every phase of economic geography, and, to a lesser extent, in other cultural aspects of geography.

To some degree, therefore, geography may be called a generalizing or nomothetic science. Both Hettner and Penck find that in this respect geography has a great advantage over history [*161, 223; 158*], though it would be an error to overlook the use in history of generic concepts and principles contributed by the systematic social sciences.

Few geographers are satisfied, however, with the present state of development of principles in geography—either in terms of their quantity or of their reliability. Insofar as more principles or more reliable principles are possible, no one, of course, should be satisfied. Professional geographers, however, may feel less discouraged if they perceive clearly the difficulties and positive limitations that are placed in their way in the effort to develop generalizations and principles. More than that, the clear recognition of those facts may prevent misdirected efforts that can only yield fallacious principles.

James seeks the explanation for our difficulty in the fact that the phenomena which we observe are so much larger than the observer that we cannot see the woods for the trees, or, to use his excellent simile "like a microbe crawling over the face of a newspaper photograph, we see only the details of the printed dots—the larger design of the photograph lies beyond our range of vision" [*286*, 84 f.]. Granted the force of this difficulty, have not geographers long since discovered the solution which he recommends, namely the use of the map to bring the larger relationships within the range of our vision? And yet we have not advanced very far in the establishment of principles.

One major difficulty lies in the fact that the integration of phenomena which we must study in areas is an integration of a large number of independent, or semi-independent factors. Consequently we seldom have to do with simple relationships—*e.g.,* rainfall to soil, temperature to crops, etc. Theoretically we might follow the logic of the systematic sciences, by assuming that all other conditions remain the same, but we have only the laboratory of reality in which to study these features, and in that laboratory the other elements do not remain the same, except perhaps in a very small number of cases, and we have no way of making them remain the same. Indeed, even if we knew the theoretical principles governing the relation of each individual factor to the total resultant, in the case of such complex resultants as cultural features, a principle which attempted to state the sum total of all the relationships, each in its proper proportion, would be far too complicated for us to be able to use. This is a general difficulty that applies not only to all the more complicated aspects of the social sciences, but also to many phenomena in the natural sciences. Even if one knew all the principles and had all the data, the solution would be involved in a mathematical equation so complicated that no finite mind could solve it. Even when geography has attained a maturity of techniques and methods, it will, as Colby suggests, hesitate to predict resultants of such a complexity of factors [*107*, 35 f.].

The second major difficulty, definitely limiting the applicability of such rough principles as may be developed in geography, results from the insecure foundation on which the principles rest, namely the generic concepts. This limitation is present, to be sure, in every branch of science, in greater or less degree. In those sciences that are able to perform controlled experiments, specimens may be selected that conform almost exactly to the generic type involved in the principle; to the extent to which actual specimens in reality differ from the generic type, the principle will not apply. In geography we seldom have a sufficient number of specimens of any generic type to permit us to discriminate in our selections. Further, as we have seen, our specimens do not conform closely to a generic type, but only within very wide limits. This wide divergence of the individual within the type is illustrated by the marked lack of agreement among geographers, not on the classification of a minority of doubtful specimens, but in regard to the very types, and systems of types, themselves. Since sound principles can only be developed if we have a sound system of generic concepts, we must consider for a moment how such concepts are developed in geography. *

We indicated earlier that many, if not most, of the generic concepts in use in geography had been brought in from other branches of knowledge. One group, perhaps the larger, are those whose expression in words of common speech reflects the fact that man studied geography before he ever heard of such a thing as science. While the concepts of common sense are likely to be crude, many of these have been found to be adaptable to scientific thought. Ocean, lake, river, mountain, plain, city, port, farm, crops—for all of these the ordinary meaning, as used in common speech, can be sharply defined for scientific use. To invent, in their place, technical terms with which to impress the outsider would be pretentious [cf. Schmidt, 7, 192]. On the other hand, many similar terms in the language are used in such different ways that it may not be possible to pin them down to definite technical meanings.

A second large group of generic concepts has been brought into geography from the systematic sciences, notably from geology. Though many of these have also proved useful, it is erroneous to presume that their eminent scientific authority assures their logical utility in geography. The common rule of scientific work that requires the students in one field to take cognizance of the findings in neighboring fields and, unless they can show errors, to assume that the facts and relationships established in the other branch of science are correct, does not apply to a classification system, which cannot be called either correct nor false, but simply more or less useful. If, for example, geologists have classified landforms into generic types based on their past history rather than on their present forms, their concept of a "plateau" need not necessarily be accepted by the geographer simply because it is a "scientific concept," in contrast, say, to the time-honored concept of the same word in common speech, and as used by geographers long before the geologists gave it a new and different meaning.

Finally, geographers themselves have developed various systems of types into which they may classify the features which they study. They have been particularly concerned to supplement generic concepts based on individual elements with generic concepts based on element-complexes. These new

* See Supplementary Note 48

systems of classification often appear in opposition to those taken over from common thought or from the neighboring systematic sciences, so that there has been a great deal of argument as to which generic concepts are proper for the science of geography to use. This is not, as many have thought, simply a matter of preference either for the "new" or for the "old." The question is more fundamental: from the particular point of view of geography, which concepts are of greatest value? Since this has caused much argument, and is clearly of great practical importance in the development of the field—in contrast perhaps with some of the questions that have required our consideration—it merits a brief analysis.

The fact that it is impossible to include all the characteristics of any object in a single generic concept means that any particular object may be classified simultaneously in many different types according to the bases of classification. But these types may be sufficiently alike so that, in view of the limitations of our vocabulary, we use the same terms for different concepts. Take the very simple case of coal. The paleobotanist, I presume, will classify coal as a vegetable product, to the stratigrapher it is simply a kind of rock, and the student of mining will say it is a mineral. Since coal is, in fact, simultaneously all of these things, it is included in each of these concepts though no one of them of course completely expresses what coal is.

Consequently the classification of objects into generic types cannot be determined simply on the naive basis of "the inherent characteristics of the objects themselves." In terms of all their characteristics, no classification is possible. Which characteristics will be considered in developing generic concepts will therefore depend on the purpose for which the generic concepts are developed. Generic concepts are not an end in themselves but merely a scientific tool whose particular purpose varies according to the point of view of the different branches of science.

In general, the purpose of developing generic concepts is to provide a single statement of a collection of common characteristics shared by objects which otherwise differ. This is useful for the primary purpose of describing reality in that it provides a shorthand method of description which likewise enables us more easily to grasp and retain descriptions. Furthermore such precise shorthand collectives are extremely useful, if not practically indispensable, in developing principles of relationships between objects. In both cases it is clear that the generic concepts should express those characteristics of objects which are of greatest importance according to the particular point of view under which they are being studied. For example, a student in the science of heat engineering may ignore, in his classification of coals, various characteristics that have no effect on the burning qualities of coal, however essential these characteristics may be to the paleobotanist.

We have, therefore, the fundamental basis for judging different systems of generic concepts in geography: which classification most completely and clearly expresses those characteristics that are of most importance from the point of view of geography, whether for geographic description or as a basis of establishing principles of geographic relationships.

To the extent that the point of view in geography is different from that of other branches of science, whether systematic or historical, there is no presumption in favor of generic concepts introduced into geography from other sciences; on the contrary it might be safer to presume the opposite. On the other hand, the generic concepts which we have found in common speech are usually developed from the same naive point of view toward reality that geography also expresses. To the geographer, as well as to the sailor, the whale is a fish or—if this be regarded as a misuse of a word—the whale is to be included under the general concept of ocean animals rather than land animals.

Many of the concepts which geography has inherited from the geologic physiography of the past century were developed by students whose primary concern was in the processes of formation of landforms, presumably as a part of earth history. It was, therefore, appropriate that they should attempt, however difficult it may be, to classify landforms in terms of those characteristics which resulted from, and therefore were the key to, their genesis.

If the geographer were likewise directly and primarily concerned with the genesis of landforms, he would, without hesitation, accept the same generic concepts; but on this basis geography would be difficult to distinguish from dynamic physiography. Conceivably, however, the geographer might require a genetic classification as the best means for expressing to the fullest the character of the landform. This, if I understand it correctly, is the line of reasoning followed by Hettner. The geographer seeks to establish types of landforms or climate "according to the entirety of their characteristics" [*161, 222* f.]. This, we found, cannot be done directly. Conceivably it might be produced by a genetic classification (though that would appear as a first step rather than a final step, as Hettner suggests). But is this possible? Let us assume different landforms, each of which represents a resultant form developed by separate processes of folding, elevation, vulcanism, and erosion. A system of classification could be constructed in which these various processes determine the major and minor subdivisions, but again we would have the unanswerable question of which processes are major, which are minor.

Penck discussed this difficulty, from the point of view of geomorphology itself, as early as 1906. He observed that there had appeared to be the finest harmony between structure and surface form, so that geomorphology seemed to have won a sure base for its concepts in tectonics. Thus "in place of the original geographic concept of a mountain area as a sum of unevennesses, many had adopted a tectonic concept according to which a mountain area was a strip of strongly folded land." But more careful morphological investigations had shown that the relations between tectonics and geomorphology were not so intimate as one had at first supposed, and one must conclude that the form of the land which the geographer is to study and explain is something different from the structure of the land which the geologist studies [128, 15 f., 35 f.]. More recently, Penck has illustrated the same conclusion by comparing the Bohemian massif and the Central massif of France. Both, he says, are products of bursting (*zerborsten*); they have great similarity in composition and full relationship in their later geological development, and yet geographically are as different as convex and concave forms [249, 6].

A complete genetic description, to be sure, would arrive ultimately at a complete description of form, but a genetic classification allowing for all the processes that may in any area be significant—in all possible arrangements in time—would not be a classification of types. It would only be an indirect way of describing an end result that could more briefly and accurately be described by direct consideration of the present characteristics of the landform.

If, then, any classification limits us to some characteristics, forces us to ignore others, which are the characteristics that are of most concern to the geographer seeking to study an area in terms of the integration of its phenomena? Surely those characteristics that are significantly related to other phenomena of the present earth surface. In other words, he is more concerned with the form or physiognomy of the landform as a functioning factor in the total complex of areal phenomena than as an end product of its own genetic causes.

No doubt there are cases where the characteristics of primary concern are the same from both points of view, volcanoes or canyons, for example. In such cases, both groups of students will use the same generic concepts. Where this coincidence of interest is lacking, geographers are under no compulsion to follow the direction of the genetic physiographer, or the geologist. In German geography the struggle for independence in scientific approach to the study of landforms took shape in opposition to the methods of W. M. Davis. According to both Ule and Bürger, the distinctly geographic point of view was most effectively presented by Passarge, as early as 1912, and, throughout his work since, he has emphasized the description of present landforms as functioning factors in the region (or landscape) [170, 497 f., 11,

77]. In passing, one may add a note of defence for Davis—if defence be needed. Lecturing at the Sorbonne, according to the testimony of Lehmann, Davis stated that "if it proves very difficult to work out the development of the landform, so that, in the problems of the past one loses sight of the landscape, the geographer will do better to describe the land simply with the help of the older orographic presentation" [113, 237].

No doubt most American geographers today would accept Sauer's conclusion that "there is no necessary relation between the mode of origin of a relief form and its functional significance" [211, 37, see also Bowman, 106, 141 f.]. Frequently, however, they fail to draw the logical corollary that type concepts based on genesis may have little significance for function. They forget, for example, that to tell us that an area is covered with "Wisconsin moraine" gives us no precise description of either its surface form or its parent rock content. Geographers have freed themselves from the genetic classification, Sauer notes, most fully in the study of climates [211, 33]. Possibly the delay in establishing a corresponding freedom in the study of landforms may be due to the fact that most geographers, in this country as well as in Germany, have been trained particularly in geology and geomorphology. (Note Fröbel's pertinent statement of 1836, quoted in footnote 36, Sec. III A.)

It should not need to be stated (but since the writer has been misunderstood in oral discussion of this point, he may be permitted to clarify his standpoint) that the suggestion that geographers need to free themselves from a geological point of view is in no sense an attack on geologists. One may feel that the latter have abused a good geographic term in calling a "plateau" what they mean to describe as a "former plateau," but there are too many difficulties in terminology to justify making an issue on that. The concepts which the geologist has developed are not subject to attack by the geographer; presumably they are entirely suitable for the purposes of studying reality from the geological point of view. All that is said here is merely that there is no reason to presume that the concepts of the geologist are suitable—and many reasons for believing the opposite—for the purposes of studying reality from the geographical point of view.

To establish sound principles of the relation between, say, cultivation or soil wash on the one hand, and landforms on the other, it is necessary to have generic concepts which express measured characteristics of landforms that are significant to cultivation and soil wash, rather than characteristics expressed by types of origin. In this case, the problem is simple at the first step, but very complicated beyond that. The essential characteristic of the landform, from the point of view just stated, is its slope, but its slope is a surface in solid geometry which varies constantly and irregularly at every point and in every direction at every point. Recent developments in the measurements of slopes by intervals, however, indicate that this problem does permit of partial solution.

In brief, in the naive point of view toward reality which geography shares with the common man, phenomena are significant in terms of their relations to other present phenomena of geographic significance rather than in terms of their origins. Generic concepts and systems of classification will therefore be more useful in geography if based on the functional rather than the genetic aspects of phenomena.

In spite of the difficulties involved, geography has been able to develop principles of the relations between the variable elements of area. To a lesser extent, as yet, have the relations between the variations in element-complexes been put into general principles. This should also be possible, even if more difficult. In general the study of element-complexes is still largely concerned with the first step, the development of types; but even in this stage—as represented by the two systems of classification of rural areas discussed in the previous section—generic relationships are at least suggested. When we consider, however, not the elements, nor the element-complexes of areas, but the areas themselves, can we hope to develop general principles? This question lies at the very heart of geography.

It is axiomatic that a necessary condition for the development of general principles is the construction of generic concepts or types. It is presumably on this account that Passarge's suggestion of a systematic classification of areas into types has made such an impression on many students: it appears to offer the possibility of developing universals in the very core of geography, thus raising the "scientific" quality of our work.

Our approach to this question will be more ingenuous if we remember that, while it is essential to science to *seek* for universals, it is in no way essential that they be found; that, on the contrary, important parts of scientific knowledge—in every branch of science—cannot be expressed in universals. By placing a special value on the endproduct, scientific principles, we are in danger of passing lightly over the fundamental bases for them. We will not be able to establish principles by first calling what are not objects, objects, and then classifying them as generic types without considering carefully whether they are generic types. We may deceive ourselves with schematic outlines of classification and with schematic explanations based on vaguely expressed principles, but the inapplicability of these principles to the reality of the world will show our substructure to be counterfeit. As

Kroeber observed—in discussing a different branch of science—"all schematic explanations seem essentially a symptom of a discipline's immaturity" [*116*, 542].

We need not repeat the reasons why any efforts to establish a world classification of areas in terms of the totality of their characteristics is doomed to failure. All the areas of the world can be classified generically in accordance with any particular element-complex, but the sum total of element-complexes varying in different areas cannot be classified in a single system 'of generic types (Sec. X, G). For our present purposes, however, it is not necessary to classify all areas; if we could find some areas that could, as wholes, be classified in generic types, we might hope to go on to generic principles.

No matter how small the number of areas we consider, however, the essential difficulty remains the same. The integration of phenomena which we must study in any area is not a complete unit integration, but rather consists of a set of somewhat related but somewhat separate element-complexes, some of which are but parts of integrations extending into other areas—in the full sense, extending over the whole world.

If we may borrow James's simile of the microbe studying the newspaper photograph, we may say—with some exaggeration—that in the geography of any area, it is as though several separate photographs, each with its own design and each of different size and shape, had been superimposed in printing, each cut arbitrarily to fit. The geographer, in the position of the microbe, may be able to reconstruct the pattern of each separate design and classify it generically, but how can he classify the sum total of the separate designs in the particular superposition in which they are found? If the same combination of element-complexes, overlapping in the same way, occurred in more than one area, one could establish a generic type of this very complex form. While it is conceivable that such repetition may occur—just as it is conceivable that a hand of bridge may be repeated—it would be necessary to find numerous cases of repetition to provide us with generic concepts of regions from which we could develop principles.

We are in danger of confusing ourselves by using the word "region" as a more convenient term than "area of a certain type." When we say that the Po Plain and the Middle Danube Plain are "agricultural regions" of the same type as the American Corn Belt, we only mean that these are areas within which we find approximately the same agricultural element-complex. Even if we found *exactly* the same agricultural element-complex, the three areas could not be called specimen areas of the same species; there is, and can be, only one "Corn Belt" in the world. Any element-complex of an area appears in other areas, but the combination of all of them in the actual mixture as actually found, occurs but once on the earth [after Banse, *246*, 41, and Hettner, *161*, 293].

The reader may feel that, while this conclusion holds for regions of such large size as those just considered—since the world can include but a relatively small number of such large areas—if we consider smaller regions, localities, perhaps, we should be able to establish types. A colleague who has been good enough to read this paper in manuscript suggests that he could select three localities, one in Mongolia, one in Patagonia, and one in the Great Plains, which are so similar that they could be compared, "if not to peas in a pod, at least to peas in pods in different gardens." Undoubtedly we have in this case very similar combinations of many of the same elements and element-complexes, but can we say that these common features are, in each case, the most important features of the area? Almost all systems of land-use types, including that developed by the colleague who made this suggestion, classify these areas in different major world types. Even if we omit the nomadic area in Mongolia and consider only the commercial grazing localities in Patagonia and the Great Plains, on what basis can we say that the marked differences in development of roads, railroads, and urban communities are of minor importance in comparison with the similarity of other features?

The previous example appears to lead to type localities because it involves areas dominated by one particular element-complex—including both cultural and natural elements. The situation becomes clearer if one takes localities of more average complexity. Thus a locality in the Po Plain may appear, in certain respects, like one in the American Corn Belt, and so one is tempted to classify both as "peas rather than as tomatoes," but in other, no less important respects, the locality of the Po Plain is like one in the Neapolitan plain, and on that basis, we must, so to speak, classify it also as a tomato. Further, in respect to all those features in the Po Plain locality that result from its location at the foot of the Alps, on main routes from northern Europe to peninsular Italy, in proximity to the Adriatic and the Tyrrhenian Seas, this locality is essentially different from all localities in other parts of the world.

In sum, the uniqueness of the region is of a very different order from the uniqueness of each human spine, or of each pea in a pod. Each of these is unique in characteristics that can unquestionably be called minor while major characteristics are identical, whereas a region is unique in respect to its total combination of major characteristics. In this sense we may agree with those who speak of the "individuality" of areas, though we may find another term used by French geographers, "personality," too suggestive of an organic whole [Musset, *93, 274* ff.; *cf.* Schlüter, *148,* 218; Penck, *158,* 49; Creutz-burg, *248;* and Finch, *223,* 16 f.].[100]

In a very interesting discussion of this problem, Gradmann has suggested that when one has studied a locality in terms of all its interrelated phenomena one has the impression of a complete and single picture which one therefore feels should be expressable in just the right word or brief expression. But "what we possess as a mental picture cannot be passed on to others in its finished state. Each one must work it out himself." The author must lead the reader through the entire analysis and synthesis so that he may himself come to the ultimate perception. His concluding comment should not be spoiled by translation: *"Damit wird reichlich Wasser in den Wein unserer jungen Begeisterung gegossen, und es ist niemandem zu verargen, wenn ihm der Trank fürs erste nicht recht munden will"* [*236,* 131 f.].

There remains, however, a method of limited application by which it might appear possible to recognize localities as specimens of the same type. Within any single large area, for example the Corn Belt, or the Austrian Alps, do we not find a definite type of locality in repeated examples which are as like as peas in a pod? Cannot these be considered, as specimens of the same type of area, defined even in terms of total characteristics? One sees at least a resemblance to the specimens of a biological species, since the features of the different localities within a major region have more or less common origins.

If we assume that the particular combination of natural elements is essentially the same in different localities of a single large area and that the cultural development was controlled by essentially the same human factors, there still remain two significant differences. Even within this limited range, we cannot ignore the significance of relative location. The localities near the center of the region will differ from those nearer the periphery in a number of ways that may be of great importance. Further, if these localities are specimens, their characteristics must include size and shape—characteristics which are not merely of academic interest, but may well affect the development of features, notably urban features, within them. But how can we

---

[100] Schlüter concluded that only on its physical side was geography *einmalige,* whereas on its human side it was more nomothetic—this, in the same article in which he emphasized that geography was not two sided, but many-faceted [*148,* 218, 145–6]! Graf also finds Schlüter self-contradictory, though for a different reason. He himself comes to just the opposite conclusions in regard to the natural science and social science side of geography, but his argument is admittedly based on the particular philosophy of the sciences which divides the natural from the social sciences on the basis of nomothetic versus idiographic—in other words, an argument in a circle [*156,* 106; note also Hettner's review and later discussion, and Kraft, *166,* 11–13].

determine the size and shape of the localities? Only on a very approximate and somewhat arbitrary basis—which would in any case give us differences as great as though the peas in a pod included objects formed like peas, pumpkins, and goose-neck squashes.

This last consideration reveals the essential error in our analogy. The localities of a larger region do not represent independent specimens of a species but simply similar parts of a whole, whose similarities are based in major part on the fact that they are but parts of a whole. Indeed that "whole," the larger region, is not really a Whole, but only a part of the single complete Whole which we have, the whole world.

In other words, the attempt to develop generic concepts *of* areas, as distinct from generic concepts *about* areas—and, on that basis, to compare areas in themselves and develop principles of their relations—rests on the fallacious assumption of the area as an actual object or phenomenon. We are misled by our terminology. When we say that certain areas belong to the same type, we can only mean that they contain one or more elements, or element-complexes, each of which is of the same type in all the areas. That is, having classified the phenomena found in areas in various systems of generic types, we then label an area that includes one or more of those types in terms of the types it contains. We are not actually classifying the area, but only one or more of its characteristics. The area itself is not a phenomenon, any more than a period of history is a phenomenon; it is only an intellectual framework of phenomena, an abstract concept which does not exist in reality. It cannot, therefore, be compared as a phenomenon with other phenomena and classified in a system of generic concepts, on the basis of which we could state principles of its relations with other phenomena. Indeed we cannot properly speak of relationships (other than purely geometric) between areas, but only between certain phenomena within different areas. Likewise, the area, in itself, is related to the phenomena within it, only in that it contains them in such and such locations.

It may be objected that we have considered the word "area" too literally, that what is meant by area is simply the sum total of all interrelated phenomena found within an abstractly limited space. This sum total is, of course, an actuality, but it is not a phenomenon: the combination of more or less related phenomena, some of which are incomplete, since parts lie outside the area, does not form a phenomenon—is not a something that has relations as a unit to similar units, other than the purely geometric relation of location. (To clarify any confusion resulting from earlier use of terms, we should note that, if we are justified by certain authorities in calling a sum of interrelated elements that form a *relatively* closed total, a *unit*—but not a Whole—we must not then consider this loose unit as having the attributes of a precise unit—*e.g.,* of forming a single phenomenon that has relations as a unit with other similar units. To return to geographic ground, we know that the relations that we may loosely speak of as relations between areal units are, in fact, nothing but relations between some of the elements in one area and some in another.)

The conclusion that areas, as such, cannot be studied in terms of generic concepts, but can only be regarded as unique in their *einmalige* combinations of interrelated phenomena, leads some writers to the conclusion that the study of regions is no proper subject for a science. Since we refuse to define a science—though we have referred the reader to the views of such students of cognition as Cohen, Barry, and Kraft—we cannot debate this question. We may repeat, however, that every branch of science is, to a greater or less extent, concerned with the unique. From the standpoint of general culture, as Granö suggests, there is scarcely anything unique that deserves so much attention as the totality of interrelated phenomena in area that forms the milieu of man [*252, 44*].

On the other hand the conclusion that we cannot consider areas themselves in generic concepts and principles does not mean that we cannot utilize generic concepts that express marked similarities in the characteristics of different areas. On the contrary, it is of great value to discover element-complexes and combinations of several element-complexes repeated in different areas. Both for the purpose of simplifying the enormous task of learning the complicated character of the different areas of the world, and in order to develop principles of relationships to aid our understanding of that world, we shall want not only relatively limited element-complexes that express some character of many areas, but also the most complete element-complexes possible to express many characteristics of perhaps few areas. These are of great value, even though no single generic concept can express all the characteristics of an area, even though any one system of element-complexes may not apply to all areas of the world, and even though a single, very complex combination of element-complexes may be found in but very few areas.

For our present purposes, therefore, we can utilize many element-complexes which we found inapplicable to a system of world division. Thus, though it is not possible logically to divide the world in terms of the combination of relief, soil, and drainage, we do find many localities in which a particular integrated combination of these factors is present—*e.g.,* "black bottoms," as a particular type of floodplain (whether we have as yet a classification of floodplains suitable for geographic purposes is another question).

Likewise we can recognize and utilize element-complexes that include both natural and cultural elements. When one has described the terraced vineyards of steep valley walls, whether of the Italian Alps or the Rhine Gorge, one has provided a major part of the description of similar localities found anywhere between those places. Similarly, one may construct a generic description of a port at the mouth of a navigable river in the humid tropics—the warehouses and factories constructed of imported materials by imported techniques in striking contrast to the background of primitive forest, the residence district of the foreign controlling group in contrast to the native quarters. In its general outline such an element-complex will be found repeated at perhaps hundreds of points on tropical coasts.

Generic descriptions of this kind are of unquestioned value, both for description and for interpretation—*i.e.*, the development of principles of relations. It is only necessary to remember that they cannot include all, even of the major, characteristics of the localities described, nor can they be arranged to form a single system that will describe the world, even in outline.

E. ORGANIZATION OF KNOWLEDGE IN GEOGRAPHY

In any field of science, knowledge must be organized into systems in order that the student of any particular problem may have ready at hand the knowledge of facts, generic concepts and principles that bear upon his problem. If all the knowledge in a field cannot be organized into a single system, it is necessary that the several systems be interrelated as far as is possible [Cohen, *115*, 106–14].

In geography, knowledge is organized into systems in two quite different ways. We may divide the phenomena of areal differentiation into major groups each consisting of closely related phenomena, and thus develop *specialized* branches of the whole field of geography. These include, then, physical geography, economic geography—which may be further subdivided into agricultural geography, the geography of mining, of manufacturing, etc. —political geography, and what might be termed sociological geography. It is, of course, an error to include historical geography in this group, for that is, in itself, a complete geography of any past period (*cf.* Section VI C).

On the other hand all geographical knowledge, including that in each of its specialized branches, may be organized according to the two different points of view required in studying the areal differentiation of the world: the view of any particular variable phenomenon in the relations of its differentiation to that of other variables over the world, and the view of the total character of all the variables within the area.   *

The organization of geographic knowledge in terms of the individual phenomena studied is called "general geography" by European geographers,

* See Supplementary Note 49

or "systematic geography" in this country. The organization according to areas is most commonly referred to as "regional geography." (The use of the term "special geography" for this form of organization is fortunately no longer common. In German geography the usual term is *"Länderkunde."*)

These two different methods of dividing the field of geography cannot be combined on a single plane. Each of the specialized branches of geography is represented in both systematic and regional geography—in systematic geography by separate studies of single elements or element-complexes, in regional geography by a *part* of a regional study, limited to a particular group of related aspects. Thus the study of the element-complexes of concern to agriculture—complexes of land-use and of natural conditions, etc.— is a study in regional agricultural geography—*e.g.*, Colby's study of the raisin production of south central California [337].

It is clear that other groupings of geographic features into special fields are possible, as in the geography of settlements (*Siedlungsgeographie*) or in urban geography. The geography of any particular city is clearly a special form of regional geography, whereas the studies of individual urban features as found repeated in many cities would be included in systematic geography. Further, because a city, like a farm, represents an actual element-complex that may be considered as a distinct unit, systematic urban geography may study the cities of the world, or any part of it, in terms of their differential character in relation to other geographic differences.

That no place is provided for "mathematical geography" and cartography may require a word of explanation. In climatology we are concerned with areal differences that result from the relation of the planet earth to the sun, but for the facts of that relationship we depend upon astronomy. Likewise, the study of the exact shape and measurement of the earth presents problems to astronomy and geodesy (see Sec. III B). The problem of projecting a sphere, or the geoid, on a plane surface, the science of projections, is essentially a problem in applied mathematics, which is of no less concern to astronomy than to geography. Consequently, the chapter on "mathematical geography" that often forms the introductory chapter of a geography text, commonly consists, as Hettner has noted [111, 274 f.], almost entirely of astronomical and other non-geographical material. Though such an arrangement may seem logical, it may be questioned whether the most effective method of introducing students to the field of geography is to begin with detailed studies from other fields—as though a text in biology should begin with a detailed study in chemistry, leading up to biochemistry. Finally, cartography, which is a technique rather than a science, is a form of knowledge of service in many sciences. Because it is more essential to

[398]

geography than in any other science, and has been developed to highest extent in geography (see Sec. VIII D), it is both natural and reasonable that it should be most closely associated with our science, but it is no more a branch of geography, logically speaking, than is statistics a branch of economics.

Geography, as the study of the world, necessarily includes a large number of different aspects that are, or may be, represented by specialized fields. Nevertheless it is a well-known fact, which can be tested by any random observation of the literature, that by far the greater part of the work in the field—whether in systematic or regional geography—consists of either physical (natural) or economic geography. Urban geography theoretically overlaps beyond these special fields, in actual practice consists of but little more. The fairly sizable literature in political geography is as yet of minor importance and of sociological geography we have almost none. This situation might be regarded as a natural result of the particular manner in which geography has developed in the past century, or it might be suggested that geographers have commenced with the more obvious and simpler problems, leaving the more complex ones for later study. Overlooking the dubious psychological assumption involved in the latter explanation, there are reasons of much more permanent validity.

In the study of the areal differentiation of the world, the interest of geography in each of the many features which contribute to that differentiation is in proportion to its relation to the total. Each of the natural features varies notably in different parts, and its variations are significantly related to those of some other natural features and many cultural features. In general, the marked differences in the natural environment of different parts of the world, and the partial dependence of most cultural features on the natural environment, is adequate demonstration of the axiom that physical geography is of fundamental importance in geography as a whole.   *

The justification for the notable concentration of geographic work of recent decades on economic geography is not so obvious. To many students this may appear to represent an emphasis on studies of presumably practical value that appears foreign to the spirit of science. Undoubtedly, much of the work in this field is stimulated by such extraneous considerations; if it is definitely designed to serve practical purposes—as in land-use surveys, etc.—it finds its justification as an applied form of geography. Undoubtedly any science receives stimulating reactions from the work of those who endeavor to apply its knowledge and methods to particular problems, but there is an essential incompatibility between the two forms of science, theo-

* See Supplementary Note 50

retical and applied.   The latter is defined and determined in character by the nature of the problem which it has to study in any particular case.   A particular, complex problem will not fit into any theoretical branch of science, but calls for all forms of knowledge and techniques applicable to it.   Even if the work be divided among a group of different scientists, the division can hardly follow, even in principle, the divisions of theoretical science.   *

Further, it may be claimed that much of the interest in economic geography represents primarily an interest in economic phenomena, which are studied more or less in their geographical aspects.   Granted that this may be true for many studies—notably those concerned with the world distribution of certain products, or with the economic situation of particular countries, even though studied in terms of a "geographic basis"—there remain, nevertheless, fundamental reasons that require all the parts of geography concerned with human phenomena to recognize economic geography as of fundamental importance.

In our examination of a long list of cultural features (see Sec. X F), we found that the cultural features whose differences in different areas were of the greatest geographic significance—*i.e.*, in terms of their relation to other areal differences, both natural and cultural—were for the most part economic features.   (They do not by any means include all economic features, since many of these are of very little geographical significance, though they may be of great economic significance.)   Consequently geographers are justified in regarding human, or cultural geography very largely in terms of economic geography.

Furthermore, since economic geography requires detailed consideration of the natural features to which economic phenomena are related, a regional study in economic geography constitutes the greater part of a full study in regional geography.   To put it simply, most people live where they live— rather than move elsewhere or die—not because they like the climate, the politics or the customs, but because they are able there to make a living; the manner in which they make a living, and consequently the manner in which they live in general, are, in large part, determined by the interrelation of economic and natural features which it is the function of economic geography to study [*cf.* Schmidt, 7, 162–200].

On the same basis we justified the major emphasis in geography on agricultural (in the sense of land-use) geography.   The features of urban areas are, to a much greater extent, undifferentiated in different areas.   Needless to say, however, this does not mean that the regional geography of even a predominantly rural area is complete if the cities are omitted or barely mentioned; the relations between the agricultural areas and the cities is of essential importance in understanding the character of the area.

[400]

* See Supplementary Note 51

On the other hand, in at least two major parts of the world—namely the two parts of greatest concern to European and American geographers—areal differentiation is represented in great part by the differential development, in intensity and in character, of urban development. For these areas in particular, geography is greatly concerned with the geography of manufacturing, upon which the urban development is largely based, and with the study of the cities themselves as the most extraordinary features of areas that man has produced.

The conclusion that physical and economic geography make up the major portion of geography as a whole does not for a moment suggest that the other parts are to be ignored. In the first place there are many features of economic geography that cannot be correctly interpreted without an understanding of their relation to areal differences in culture, in the narrower sense of the word, and in political organization. These are therefore geographically significant features, and, in order that their relationships to others may be known with certainty, they need also to be studied systematically.

That major areal differences in culture may be of great importance in relation to other geographic features is familiar to anyone who has traveled in eastern Europe, or has crossed the Rio Grande, or been to mid-latitude South America, not to mention the vast differences to be observed in areas of Sino-Japanese and Indian culture. The differences of culture of a secondary order, however, appear to have but minor effect. A careful study of the maps of crop production in Europe reveals the importance of a boundary running from the North Sea to the Alps, east of which rye and potatoes are far more important than in areas of similar natural conditions to the west; the line follows approximately the Franco-Germanic cultural (not national) boundary. The difference mentioned, however, is clearly minor. Likewise, the more obvious differences in architecture, in customs, etc., can hardly be regarded as geographic differences of the first magnitude. Consequently, a detailed systematic study of the geography of peoples, whatever its value for other purposes, would appear to offer a minor contribution to geography as a whole.

On the other hand, just as the importance of the geography of mining is found, not in itself, but in its significant relation to the far more important geography of manufacturing, so the geography of peoples, even in great detail, is of major concern to the geography of states. While the study of states—in the sense of independent political units—is perhaps but a lesser part of political science, the geography of states constitutes the major part of political geography. This conclusion results from a particular characteristic * of the state that is often lost sight of in the discussions of the place and function of political geography as a part of the field as a whole.

[401]

* See Supplementary Note 52

As a social organization any political form, whether a state, a government, or other commonalty, is a feature that differs, to be sure, in different parts of the world, but the differences have but little relation to those of other features. The proposition that the mountainous islands of the Hellenic archipelago constituted the necessary background for the development of the democracies of the Greek city-states is a proposition in the geographic aspects of political science; it would not be suggested by a systematic study in the geography of mountainous islands or of democracies. If the "mother of parliaments" is located on a fair-sized island, close to, but separated from the mainland, an island of certain conditions of climate, relief and soil, her daughters appear to thrive in areas of radically different natural environment and of very different economic geography. In the study of political organizations considered from this point of view, geography, as Penck suggests, may have little place other than to offer supplemental suggestions to the political scientist [*163*, 51 f.].

On the other hand, the geographer has a direct interest in the state as a division of the earth surface. The fact that that division, as Penck insists, is made by man and is not inherent in the nature of the earth, is immaterial to us, since we have found that any division of the earth can only be made by man. Neither can we accept his description of the earth surface of a state as "merely the stage of man's (political) action, to be sure a stage that influences it"; such a description represents only the political scientist's point of view of the state-area. For geography, the state is an area in which certain conditions are universally true in contrast with those of other state-areas. It is, therefore, an area of homogeneity in certain very important respects, and so forms the simplest as well as the most definitely delimited of all geographic phenomena. Further, as we observed in our examination of the concept of regions as units, the state is the only area larger than a city which is organized as a functioning areal unit Whole (Sec. IX F). Unlike the abstract concept of the "region," the state area is, in many respects, a concrete unitary object; it is a piece of the earth's surface sharply defined and separated from other pieces, with which its relations are, in many respects, the relations between whole units. Like other concrete objects, the state-area has size, form, and structure. Indeed, it is in the consideration of the structure of a state that our concept of regions becomes of practical rather than merely academic importance.

To many geographers, particularly in this country, the concept of the state as an areal unit may appear remote from geography, not merely because it can be observed in the visible landscape only with difficulty, if at all, but also because of a wide-spread impression that man's political structure is something extraneous, essential neither to the earth nor to man. On the contrary, as Schlüter has noted, "the state (in the widest sense) is no younger than man, but rather older. Man could not become a human being without the protection of an association which contains the seed of the state. Man has been from the beginning the *zoon politikon* of Aristotle" [*134*, 410]. For man, as a political animal, it has been just as natural to create a state as a farm or a city [*cf.* Vogel, *271, 5*], and the state which he creates, as East concludes, "whatever else it is . . . is additionally and inevitably a geographical expression and as such forms part of the subject matter of geographical science" [*199*, 270; *cf.* also *216*, 802–4].

Since there has been considerable controversy over the relation of this part of geography to the field as a whole, we may examine briefly the significance of the state-area as a subject of geographic study. The notable differences in size and form of the state-areas of the world is one of the most obvious facts of areal differentiation of the earth surface. It is, therefore, one of the characteristics of the world which needs to be understood in a subject devoted to the study of the differences in different parts of the world. That it could be considered as a fact of minor importance is refuted by the reality of the power of states in controlling not only the political, but also the economic life within their areas. Finally, this important fact of areal differentiation is significant in geography if it is significantly related to other features of areal differentiation.

It is of course false to conceive of the division of any part of the earth into state-areas as "natural"—*i.e.*, as determined by the natural conditions; since the phenomenon itself is cultural, neither the state nor any of its elements, such as boundaries, can be natural [*357*]. But one cannot compare the political map of Europe or Asia with the corresponding relief map without realizing that there are close relations between the two forms of differentiation. Even where the political map may seem highly arbitrary, as in South America, a consideration of the map of population density, and then of the maps of natural elements that explain it, will show that it is only the outer boundaries that are arbitrary; the essential division into states is very significantly related to the natural geography of the continent.

The state-areas differ not only in the obvious features of size, form, and locational relations to each other but also in their internal structures. The differences in structure in different state-areas is directly related to the regional structure of the area in terms of physical, economic, and ethnological geography.

In the reverse direction, the economic differentiation of the world with which we are concerned in economic geography is notably affected by the efforts of all states—more marked in some than in others, but present in all —to organize all the cultural features of its area into a more or less homogeneous and closed unit.

The most obvious product of these efforts is tariff walls, which, though visible only in the minute form of boundary stones and frontier stations, may have a greater effect on the geography of production and trade than the highest mountains or the widest oceans. Many of the less obvious effects are perhaps, in total, even more important. Thus, the regional geography of the Paris Basin can in itself provide but a minor explanation of one of the most important features within it, the city of Paris, in size and character out of all proportion even to the large fertile plain of Northern France. Only the successful effort of the state of France, in past centuries, to bring all its regions, from the North Sea to the Mediterranean, into an organized unit with Paris as the center can account for that particular geographic phenomenon. Obversely, the agricultural and industrial development of each of the regions of France has been notably affected by its inclusion in this political-economic areal unit.

The position of political geography, as primarily the geography of states, is therefore not to be considered as a remote peripheral location, but rather as one in close relation to the major aspects of geography. On the other hand, the special field of *Geopolitik,* which has had such a notable development in post-war Germany, represents a very broadly defined—or quite undefined—field in which geography, in terms of political geography, is utilized for particular purposes that lie beyond the pursuit of knowledge. It represents, therefore, the application of geography to politics [Hassinger, *165, 23*] and one's estimate of its value and importance will depend on the value that one assigns to the political purpose it is designed to serve. Since it is designed to serve national politics from the German point of view, its positive value from that point of view may be considered as offset by its negative value from the point of view of other countries. In its influence on the world's thinking, this branch of geography, which produced the concept of the *"Lebensraum"* of the state, would appear, at the moment, to be the single most influential branch of geography. (The history and methodological problems of *Geopolitik,* as well as of political geography, proper, have been treated previously by the writer [*216*]. Since then, Ancel, in France, has attempted to rescue the term geopolitics (*Geopolitique*) for international science [*187*], and East, in England, has published a significant study [*199*]; the most comprehensive study of political geography is that of Maull [*157*].

When one considers the theoretically possible field of sociological geography from the point of view which we have been following, it may appear doubtful that the development of studies of that character can make important contributions to geography.   In areas of primitive development one might study the geography of clothes, or implements, or conceivably of manners and customs, and religions.   For the important parts of the world, however, such studies would apparently have little geographic significance.   Men wear hats in Chicago that allow their ears to freeze because the winters are not cold in London.   The urban architecture of Midwestern United States is significant chiefly in revealing the lack of indigenous culture.   Areal differences in religion are of some importance in political geography and perhaps in a few cases in economic geography, but in general this is a phenomenon that shows relatively little relation to other features of areal differentiation. Even if we take the case most commonly cited, that of Mohammedanism as the religion of the nomads of the steppes, we find that it flourishes also under tropical rains in the intensively cultivated paddy fields of Bengal or Java.

Prejudgments of future developments in science, however, are among the most dangerous of predictions.   Many aspects of culture, other than those of economic and political facts, are significantly related to regional differentiation.   Certainly it would be unwise to dismiss this field on the basis of examples selected to prove the lack of significant relation.   The studies of settlement forms, house types, etc., that have been made in Europe and in some parts of this country—and with particular enthusiasm recently, I am told, in Japan—indicate the possibility of systematic contributions to geography as a whole (cf. particularly Schmidt [180, 54–80]; see also the previous discussion in Sec. VII F).

In each of the specialized branches of geography which we have discussed, the pursuit of knowledge is dominated by the same ideals that we found applicable to geography as a whole.   The degree to which these ideals can be attained varies in the different fields as it does in the corresponding fields of systematic science.   In general, no doubt, the degree of accuracy and certainty may be highest in physical geography and decreases in the various branches of cultural geography, more or less in the order in which we discussed them.   Such a comparison however is by no means universally true. Some economic facts can be established with far more accuracy and certainty than many facts of natural phenomena, and few facts in geography can be established with such a high degree of certainty and exactness as those concerning the extent of state-areas.   *

In the development of universals the greatest contrast in geography is

[405]

not found among the special fields but rather between systematic geography in all its branches, and regional geography, whether partial or complete. It is particularly on this account that the separate development of systematic geography is so important, both in itself and in its relation to regional geography.

The division between these two different points of view is made necessary by the very nature of geography. The areal differentiation of the earth surface, which geography is to study, is a differentiation expressed in a great number of individual features whose variations are in part related to each other, in part independent. Geography, therefore, must study the areal differentiation of each of these features over the world, not as a part of the systematic study of that particular object, but as a study of one form of areal differentiation of the earth surface; this is systematic (or general) geography. It is clear, however, that a full understanding of the differences between areas cannot be obtained by simply adding together the appropriate sections of systematic geography. It is necessary to study the totality of interrelations of all geographic features found together at one place; this is regional geography. In the historical section we saw that this difference between the two points of view within geography, first stated by Varenius, was present in the work of Humboldt, and formed, as Wagner noted, an inevitable dualism in the field—not in respect to materials, but in respect to methods of approach. More recently the philosopher Kraft, in examining our field, has substantiated this statement [166, 4 f.]. In general, in modern geography there appears to be agreement on the distinction between the two points of view, as outlined by Richthofen [73] and following him, more clearly, by Hettner [140; 161, 218, 398–403; cf. also Penck, 163, 44]. The relative importance of the two aspects, however, remains as it has always been, a subject of continuous controversy.

The relation between the two methods of organization has been illustrated by Hettner in the following manner. If we consider the variations of the geographic elements as though contained in surfaces parallel to the earth's surface, then all of them together would form a series of surfaces at different levels above the earth's surface. In any particular part of systematic geography, one studies all the variations in a single surface, in relation, it may be, to variations in the surfaces of the other elements. In regional geography, however, we strike a limited section vertically through all the surfaces in order to comprehend the totality of their characteristics in a single area.   *

These two methods of organizing geographic knowledge are not only interconnected in every part but are also by no means so distinct in practice as is frequently supposed. A systematic study need not cover the whole

* See VanCleef's diagram of this simile, as anologous to stratigraphy [433].

world but may be limited to a continent or to any area within which there are variations in the feature studied. (Consequently the term "general geography" is unfortunate.) If, then, all the features of a small area are studied systematically in succession, the mere addition of the total series of these systematic studies does not form a study in regional geography, but simply the systematic geography of a limited area. The essential difference is in the point of view. In regional geography the study is focussed on the individual localities or districts, which, whether smaller or larger, are conceived (arbitrarily) as units. The purpose then is to comprehend the particular geographic character of each of these units—that is, the particular manner of formation of all the geographic factors in their causal connections [Hettner, *126, 672*].

As we have implied a number of times, the terms inherited in geography for these two divisions of the field have been found unsatisfactory by many students. For work in which individual categories of phenomena are studied over extensive areas, or the whole world, Varenius' term *"geographia generalis"* is almost universally used in Germany (*allgemeine Geographie*) as well as in France (*géographie générale*). As Hettner and others have found, however, this is misleading, not only because such studies may be limited to but a part of the earth, but particularly because they consider the features of the earth surface by individual categories. Richthofen at one time suggested the term "analytic geography" [*73*, 41], but Hettner rightly objects that both analysis and synthesis are required in both forms of geography [*126, 675*; or *161*, 400]. Schlüter nevertheless appears to have adopted the term [*247*]. In his earliest considerations of this question, Hettner, we noted, endeavored to express more clearly the close relation of the two aspects of geography by using in place of "general geography" the term *"vergleichende Länderkunde"*—taken, he tells us, from the title of a course given by Richthofen. Though he did not use this term in his subsequent methodological treatments, Hettner has recently returned to it as the title for a comprehensive work covering the whole of general geography, except for the omission of the seas [*363*]. Whatever conclusion German students may ultimately come to, American geographers will no doubt agree with Penck in finding this usage unfortunate [*90*]. On the other hand, the term now widely used in the American literature *"systematic geography,"* finds ample precedent in the writings of many German geographers. Thus Richthofen described this form of geography as organized *"auf Grund systematischer Principien"* [*73*, 41], Hettner called it *"die systematische Darstellung"* [*126, 675*], and Penck has called it the method *"nach systematischem Gesichtspunkt"* [*163*, 44]. Unfortunately the fact that the term is not of Germanic origin, and its German equivalent is unsuitable, apparently excludes its use as the title for a major part of the field of geography in that country.

The other term introduced by Varenius, *"geographia specialis,"* has been very largely replaced, in Germany by *"Länderkunde,"* or more recently by *"Landschaftskunde"* (ignoring here any differences between the two), and in practically all non-German lands by the appropriate form of "regional geography." Any number of German students however have found *"Länderkunde"* unsuitable, both because it excludes the seas and because it suggests areal divisions much larger than those now commonly studied as regions. As early as 1831, Fröbel suggested "region" in place of "Land" [*54*], Supan wished to return to *"Spezialgeographie"* (or *"spezielle Geographie"*) [*78, 153*]; Waibel would prefer either that term or *"regionale Geographie"* [*266, 198*]; and Lautensach favors *"regionale Geographie"* [*173, 29 f.*]. Since neither of these terms, however, is of German origin, there is little likelihood that German geographers of the present period will change to them.

The fact that the field of geography may be divided in two different, and intersecting, directions frequently leads to confusion. We may clarify the situation by classifying a few examples; if we limit these to studies by the present writer there will be no danger that anyone be offended. The study of types of political boundaries is a systematic study in political geography in which the attempt is made to construct generic concepts and to suggest some general principles [*357*], whereas the study of boundaries in Upper Silesia is a partial regional study—regional political geography [*355*]. Similarly the study of the Upper Silesian Industrial District is a partial regional study limited largely to a part of the economic geography of the area [*356*]. Finally, the study of Austria-Hungary is a study in regional political geography of a past period, that of the beginning of this century— *i.e.*, a study in historical geography, limited to political regional geography [*358*]. In other words, we recognize no "boundary between economic and regional geography," such as Pfeifer apparently would have us observe [*109, 108*]. Each overlaps the other. But likewise it follows—as possibly he intended to say—that regional geography is not to be thought of as complete if it is limited to economic regional geography.

F. COMPARISON OF THE ORGANIZATION IN GEOGRAPHY WITH THAT IN OTHER INTEGRATING SCIENCES

The conclusion that systematic and regional geography are two coordinate forms of organization of knowledge in our field may have raised a question in the reader's mind—as it did in the writer's—concerning the position among the sciences to which we have assigned geography in general. If geography is to be classified as a chorographic science along with astronomy, and these are to be included with the historical fields as integrating sciences, should we not logically expect a corresponding division in each of these fields between the systematic and the sectional (regional or periodic) approach? At first glance, at least, one might suppose that almost all of astronomy would be included under the systematic approach, almost all of history in the study of periods. Insofar as this is the case, does it cast doubt on our thesis of the logical similarity of geography to these fields?

Although astronomy and geography are both concerned with the study of spatial integrations—things related to each other in space—the character of the spaces that they study and the things interrelated within them are so completely different that no amount of logical similarity should lead us to expect similar results in the developments of the two fields. For much the greater part of his work the astronomer may consider celestial space as extraordinarily simple, consisting on the one hand of homogeneous ether—which in much of his work, he may regard simply as empty space—and on the other hand of masses of inanimate matter, most of which are sharply defined units widely separated and therefore related to each other essentially as whole units. Within any celestial region—*e.g.,* that occupied by the solar system—the problem of integration is little, if anything, more than the systematic problem of the relations between these unit masses, a problem primarily concerned with their effects on each other's position and motions, studied in terms of but two factors, gravity and free motion in space. Even within systematic astronomy, however, not all the findings can as yet be reduced to scientific laws. Though the motion of the sun in reference to other stars has been measured, the laws determining that motion have not been constructed, possibly never can be.

Astronomy does include studies corresponding to those of regional geography. These are represented most clearly by the detailed examination of those units in the solar system that are near enough so that differences between different parts may be observed. In the same category, though different in character, are studies of the groupings of stars in our universe, and the detailed examination of individual stellar nebulae.

If astronomy is largely concerned with systematic studies because of the relative simplicity of its subject matter, exactly the reverse is the case in the historical sciences. Of this group, historical geology shows most clearly the distinction between the systematic and the periodic approach. The study of climatic changes in past ages, of changes in mountain development, and, in general, the changes in the continental landforms, or the evolution of the horse—all represent systematic studies in the history of the earth. In contrast are studies which attempt to provide a generalized picture of associated phenomena of climate, landforms, and vegetable and animal life of the Upper Mississippian or any other past period in the history of the earth.

The comparison to which we have most repeatedly referred throughout our discussion of the nature of geography is, of course, that with history in the ordinary sense of the history of "historic times." Various students, however, have suggested that the comparison can only be related to regional geography, that history lacks systematic studies [cf. Penck, 158, 48–50].

We should not be misled by the fact that history is commonly taught only in terms of what we may call "periodic history." In their research problems historians frequently concentrate on the development and changes in some very restricted group of phenomena through a succession of years. Such studies may treat the development of a particular form of constitution, the growth of labor legislation, the changes in the price of wheat in England, or the development of roads in Minnesota.

Nevertheless, so far as an outsider may judge, the work of this character has by no means the importance to the field of history as a whole that systematic geography has to geography. In particular it has not yielded to history generic concepts and principles that are nearly as definite as those developed in systematic geography.

If one compares the particular problems studied in the two fields, as shown in their publications, it is obvious that historians are concerned with phenomena whose interrelations are far more complex than those commonly studied in geography. The logical basis for this difference is not so obvious; indeed our fundamental assumption of the relation of the two fields leads logically to the conclusion that the same phenomena may be studied in each field: history may consider areal phenomena and geography may consider historical events.

Neither history nor geography, however, need consider all the phenomena that are found in the sections of reality which they study, but only those phenomena which differ significantly in different sections of time or space, respectively. In each case the attention is chiefly focussed on those phenomena which differ most and whose differences are most significant to the total differentiation. In the total reality with which both history and geography are concerned—namely the phenomena of the world in historic times—there is one major group of phenomena, the natural phenomena, which are causally of fundamental importance to all the other phenomena, but which, while differing markedly in different areas of the world, differ but slightly in different periods of historic time. This, of course, constitutes a great difference between history, in the narrower sense, and pre-history, not to mention paleontology.

In consequence, the areal differences that are of greatest importance in geography are either differences in the natural features themselves or in cultural features which are closely related to the natural features. We would have a similar situation in history only if such features as climate and landforms had varied as radically through historic times at the same place, as they vary over the world at the same time. In other words, if the natural environment of England since the time of Caesar had varied from humid to arid, from polar to tropical, from plains to mountains, the agricultural history of England would represent the most important part of its history, and history would long since have developed the systematic branches of climatic history, landforms history, etc. Indeed, if Ellsworth Huntington's thesis of the historical importance of even minor variations in climate should be substantiated, it would be not only logical but necessary for history to develop a systematic study of climatic history—the study of the relations of climatic changes to other historical features.

In any case, the exceptional character of the example just cited tests the rule: the relative fixity of natural conditions during historic times results in a notable degree of constancy in those cultural features which are most closely related to natural conditions. The manner of land use in any area may remain much the same for centuries, in China, for millenia. Cities do not pass through a generic process of youth, maturity, and old age to death; they may continue in approximately the same condition for indefinite periods of time.

Consequently the phenomena which show the most notable differences in relation to time are cultural phenomena less closely related to natural conditions—commonly, therefore, phenomena of much more complicated character—such as manners and customs, political organization, inventions, etc. Furthermore, not only are these phenomena in themselves more complex than those with which geography is most concerned, but their interrelations through different periods of time are more complex than the interrelations of the principal geographic phenomena in different areas. Indeed, in most cases the character of one period of history largely determines the character of the next, whereas the character of one area in geography has commonly but minor effect on the character of its neighbors. It is not surprising, therefore, that historians are more clearly aware of the fictional nature of their divisions of time than are geographers of their corresponding divisions of area.

On the other hand there are some sudden breaks in historical development that produce changes almost as great as the change from sea to land in geography, namely, when new discoveries or inventions, or the migration of peoples, introduce a new culture into an area. The frontier of settlement in America during the past several centuries was not only a line marking great geographic contrasts but, as it passed through any region, it represented an historical revolution in the adaptation of man to nature. The historian of this revolution, therefore, must understand the principles governing the relation between cultural and natural features in order to study history. Much the same is true of such historical problems as the industrial revolution, and the associated agricultural revolution in Europe.

The historian who is concerned with these problems, involving less complex features than those that form most of the material of history, presumably will not hesitate to make systematic studies wherever possible and to explain the relationships where he can. The fact that most historical events may be too complex to permit of definite explanation should not lead to a dogma that no historical events can be interpreted. Unfortunately, however, situations comparable to those mentioned above, in which the fundamentals of man's adaptation to nature are notably changed, are relatively few in history and most of them took place at such an early date that the historian has scanty reliable data from which to study them. Thus, a systematic study of the history of "frontiers"—in the sense of a border of progressive settlement— should consider not only the frontiers in the New World and in Siberia, but also the frontier of German settlement in central Europe in the Middle Ages and the still earlier frontier of Anglo-Saxon settlement in Great Britain. It is obvious that, even were data available, such a problem would be extremely complicated, since it involves not merely different periods of world history but also different areas of the world of radically different character.

In general, the problems which must be handled in systematic studies in history are far too complex, and involve factors too difficult to observe and measure, to permit of the development of generic concepts and principles similar to those developed in systematic geography. There are, to be sure, some students of history—chiefly non-historians—who assume that it is possible to develop scientific laws concerning the rise and fall of states, the causes of revolutions, or the development of particular social movements, but their theses are more notable for the ardor with which they are advanced than for the evidence which has been brought to support them. Most professional historians are sceptical of the possibility of developing a systematic history in which the phenomena important in history may be classified in generic concepts leading to principles. The rather naive belief of some geographers that geography can provide this deficiency in history has not, as yet at least, been substantiated.

This contrast, therefore, between history and geography results from the fact that the interrelations of the phenomena that vary most notably in historic times are far more complex than the interrelations of the phenomena that vary most notably in the earth surface. It does not affect the logically common nature of the two fields, as sciences that attempt to integrate phenomena as they are found in reality.

G. THE CHARACTER OF SYSTEMATIC GEOGRAPHY   *

The simplest form of study in systematic geography is the consideration of the differential character of the earth surface in terms of any single geographic factor. In the past such studies were in large degree limited to the natural factors—the climatic factors, landforms, soils, etc.—but, as many students have pointed out, if geography in general is to consider human or cultural features, they must be studied in systematic as well as regional geography [cf. Hettner, 126, 672; Hettner's theoretical treatments of systematic geography will be found in 140; 152, 46–48; 161, 398–404; and 167, 281–86; his detailed survey of the land areas of the world on this basis, in 363]. Most students today recognize that in this respect the development of systematic geography to date has been one-sided. An extreme illustration of the contrast is offered, as we noted earlier, in Finch and Trewartha's "Elements of Geography." In contrast to nearly six hundred pages of masterly treatment of the systematic geography of the natural elements, they present hardly a tenth of that amount on the much more complicated problems of systematic cultural geography [322]. This difference, to be sure, hardly does justice **

to the present development of systematic cultural geography, even in the literature of this country; it is not clear why the authors should appear to disregard a number of excellent studies, including some by Finch himself. Further, a large number of systematic studies have been made in cultural geography by many German writers, including notably Schlüter and his students [cf. Waibel's discussion, 266, 201 f.; Schlüter's model example, the study of bridges, has been noted, 247; for an example of a full detailed survey of systematic human geography see Hassinger's volume, 360].

With the increasing interest, in both Germany and America, in full regional studies, necessarily including cultural as well as natural features, geographic research suffers not only from the lack of adequate foundation in systematic cultural geography but also, as Broek notes, in the lack of training of most geographers in the social studies [108, 252]. If students are to be prepared to make full studies in regional geography—and there is fairly general agreement that all geographers should do at least some work in this field —it would be logical to require supplementary training in the related social

[413]

---

sciences no less than in the related natural sciences. This is the case in few, if any, of our departments of geography. Since, in addition, the academic relations of geographers are commonly closer to geology and other natural sciences than to the social sciences, the individual geographer knows that his work will be subjected to careful criticism so far as it touches natural science. For the same reasons he has been free to indulge in almost any sort of economic theorizing or political speculation that occurred to him, with little or no risk of being called to account.

On the other hand, the relatively recent shift from the emphasis on a geomorphology closely related to geology to the emphasis on human geography, has resulted in a tendency on which Krebs has very recently commented: much material that is essentially economic, historical, or sociological is taken over and presented without digestion in presumably geographic studies [91, 244]. American geographers appear also to be aware of the need for independent research to develop concepts and principles of systematic cultural geography—indeed there is evidently a widespread feeling that this is the single most pressing need in geography today.[101]

It is desirable, therefore, to consider carefully the distinction between systematic studies in geography and the studies made in the related systematic sciences. Especially is it necessary, as Schmidt insists, for those who work on the borders of geography to keep the distinction in mind. Though they must be familiar with the concepts and methods of the neighboring sciences and may use these in their work, they must use them for purposes dictated by the point of view of geography as distinct from that of the related systematic science [7, 162 ff.]. In particular, Schmidt has contributed a very thorough and valuable study of the relations between economic geography and economics, as well as a detailed study of systematic economic geography [7; 386].

The divisions of systematic geography, as we noted earlier, correspond to the divisions of the systematic sciences and there is inevitably close relationship between each branch of systematic geography and the corresponding systematic science. This relation is not accurately expressed by the phrase "neighboring sciences," since geography is not a branch of science situated beside the systematic sciences, but represents a point of view in science which cuts through all the systematic sciences (Sec. IV A). There is, therefore,

---

[101] As indicated in answers received to a questionaire distributed to a wide group of American geographers by the Geographic Section of the Committee on Research in the Earth Sciences of the National Research Council. A general acknowledgement may be made here for a number of suggestions that have been taken from those answers and utilized in this paper.

no line separating systematic geography from the systematic sciences, but there is an essential difference in point of view which must be maintained by the individual geographer who wishes to do geographic work rather than work in some other branch of science.

The distribution of a particular kind of phenomena is significant both in geography and in the systematic science concerned with that kind of phenomenon, with this difference: in geography the focus of attention is concentrated, not on the phenomenon—one of whose aspects is its distribution—but on the relation of that distribution to the total areal differentiation of the world.

This contrast in point of view may be illustrated in the case of the production of corn (maize). The totals of production by countries and the resultant effects on national and international markets are presumably of concern in economics, but not in geography. (One may be confused here by the fact that economists have generally been willing to leave instruction in the geographic aspects of economics to economic geographers, with the result that geographers have frequently attempted to do research in what is essentially a part of economics.) Likewise the relation of annual variations of corn production to annual variations in rainfall is of great concern to the student of agriculture but is not of direct concern in geography, whereas the fact that the variations in rainfall in Nebraska have a greater effect on corn yields, than the same degree of variation in Pennsylvania, is of geographic concern. Geography is concerned with the marked areal variation in corn production, since this represents a part of the total areal differentiation, in which it is associated on the one hand, in its relation to differences in climate, soil, relative location, or cultural conditions and on the other hand, in its relation to differences in the total crop-livestock element-complex, the character of barns, the presence of grain elevators, etc.

In other words, the facts of distribution of corn production are not in themselves "geographic," not even when shown on a map. It is what is studied concerning those facts that is significant for geography. Merely to describe and analyze the facts of distribution of physical and social phenomena found in different areas is to produce a compendium, not a geography, either systematic or regional. The facts concerning the areal differences in these phenomena must be studied in their *areal relations,* that is, their significance to the area as determined by their relations to other phenomena of the same place, and by their spatial connections with phenomena in other areas.

In a study in systematic economic geography, for example, unless the geographic point of view is clearly maintained from the beginning, the work may turn out to be a study in the geographic aspects of economics. The reason for this, of course, is that both of these kinds of studies start with the same step, namely, the establishment of the facts of distribution of the particular phenomena studied. As this first step focusses the attention of the student on the phenomena themselves, it often results in his continuing the work with that point of view, thus producing a study of those phenomena— *i.e.,* a study in a systematic science.

Because the botanist or economist is concerned with the location of his phenomena in only some of his studies, whereas the geographer is always concerned with the location of facts, it is often supposed that the determination (and interpretation) of the "Where" of things is exclusively a function of geography, if not the whole, of geography—*i.e.,* geography as the science of distributions. But it would be both presumptuous and contrary to what actually takes place for us to claim that the zoologist, geologist, or economist concerned with the distribution of his phenomena must look to the science of geography for the answers. Likewise the fact that in such studies the students of other sciences may use the geographic technique of mapping does not make them geographers; the economist or political scientist may often use the historical method in determining the "When" of past events, but their work does not thereby become history. Similarly all the students of the systematic sciences may use geographic methods in presenting the distribution of their phenomena—whether particular kinds of plants, animals, or factories—without depending upon the science of geography (see Sec. III D).

In the reverse direction, geographers have, in fact, long been accustomed to looking to certain of the systematic sciences for their knowledge of the distribution of certain kinds of facts. For the location of mineral deposits and different kinds of surface rocks, we depend upon geology; for the occurrence of soils, on soil science; for the distribution of native plants and animals, on botany and zoology. As Hettner, in dependence on Wallace, has insisted, these latter represent geographic studies in botany and zoology, as distinct from studies in plant and animal geography in which the interest is focussed on areas, studied in terms of their plant and animal contents. It is only in those fields in which the systematic sciences concerned have given little attention to the geographic aspects—notably in economics—that the geographer has been forced to do his own spade-work in determining distribution. It is significant, however, that once geographers had introduced their technique of mapping into the study of domestic crops and animals, agricultural economists have taken over this work as an integral part of their field.

The study of the distribution of phenomena presumes a classification of objects into types. In many cases the objects are sufficiently simple so that the classification is both obvious and acceptable to all the sciences concerned —*e.g.,* the classification of cultivated plants into different kinds of crops: corn, oats, wheat, etc. If less simple phenomena are involved, however, we noted that the classification will depend on what aspects are selected as most important for the particular study. Consequently two sciences concerned in studying the distribution of the same phenomena may differ, even in the presentation of facts of distribution; though this difference may in itself be slight, it may be of major significance in later stages of study.

One should not forget, however, that economy of effort is an axiomatic *desideratum* in scientific work. In any case where the classification and establishment of facts concerning the distribution of phenomena that have been developed in another science are found to be suitable for the purposes of systematic geography, there is no call for the geographer to do the work over again in a different way. Nevertheless, as we noted earlier, in the consideration of relatively complicated phenomena, such as land forms, the facts established for the purpose of another science have been found not to be the facts needed in geography; consequently the systematic study in geography must begin anew at the first step (Sec. XI D).

Any presentation of facts in science calls for interpretation. Consequently, geographers have often presumed that, in presenting the facts of distribution of any phenomenon, it was also the function of geography to study the causes of that distribution. But in every branch of science facts are presented and utilized whose interpretation is the function of some other branch of science. In this case, namely the distribution of any phenomenon, does interpretation of the facts of distribution fit logically into geography or into the systematic science?

This question is not raised here in order to argue over the location of a borderline between sciences, certainly not with the idea of establishing any rules of conduct for geographers. In considering this question we will, I believe, come to a clearer understanding of the whole relation of systematic geography to the systematic sciences.

One point appears clear. Whichever student is to interpret the distribution of a particular phenomenon will study that phenomenon in terms primarily of those aspects which indicate its causal development. If the distribution be measured in terms of other aspects they must first be referred back to the genetic aspects in order to provide interpretation. One of the essential contrasts between geography (and history) on the one hand, and the systematic sciences on the other, is that the former are interested in the integration of phenomena, the latter in analyzing the *processes* of particular kinds of phenomena. The fact that studies of processes involve the time element does not make them history, as Kroeber has emphasized [*116*, 545 f.] ; neither does the fact that distribution involves the element of space make its study a part of geography. The explanation of the world distribution of a particular kind of phenomenon would appear to be an end resultant of the study of the processes of development of that phenomenon; it is therefore the proper subject of study in a systematic science. In systematic geography, however, it represents the world picture of an element with which one is concerned in its functional relation to the differential character of the areas of the world. In other words, though geography must know where things are, the study of the "Where" is not geography nor an integral part of geography, and it is therefore not the function of geography to explain the "Where"—that is, to give the full explanation of why a phenomenon is found where it is found. Consequently, systematic geography is free to overlook generic concepts based on *genetic* aspects of phenomena in order to develop generic concepts based on aspects that are functionally significant.[102]

Although we conclude that it is not the function of the geographer to explain the distribution of any phenomenon, it is at the same time clear that he may be concerned with such an explanation in order to interpret the relations of that phenomenon to other geographic phenomena. For example, in the geography of soils, the interpretation of the relation of the soil of any area to the character of its climate and bedrock, necessitates an understanding of the whole development of soils; but it is the function of the soil scientists to provide the explanation of soil development in terms of all its factors and processes.

The systematic geography of any particular phenomenon depends we conclude, for the principles governing its distribution, on the systematic science concerned with that phenomenon. In many cases, however, the geographer may find that the students of the appropriate field have not been interested in developing such principles. In such cases he can hardly be expected to wait indefinitely, but may have to undertake the study himself. If, however, he does that without realizing that he is shifting his point of view, he may later discover that he has definitely passed over into a field in which he may not be adequately prepared.

---

[102] It may be noted that Hettner, who repeatedly and vigorously refutes the concept of geography as the study of distributions, the "Where" of things, has apparently not considered the line of reasoning here followed, since he assumes that classification of phenomena in geography should be genetic [*161*, 223], just as he claims that the genetic principle is essential in a logical system of regions. The fact that Hettner, whose "methodological masterpieces" underlie the thought throughout this paper, should not have come to the conclusion stated above has caused me to re-examine it repeatedly and critically, but, as yet at least, without finding any error in the reasoning or in the conclusion.

Since the previous paragraph might seem to be pointed at individual geographers, it may be appropriate to illustrate it from the writer's own experience. Geographers had long recognized that the concentration of iron and steel mills in certain areas was somehow related to the presence of coal and iron mines in the same or other areas, and every text in economic geography attempted to state that relationship. None of these statements, however, were found to be adequate and the reason for this was, no doubt, the failure of students of the economics of industry to study the problem of the distribution of iron and steel mills. The geographer wishing to interpret the character of areas found to have, as one of their major characteristics, intensive development of this industry, must be able to explain the relationship of these factories to other geographic features. The writer, therefore, undertook to develop the principles governing the location of the iron and steel industry [352], on the basis of which that industry could be studied as a part of the systematic geography of the United States [353]. The interest developed by the first study led the writer into similar studies of the principles governing the location of other industries, and of industry in general. It has since become clear to me that only in the study of the iron and steel industry in the United States was my attention focussed on the areal significance of the industry—as a particular characteristic of certain areas; in the others the center of attention was the industry as a phenomenon of which one aspect, namely its location, called for explanation. It is not surprising, therefore, that any interest shown in these studies has been confined almost entirely to economists.

The conclusion which the writer has drawn from this personal experience may have general application. The study of the *"Standorts"* problem—the determination of principles governing location of units of production—not only requires more training in economics than in geography, but also requires a full concentration of interest on the problem for the sake of the problem itself, rather than for the sake of the results; it is the economist who is interested in the problem, the geographer in the results [*cf.* Tiessen, *160, 8*].

On the other hand it might be claimed that, regardless of a logical division of problems between the sciences, geographers had, in fact, developed this particular subject to such an extent as to justify their retaining it as a part of their field, by right of cultivation, regardless of the logical division of work in science [*cf.* Kraft, *166, 7*]. Undoubtedly geographers have made contributions to the location of economic activities, but we have hardly pursued the problems with sufficient consistency and system to register a valid claim based on thorough and successful cultivation. The world of knowledge as a whole does not look to us to supply the principles governing these phenomena.

Confirmation of this conclusion may be drawn from critical survey of the work of geographers in this field made by the Swedish economist Palander, in the introduction to his exhaustive study of the theory of the *Standorts* problem [*372*]. It is only fair to add, however, that geographers entered this field only because the results which they needed in their work had not been adequately developed by economists. American economists in particular have shown little interest in this field and at the time the writer was concerned with it were hardly aware of Weber's work [*396*], which, in any case, Palander has shown to be impracticable.

Geographers, therefore, will welcome the attention which economists are now giving to this problem. Geography will not be merely receptive in this relation, however, for even though we may agree that the problem comes logically under the point of view of economics, it certainly represents a geographic problem in economics, which requires some understanding of the geographic point of view, and for which geography can continuously contribute both positive materials and effective criticism.

Since much has been said of the dependence of geography on the systematic sciences, we may appropriately note one or two significant suggestions that geography may contribute to the problem of interpreting the distribution of economic features.

The first step toward an interpretation of the distribution of any phenomena is, of course, to portray that distribution. Students of the systematic sciences who have acquired something of the geographic point of view, will realize that the only language in which the location of things on the earth's surface can be portrayed intelligibly is the map, and that reliable interpretations require maps more detailed than cartograms of units as large as our States. Although this proposition is axiomatic in geography and geology, and is now thoroughly recognized in agricultural economics, in other branches of economics it is frequently overlooked. When economists attempt to interpret the location of the steel industry in the United States in terms simply of the amount of development in the States of Pennsylvania, Ohio, Indiana, and Illinois, it is not surprising that they should reach a defeatist conclusion as to the possibility of principles of location. Even though our census figures by counties are less complete than those by States, they must be used to gain an approximately accurate measure of the development of the steel industry in the areas of Southeastern Pennsylvania and Maryland, the Pittsburgh-Youngstown region, the Lake Erie Ports, and the Calumet District [*353*]. Likewise one cannot hope to interpret the contrast between the industrial development of Wisconsin and that of Minnesota until one has seen the facts portrayed on a detailed map and observed that the concentration of distinctively manufacturing cities is not to be considered in terms of Wisconsin versus Minnesota, but in terms of proximity to the west shore of Lake Michigan [326].

To the student of economics who has not been trained in geography, even a detailed map of distribution of an economic feature may appear to present a comparatively simple problem in comparative location—that is, he is apt to think almost exclusively in terms of relative location, considered purely geometrically, and to ignore other variants of areas. Thus, in many economic texts the consideration of the distribution of different types of agricultural production has long been dominated by Thünen's simple picture (constructed, we may note, by a writer living in the relatively homogeneous North German Plain) of concentric belts of differential production surrounding a city center [for an outline of the theory, see Jonasson, 313 (1925), 284-6]. Whatever validity this analysis may have had in earlier periods has been very largely destroyed by the development of modern commercial facilities that have made relative location a factor of secondary importance in determining land use. O. E. Baker showed, some years ago, that, with such well-known exceptions as market-gardens and fluid-milk farms, the location of different types of agricultural production is far more dependent on climate, relief, soil, and drainage than on relative location; this writer has demonstrated the fact in detail in the agriculture of Europe [324]. On the other hand, this does not mean that the factor of relative location may be entirely ignored in such problems, as is often the case in studies in agricultural geography. (Waibel has recently considered Thünen's law in full detail in the light of the radical changes in conditions since it was first stated [395, 47-78].)

This discussion of the importance of the geographic point of view to the problem of interpreting the distribution of any phenomena may appear to suggest that, in spite of the logic of classification, the geographer is best equipped to handle the problem. Before any geographers accept that conclusion, they should first examine the specific problems treated in Palander's masterly treatise—in particular the enormous amount of economic detail required in the handling of the transportation problem; they should observe the technique of economic analysis developed in a study of the location of manufacturing in the United States by Garver and associates—even though that study provided the examples of lack of map-mindedness to which we referred above [347]; and finally they should consider the studies made by economists who have had some geographic training, for example, Hoover's study of the boot and shoe industry [364].

All that we have shown is that the study of geographic aspects of any field of science, such as economics, requires something of the geographic point of view. That historical problems in economics require something of the historical point of view requires no emphasis, because all economists have no doubt had training, in one way or another, in history; but relatively few have in geography. Every science, Schmidt concludes, "that concerns itself with the study of the areal distribution of its objects on the earth is necessarily led to the geographic method; it must interpret the differences of its objects in relation to area, and so must make use of the method of thinking in geographic comparisons as one of the most important means of attaining general concepts in its own field and of penetrating into the character of the scientific objects of its own science. Thus, every research worker in economics must be a geographer [in the sense that he must use the geographic method] whether he will or no; the sooner he wills it and knows it, the better for him and his research" [7, 4].

In brief, we conclude that, both in terms of the logic of the classification of the sciences, and in terms of the professional equipment of the students—in techniques and in knowledge of the literature—the problem of principles of distribution of economic phenomena can best be studied by the student who is primarily an economist, but it is necessary also that he be in some degree a student of geography.

To avoid any misunderstanding it may be necessary to add that throughout this discussion—indeed throughout this paper—the term "geographer" is to be understood as an abbreviation of "student of geography." Any individual person may presumably be simultaneously a student of geography and a student of economics and may study wherever he finds himself interested and believes himself competent. Undoubtedly individual geographers have made, and may continue to make, important contributions of thought to the work in related fields. Indeed, such personal interconnections between the different fields of science are not to be regarded as merely permissible, but rather, as Penck properly insists in his own defense, are greatly to be desired [147, 124 f., 134]. If this is true of science in general, it is particularly true of geography, which not only is related to other sciences along border zones that "would be left fallow if scholars always limited themselves to a single science," but in every part of its field intersects the studies of the various systematic sciences. It is fortunate, therefore, that "the boundaries of the sciences are not insurmountable walls," as Penck writes in discussing the same question later, and one might ask him to modify slightly the analogy that he does suggest, of "the boundaries of states, which one can cross if one has the necessary pass, in our case, capacity." Perhaps Professor Penck's own experience on certain international frontiers would persuade him to agree that there should be no border guards along the boundaries of science, but that each student is to be his own judge of his pass, subject always to the ultimate verdict of those qualified to judge in the field into which he crosses. In any case, all will agree that "he who works across the border areas of geography must be able to ride in several saddles" [90, II, 36]. Further, as Penck has elsewhere indicated, the requirement that the student should feel himself competent wherever he works, requires that he should himself know in what field he is working at any time. And in order that geography may maintain clearly its own fundamental point of view, any cross-fertilization should be recognized for what it is, and not accepted as an extension of our field.

On much the same basis we may be permitted to dispose of the difficult problem of the relation of geomorphology to geography—without attempting to solve it. This question has long been a matter of controversy, particularly in the English-speaking countries. As early as 1908, Chisholm, in agreement with Geikie, expressed the view that if the study of landforms follows genesis it leads to geology, and many others have echoed that view since. At the same time, however, a large part of the work in that field—in America, if not in England—has been carried on by geographers—particularly of course as a result of the work and influence of Davis [cf. D. W. Johnson, 103]. In Germany, the course set by Richthofen, and followed particularly by Penck and his students, has made geomorphology so definitely a part of the field of geography that few, if any, question its permanent inclusion. (In the Netherlands, in contrast, geographers apparently distinguish more clearly between geography and geology in this field [92, 294].) Examining geography from the point of view of knowledge as a whole, the *Erkenntnistheoretiker* Kraft concluded that the study of geomorphology, including the genesis of landforms, disrupts the logical unity of the field, but that, as a result of historical evolution, this field is in fact included in geography— in Germany at least [166, 7]. Its inclusion cannot, therefore, be questioned, he concludes—so long as, one may add, the geomorphologists continue also to be geographers. In other words, the geographer in Germany is, by his training, a geomorphologist as well, and geography, therefore, as a division of labor within the sciences includes that special field.

Whether the same conclusion, based on the history rather than the logic of the field, applies in this country, the writer would not attempt to judge. It is important to note, however, that the close association of geomorphology with geography has brought the latter not only undoubted advantages, but also certain disadvantages. If geomorphology is primarily concerned with

[423]

landforms as objects to be studied in themselves, as the botanist is concerned with plants, then, as Michotte notes, the point of view is that of a systematic science, in contradiction to that of geography as a chorographic science [*189, 26*]. A general result of this contradiction is to be found in the difficulty that many geographers who have been trained primarily as geomorphologists have experienced in maintaining consistently the geographic point of view, as Penck himself has observed [*90*, I, 38 f.], not to mention the confusion that many of them have introduced into methodological thought in geography. A more specific result has been suggested in an earlier connection (Sec. XI D). The study of landforms as objects in themselves, leads logically to a classification of them as individual objects of a systematic science rather than to a classification of the areal character of landforms—"the character of the various morphologic areas of the world," as Michotte puts it. While geographers have felt free to classify climates, natural vegetation, or farm types independent of the classifications of the corresponding systematic sciences, as long as geomorphology was regarded as an integral part of geography, they were inhibited from developing a different classification of landforms suitable for chorographic description. In many cases, to be sure, the types of individual landforms were suitable, but the attempt to make them usable in all cases led, for example, to that paradox of areal terminology, the description of the White Mountains as "a collection of monadnocks."

Whatever conclusion may be drawn with respect to the relation of geomorphology to geography, it is necessary to note that if, as Kraft holds for German science, the facts of historical development make it a part of the field of geography, that conclusion does not provide an argument for the inclusion of logically analogous problems in other parts of geography in which they have not established themselves in the past. Thus, the claim made by Maull, and more recently by East, that the processes of evolution of state-areas are as properly a problem in geography as the study of the evolution of landforms, would be valid only if it could be shown that such studies in the geographic aspects of political history have in fact been developed primarily by geographers, rather than by political scientists or historians [*157; 199, 270; cf. 216*, 956 f.].

We may summarize briefly our examination of the relation of systematic geography to the systematic sciences. *Ideally,* systematic geography receives from other sciences, or from general statistical sources, the necessary data concerning the distribution of any phenomenon; it classifies the various forms of that phenomenon in any way that is suitable for geographic purposes—*i.e.,* in terms of characteristics significant to regional character—whether or not such classification is available from other sciences. Further,

[424]

ideally, it receives from the systematic sciences the explanation of the distribution of the phenomenon, that is, its genesis. Whether it be landforms, forests, crops, steel mills, or political states, the principles of development and the causes of distribution, as such, are the concern of the appropriate systematic fields. Geography starts with those facts and principles—assuming always, of course, that the systematic sciences concerned have provided them—as frankly borrowed material.

We have dwelt on the preceding question at some length because it is particularly in systematic geography that the student is likely to lose his sense of the geographic point of view, so that he may, as Lehmann has suggested, give a false picture by disproportionate consideration of phenomena that are the objects of the systematic sciences [113, 237 f.], or he may leave geography and enter entirely into other fields. It is doubtful if this can be prevented merely by drawing boundaries, however sharp and clear they may be. The reader may already have objected that we have drawn no clear boundary between systematic geography and any systematic science. No such attempt has been made, and if we remember that the relationship involved is not the borderland of neighboring fields but rather the intersection of fields lying in different planes, no such boundaries are needed. The distinction is in the point of view: that of the systematic science is focussed on the particular phenomena, which are studied in terms of distribution; that of systematic geography on the part which that distribution plays in forming areal differentiation. In many studies, the geographer may find it necessary to make excursions in the plane of the related systematic science, away from the common line of intersection. If he has the geographic point of view clearly in mind, he will need no boundary stones to remind him that he is making an excursion out of his field, but will return, as soon as he has established the necessary data or conclusions, to the geographic plane (See Fig. 1).

To maintain the geographic point of view in systematic geography, it is necessary for the student to refer his work constantly to the field of geography as a chorographic science. Most writers agree that this view is most clearly indicated in regional geography. Consequently, many have agreed with Penck that every geographer, no matter how great his interest in specialized systematic branches of the field, should make regional studies [129, 639]. In any case, it is essential, as Lehmann insists, in making any study in systematic geography, constantly to consider the relation of that study to regional studies [181, 49]. If a systematic study is considered from this point of view it is immediately clear that the interest of the geographer is not in the phenomena themselves, their origins and processes, but in the relations which they have to other geographic features (*i.e.,* features significant in areal differentiation).

On first thought it might be supposed that the conclusions to which we have arrived would result in the elimination of most, if not all, of the work in systematic geography. On the contrary, they free that part of the field for its essential function of providing systematic study of the relation of the differentiation of specific kinds of phenomena to total areal differentiation. The areas of the world differ from each other in terms of a mutually interrelated complex of heterogeneous features, each of which is different in the different areas. The complete interpretation of an individual area requires that, at some level of size, we break it down, mentally, into the component parts formed by the specific categories of phenomena. As Michotte puts it, we must study the vegetative character of the area, its geomorphological character, the character given it by each of the major cultural features, and so on [*189,* 17–33]. Further, the comparison of such completed individual studies would not give us a complete understanding of the areal differentiation of the world. It is necessary to know also how these areas, considered solely in terms of their natural vegetation, landforms, or each of various cultural features, differ from, and are related to, each other. Michotte speaks of these comparative studies as "comparative plant geography," "comparative morphological geography," etc. Similarly Hettner's title for his several-volume study of systematic geography, *"Vergleichende Länderkunde,"* though unfortunately misleading, as we have seen (Sec. II D), is not so inappropriate as might be thought. Likewise, we may add, it is significant that the physiographer, Fenneman, who first presented the chorographic concept to American geography should demonstrate that viewpoint, in systematic geography, in his masterly studies of the regional physiography of the United States.

It is particularly, however, the students whose interest in regional geography has motivated them to make systematic studies who have most clearly indicated the type of work that systematic geography should undertake. Recognizing that the relation of any specific feature to the character of an area is to be measured particularly in terms of its relation to the other factors in that total character, they have perceived that absolute mueasurements of individual elements are less valuable than relative measurements, or ratios, of elements in reference to each other.

One of the most important advances in making studies in systematic geography more geographic in character has been the development of the isoplethic method of mapping ratios. Based on the work of Engelbrecht,

[426]

this method was developed and effectively presented in this country by Wellington D. Jones [283] and is now in widespread use. Compare, for example, the utility, for a study of agricultural differences in different parts of China, of the maps of crop ratios that Trewartha has recently published [392] with the dot maps showing absolute values upon which we previously had to depend. That this method may be carried further, so as to show simultaneously two significant ratios concerning the same phenomenon, is indicated by the writer's isoplethic map of the dairy areas of the United States [325; the demonstration is very inadequate due to the small scale on which the map is reproduced]. This study (an extension and amplification of Jones's map of a smaller area) actually portrays the differences in areal character of land use resulting from varying degrees of intensity of dairy development, which can be inferred only indirectly, and in many areas incorrectly, from the ordinary census maps showing distribution of dairy cows or milk production. In both of the systems of world division of rural areas discussed in the previous section, the determination of agricultural types and the delimitation of "agricultural regions" depended on the construction of a large number of isopleth maps (not published) showing areal differences in ratios of individual crops to total crops, of cropland to total land, of live-stock units to cropland, etc.

The ratio method in systematic geography is not limited to studies in agricultural geography. In addition to ordinary "relief" maps, which actually show directly only elevation, maps of "relative relief" have been constructed after Partsch, by various European geographers, and, in this country, by Guy–Harold Smith [see particularly James' survey in 294, and Cressey's recent example, 338]. In "sociological geography," Kniffen has used the method in mapping house-types and the writer has used it to show the areal differences in racial construction of the population of the United States [359].

Even where the character of the distribution does not permit of isoplethic mapping, as in the treatment of characteristics of cities, the principle of measurement by ratios rather than by absolute figures may be used to bring out the differential character of cities—*i.e.,* that character, other than size, which is most significant in the comparison of cities, and in the comparison of regions in terms of their urban development. This is illustrated by the writer's study of the manufacturing belt of North America, in which the manufacturing functions of cities are measured in relation to their total functions, rather than in absolute values [326].

A much more complicated technical tool for work in systematic geography has but recently been presented by John K. Wright, under the title of "Some Measures of Distribution" [293]. The fact that it involves rather complicated mathematical formulae should not prejudice geographers either for or against its use. The various small examples which Wright offers suggest that we may have here a new technique of great value in enabling systematic geography to arrive at conclusions useful in regional geography that will be far more accurate than those now available. While this possibility is suggested by the examples which he gives, for the writer, at least, they seem inadequate to establish the utility of his technique. It is to be hoped, therefore, that some student will be interested in applying the technique to some actual problem to see what results it may produce.

In their simplest form systematic studies in geography are confined to single elements. We have previously noted, however, the importance of the concept of "element-complex" in geography—*i.e.*, an interrelated association of various elements, regardless of kind. If approximately the same element-complex is found repeatedly in different areas and its distribution is geographically significant, it may also be studied systematically—over the whole world or any large area. Such studies, interconnecting different branches of systematic geography, may be considered as stepping stones from the study of single elements to the study of the total complex of a particular area in regional geography.

A single element-complex may represent an interrelation of elements at a single point—*e.g.*, rainfall, temperature, slope, soil, drainage, and vegetation—in which case we may speak of a vertical complex of indefinite horizontal extent. On the other hand, the elements may be situated at different points so that their interrelation constitutes an areal form of more or less definite horizontal extent. Thus a longitudinal **U**-shaped Alpine valley in its primeval condition was a natural-complex in which the factors of slope, soil, drainage, and vegetation varied in a definite manner from the mountain shoulder on one side to that on the other. It is an areal form fairly definitely determined in the transverse direction though its limits in the longitudinal direction are indefinite so far as the concept itself is concerned— *i.e.*, are determined only in each specific case. A *polje,* in contrast, is a similar complex areal unit definitely limited in all directions.

The Estonian geographer, Markus, has contributed an interesting and suggestive study of element-complexes (*Naturkomplexe*) confined largely, if not entirely, to combinations of natural elements [239]. He notes that geographers have long recognized certain more obvious cases of element-complexes by such terms as "tundra," "high moor" and "low moor," "grass-rich depressions in steppes," etc. Noting that changes in any factor in a complex do not cause immediate change in the others, but rather that these adjust themselves to the new conditions at different rates of speed, he distinguished between "normal complexes," in which all the elements correspond to each other completely, and "abnormal" in which the adjustment has not yet reached completion. These terms give a clearer description than the distinction between "harmonious" and "inharmonious" that other writers have suggested in a similar connection (see Sec. IX D). Markus speaks of a "positive shifting" of an element-complex where it is pushing into an area of another complex that requires a lesser amount of any particular factor—as in the advance of forest into steppe—and negative shifting in the reverse direction. Further, he projects a complete classification of element-complexes in which real complexes are reduced to abstract species or types —by consideration of their essential common characteristics—and in which these are arranged in families, orders, etc.

This ultimate object—establishment of a Linnean classification of types, even of abstract types, of element-complexes—faces essentially the same insurmountable difficulties as we have met in the attempt to arrange regions in a single system of classification. Forested mountains, forested plains, mountain steppes, and plain steppes are four distinct types of element-complexes that cannot logically be arranged in any unilateral system of classification, since we have no method of deciding objectively whether the difference between mountain and plain is more important than the difference between grass and forest. Likewise, we cannot accept Markus' further implication that a geographic region (*Landschaft*) can be expressed as a single type of element-complex; if we consider all the elements involved in the complex of which the region consists, we arrive at the unique case, not the type (see Sec. X E). Nevertheless, though we cannot accept the more optimistic conclusions that Markus draws, we can expect valuable results from the systematic study of particular types of natural element-complexes, each of which is to be regarded as expressing more of the character of any area than a single element, though not its full character, not even in outline.

Our previous discussion of regional division indicated that we may expect much more useful results from the study of the many element-complexes, extending in many cases over wide areas, that have been produced by the organizing hand of man. These complexes are of a different order from the natural element-complexes that we have discussed, in that they are not merely the sum total result of the interaction of forces accidentally placed together—any one of which may be understood by itself in its relation to the others. These cultural element-complexes have been purposely created by man for the sake of the ultimate result, and the presence of any one element is to be understood not in terms of its relations to the others but in its relation to the ultimate result. For example, the importance of oats in the Corn

[429]

Belt is to be explained in terms of its significance to the total crop and live-stock association that man has organized on Corn Belt farms, or rather, if one will, in terms of his ultimate purpose of securing the highest monetary return with the least expenditure of labor, capital, and land. Consequently, as noted earlier, these cultural element-complexes are, to a considerable degree, organized as unit Wholes, an understanding of which should be a first step in the development of cultural regional geography.

The relatively small units of cultural element-complexes—*e.g.*, farms—involve a much larger number of factors than those commonly found in natural element-complexes and include both material and immaterial elements. O. E. Baker has recognized that contrasts in types of farms in the United States involve contrasts in the character of the farm population: farmers represent farm elements just as definitely as do the livestock and crops [*312*].

Though the cultural complexes commonly form but small areal units they may be found to be organized together, by man, into looser areal complexes that, individually, cover relatively large areas.

Finally we may recognize complexes involving both cultural and natural elements. Because man, in many cases, has developed the same cultural element-complex in areas of similar natural conditions, we may expect to find a number of complexes consisting of cultural element-complexes in interrelation with certain natural elements. Since man, however, has been far from consistent in his form of adjustment to natural conditions, we must expect these compound element-complexes to have relatively restricted applicability.

Geography finds that certain element-complexes with which it is concerned have been studied by other sciences. If the complex includes only elements of the same general category—*e.g.*, the natural vegetation as a complex of different plants, or an iron and steel works as a complex of blast-furnaces, steel furnaces, rolling mills, fabricating mills, storage yards, etc.—one of the systematic sciences will presumably be concerned with the study of the complex in itself and in its distribution over the world. The more complex forms however—involving combinations of heterogeneous phenomena—may be of concern to the geographer alone. In either case, the classification of the types of element-complexes for use in geography must be adapted to geographic purposes. If economists have produced a classification of farms suitable for geographic purposes, the geographer will utilize that classification; but if not, he is free to develop his own. In this particular case it appears likely that both groups of students working in cooperation may develop a classification suitable from the point of view of both sciences [*cf.* *320* with *319* and *324*]. But if economists' classification of manufacturing industries offers little of value to geography, geographers must develop their own.

Any study in systematic geography, whether of an element or an element-complex, concentrates on one particular kind of phenomena or phenomenon-associations. It naturally leads, therefore, to the establishment of generic concepts; that is, for each element, or element-complex, a logical system of types may be established. On this basis the relations of the feature studied to other geographic features, for which types have also been established, may be stated in the form of principles, however limited or inaccurate in application. Whether one considers rainfall, soils, stream deposits or erosion, crop-animal associations, steel mills, or political boundaries—in each case over wide areas if not the whole world—it should be possible to establish principles of relationships between the feature studied and other features geographically significant.

It is by no means possible, however, to express all the findings of systematic geography in generic terms, whether of concepts or principles [Hettner, *167, 283*]. In the systematic study of volcanoes, Krakatoa cannot be adequately treated solely as an example of a type. Likewise, in the relations of one geographic feature to another, innumerable cases will be found, each of which is unique. Nevertheless a very large part of the work in systematic geography does deal with universals and leads to the development of principles. Do these provide geography with that precious power that is often regarded as a hall-mark of "science," the power to predict?

The essential characteristics of that form of "knowing" which we call science—to use the term suggested by the physicist, John T. Tate[103]—are not determined by the character either of the knowledge or of the capacities acquired; these are rather resultant products of that manner of pursuing knowledge which is science. The ability to predict in any branch of science represents the attainment of such a degree of certainty of knowing that, by deduction from principles, the future outcome of a combination of present factors is known almost as certainly as it can later be known as an observed fact. The qualification "almost" represents more than a margin of error, of inaccuracy in measurements: in every field there is an ever-present margin of uncertainty, of not-knowing, which cannot be eliminated even in the physical sciences—nor do modern physicists expect that it may ever be eliminated.

[103] In an informal talk to a group of colleagues at the University of Minnesota, on "What I think about Knowing."

The ability of any science to predict is therefore the result and outward evidence of a high degree of attainment of the ideals of accuracy, certainty, universality, and system. It is not the test of a science, but only the test of success in "knowing" in any science. That success is not to be attained by striving directly for its result, the power to predict, but rather by striving for the highest degree of attainment of the fundamental ideals of "knowing."

In plain words, we will not learn to predict in geography by attempting to predict. It is a corollary of scientific principles that we should seek to know to what extent our knowledge is incomplete or uncertain. To attempt predictions in situations for which we know we lack the necessary knowledge is to be unscientific. Science does not require that we be able to predict. The sound demonstration of a low capacity for prediction in any particular field of knowledge is not evidence that that field is not a branch of science, but, on the contrary, is a scientific conclusion testifying to the scientific character of that field.

No professional geographer, I presume, would claim that research in systematic geography had as yet reached such a high degree of attainment of the requisite ideals as to enable it to make predictions of high degree of certainty. Though a more maturely developed geography should show far higher attainments, we must recognize certain insurmountable difficulties and limitations that will always be present in geography.

We know, in the first place, that the nature of most of the phenomena that must be measured in systematic cultural geography, and of many of those in systematic natural geography, will never permit of such accurate and certain measurements as are possible in some of the natural sciences. This difficulty geography shares with many of the systematic sciences, notably, of course, the social sciences.

We know that our knowledge of phenomena and their interrelations, in every branch of systematic geography, can only incompletely be contained in generic concepts and principles, and that there is inevitably, therefore, a margin of uncertainty in prediction. While this margin is present in every field of science, to greater or less extent, the degree to which phenomena are unique is not only greater in geography than in many other sciences, but the unique is of the very first practical importance. This is true not only of geography and the social sciences but likewise of human physiology and psychology—from the point of view of the individual and his family, at least —and of certain aspects of meteorology and geology. To predict that the islands of Japan will experience innumerable earthquakes is of little value; who will predict the date and location of the next major earthquake?

We know further that the complex interrelations of phenomena that we

* See Bowman [106, 17, 31-33].

study in systematic geography cannot be taken into the laboratory where some of the variables may be controlled in experiments, so that we could learn the exact significance of each factor. This handicap, again, geography shares not only with the social sciences but also with human physiology, with most of the branches of geology, and with astronomy. The students in all these fields can work only in the laboratory of reality, can observe only those experiments that reality chooses to perform for them. Actually the same limitation applies to the physicist studying the actions of electrons in the laboratory; he cannot control the individual electrons which he is attempting to study. There is "no hard and fast line between observation and experiment," Cohen concludes [*115*, 111].

In the performances that reality presents us as substitutes for laboratory experiments, we know that geography is handicapped in two ways. Whereas some fields are presented with thousands or millions of repetitions of nearly similar cases, in geography, as well as in geology and astronomy, and in parts of all the social sciences, there may be only hundreds of similar cases, or only a handful, or often but a single case. Where the number of factors involved in the relationships is relatively small, as in astronomy, and the clearness and exactness of observation may be at least as fine as in the laboratory, a few cases may suffice to provide an adequate basis upon which to develop scientific laws of high degree of certainty; but in geography, geology, and the social sciences, one or both of those conditions are lacking.

One further difficulty remains in any branch of science that must deal with extremely complex functions of a large number of more or less independent variables. If some divine power should present the scientist with complete statements of every one of the interrelations involved, expressed (if that were possible) in mathematical equations of the greatest complexity, and if then, in any particular situation, divine power should also provide complete and accurate knowledge of the individual factors, the complexity of the problem which must then be solved to arrive at certain knowledge of the outcome would be beyond the ability of finite minds. *

In conclusion, therefore, geography is, by its nature, one of the branches of science from which we are to expect relatively little knowledge of the future of such a degree of certainty as to justify the word "prediction." ** Undoubtedly one could postulate many cases where such certainty was possible—even in our present state of development. If large deposits of high-grade iron ore should be discovered in West Virginia we could not only predict the mining development that would result, but we could no doubt predict, in a general way, notable changes in the iron and steel industry in the Pittsburgh and Calumet Districts. The reader will observe that this is

[433]

* A conclusion suggested by the philosopher, Charles Hartshorne.
** This is an over-statement. See Supplementary Note 55

not only an extreme case but one in which the relationships are unusually simple, since but three variables are of major importance [*352*]. For the most part, the knowledge of the future that systematic geography can provide is limited to that lesser degree of certainty that we express by such terms as "trends" or "likelihoods," and must further be qualified by many uncertain factors due to the more or less arbitrary action of individual men or groups of men.

In sum, we may justifiably predict that a mature geography developed to the maximum, cannot attain more than a very restricted capacity for prediction [*cf.* Schmidt, *7,* 210–13; Colby, *107,* 35 f.; and Finch, *223,* 19].

While we may dismiss the question of ability to predict as not fundamentally relevant, the pursuit of universals, generic concepts, and principles, must be regarded as of major importance for the development of any science. In geography, the greatest opportunity to develop generic concepts is in systematic geography. A large part of the work in each section of systematic geography is concerned with phenomena and relations between phenomena that repeat themselves in similar specimens in different parts of the world, so that it is possible to express them in universals and thereby to develop principles. Consequently, for those among the ranks of geographers who by reason of temperament, ability, or training, prefer to study the generic, with the opportunity to develop scientific principles or laws, there is plenty of work to be done in geography. Since such work is not merely an integral part of geography, but forms the necessary base for the studies of regional geography, no geographer need berate these students as deserters from the field. Those who use Fenneman's picture of the field of geography and speak of regional geography as the center, should not overlook his qualification: "There is no intention of assigning more dignity to one part of the field than to another, nor of asking any man to turn aside from that which interests him to something else. There is no more inherent worth in a center than in a border" [*206,* 10].

Our examination of the character of systematic geography emphasizes the inescapable comprehensiveness of geography, a condition that would not be reduced in the slightest if one were to omit regional geography; on the contrary, the one method by which the diversity of interests is brought into unified study would be lost. Even if one should attempt to reduce systematic geography to the study of natural, non-human, elements, it would still be concerned with a heterogeneity of phenomena as great as that of all the systematic natural sciences put together, and the elimination of human factors would make it impossible to unify this diversity in the study of actual regions. Any attempt to arrive at a unified field by further whittling can change the situation only relatively: if one throws out plant and animal geography, one still has subjects as different as the study of climates and of landforms. As these are both physical sciences, and are both concerned with the earth, they can logically be combined, either from the point of view of physics, or of the earth. From the point of view of physics they are widely separated fields that are not brought into logical combination by the incidental fact that they both concern the earth. If the earth forms the unifying framework they are combined only in the earth surface, as broadly conceived, where they are inextricably intermixed with the elements studied in plant, animal, and human geography. Only in the study of all earth surface features in their actual interrelated combinations in regions, can the heterogeneity of systematic geography be unified into one science. "We need not be frightened away by the fullness and breadth of the problems," Richthofen concluded. "The field is great. But the work can be divided among many. No one today can do research in all the parts of geography. But he who devotes himself seriously to geography, can master it sufficiently to follow advances in all branches; and he who, through modest limitation, is fortunate enough to investigate productively in one part, should always strive to comprehend the relation of that part to the rest and never to lose sight of the interconnection of the whole" [73, 67–70].

At the same time, Richthofen felt that the individual geographer who wished to contribute research to the advance of geography, "the higher he sets his goal, the more should he concentrate his preparation on one part of (systematic) geography and the particular systematic science that forms its foundation, without neglecting instruction in the other parts." It was natural for Richthofen to emphasize geology as "the surest foundation," since that had been his own, but both Oberhummer and Gradmann are drawing the logical conclusion from his general principle when they state that individual geographers may just as properly select some other science as their principal supplementary field—whether meteorology, botany, economics, or some other [124, 11; 251, review, 552].

An individual geographer who specializes in a branch of systematic geography, and is adequately equipped in the corresponding systematic science, will no doubt have occasion to make studies in that other field as well as in systematic geography. Just as it has always been regarded as appropriate for individual geographers who were adequately equipped therefor to do research in geology, it may similarly be appropriate for individual geographers—under the same condition—to do research in anthropology, economics, or political science. Inasmuch, however, as geographers are not capable of judging the research in other fields, it seems logical that such research should be presented, not to geographers, but to the workers in those other fields.

Need it be added that such transfers of point of view may equally well be made in the opposite direction? The student of a systematic science, interested in the geographic aspects of his field, will frequently be able to contribute to the field of geography, and one trusts that in this exchange there need be no grumblings of trespassing on either side.[104]

H. THE CHARACTER OF REGIONAL GEOGRAPHY

The development of geography during the past thirty-odd years has been marked by an increasing interest in regional geography. Under the leadership of Vidal in France, of Hettner, Penck, Gradmann, Passarge, and many others in Germany, European geographers gradually shifted away from the concentration on systematic geography, which had been a natural result of the emphasis on universals in all science. Likewise, in this country, the programmatic papers of Barrows and Sauer, however divergent in other respects, agreed in the emphasis on regional studies as the core of geography [208; 211]. Though Pfeifer is correct in noting the similarity of these, the two most influential methodological statements in current American geography, he over-estimates their importance in determining the course of current thought in American geography [109, 96 ff.] by failing to note the major degree to which, like the earlier methodological pronouncements of the presidents of this association, they simply "mirror . . . geographic opinion in America" [94]. As Platt has pointed out, the roots of the current movement, in particular of the tendency for detailed studies of small areas, reach back to geological field courses before the World War and military mapping during the War.[105] It is neither possible nor necessary to determine even approximately what forces or what individuals have been responsible for this development. Mention should certainly be made of the influence that Bowman, as Director of the American Geographical Society, exerted towards intensive regional studies [cf. 106]. Possibly most important of all has been the personal influence exerted by the group of Midwestern geographers

---

[104] It may be admissable to add that a logical corollary of this situation is to be found in the character of membership of this Association. The relatively large number of specialists from other fields, included as geographers, does not represent the normal overlapping along an actual border line between sciences, but rather the fact that geography, by cutting through the systematic sciences, in a sense includes all of them.

[105] In a paper read before the association at the recent meetings, 1938. Specifically Platt notes that the first publication cited by Pfeifer as containing "proposals made by Sauer" [footnote 12] actually consisted of proposals, presented without distinction of authorship, of both its co-authors (as well as of other unnamed members of a seminar group at the University of Chicago): W. D. Jones and C. O. Sauer: "Outlines r Field Work in Geography," BULL. AM. GEOGR. SOC., 47 (1915), 520–5.

whose annual field conferences, in the years 1923 and following, concentrated the attention of a much larger number of workers on the problems of regional mapping [note, for example, the report of the joint conclusions of this group (see the bibliography for its members), which Jones and Finch published in 1925, *281,* as well as the significant studies listed as *282–290*, incl.]. *

If geography, in America as well as in Europe, may be said to have returned, in a certain sense, to the point of view that was common with Humboldt and Ritter (see Sec. II D), its long period of concentration on systematic studies has enabled it to return far better equipped with generic concepts and principles with which to interpret the findings of regional geography— though unfortunately this equipment is relatively deficient in respect to human or cultural features, both in geographic literature and in the training of most of its students.

Many geographers who have accepted this shift in emphasis evidently have done so under the provisional assumption that regional geography is to be made as "scientific" as systematic geography has been, that somehow it must be raised to the plane on which scientific principles may be constructed. We have noted a number of difficulties into which this ambition has led. In our final consideration of regional geography it is necessary to understand clearly certain limitations imposed upon the student that are not found in systematic geography.

After a number of unsuccessful attempts to express the special nature of regional study in words, I find it can be most clearly presented if we may use mathematical symbols, though we shall not, of course, find it possible to express such complicated problems in any real mathematical formulae or equations.

Any particular geographic feature, z, varying throughout a region, might theoretically be represented as a function, $f(x, y)$, x and y representing coordinates of location. As a function of two variables, any such feature that we are able to measure mathematically—such as slope, rainfall, or crop yield —can be represented concretely by an irregular surface. Such a surface would then present the actual character of that feature for the whole region; it would, theoretically, be correct for every point, and for every small district. Furthermore, if the function involved were not too complicated, the theory of integral calculus would permit us to integrate the total of that feature for any limited section, as well as for any individual point. In a sense, part of our work in systematic geography corresponds to this form of presentation.

Likewise, the relation of any two or three geographic factors to each other

[437]

within a region—*e.g.*, the relation of crop yield to rainfall and humus content of soil—might be represented as a functional equation involving that many variables: $z_3 = f'(z_1, z_2)$. The concrete representation of this relation would require again a surface form. More commonly, in systematic geography, we consider only the relation of one factor to but one other, which we may then represent as a curve on a plane surface. Each of these factors, z, is of course a different function, $f(x, y)$, and the more complex equation, $z_3 = f'(z_1, z_2)$ holds true only if z_3 is unaffected by other z factors, or if those which affect it are constant throughout the region under consideration. Neither of these conditions is strictly true: almost any geographic element we may consider is affected by more than two of the natural elements, and may also be affected by incommensurable, or quite unknown, human factors; and all of the factors considered vary to some extent no matter how small the area considered. Consequently, we have introduced a degree of distortion of reality even at this step in systematic geography.

We may introduce a further step by establishing element-complexes, u, each representing functions of many z elements, varying, by more or less regular rules, with the variations in a smaller number of those elements. Thus, given certain conditions of soil, slope, temperature, and rainfall we may presume within a wide margin of both inaccuracy and uncertainty, certain conditions of natural vegetation and wild animal life, and we may express the total of all these z elements by one u element-complex. If it were conceivable that we could express this feature, u, arithmetically, its character over an area would likewise form an irregular surface that would indicate its character for any limited part. From the nature of these element-complexes, however, it is obvious that any such representation would have a high degree of unreliability.

In regional geography, however, we are concerned with a vastly more complicated function of the location co-ordinates. It cannot be expressed as the function of any one element or element-complex, but rather of various semi-independent element-complexes, u, and of additional semi-independent elements, z'. Thus, the total geography, w, at any point, might be expressed by the function, $F(u_1, u_2 \cdots u_n, z'_1, z'_2 \cdots z'_n)$. If we could have accurate and complete information concerning the form of the function, F, and every one of the element-complexes, u—each as a function of various z elements—and of the semi-independent elements z', the function would be so complicated that we could not hope to represent it by any concrete form, even in terms of n-dimensional space. We would have a function that could be solved only for each point, x, y, in the region, but could not be correctly expressed for any small part larger than a point. In other words, we could study the geography of the area only from the study of the geography of the infinite number of points within it. This task, being infinite, is impossible. The problem of regional geography, as distinct from a geography of points, is how to study and present the geography of finite areas, within each of which the total complex function involved depends on so many complex functions, complexly interrelated, as to permit of no solution by any theory of integration.

Consequently we are forced to consider, not the infinite number of points at each of which w is in some degree different, but a finite number of small, but finite, areal divisions of the region, within each of which we must assume that all the factors are constant. In order, then, to cover an entire region we will need but a finite number of resultants, w, each representing the geography of a small unit of area rather than of a point. This method is legitimate only if one remembers that it inevitably distorts reality. The distortion can be diminished by taking ever smaller unit areas, but it cannot be eliminated entirely; no matter how small the unit, we know that the factors which we assume to be constant within it are in fact variable. In practice, the smallest units that we can commonly take time to consider are sufficiently large to permit of a marked degree of variation, and therefore of a significant distortion of reality in our results.

To express our conclusion in more common terms, in any finite area, however small, the geographer is faced with an interrelated complex of factors, including many semi-independent factors, all of which vary from point to point in the area with variations only partially dependent on each other. He cannot integrate these together except by arbitrarily ignoring variations within small units of area, *i.e.*, by assuming uniform conditions throughout each small, but finite unit. He may then hope to comprehend, by analysis and synthesis, the interrelated phenomena within each particular unit area.

Although the studies of all the unit areas added together will constitute an examination of the entire region, this does not complete the regional study. As Penck has emphasized, it is not sufficient to study individual "chores" (approximately homogeneous districts) and to establish types of chores. "Above all geography must consider the manner in which these are fitted together to form larger units, just as the chemist does not limit himself merely to studying the atoms, but investigates also the manner of their situation beside each other in individual combinations. The comprehension o geographic forms (*Gestalten*) has scarcely been taken into consideration by the new geography." Just as a mosaic cannot be comprehended, Pencl continues, by classifying and studying the individual stones of which it is made, but requires also that we see the arrangement and grouping of the individual pieces, so the study of the arrangement of the "chores"[106] will present different structural forms of significance [*163*, 43 f.; in part also in his address given in Philadelphia and published in English, *159*, 640].

Our second step—in a theoretical approach to regional geography—is to relate the unit areas to each other to discover the structural and functional formation of the larger region. Since all the factors concerned, and therefore the resultants, have been made arbitrarily constant for each small unit, it may be permissible to speak of functional relations between one factor in one unit and another in another unit, as though these were functional relations between the units themselves—provided that we understand that this is not strictly true. Further, the regional structure produced by this method will have the character of a mosaic of individual pieces, each of which is homogeneous throughout, many of them so nearly alike that in any actual method of presentation they will appear as repetitions in different parts of the region. But we are not to be deceived into regarding this mosaic which we have made as a correct reproduction of reality. It is simply the device by which finite minds can comprehend the infinitely variable function of many semi-independent variable factors. The fiction involved is threefold: we have arbitrarily assumed each small unit area to be uniform throughout; we have delimited it from its neighbors arbitrarily, as a distinct unit (individual); and we have arbitrarily called very similar units identical in character.

There are certain other fundamental limitations that must be insisted upon if we are to compare the face of the earth, even in the more or less distorted form in which the geographer must present it, to a mosaic. We may say that there is a similarity in the detail of technique but, unless we are to return to some teleological principle, we cannot liken the face of the earth to any work of art, for we cannot assume that it is the organized product of

---

106 This is the word that Sölch introduced as a term for a unit area [*237*]. As he defined it the concept is independent of size; the chore is simply an area of land determined by the relative degree of homogeneity of all geographical factors—"geofactors." A chore established on any particular scale could be divided into smaller chores each of which would presumably show a higher degree of homogeneity; the limit of such a process is, of course, the perfectly homogeneous unit, which can only be a point. In adopting this term Penck has used it in a different meaning, according to which the "chores" appear as the smallest land units, indivisible cells, so to speak, which he adds up to form larger "forms." We do not follow this usage, not only because it changes the meaning of the term as the inventor defined it, but also because there can be no smallest land units. As Penck himself elsewhere has recognized, we may continue the process of division indefinitely and our subdivisions are no less (and no more) real units than those we divided.

a single mind. On the contrary, if we may transfer Hettner's analogy of a building built by several architects working independently to Huntington's picture of "The Terrestrial Canvas," we may say that the face of the earth has been produced by the interrelated combination of different color designs each applied by different artists working more of less independently, and each changing his plan as he proceeded. In systematic geography one might say, we attempt to separate each of the individual designs in order to understand its form and its relation to the others and, thereby, to the total picture. Since the total pictures were not produced simply by superimposing different color plates in printing, but are, to some extent, causally related to each other, this separation involves the analysis of the causal and functional relations of each design to the others. In regional geography we first reduce the subtle gradations which the different artists of nature have applied and intermixed on the face of the earth, to the stiff and arbitrary form of the mosaic technique. When we then survey the formation of the mosaic pieces, we are not to expect some unified organized pattern such as every work of art must have. On the other hand, neither need we expect mere chaos, or a kaleidoscope; for we know, from our studies in systematic geography, that there were principles involved in the individual designs, and if our determination of the unit areas of homogeneity has not been purely arbitrary, but has been based on the combination of careful measurement and good judgment, we may expect the combinations of these designs to show more or less orderly, though complex patterns. Further, whatever the explanation of these patterns may be, their form is significant to each of the parts, since the development in each unit part is affected by that in the others.

The last thought leads us finally to another major respect in which any analogy of the earth surface to a work of art is inadequate, namely, the fact that, while the latter is static, consisting of motionless forms, the face of the earth includes moving objects that are constantly connecting its various parts. (To attempt to introduce the artist's special use of such terms as "lines of force," "movement," "opposite forces," etc., would merely add to confusion here.) In other words, the geographer must consider function as well as form. In establishing our arbitrary small unit areas we not only assume that each is uniform throughout in character, but also in function. Likewise, in combining these units into larger regional divisions our problem is complicated by the fact that we must consider the functional relations of the units to one another as well as their form. For example, if two neighboring areal units are so similar that we have painted them as much alike as two pieces of mosaic of the same color, but one of them is functionally related to a city center in one region, the other to a city center in another, are we to include them in the different regions, or, if in the same region, in

[441]

which? Any answer to this question can only be more or less intelligent: there can be no one "correct answer."

Just as it is necessary to know the arrangement of unit areas in a region, it is likewise necessary to understand the arrangement of regions to each other. Both Penck and Granö (who follows a similar line of thought [252, 28–31]) would carry the process on to larger units; the size of the areas concerned is immaterial. Regional geography, therefore, studies the manner in which districts are grouped and connected in larger areas, the manner in which these larger areas are related in areas of greater scale, and so on, until one reaches the final unit, the only real unit area, the world.

There is, however, one important difference at the different levels of integration. Both Penck and Granö appear to ignore the fact that the small, but fundamental, element of fiction in the assumption of homogeneity of the smallest units of area increases progressively as one advances to larger divisions. Consequently, the determination of these larger divisions requires increasingly arbitrary distortions of fact.

Assuming the first step, the establishment of "homogeneous units" of area, we may proceed to the second by enclosing in a continuous area which we call a region, the greatest possible number of "homogeneous units" that we judge to be nearly similar, together with the smallest number of dissimilar units. Our judgment of similarity will involve subjective judgment as to which characteristics of the homogeneous units are of greater importance than others, so that, at best, the determination of the region is in a sense arbitrary.

Furthermore we seldom find in reality such a simple solution as that described. Though some geographic features vary but gradually from place to place, the irregular and steep variations of others—such as soils, slopes in mountainous areas, urban settlement, and all the features of essentially linear form, rivers, roads, and railroads—will force us to include in any region, "units" of quite different character. It is necessary therefore to determine which kinds of units are, either in actual interrelation or merely in juxtaposition, characteristic of the region as approximately considered, and then so determine it as to include the greatest number of those several kinds of similar units, with the smallest number of units of other kinds.

In considering any large area in which we have first recognized "homogeneous units" and are attempting to form them into regions, which we can briefly characterize in terms of similarities or relations among some of those units, we may find the task relatively simple in parts of the area, where perhaps the great majority of the units are notably similar. But it may be extremely difficult in parts between these, which may be characterized by units that are, in some respects, similar to units on one side of them, in other respects, to units on another side. Further, we will find areas containing such a variety of different kinds of units that we cannot see where to include them. In some cases, to be sure, we may recognize such areas as transition zones, but that merely postpones the fundamental problem without solving it. Likewise, to call them "characterless areas," or areas of "general" or "mixed" types is simply to dodge the problem entirely (see Sec. IX E).

The individual student, no doubt, would gladly wipe such troublesome areas off the map, but he is not granted that privilege. Neither is a science which seeks to know what the world is like permitted to ignore more difficult areas and confine itself to those easier to organize into its body of knowledge. Since these doubtful areas are commonly not merely narrow borders of transition, but areas of wide extent, perhaps as great or greater than those more clearly classified, there is no basis for assuming that they are of less importance in the total picture of the larger area, or of the world, than the areas whose character we can more readily describe. Fenneman's statement with reference to the different parts of geography applies even more literally to parts of an area—"there is no more inherent worth in a center than in a border."

Consequently, when we divide any given area into parts which we call regions, so determined that those characteristics that we have judged to be most important may be most economically stated for each region, we cannot avoid many decisions based on judgment rather than on measurement. We must, therefore, acknowledge that our regions are merely "fragments of land" whose determination involves a considerable degree of arbitrary judgment. On the other hand, if all possible objective measures have been used, and the arbitrary decisions are based on the student's best judgment, we may properly regard his regions as having more validity than is expressed by the bare phrase "arbitrarily selected." On the other hand, the view of various writers previously noted, that geographers could be expected to come to approximate agreement on the specific limits of regions—or even on their central cores—appears, in view of all the difficulties listed, overly optimistic.

It hardly needs to be added that the conclusion that geography cannot establish any precise objective basis for regional division does not permit it to shirk the task of organizing regional knowledge into areal divisions determined by the best judgment possible. In order to utilize the generic concepts and principles developed in systematic geography to interpret the findings of regional geography, the latter must be organized into parts that are as significant as is possible. In the present state of development of the field —if not indefinitely—we do not have what would be the simplest solution, namely, a single standardized and universally accepted division and subdivision of the world into regions. Therefore, each student of regional geography has imposed upon him the task of standardizing his own system of regional division—unless he can utilize that of some colleague. "Standardized" is used here to indicate that the regional system is based on certain standards specifically stated, so that other students may know precisely what the organization is.

The complete organization of regional knowledge in geography requires —whether as a final or as a primary step—the division of the whole world. In whichever direction the process is carried on—and we noted that it requires consideration in both directions (Sec. X A)—the completed system must provide a regional division of the world in which our knowledge of each small part may be logically placed. For this extremely difficult problem we found two different methods of solution. Geographical knowledge may be logically arranged in systems of areas classified according to certain characteristics of the areas. Though this method has distinct utility for comparative purposes, it does not permit organizing all regional knowledge into one system, but requires several independent systems. Furthermore, it does not present the actual relations of areas as parts of larger areas. These relations can be included only in a realistic division of the world into a system of specific regions, in which all regional knowledge may be incorporated in a single logical system. Such a system unfortunately is not provided the geographer by any natural division present in reality, nor by anything corresponding to the simple division of organic forms. It must be developed and constantly modified by geographers as a result of research, at the same time that it is being used, always in tentative form, as the organizing structure of regional research.

We have suggested, in very general terms, the manner in which the problem of delimiting regions may be met, in order that geographic knowledge may be organized intelligently in regional units. What kind of knowledge is to be included within the regional study itself? So far as the nature of the material is concerned, we have previously indicated that a complete geography of a region includes all the kinds of phenomena that are included in systematic geography—insofar as they may be present in the particular region. The only field of geography that is not included in regional geography, as well as in systematic geography, is historical geography. As there was a different geography in every past period, there may be any number of independent historical geographies, each including its own systematic and regional divisions.

[444]

The kinds of phenomena present in regions, the particular manner in which they are present, and the nature of their interrelations, both within each unit area and across unit divisions, determine the particular forms and the functions of the area. Though most students agree in theory that these are of coordinate importance, much of the work of recent decades tends to emphasize the study of forms to the neglect of functions. We found this to be particularly pronounced in the work of the "landscape purists" (Sec. VII E). On the other hand, Granö finds that many students, like Spethmann in particular, conceive of an area as "the field of forces, as a dynamic complex." Geography, Granö insists, is not the study of forces, of interrelations, but the study of things in interrelation in areas. Judging by the major example which he has presented in German, Granö himself tends to emphasize physiognomy and gives but little attention to the functions of areas [252, 114 f.; cf. Waibel, 266, 204].

When we speak of the functions of areas, we are not to forget that in reality the area is not a thing that functions, it is only certain things within it that have functional relations to things in other areas. If our fiction of the small homogeneous areal unit, uniform in both form and function, permits us to speak figuratively of the unit area as having a functional relation to other unit areas, we are not to ignore the fictive character of this concept by attempting to consider areas as having, in themselves, functional relationships.

In particular, it is necessary to note that the concept of the small areal unit breaks down when we attempt to study "the genesis of an area." When we study the previous historical stages in the geography of the area, we find that any one of our small unit areas of homogeneity may not have had in the past even that incomplete degree of validity that we may grant it today. That is to say, since areas, no matter how small, do not grow as units, but change only as a result of the differential change of different things within them, the unit area of today was probably not a unit area in an earlier stage, and will probably not be in a future stage. The very concept of mosaic is incompatible with the concept of gradual and differential change. Consequently, the study of genesis in geography can only be undertaken in the form of systematic studies: the study of "the genesis of an area" can only be broken down into studies of the genesis of each of the various objects contained within it. These are therefore studies in systematic geography; to what extent they may be desirable for an understanding of the geography of any region is a controversial question which we touched on earlier, and need not reconsider here (Sec. VI B).

[445]

We should now be in a position to answer the question that is of greatest importance in contemplating the possible development of regional geography: may we hope to progress in this branch of our field to the construction of universals, of generic concepts, and scientific laws or principles?

One form of generalization used in regional geography we have already described: the construction of regions from small unit areas. The philosopher, Kries, has distinguished such generalizations of heterogeneous and semi-independent parts as a third type of scientific description, together with type concepts and the description of the unique [according to Graf, *156*, 57–62, 105]. The importance of the distinction lies in the fact that this form of generalization offers no basis for establishing general principles; for that we must have type concepts.

It is obvious that any universal principles that we might attempt to construct on the basis of the fictive areal units set up for the purposes of description, could have no more validity than the units themselves. Unless these are taken as extremely small units, the margin of error introduced by our personal judgment would lead, in any principles we might set up, to a degree of error so great as to render them of very doubtful value.

Regardless of that essential difficulty, however, we found that even these arbitrary units, each involving a complex combination of associated forms, cannot be classified into a system of types based on the sum totals of its varied and semi-independent factors. Though in any one region we find unit areas so similar that we may, with but a minor degree of error, call them alike, we do not find unit areas of that kind of similarity in other regions of the world. A small district somewhere in the Upper Rhine Plain may be very much like many other such districts in the same region, but no matter how small a district we take, it is fundamentally different from any unit area in any other world region (see Sec. XI D).

We arrive, therefore, at a conclusion similar to that which Kroeber has stated for history: "the uniqueness of all historical phenomena (meaning, I take it, the particular combination of phenomena at a particular time) is both taken for granted and vindicated. No laws or near-laws are discovered" [*116*, 542]. The same conclusion applies to the particular combination of phenomena at a particular place.

One is not to suppose, however, that regional geography is studied without the use of generic concepts and principles. On the contrary, the interpretation of the interrelations of phenomena within each region depends upon the type concepts and principles developed in systematic geography [*cf.* Schmidt, *7*, 194]. In other words, for the individual items included in regional geography, and the simpler relations between them, we depend constantly on universal concepts supplied from the systematic studies, but the total interrelated combination of each areal unit represents an essentially unique case for which we can have no universals.

An objection may be made that one form of study used by many geographers in the consideration of regions represents an approach to the construction of scientific laws—namely what has been called "comparative regional geography," the comparison of regions of notable similarity. As a current example we may cite Maull's effective comparison of the Amazon, Congo, and Insulindia areas [*179*, 184–6]. The fact that in other sciences "comparative studies" have marked an adolescent period preceding the flowering of a nomothetic science, has led many to suppose that regional geography might be expected to grow out of its youth by progressing from comparisons to scientific principles.

The essential idea involved is nothing new in geography. Introduced by Humboldt—if not by earlier writers—it was used, according to Hettner, by Brehm, Nehring, and particularly by Richthofen [*161*, 403 f.]. Plewe * found, however, that these represented merely occasional examples, that our literature contained no comparative regional geography as a branch of the field [*8*, 46–55, 77]. Such occasional comparisons, he noted, are used in all sciences, citing as an example, Th. Litt's comparative study of Kant and Herder (Berlin, 1931). Historians, we may add, frequently find it valuable to compare the developments of any two or more periods that are significantly similar in certain respects. These examples should make us sceptical of the likelihood of our discovering anything that could be called laws, or near-laws, of regional geography.

Passarge recognized the limitations that prevent a comparative *Länderkunde* from developing universal concepts, but still (in 1936) believes that these can be avoided or overcome in a comparative *Landschaftskunde* [*272*, 61]. In order to discover the laws of regions it is necessary, he says, to have a *tertium comparationis* and this, he believes, is provided by his system of abstract types. As we saw in our previous discussion (Sec. X E) he has, in part, merely reduced the difficulties, by reducing the size of areas concerned, and for the rest he has simply dodged the limitations by setting up types that are not even in outline complete abstractions of real areas. The difference between the real *Land* and the real *Landschaft* (as area) is only a difference in size; a *tertium comparationis* is equally impossible in both cases. We may go on comparing areas of whatever size forever with no hope of discovering regional laws.

Plewe concludes, therefore, that the comparative study of regions is neither a preparatory step to a nomothetic regional geography nor an inde-

[447]

* See also Jessen's paper on the use of comparisons [*440*].

pendent branch of geography. Ritter's introduction of the concept, over a century ago, represented a transfer from a quite different kind of science; he never clearly defined his concept, and others who have taken it up have used it in many different ways but without leading to any important development [8, 82 f.].

Nevertheless the use of this method, as a supplementary device, appears to offer certain distinct advantages. If widely separated regions are in many respects similar, so that, in respect to certain elements or element-complexes, they may be classified as of the same type, the comparison of their similarities, and particularly of their differences, may well serve as a check on the interpretations we place upon the relation of phenomena within each one of them.

Even more useful is the employment of this method in the comparison of localities within a major region, where there may be a much larger number of element-complexes of the same type. By selecting those localities that are alike with respect to the greatest number of features, and comparing them with those that are like them in many, but not all, of these features, we may have a key to the significance of specific features for the area as a whole.

To take a well-known example: the consideration of the major characteristics of the Cotton Belt as a whole might lead one to suppose that—taking certain cultural conditions for granted—the importance of cotton in the area was to be explained simply in terms of climatic conditions. We have learned, however, by contrasting the localities in which cotton is the all-important crop, with those where cotton is of minor importance though the climatic conditions are the same, that the cotton crop of the South as a whole is not to be understood without considering the character of the soil.

Likewise, American geographers, at least, have long realized what is not so clearly recognized in popular thought—or even by many European geographers—that the climatic conditions of the South do not directly explain that feature which is of greatest importance in the contrast between North and South—namely the high proportion of Negro population. By the same method of comparison of localities one finds that this element—and all the cultural elements associated with it—cannot be understood without considering the combination of climatic and soil conditions that are necessary for cotton. This conclusion, however, is incomplete: in the cotton district of greatest importance today, in central Texas, the proportion of Negro population is low. The complete explanation can be reached only when one also compares the localities which were developed for plantation crops—including tobacco as well as cotton—before the end of the slavery period, with those localities developed for the same crops since that time [359].

This method of comparing localities within the same larger region might appear to lead to generic principles. But it can lead to conclusions that are applicable only to the single larger region concerned. If we should add to the districts of the Cotton Belt a district in the Yangtse Valley and a district in the Bombay province, we could not include them all under any generic concepts of districts. In the comparison limited to districts in the Cotton Belt, we are not, as we noted in an earlier connection, comparing separate units, but only similar parts of a single larger region, parts whose similarity is simply a result of the fact that they are parts of the same region. Valuable as the device may be for checking our interpretations, it leads to no universal concepts or principles.

Regional geography, we conclude, is literally what its title expresses: the description of the earth by portions of its surface. Like history, in the more common sense of periodic history, it is essentially a descriptive science concerned with the description and interpretation of unique cases, from which no scientific laws can be evolved. Though this is undoubtedly a disadvantage, making the interpretation of findings far more difficult than in those fields that are able to develop general laws to explain individual cases, it does not mean that regional geography lacks any scientific goal. As previously noted, the construction of scientific laws is not the purpose of science, but a means toward its purpose, the understanding of reality. To any "who find the title 'earth description' (*Erdbeschreibung,* or *geographia*) insufficiently learned and scientific," Heiderich has answered, "description is the last and highest goal of scientific work, to be sure, not a mere outward external description that remains on the surface of the object, but a description that aims . . . to comprehend synthetically all that has been learned analytically from the characteristics of the object" [*153,* 213]. All that science requires is that, in order that the interpretive description may have a maximum degree of accuracy and certainty, universals shall be constructed and used wherever possible. Regional geography utilizes all the appropriate generic concepts and principles developed both in the systematic sciences that study particular kinds of phenomena, and in systematic geography which studies their relations to each other over the earth.

The conclusion to which we have arrived concerning the nature of regional geography may enable us to answer one or two questions that have been raised by a number of students in very recent years. The course of thought among American geographers concerning regional studies has been discussed in two articles published in Germany during the past year, by Broek and by Pfeifer [*108; 109*]. From these surveys, and the critical articles to which they refer, the reader might suppose that after a period of enthusiastic concentration on regional studies that was introduced by the methodological papers of Barrows and, more particularly, of Sauer, American geographers had now begun to doubt whether much was to be expected from regional geography after all. It may be that the testimony has been exaggerated in the echoes back and forth across the Atlantic; possibly we are presented with a revolt within a single university department that has reverberated among its present and former members here and in Germany.[107] Undoubtedly, however, other American geographers in oral discussions have expressed a note of scepticism concerning the results to be expected from regional studies.

In a number of cases, the sceptics have spoken, or written, as though after a long and earnest attempt to advance geography by regional studies, we had discovered that the works produced did not add up to, or yield, significant general results. It is difficult to believe that this argument is meant seriously. American geography has concentrated its efforts on regional studies for scarcely twenty years and never completely. During that time, perhaps a

---

[107] In a critical comment on these studies, Platt speaks of the misunderstanding of methods and purposes on the part of writers who have not been eye-witnesses of the type of field work involved nor participants in discussions current during the past fifteen years among those experimenting in that work [224, 125]. To explain this difficulty, it is important to remind ourselves of the major distinction in attitude toward methodological discussions between American geographers, as a group, and the Germans. In marked contrast with the latter, American students seldom regard such problems as appropriate for research studies prepared for publication. On the contrary, such problems are more often regarded as matters of opinion, on which individuals may express their personal views in more or less informal symposia and, particularly, in oral discussions "out-of-meeting." Only in the "mature" pronouncements of the association's presidents do such views commonly attain formal presentation—and then usually long after they have been most influential. The few exceptions, it is significant to note, have been contributed by students influenced by the German attitude. Since these have come largely from one institution, the development of current methodological thought in American geography may well appear to a foreign student to be largely the development of thought in California [e.g., Dickinson, 202]. Both Broek and Pfeifer attempted to escape this limitation, but were hampered by the fact that the methodology of other American students appears, in publication, only in reports of fragmentary statements in symposia or in even more fragmentary explanations included in their actual research studies. There is no literature available, comparable to that in Germany, in which one may directly trace the development of the methodological views of American geographers; the task is therefore exceedingly difficult for any students remote from the actual course of development—which in large degree, unfortunately, include the group on our Pacific Coast. Probably a more reliable source than the few methodological treatises is to be found in such thorough studies of the general development of American geography, as that of Colby [107].

score or so of research students have each made one, two, or three regional studies in areas scattered from the Peace River Country to São Paulo, from Europe to China. Since neither of the two principal American promoters of the regional concept in theory has presented a concrete example of a full study in (present) regional geography, each of the individual research students has had to work out more or less independently his own methods of determining his region, of selecting the phenomena for consideration in it, and of presenting his results. Would anyone seriously consider that we have had a fair test of the possibilities of developing general results from regional studies? Even if all the work had been carried out under standardized procedures, such a small number of cases scattered over more than half the world could hardly be expected "to add up" to any general results, or to provide the basis for generalizations.

It seems more likely that many students have begun to suspect, for other reasons, that no matter how many regions are studied, no matter by what methods, no scientific laws will be forthcoming. This conclusion we have found can be thoroughly demonstrated in theory, so that we can agree that any who have made regional studies with that ultimate purpose in mind have been following a will-o'-the-wisp; the sooner it is abandoned the better for all concerned.

If, however, the purpose of geography is to gain a knowledge of the world in terms of the differential development of its different areas, the task of studying regions as areal divisions of the world, is not subject to question in geography. Neither need the workers in any science feel discouraged if the efforts of a relatively small number of workers over a period of less than twenty years have produced less than enthusiasts led them to expect. Though the object of geography, the world, is large, it is limited in size, and we must assume that geography has a long life ahead of it. No doubt the group efforts of American geographers would show more productive results if they could concentrate the attention of all or most of their members on some limited part of the world—as French geographers have done on their own country. But the many factors that persuade students to travel far afield are not to be restrained, even if it were desirable. One can only hope for a larger total number of workers, and possibly, for increasing concentration within this country of the work of particular groups—as at Wisconsin—on the regions of a relatively limited area. In particular, as Finch notes, we should not expect results of far-reaching scientific value from the practice of "skimming the cream of the more clearly given from a region and its abandonment for another area" [223, 26]. The value that studies of this kind may have for teaching purposes may justify the time and effort spent on them—provided that the areas concerned are significant for class instruction. Lasting progress in research in regional geography will require a much greater amount of concentration of the individual's time—whether or not one would go as far as Finch and consider one region as sufficient for one student's life work.

On the other hand, these considerations do inevitably raise the question of what size of area should be considered worthy of research in regional geography. The regional studies launched under Vidal's leadership in France formerly examined areas the size of a province, but increasingly smaller areas have been selected. Demangeon feels that the extreme limit of "microscopic" study has been reached by Allix, whose examination of "L'Oisans," a part of an Alpine valley in the Dauphiné smaller than an *arrondissment,* requires 915 pages, with a bibliography of 861 works [*329*]. This averages, Demangeon reckons, a little over a page per square kilometer or for 12 inhabitants. American geographers, by comparison, hardly seem justified in applying to their work the word "microscopic."

The question raised admits of no simple answer. Historians welcome extremely detailed studies of very short periods, in addition to less intensive studies of an extensive series of periods. The criterion, in either field, is the same—namely the significance of the study—but that is a criterion for which we have no objective measure. We have previously suggested two major [*] considerations—namely, the significance of the area in itself,[108] and its possible significance as representative of a large area, or a large number of similar small areas. Outside of the proper interest of the citizens of L'Oisans itself in the geography of their own district, we may assume that the world of knowledge in general has little need for such an exhaustive study of this small and unimportant district. On the other hand, if we had but little knowledge of the valleys of the French Alps, and reconnaissance had shown that this particular district was in large degree representative of hundreds of others, such a study might provide us with an approximate view of the regional geography of the entire area—or of a large part of it. Presumably, however, such a study would be limited by the desire to express primarily

---

[108] In the previous discussion we considered the significance of an area in itself merely in terms of its relative importance in the actual world. Finch reminds us however that an area may have a special significance to our science of the world if it includes some unanswered question, if it has some peculiar association of features [*223, 23*]. I find it particularly desirable to add this criterion since it would seem to offer the only justification for an American geographer to have occupied himself with such a small, remote area as Upper Silesia, representative of almost no other areas near it [*355; 356*].

[452]

* "previously"; on pages 288 f.

those characteristics that were representative, and one might question whether that would require a thousand pages. Demangeon finds much of the study superfluous because it merely duplicates findings that Blanchard and others have presented in works on similar districts. Insofar as Allix's work has served to corrobarate that of his predecessors, that fact might have been more briefly presented. On the other hand, another competent critic[109] finds that Allix has contributed a much more thorough treatment of the problems representative of the French Alps than have any of his predecessors.

It is particularly against such "microgeographic" studies—to use Platt's term—that criticism of regional geography in America has been directed. Recognizing that it would be impracticable, in any reasonable length of time, to cover the whole land area of the world by the total addition of such small studies—and that the total might be indigestible if it could be attained—critics have feared that we would have but a miscellaneous collection of scattered pieces selected at random [cf. Leighly, 220]. More particularly, however, the critics have asked what general principles can we expect to derive from such minute and scattered studies. Even some who have made such microgeographic studies, like James, have given expression to a later feeling that "the more detailed and specific is the study the more insignificant are the results" [286, 84].[110]

To these attacks, Platt, in particular has replied vigorously, both in two published papers [221; 224] and in unpublished statements read before this Association. Microscopic geography, he observed in one of the latter, developed "as a rational and timely drive against the limitations of armchair compilation from promiscuous data, of subjective impressions from casual travel, and of environmental theory not founded on data." To attain these purposes, geographers took to the field and "in the field all geographers are microscopic." There "they face the geographer's dilemma in trying to comprehend large regions while seeing at once only a small area." They do not, he insists, plunge into detailed studies of minute areas because the methodological conclusions of others have led them to believe that thereby something will ultimately be gained for geography. On the contrary, their

---

[109] J. Sölch, in personal communication.

[110] This statement may possibly account for Pfeifer's conclusion that James "questioned even whether the 'microscopic method' represented any advance at all" [109, 115 ff.], a conclusion that is certainly not consistent with the general view expressed in such statements as "the detailed study of the small area becomes significant in so far as it contributes to the more accurate generalization of this detail on chorographic (mesochoric) or geographic (macrochoric) maps"; or the conclusion: "topographic (microchoric) studies are vital parts of the chorographic (mesochoric) or geographic (macrochoric) investigations" [286, 85 f.]

own efforts to comprehend areas of larger extent has led them to the reasoned conclusion that, in addition to general reconnaissance studies and detailed systematic studies covering large areas, accurate generalizations for larger regions require an examination of the total fundamental complex of inter-related features that can be examined, in detail, only in the small area.[111]

Platt's defence of microgeography, however, is based less on theoretical discussion than on the actual work that he has been carrying on for some years in Hispanic America, which forms the most significant series of micro-geographic regional studies in American geography [listed in *221*, 13; to that list should be added *224*]. Of the sceptical questions that have been raised concerning the value of such a series of studies, many appear irrelevant to its purpose. This, I take it, is simply to increase our organized, objec-tive and reliable knowledge of the lands south of the Rio Grande. That such knowledge of the different parts of the world is desirable and requires the research of trained workers is, we repeat, the fundamental justification for the field of geography. That our present knowledge of the Hispanic American area is inadequate is obvious to anyone who has attempted to gather the materials necessary even for an elementary course concerned with that part of the world. Consequently, we are not to test the value of such a series of detailed studies of scattered districts by asking whether they can yield us any "scientific principles," or whether they will aid us in drawing conclusions concerning "the larger relationships" of which Pfeifer speaks [*109*]. So long as Platt does not claim that all geography should consist of such "microgeographic" studies, or of regional studies in general, these questions are irrelevant. The relevant question is, granted that we want more adequate knowledge of the geography of South America, is his method of study appropriate to produce such knowledge?

Few will question the inadequacy of the general surveys of South Amer-ica now available. In his most recent study, on coastal plantations in British Guiana, Platt has noted that the best available, generalized maps of the con-

---

[111] By way of illustration, Platt notes that Finch has contributed not only the extremely minute study of Montfort [*285*], but also systematic studies covering the agriculture of the world [with Baker, *343*], to which might be added the many essen-tially research studies included in the more recent text written with Trewartha [*322*]; Whittlesey not only surveyed a small district in Wisconsin ["Field Maps for the Geog-raphy of an Agricultural Area," *Ann. Assn. Am. Geogrs.*, 15 (1925), 187–91], but has endeavored to establish the major agricultural regions of the world [*319*]. Further-more, we may add, the latter finds as a major difficulty in interpreting the findings of such a world survey, the lack of detailed studies of small, representative, districts, such as Dicken has offered for the Mexican highlands [*340*], and Platt has presented for several districts in Hispanic America. Finally, in his most recent study, Platt has shown directly the relation of microgeographic work to the broader purposes of reconnaissance [*224*].

tinent give erroneous impressions of the soils, vegetation, and population density of the specific districts he studied. Even if we had accurate detailed information on the climates, land forms, soils, crops, races, and commerce of South America, these would not add up to the geography—the areal differentiation—of the different parts of that continent. In studies limited even to provincial scale, the American student is frequently baffled because he lacks the detailed knowledge of the cultural element-complexes that are basic to the cultural geography of the region. For areas in United States or Europe, he may have acquired that knowledge unconsciously, whether as a by-product of field work, or merely from his general knowledge. These essential features must be studied first in relatively small areas—particularly in a world area where there is lack of cultural homogeneity. If, then, one has acquired an understanding of a particular ranch in Panamá and may be permitted to assume somewhat similar features scattered through a large area, one has a more correct picture of the geography of the larger area concerned than can be acquired by any small-scale measures [see Platt, *221*].

The essential assumption in this proposition, of course, is that the minute district studied is in fact representative of others; as Finch notes, it can hardly be typical in any full sense [*223*, 24]. If it is representative, however, it presumably will be typical in certain limited respects, and it is important that we know in what respects it is approximately typical. In areas that are adequately covered by census and climatological data, geological, topographic, and soil surveys, it may be possible to give approximate answers to these questions from the study of such data. The utility of element-ratios and isopleth maps in this connection has been suggested previously. In other areas, one can only depend on the student's judgment formed from reconnaissance. Though such judgment can only give answers that are far removed from scientific certainty, they are better than no answers at all, and should therefore be provided—even at the risk of being shown erroneous by later work by the same or other students.

Perhaps only in his most recent study has Platt clearly demonstrated the relation of these detailed studies of small districts to the reconnaissance study of large areas. Though the microgeographic area which he studied in detail is not, in this case, "typical of broad regional types," it is shown to be "a normal feature of a coherent plantation district, which in turn has a consistent place in the intricate geographic pattern of South America" [*224*, 123 ff.]. No doubt the significance, to broader regional knowledge, of his previous studies in small and widely separated districts, apparently chosen at random, will be made clear in the ultimate publication of his "reconnaissance study of Hispanic America," of which these detailed unit studies are to form integral parts. **

* See Supplementary Note 57

** Published in 1942, Latin America: Countrysides and United Regions.

In sum, the student who presents a study of a small area of no special importance in itself, needs to keep in mind that the purpose is not to present the area in itself, but to provide an accurate illustration of the representative character of a larger region, too large to permit of such intensive study. So long as he keeps this broader purpose in mind, there are no grounds apparent on which we can prescribe the minimum size of area that may be studied.

I. THE INTEGRATED DUALISM OF GEOGRAPHY

The final question raised by our examination of the nature of geography, as presented to us both by its historical evolution and by the logical consideration of its position among the sciences, is the same question that has provoked so much controversy throughout almost the entire history of modern geography—certainly ever since Bucher raised the issue in 1827. If geography studies the areas of the world according to the differential character of their phenomenal contents, either according to a systematic point of view, category by category, or, on the other hand, according to an areal point of view, each area in terms of all its heterogeneous phenomena, how can these two points of view be related to each other in a unified field of geography?

Our historical survey showed that, while modern geography from its beginnings has included both of these points of view—in theory even in Varenius's outline—it has experienced notable shifts in emphasis from one to the other. Whereas the work of Humboldt combined both points of view, under the influence of Ritter, systematic studies were placed in a subordinate position and easily lost sight of. Though the protests of Bucher and Fröbel were of no avail at the time, a later generation following Peschel, and motivated by scientific standards developed in such fields as geology, swung the center of interest the other way. As late as 1919 Hettner found that in Germany systematic geography was generally regarded as "something higher, more distinguished" than regional geography. He therefore repeated the arguments that he had presented at various times during nearly a quarter of a century to show that the two parts of the field were scientifically on the same level [142, 22 f.]. Less than a decade later, however, he found it necessary to present opposite arguments to urge the same conclusion; for "youth, which is given to exaggeration, has turned far too much away from systematic geography" [161, 401]; (we can hardly suppose that Hettner was unaware of the fact that some of those concerned were not much younger than he). The reaction that had taken place earlier in France, under Vidal, swept German geography in the post-War years toward an increasing emphasis on regional geography, as the real goal of geographic work. Thus Obst, believing that one could develop a science of *"Länderkundliche Typolo-*

* See Supplementary Note 49

gie" wished to shift the time-honored term of "general geography" to the study of regional types as the goal of geography, and placed systematic geography (as *allgemeine Erdkunde*) in the subordinate position of a necessary propaedeutic [*178*, 6–9] ; somewhat similar views have been expressed by Braun [*155*, 5], Volz [*151*, 247], Ule [*170*, 486], and Gradmann [quoted in *166*, 13].

Likewise in this country, the emphasis that Barrows, and particularly Sauer, had placed upon the study of regions (in the latter case, "landscapes") led some to regard systematic studies as necessary only for instructional purposes but inappropriate for geographical research.

On the other hand, such veterans as Hettner and Penck have never wavered in their insistence that both points of view were of equal importance in geography [Hettner has said as much in almost every methodological treatment he has written; for Penck, see *129*, 639; *137*, 173–76; *163*, 44]. The very fact that geography has experienced these successive shifts in emphasis from one side to the other is in itself indirect evidence that both are of coordinate importance in the field [*cf.* Hettner, *2*, 306].

In his critical investigation of geography as a single, unified field of science, Kraft finds that, while one could dismiss the charge of dualism of content—natural and human features—as invalid, the inclusion of the systematic and the regional points of view was an unquestionable form of dualism. He agrees with Hettner, however, that this dualism cannot be expressed simply as the combination of a nomothetic and an idiographic science; systematic geography must include the study of unique cases, and regional geography must use generic concepts and principles. In any case, neither construction of laws nor the description of the unique represents the purpose of geography, or of any other science. The purpose of geography is the same in both branches, the comprehension of the areal differentiation of the earth, and this purpose cannot be solved either by systematic studies alone nor by regional studies alone, but requires both approaches. Consequently, he concludes, this dualism in approach is justified as necessary for the single aim which makes geography a unified science [*166*, 11–13].

This view, we may add, is further supported by the fact, stressed by Hettner, that it is frequently difficult to classify particular studies under one heading or the other. The difference is not in the substance, but in the point of view, and in certain kinds of studies these may be combined. For example, the systems of land-use classification previously discussed (Sec. X F, G) are intended to provide backgrounds for agricultural regional geography and they involve, in outline, a major part of the regional study of any area. At the same time, however, they represent systematic studies of particular element-complexes in their world distribution, so that it is by no means clear whether they belong more in the one or the other of our two major divisions.

Finally, if one agrees that both regional and systematic studies are included as essential parts of geography, we may perhaps dismiss any question of relative importance as irrelevant. For systematic geography, regional studies provide, not merely a source of detailed factual information that otherwise would hardly be available, but they also indicate problems of relationships that might easily be overlooked in systematic geography, and they provide the final testing ground for the generic concepts and principles of systematic geography. On the other hand, it is even more obvious that progress in interpretation of the interrelated phenomena of regional geography is constantly dependent on the development of such universals by systematic studies. Any assumption that these studies can be left to the systematic sciences concerned with each particular category of phenomena has been shown by experience to be unwarranted. The aspects of these phenomena with which geography is concerned—their relation to other earth phenomena in different parts of the world—are not of direct concern to those systematic sciences and are more commonly left unstudied, unless geographers study them, as Lehmann has shown. Systematic geography, he therefore concludes, is not to be thought of as a border area of geography, or merely as a propaedeutic, but represents "organs vital to the growth of geography, without which its regional crowning can as little exist as a real tree without its roots" [113, 236 f.].

Further, Lehmann suggests, the point of view developed in systematic geography is different from the general point of view in regional geography but at the same time of such value to it that every regional geographer should work productively in some systematic branch or branches (he recommends two or more). On the other hand, Penck, whose most notable contribution has no doubt been the systematic study of landforms, urges that "the cultivation of regional studies is indispensable for the geographer; they form for him the touchstone of his whole concept of geography, of his geographic system" [129, 639; cf. also Graf, 156, 82].

The mutual dependence of the two interconnected points of view in geography has been consistently maintained by Hettner from his earliest methodological treatment of more than forty years ago to the present time. The development of sound universal concepts in systematic geography is the essential basis for progress in regional geography, but since systematic geography is in method similar to the systematic sciences, "the geographer who works only in it and does not cultivate regional geography runs the risk of leaving the ground of geography entirely. He who does not understand regional geography is no true geographer. While regional geography alone, without systematic geography, is incomplete, it remains geographic; systematic geography without regional geography cannot fulfill the full function of geography and easily falls out of geography" [*142, 22* f.].

We may assume, therefore, that there is plenty of work to be done in the field of geography by both methods of approach. It is not for any student specializing in either approach to speak with scorn or condescension of those who are working in the other. "Differences of approach," as Kroeber suggests, "are probably at bottom largely dependent on differences of interest in individuals" [*116, 569*]. Paraphrasing his statement further, we may conclude that it is perfectly legitimate to confine one's interest to the specific approach of systematic geography, or to the integrating approach of regional geography, or to use alternately one or the other according to occasion. But sympathetic tolerance is intrinsically desirable and certainly advantageous to understanding: to *scientia*.

XII. Conclusion: The Nature of Geography

Our examination of the great variety of different ideas that have been suggested for geography has repeatedly led us into sidetracks that proved to be blind alleys or routes leading outside of geography. No doubt also we have lingered at other points along the way to investigate in detail certain important problems within the field. It may be well therefore to summarize briefly the positive conclusions to which we have arrived concerning the nature of geography.

In its historical development geography has occupied a logically defensible position among the sciences as one of the chorographical studies, which, like the historical studies, attempt to consider not particular kinds of objects and phenomena in reality but actual sections of reality; which attempt to analyze and synthesize not processes of phenomena, but the associations of phenomena as related in sections of reality.

Whereas the historical studies consider temporal sections of reality, the chorographical studies consider spatial sections; geography, in particular, studies the spatial sections of the earth's surface, of the world. Geography is therefore true to its name; it studies the world, seeking to describe, and to interpret, the differences among its different parts, as seen at any one time, commonly the present time. This field it shares with no other branch of science; rather it brings together in this field parts of many other sciences. These parts, however, it does not merely add together in some convenient organization. The heterogeneous phenomena which these other sciences study by classes are not merely mixed together in terms of physical juxtaposition in the earth surface, but are causally interrelated in complex areal combinations. Geography must integrate the materials that other sciences study separately, in terms of the actual integrations which the heterogeneous phenomena form in different parts of the world. As Humboldt most effectively established, in practice as well as in theory, though any phenomenon studied in geography may at the same time be an object of study in some systematic field, geography is not an agglomeration of pieces of the systematic sciences: it integrates these phenomena according to its distinctive chorographic point of view.

Since geography cuts a section through all the systematic sciences, there is an intimate and mutual relation between it and each of those fields. On the one hand, geography takes from the systematic sciences all knowledge that it can effectively utilize in making its descriptions of phenomena and interpretations of their interrelations as accurate and certain as possible. This borrowed knowledge may include generic concepts or type classifications, developed in the systematic sciences; but, where these are found unsuitable for geographic purposes, geography must develop its own generic concepts and systems of classification.

In return, geography has contributed, and continues to contribute, much to the systematic sciences. In its naive examination of the interrelation of phenomena in the real world it discovers phenomena which the sophisticated academic view of the systematic sciences may not have observed, shows them to be worthy of study in themselves and thus adds to the field of the systematic studies. Further, geography constantly emphasizes one aspect of phenomena which is frequently lost sight of in the more theoretical approach of the systematic fields, namely, the geographic aspect. It serves, therefore, as a realistic critic whose function it is constantly to remind the systematic sciences that they cannot completely understand their phenomena by considering them only in terms of their common characteristics and processes. They must also note the differences in those phenomena that result from their actual location in different areas of the world. In order to interpret these differences correctly, and to interpret the resultant world distribution of their phenomena, the systematic sciences take from geography something of the particular techniques which its point of view has required it to develop—notably the techniques of maps and map interpretation.

Geography, like history, is essential to the full understanding of reality. The naked, schematic study of the systematic sciences divides up reality into academic compartments, and thereby necessarily destroys something of its essential character:

> "Ach, von ihrem lebenwarmen Bilde
> Blieb der Schatten nur zurück."

Geography adds, as Vidal said, "the aptitude of comprehending the correspondence and correlation of facts, be they in the terrestrial milieu which includes them all, be they in the regional milieu in which they are localized" [*183, 299*].

It is a corollary of this proposition that, in the application of science to society, as Finch observes, the chorological science of geography can function directly, since many of the problems of society—notably those concerned with the most efficient organization of land use—are, in fact, regional problems. But that statement does not mean—and I assume that Finch did not intend it to mean—that the chorological point of view requires the justification of utility [*223,* 21 ff.]. On the contrary, whatever value geography has in relating science to the problems of society, merely confirms the fact that in pure science itself—the pursuit of knowledge for the sake of gaining more knowledge—there is need for a science that interprets the realities of areal differentiation of the world as they are found, not only in terms of the differences in certain things from place to place, but also in terms of the total combination of phenomena in each place, different from those at every other place.

Geography, like history, is so comprehensive in character, that the ideally complete geographer, like the ideally complete historian, would have to know all about every science that has to do with the world, both of nature and of man. The converse of this proposition, however, is that every student of a systematic science is somewhat at home in some part of geography. Further- more, both geography and history endeavor to describe and interpret actual sections of reality as they exist, and in these sections they observe phenomena by methods that, in a general way, are available to the common man. Con- sequently geography, like history, is a field apparently open for layman to enter. Whereas the study of history, other than current history, at least requires the degree of learning sufficient to utilize the records of the past, geography may be studied by any one who has the opportunity to travel and the ability to describe what he sees. Consequently, geography was in fact studied by laymen long before any organized subject of geography was con- structed, and countless non-professional travelers since have contributed more or less useful data to its literature. This characteristic, likewise, it shares, for good or ill, with history.

In consequence, Richthofen has noted, "many have the delusion that geography is a field in which one can reap without sowing. Because a great part of that which the serious research students have won in it is easily under- stood, one thinks that he can work successfully in it without preparatory training, and can win laurels by the easy means of describing fleeting travel observations or by uncritical compilations. An endless flood of superficial literature, which, in spite of its deficiencies, may not be denied the service of popularization, has been able to obscure the judgment of a great part even of the educated public concerning the scientific content of geography. But, just as with history, the apparent ease with which a great part of the facts secured can readily be understood, stands in contrast to the difficulty of sound research" [73, 68]. Allen Johnson, among many others, has discussed the importance of the same contrast in history [117].

Since the vulnerability of both geography and history to occasional tres- pass by wandering laymen is a result of the fundamental character of the field in each case, little would be gained by attempting to set up barbed wire fences in the form of erudite technical terms designed to bar trespass. Few geographers, presumably, will wish to have their subject strive for prestige by hiding its knowledge behind smoke-screens. On the contrary, in a subject in which the field includes vast areas that few professionals will have the opportunity to explore, the assistance of the interested amateur may heartily be welcomed. The sole provision that we might like to suggest is that the amateur, as in any activity of life, should recognize his need of securing as much knowledge and training from professionals as is possible for him, so that his efforts may produce results of greater accuracy and interest in themselves and of more lasting value for the science of geography.

Geography and history are alike in that they are integrating sciences concerned with studying the world. There is, therefore, a universal and mutual relation between them, even though their bases of integration are in a sense opposite—geography in terms of earth spaces, history in terms of periods of time. The interpretation of present geographic features requires some knowledge of their historical development; in this case history is the means to a geographic end. Likewise the interpretation of historical events requires some knowledge of their geographic background; in this case geography is the means to an historical end. Such combinations of the two opposite points of view are possible if the major emphasis is clearly and continuously maintained on one point of view. To combine them coordinately involves difficulties which, as yet at least, appear to be beyond the limitations of human thought. Possibly one approach to such a combination can be made in geography by the lantern-slide method of successive views of historical geographies of the same place. An attempt to develop a motion picture would produce a continuous variation with respect to both time and space which would, of course, represent reality in its completeness, but which appears to be beyond our capacity even to visualize, not to say, to interpret.

Though the point of view under which geography attempts to acquire knowledge of reality is distinct, the fundamental ideals which govern its pursuit of knowledge are the same as those of all parts of that total field of knowledge for which we have no other name than science.

Geography seeks to acquire a complete knowledge of the areal differentiation of the world, and therefore discriminates among the phenomena that vary in different parts of the world only in terms of their geographic significance—i.e., their relation to the total differentiation of areas. Phenomena significant to areal differentiation have areal expression—not necessarily in terms of physical extent over the ground, but as a characteristic of an area of more or less definite extent. Consequently, in studying the interrelation of these phenomena, geography depends first and fundamentally on the comparison of maps depicting the areal expression of individual phenomena, or of interrelated phenomena. In terms of scientific techniques, geography is

[463]

represented in the world of knowledge primarily by its techniques of map use.

There are no set rules for determining which phenomena are, in general, of geographic significance. That must be determined, in any particular case, on the basis of the direct importance of the phenomenon to areal differentiation, and of its indirect importance through its causal relation to other phenomena. In order to determine his findings as accurately as possible, the individual student, in any particular case, must depend upon those among the significant phenomena for which he is able to secure some sort of measured data. Non-measurable, but geographically significant, phenomena must be studied indirectly, by whatever measurable effects they have produced.

These general principles lead to no general exclusion of any kind of phenomena, nor of any aspect of the field. In any particular study in systematic geography or in any partial study of a region, particular kinds of phenomena may logically be excluded only if they are not significant to the interrelations of those that are being studied. Finally, the ideal of completeness requires geography to consider not only those features and relationships that can be expressed in generic concepts but a great number of features and relationships that are essentially unique.

In order to make its knowledge of interrelated phenomena as accurate and as certain as possible, geography considers all kinds of facts involved in such relations and utilizes all possible means of determining the facts, so that results obtained from one set of facts, or by one method of observation, may be checked by those secured from other facts or from other observations.

With the same ends in view, geography accepts the universal scientific standards of precise logical reasoning based on specifically defined, if not standardized, concepts. It seeks to organize its field so that scholarly procedures of investigation and presentation may make possible, not an accumulation of unrelated fragments of individual evidence, but rather the organic growth of repeatedly checked and constantly reproductive research.

In order that the vast detail of the knowledge of the world may be simplified, geography seeks to establish generalized pictures of combinations of dissimilar parts of areas that will nevertheless be as nearly correct as the limitations of a generalization permit, and to establish generic concepts of common characteristics of phenomena, or phenomenon-complexes that shall describe with certainty the common characteristics that these features actually possess. On the basis of such generic concepts, geography seeks to establish principles of relationships between the phenomena that are areally related in the same or different areas, in order that it may correctly interpret the interrelations of such phenomena in any particular area.

[464]

Finally, geography seeks to organize its knowledge of the world into inter-connected systems, in order that any particular fragment of knowledge may be related to all others that bear upon it. The areal differentiation of the world involves the integration, for all points on the earth's surface, of the resultant of many interrelated, but in part independent, variables. The simultaneous integration all over the world of the resultant of all these vari-ables cannot be organized into a single system.

In systematic geography each particular element, or element-complex, that is geographically significant, is studied in terms of its relation to the total differentiation of areas, as it varies from place to place over the world, or any part of it. This is in no sense the complete study of that particular phenomenon, such as would be made in the appropriate systematic science, but the study of it solely in its geographic significance—namely in its own areal connections, and in the relations of its variations to those of other features that determine the character of areas. Although the study of any single earth feature is thus organized into a complete system in systematic geography, it is clear that at every point on the earth it is connected with the coordinate systems concerned with the other features.

In regional geography all the knowledge of the interrelations of all features at given places—obtained in part from the different systems of sys-tematic geography—is integrated, in terms of the interrelations which those features have to each other, to provide the total geography of those places. The areal integration of an infinite number of place-integrations of factors varying somewhat independently in relation to place, is possible only by the arbitrary device of ignoring variations within small unit-areas so that these finite areal units, each arbitrarily distorted into a homogeneous unit, may be studied in their relations to each other as parts of larger areas. These larger areas are themselves but parts of still larger divisions—ultimately divisions of the world.

The problem of dividing the world, or any part of it, into subdivisions in which to focus the study of areas, is the most difficult problem of organiza-tion in regional geography. It is a task that involves a complete division of the world in a logical system, or systems, of division and subdivision, down to, ultimately, the approximately homogeneous units of areas. Difficult though the task may be, the principles of completeness and organization demand that geography seek the best possible solution.

One method of providing such an organization represents perhaps an intermediate step between systematic and regional geography. On the basis of any one element or element-complex—which latter may represent a great number and variety of closely related elements—we may construct a logical system of division and subdivision of the world according to types. Each of these systems of division determined on the basis of generic concepts of element-complexes, may be carried through by objective decisions based on measurement. Possibly as few as three such systems—each based on a cultural complex of many elements—may be adequate to provide outlines into which to organize most of our regional knowledge of the world. In each case, however, we are organizing separately different aspects of the geography of regions, we are not organizing the complete geography of regions.

A single system in which to organize the complete geography of the regions of the world must be based on the total character of areas, including their location as parts of larger units. Such a system of specific regions requires the consideration of all features significant in geography, some more significant in some areas, others in others. The determination of the divisions at any level involves, therefore, subjective judgment as to which features are more, which less, important in determining similarities and dissimilarities, and in determining the relative closeness of regional interrelations. At any level therefore, the regions are fragments of the land, so determined that we may most economically describe the character of each region, —that is, that in each region we will have a minimum number of different generalized descriptions of approximately similar units, each description involving the maximum number of nearly common characteristics and applicable to the maximum number of similar units.

Although all the fundamental ideals of science apply equally in all parts of geography, there are differences in the degree to which they can be attained in the different parts. These differences among the special divisions of geography—physical, economic, political, etc.—are differences in degree corresponding to the similar differences in degree to which the various systematic sciences are able to attain those ideals.

The greatest differences in character within geography are found between the two major methods of organizing geographic knowledge—systematic geography and regional geography—each of which includes its appropriate part of all the special fields. In addition to the difference in form of organization in the two parts, there is a radical difference in the extent to which knowledge may be expressed in universals, whether generic concepts or principles of relationships.

Systematic geography is organized in terms of particular phenomena of general geographic significance, each of which is studied in terms of the relations of its areal differentiation to that of the others. Its descriptive form is therefore similar to that of the systematic sciences. Like them, it seeks to establish generic concepts of the phenomena studied and universal principles of their relationships, but only in terms of significance to areal differentiation. No more than in the systematic sciences, however, can systematic geography hope to express all its knowledge in terms of universals; much must be expressed and studied as unique.

While there are no logical limitations to the development of generic concepts and principles in systematic geography, the nature of the phenomena and the relations between them that are studied in geography present many difficulties preventing the establishment of precise principles. These difficulties are of the same kind as are found, in differing degree, in all parts of science. In many of the systematic sciences, both natural and social, the degree of difficulty is as great, or greater, than in geography. In that field which is most nearly the counterpart of geography, namely history, the difficulties are in almost every case far greater. Systematic geography is therefore far more able to develop universals than is "systematic history." Nevertheless the degree of completeness, accuracy, and certainty, both of the principles established and of the facts known in regard to any particular situation, seldom permit definite predictions in geography. This characteristic, geography shares not only with history, but also with many other sciences, both natural and social.

Regional geography organizes the knowledge of all interrelated forms of areal differentiation in individual units of area, which it must organize into a system of division and subdivision of the total earth surface. Its form of description involves two steps. It must first express, by analysis and synthesis, the integration of all interrelated features at individual unit places, and must then express, by analysis and synthesis, the integration of all such unit places within a given area. In order to make this possible, it must distort reality to the extent of considering small but finite areas as homogeneous units which can be compared with each other and added together in areal patterns of larger units. These larger, likewise arbitrary units, are so determined as to make possible a minimum of generalized description of each unit "region," that will involve a minimum of inaccuracy and incompleteness.

Since the units with which it deals are neither real phenomena nor real units but, at any level of division, represent distortions of reality, regional geography itself cannot develop either generic concepts or principles of reality. For the interpretation of its findings it depends upon generic concepts and principles developed in systematic geography. Furthermore, by comparing different units of area that are in part similar, it can test and correct the universals developed in systematic geography.

The direct subject of regional geography is the uniquely varying character of the earth's surface—a single unit which can only be divided arbitrarily

[467]

into parts that, at any level of division, are, like the temporal parts of history, unique in total character. Consequently the findings of regional geography, though they include interpretations of details, are in large part descriptive. The discovery, analysis and synthesis of the unique is not to be dismissed as "mere description"; on the contrary, it represents an essential function of science, and the only function that it can perform in studying the unique. To know and understand fully the character of the unique is to know it completely; no universals need be evolved, other than the general law of geography that all its areas are unique.

In the same way that science as a whole requires both the systematic fields that study particular kinds of phenomena and the integrating fields that study the ways in which those phenomena are actually related as they are found in reality, so geography requires both its systematic and its regional methods of study of phenomena and organization of knowledge. Systematic geography is essential to an understanding of the areal differences in each kind of phenomena and the principles governing their relations to each other. This alone, however, cannot provide a comprehension of the individual earth units, but rather divests them of the fullness of their color and life. To comprehend the full character of each area in comparison with others, we must examine the totality of related features as that is found in different units of area—*i.e.*, regional geography. Though each of these methods represents a different point of view, both are essential to the single purpose of geography and therefore are properly included in the unified field. Further, the two methods are intimately related and essential to each other. The ultimate purpose of geography, the study of areal differentiation of the world, is most clearly expressed in regional geography; only by constantly maintaining its relation to regional geography can systematic geography hold to the purpose of geography and not disappear into other sciences. On the other hand, regional geography in itself is sterile; without the continuous fertilization of generic concepts and principles from systematic geography, it could not advance to higher degrees of accuracy and certainty in interpretation of its findings.

It would be an error to interpret the current interest in methodological discussion as a sign that American geography had entered a period of unusual dissension. In large part, no doubt, it represents the crystallization in print of disagreements hitherto held in the more liquid solution of oral discussion, but which have been suddenly precipitated by the first challenge to fundamental principles in geography to appear in print in more than a decade. Although the basic position of geography as a chorographic science has been questioned, the challenge does not appear to have produced dissension. On the contrary, it has revealed that those who were accustomed to find themselves in opposite camps in methodological discussions, were actually in opposition only on secondary questions and were fundamentally at one on the major function of geography among the sciences.

More than at any previous time in the development of American geography, there is notable agreement, in practice as well as in theory, on the importance of studies in regional geography, while at the same time there is a continued drive to develop the various aspects of systematic geography. Further, the apparent gulf between these two aspects of the field is being narrowed: students of regional geography depend increasingly on studies in systematic geography and those making systematic studies have recognized that their value to geography as a whole depends on the extent to which they are correlated with the viewpoint of regional geography.

If American geography is approaching that major degree of common understanding on the fundamental nature of its field that was attained in Germany two or three decades earlier, and likewise underlies—even though less definitely expressed—a great part of the work in French geography, we may hope that the immediate future in this country will show a period of correspondingly rich production along a wide, but common front. Agreement on methodological questions, as on any others, can be attained, by those who are free to think for themselves, only by thorough examination of the problems involved, with adequate and fair consideration of the divergent views expressed by other students, past as well as present. By directing on current methodological problems in our field a critical review organized out of the rich literature of more than a century of geographic thought, I hope to have contributed to a more general understanding of our fundamental purposes and problems.

University of Minnesota.
June, 1939.

SUPPLEMENTARY NOTES

Page   #

4   1.   With the exception of Strabo's work, the works in this section are listed in the language of the original. Humboldt's major works (listed here as Nos. 45-47, 59-60) have been republished in many languages, but the reader cannot rely fully on correctness in the translations. This is even more true of the two volumes of Ritter's essays (Nos. 50 and 61) which were translated into English by W. L. Gage and published as *Geographical Studies,* Boston 1861 and *Comparative Geography,* Philadelphia, 1865.

35   2.   On viewpoints concerning geography in ancient times, the standard reference work in English is Bunbury's *History of Ancient Geography (402),* which has recently been republished. A shorter study, recently revised, is that of Tozer *(403).* Berger's analysis of the character of the work done in different schools is more nearly comparable with the classifications familiar to modern geographers *(404).* For the entire period from classical times to the early nineteenth century, there is no counterpart in English for Peschel's richly documented history *(401),* which however is less pertinent for the purpose of this chapter than Wisotski's detailed study of geographic thought in the late eighteenth century *(1).*

57   3.   The corrections of translations and paraphrases in this and a number of the following footnotes were deemed necessary because of errors published in the *Annals* the previous year. In the period since, no one, to my knowledge, has challenged the reliability of these corrections.

58   4.   Much of this discussion of the work of Humboldt and Ritter has subsequently been confirmed and amplified by George Tatham, "Geography in the Nineteenth Century," in *"Geography in the Twentieth Century,"* edited by Griffith Taylor, New York, 1951, pp. 42-59; and, with respect to Humboldt, by Rayfred L. Stevens-Middleton, in *La Obra de Alexander von Humboldt en Mexico,* publ. no. 202, Instituto Panamericano de Geografia e Historia, Mexico, D.F. 1956, pp. 199-246. The commemoration in 1959 of the hundredth anniversary of the death of both of these "founders" of the field led to publication of numerous studies of their work by German geographers.

67   5.   The last two sentences of the paragraph are in error. While Humboldt included man, as the highest organism, in his "physical geography," he did not include the realm of

the intellect or art, even though he recognized that the distinction was in a sense unreal (6̶0̶, 69, 386); see *Perspective. . .* , footnote to page 68.

68   6.   Ritter's writings are somewhat inconsistent in use of the term "nature"; see *Perspective. . .* , footnote to page 48.

74   7.   Leo Waibel has subsequently shown the soundness of Ritter's theory of the origin of the plantation system, in contrast to theories of several later students, *Geographical Review* 32 (1942), p. 307-310.

83   8.   Ritter was much more definite than these sentences indicate in restricting the scope of geography to the earth shell; see *Perspective. . . .* footnote to page 22.

85   9.   The statement concerning Mary Somerville was based on the source named in the next sentence, Spörer. J.N.L. Baker finds this doubtful, in his study of "Mary Somerville and Geography in England," *Geographical Journal*, CXI (1948), 207-22.

85.   10.   Guyot, though of French Hugenot parentage, had received the greater part of his advanced training in Germany, particularly in Berlin where, almost certainly, he studied under Ritter. At one time he had started to translate several portions of Ritter's work into French, but evidently never carried it through. Political difficulties in 1848 caused him to leave Switzerland and come to America. The Lowell Lectures which he delivered in French at Harvard, in 1848-49, on "Comparative Physical Geography in its Relation to the History on Mankind," were translated into English and published in 1849 as *The Earth and Man (64)*. He was long employed by the Smithsonian Institute and in 1854 was appointed Professor of Geology and Physical Geography at Princeton, retaining that position until his death in 1884. In addition to an extensive series of *Meteorological and Physical Tables*, prepared for the Smithsonian Institute (his work in this field stimulated the establishment at the U.S. Weather Bureau), he published a series of school geographies. See the "Memoir" by James D. Dana, *(410)*.

90   11.   The most thorough-going and authoritative study of Ratzel's concepts and their origins, has subsequently been made by Johannes Steinmetzler in a doctoral dissertation, "Die Anthropogeographie Friedrich Ratzels und ihre ideengeschichtliches Würzeln," published in *Bonner Geographische Adhandlungen*, Heft 19, 1956.

94   12.   Under the Nazi regime, Hettner was not permitted to publish, but his *Allgemeine Geographie des Menschen* was published posthumously, in 1947. It includes a detailed sketch of his professional background and career, by Schmitthenner.

Page	#

100   13.   DeMartonne, writing in 1924, found that geography in France had developed as an emancipation from the dominance of history *(412;* see also *415)*. In the present work, views of geography by French students are limited largely to the writings of Vidal de la Blache, Brunhes, and Vallaux; a much larger number are considered in *Perspective.* . . . pages 4, 188–89.

100   14.   Thanks to the common practice in British geography of discussing methodological operations without reference to sources, a number of important papers did not come to the author's attention. Particularly notable are early papers by Mackinder *(422* and *423)* and the collection of Herbertson's notes published after his death in 1915 *(424)*. Keltie's brief outline of "The Position of Geography in British Universities" in 1921 contained much information on the background of geographic thought in Britain, as well as direct quotations of views held at that time *(411)*. The relatively larger number of methodological papers published in recent years by British geographers are considered in *Perspective.* . . . pages 4–7, 189–91.

112   15.   The paper by Penck was published in this country, in *Congress of Arts and Science,* Universal Exposition, St. Louis, 1904, Howard J. Rogers, editor, Boston, 1906, but the passage mentioned here was not included in the American publication.

119   16.   For further discussion of the concept of the earth shell as defining the physical scope of geography, see *Perspective* . . . pages 22–25.

122   17.   Concerning the position of Ratzel and Semple on the question of "geographic determinism," however, see *Perspective.* . . . , footnote 5 on page 56.

122   18.   The most vigorous later-day upholder of the "environmentalist concept" is Griffith Taylor *(432, 434)*, who provided a record of his own development in geographic work, and that of Australian geography in general *(451);* see also *Perspective.* . . . , pages 58, 62.

123   19.   Although Barrows' presidential address was much discussed in this country at the time, and has been frequently cited in methodological writings of foreign students, its influence among American geographers was short-lived, if we except the work of George Renner *(436)*.

134   20.   Since Kant's statement about geography has subsequently appeared in several articles in somewhat different form, it may be noted—as explained in footnote 3 on pp. 38 f.— that (1) the form here presented is probably the most nearly correct presentation of what Kant actually said; (2) any differences in the several sources are of little

significance so far as this particular section of his lectures is concerned; and (3) in contrast to other portions of
the published versions of Kant's lectures, which are either
of doubtful authenticity or were based on what he presented
in the first few years of his teaching, the section here
translated is based on a full and fairly reliable record of
his lectures in 1775, in mid-career.

Whatever the merit of Kant's statement, it cannot be
assumed to have had significant influence on the subsequent
development of geographic thought; see *Perspective. . . ,*
page 180, and my article on "The Concept of Geography as
a Science of Space, from Kant and Humboldt to Hettner,"
Annals, Association of American Geographers, XLVIII
(1958), 97–108.

135   21.   Additional statements of the views of Humboldt and Ritter
on this topic are included in a previous chapter, page 77,
and more particularly in my paper of 1958, cited in the
preceding note, including full translation of his basic statement of 1793, from the original Latin.

139   22.   Almost simultaneously and evidently independently, Krebs
expressed concepts of geography which, as he noted, were
very similar to Hettner's *(416).*

140   23.   That Hettner was not aware of Humboldt's statement on
this topic was confirmed by him in personal letter to me.
Subsequent investigation has failed to find any clear evidence of connection between Kant and Humboldt or between
either of them and Hettner on this concept; see my article
of 1958, cited in Note 20.

140   24.   This concept of the logical position of geography among the
sciences was included here, and is repeated with minor alteration and amplification in the more recent study, because it seems to explain characteristics of geography as
determined from empirical evidence; it is not to be considered as a basic and necessary proposition on which
those findings depend. See *Perspective. . . . ,* pages 10 f.,
173–82.

142   25.   Ritter's expression is explained more fully on page 57,
and still further in *Perspective. . . ,* footnote on page 66;
see also, in the same work, the footnote on page 116.

144   26.   This statement is found to be an inadequate answer to
Kraft's question, which is re-considered in *Perspective. . . ,*
pages 41–47.

175   27.   This section on "The Relation of History to Geography"
might more logically have been placed later, and the student may find it less disrupting to his thought to postpone
reading it till after Section X. Its brevity, in comparison
with subsequent sections, is not to be taken as a measure

of the importance of the topic, but as a measure of the small degree to which it appeared, at the time, to constitute a topic of controversy among geographers in this country. Likewise it reflects the fact that the writer found himself in complete agreement with Hettner's treatment of the question which was followed consistently throughout the section (with one exception specifically noted on page 186 f.). The subject is reconsidered, in Chapter VIII of *Perspective. . . . ,* with a change noted below.

177   28.   These views of Hettner were vigorously opposed by Winkler, *(418, 419)* who argued that geography must be a science of time as well as of space, that landscapes form objects of study in geography (see Chap. IX in this volume) and therefore an essential theme in geography is the history of the cultural landscape *(Landschaft).* See also note 38 below.

178   29.   This study by Broek should rather have been cited as an appropriate example of "comparative historical geography" discussed later, on page 187.

183   30.   Meigs' study utilizing the relative number of young nonbearing fruit trees as an index of regional trends illustrates the thought here. These trees, of which no account is taken in production statistics, are a significant feature in the present working economy of a region, and at a future date will be a significant feature in a different way. He is interested in the situation at each of those times, not primarily interested in the process of growth of the orchards *(450).*

184   31.   Mackinder spoke similarly, in 1931, of the "danger that we should mix history with geography without seeing clearly what we are doing." Perhaps he has offered a key in the following distinction: "Geography should, as I see it, be a physiological and anatomical study rather than a study in development . . . it should be a description, with causal relations in a *dynamic* rather than *genetic* sense." *(429,* 268; italics added; see also the similar discussion in this volume, page 358).

187   32.   The consideration presented in the remainder of this chapter is re-examined, with significant change in conclusions, in *Perspective. . . .* pages 102-106.

189   33.   At the time of writing, in 1939, the restriction considered in this chapter was the subject of vigorous discussion in the methodological writings of German and American geographers. I do not know of any American geographers who now support it.

195   34   These statements made by Finch in a symposium should not be presumed to represent his considered judgment;

Page	#

they were used here because they present clearly the thinking of many who maintained the limitation.

237   35.   The term "areal differentiation," as used here and in innumerable other places in this volume, has been found to lead easily to misunderstandings, so that the author now uses rather the term "areal variation"; see *Perspective* . . . . pp. 12-21.

242   36.   The author has since concluded that geographers, in fact and for logical reasons, depend on an additional criterion in measuring the significance of areal variations, namely, the significance to man; see *Perspective* . . . pages 41-47.

252   37.   While no American geographer today, so far as I know, asserts that a region is an object, discussions of the concept of regions are nonetheless frequently based on assumptions which in fact imply such a belief.

263   38.   On the basis of similar assumptions, Winkler, writing in 1938, had urged the logical reorganization of general (systematic) geography, in terms of *Landschaft* morphology, physiology, chronology, chorology, and system of types. Geomorphology, climatology, etc., were dismissed to other disciplines *(442)*.

285   39.   The consideration of regions in this chapter and in portions of the following chapter, requires significant modifications in the light of subsequent developments in concepts of regions; these are analyzed in *Perspective* . . . , pp. 129-43.

294   40.   In 1914 Joerg presented an illuminating analysis of six published maps of "natural regions" of North America and of fifteen other divisions of the continent, each based on a single natural element—physiography, climate, or vegetation *(443)*. The study was based on thorough examination of the European literature, including, in addition to the studies of Herbertson, Hettner and Passarge, a major example of the school of Vidal de la Blache, by Gallois *(437)*. It is of historic interest to note that a committee of American geographers at that time decided to postpone, as "a work of a different and higher order," the construction of a map of "natural regions" of the United States, which would supply "the logical units of regional investigation" *(444)*. Instead, they decided to concentrate efforts on the division of the United States into physiographic units—the work ultimately carried through by the late Professor Fenneman.

299   41.   The thesis of this paragraph, which is assumed in numerous subsequent discussions as it has long been assumed by geographers generally—namely, that it is necessary in geography to distinguish among all the factors of environment between those of human origin and those of natural origin—is opposed in *Perspective* . . . , pp. 48-64.

Page	#

approaches. Reconsideration of the question has led to the conclusion that what we have in fact is a continuum of gradations from the more specialized topical studies to the more nearly completely regional; see *Perspective.* . . . pp. 113-29.

399   50. The two previous paragraphs, as well as others in the following pages were based on the widely-held assumption that all the various "natural" aspects of geography constituted one branch of the field, physical geography, in contrast to the other main part, human geography. The author has opposed this view in oral discussions and correspondence for more than a decade, and has recorded the objections in *Perspective.* . . . pages 65-80.

400   51. A brief paper by Zuber, in 1937, compares the academic relations of the sciences concerned with planning to their relations in planning work *(456; see also the papers listed as ##s 451-460).*

401   52. While the conclusion is surely correct as an empirical statement based on the works which their authors call "political geography," this may represent failure of geographers to recognize the need for other types of studies which should be included under this head. Subsequent development of views on this field are discussed in the following studies by this author: "The Functional Approach in Political Geography," *Annals,* Association of American Geographers, XL (1950), pp. 95-130; "Political Geography" in American Geography: "Inventory and Prospect," Preston E. James and Clarence F. Jones, eds. Syracuse 1954, pp. 162-225; and "Political Geography in the Modern World," *Conflict Resolution IV* (1960), pp. 52-66.

405   53. The fact that little consideration is given in this volume to the work of geographers in governmental agencies is due to the nature of such work, in which the author subsequently participated for several years, as applied geography. Such studies, as previously noted (pages 399 f.), cannot readily be classified in any theoretical branch of geography, or indeed within science itself. The nature of the work in practical problems is dictated on the one hand by the problems themselves and on the other by the particular abilities of the individuals studying them.

423   54. The question of the relation of geomorphology to the field of geography, which is avoided here as it is throughout this volume, is examined at length in *Perspective,* . . . pages 84-96.

433   55. This statement assumes much too high a degree of certainty as necessary for the term "prediction"; see *Perspective.* . . , footnote 10 on page 165.

Page	#

437 56. This list should have included Wellington Jones' short paper of 1930 *(448)*. That was the first presentation of the "fractional method" of field analysis worked out by the spring field group, later carried out more fully by Finch *(285)* and subsequently adapted for use in planning surveys, as described by Hudson *(458, 459)*.

455 57. This sentence, which might well be emphasized, is expanded further in *Perspective*. . . . page 164.

AUTHOR INDEX *

Figures in italics, in brackets, refer to title numbers in the bibliography (pages 2–21)—either publications of the author or those primarily concerned with his life or work.

Figures in roman type refer to pages of this text (as marked at the bottom of pages). Those in boldface type indicate major discussions.

[479]

* Page numbers may be incorrect by plus-or-minus one.

* Delete last line of numbers and add: 388, 392, 394, 398, 406, 407, 413, 416-18, 426, 431, 436, 441, 447, 456-59

* Add 24, 25; delete 186, 413
** Change to read: Stamp, Josiah C., [200]; 99, 100
Stamp, L. Dudley, [310]; 293–94, 312

SUPPLEMENTARY AUTHOR INDEX

SUBJECT INDEX *

All figures refer to numbers at the bottom of pages in the text. Those in bold-face type indicate pages in which the subject is discussed at length.

accuracy, in science, **374–377**, 405, 432
"actual" systems of regions, 294
adjustments, 123
aesthetic geography, 47, 67, 82, 97, 103–104, 121, 215–216, **218–219**
agricultural geography, 335, 397, 398, 400, 421
agricultural regions, 319, **338–347**
analysis and synthesis, 69, 355
animal geography, 78, 111, 128, 416, 435
anthropocentric, 43, 105
anthropogeography, **90–91**, 121, 123, 128
anthropological geography, 70
anthropology, relations with geography, 178, 211, 435
applied science, 400, 404, 461
"arbitrary divisions" of the earth surface, 442, 443
area, 44, 131, 154, 168, 265, **393–396**; see also: region, character of area
areal correspondence, **329–330**
areal differentiation, 92, 98, 143, 218, 237, 240, 242–245, 251, **334–337**
geography as, 88, 92, 98, 144, 218, 237, 240, **243–245**
areal extent, 192, 199–200, 220, 242, 245–246
"areal organism," 275, 280–281; see also: organism
areal relations, 415
arealy, 77
"areas of certain type," 312, 392
"arm-chair geography," 55
art, relations with geography, 132, 219
artificial systems of regional division, 294, 298
astronomy, compared with geography, 141, 371–372, 379, 409, 433
relations of, with geography, 77, 398

botany, compared with geography, 78
relations of, with geography, 416; see also: plant geography
boundaries; see: linguistic boundaries; political boundaries; region, boundaries; sciences, boundaries between the
boundary girdle, 269, 291

boundary zone, 267; see also: transition areas
bridges, 220

cartography, **247–249**, 398
certainty, in science, **374–377**, 405, 431–432
character of an area, 36, 57, 62, 79, 96, 130, 133, 142, 191, 195, 200, 215, 228, 230, 246
"chore," 170, 440
chorographic science, geography as a, 91, 101, **130–144**, 241, 253
"chorography," 41, 92, 101
chorological concept, 56–57, 77–78, 91, 101, 240, 371
chorological sciences, 141
"chorology," 93, 101
circulatory phenomena, 270–271
cities, 279; see also: urban activities
classical geography, **48–64**
classification of objects, 416–417
climate, 94, 204
climate, in landscape, 155, 157, 165, **191**
climatic changes, 409–411
climatic classification, 292, 308, **321–323**
climatology, 79
cognition, theory of, 30
commercial versus subsistence farming, 342–343, 352–353
common-sense knowledge, 386
"comparative geography," 59, 73, 426
comparative regional geography, 87, 311, **447–448**
comparative systems of regions, 305, **314–364**
compendium, geography as a, 71
continents, 60, 62, 69, 285, 306
conversion of natural landscape to cultural landscape, 177, 180, 348
cores of regions, 305
correlations, 329
cosmography, 77
crop-animal association, 336, 341–343, 352
cultivated lands, 336–337, 348–349
cultivated landscape, 172, 173, 348
cultivation methods, 339, 342, 352–353

[485]

* Page numbers may be incorrect by plus-or-minus one.

ON THE MORES OF METHODOLOGICAL DISCUSSION
IN AMERICAN GEOGRAPHY*

RICHARD HARTSHORNE

University of Wisconsin

AMERICAN geographers reveal a paradoxical attitude toward questions concerning the nature of their field of work, its purposes and general methods of procedure. If one were to judge from published studies, one would conclude that they are not greatly concerned to examine these questions. But just the opposite conclusion would be drawn if one judged from the amount of controversial discussion in which they engage over these questions.

Surprisingly little controversy develops over the content of substantive works. Our field is so vast in proportion to the number of workers in it that our paths cross all too seldom. More often we are working along more or less parallel lines in different areas. Methodology therefore constitutes the principal common factor in our work, the one aspect on which all feel more or less competent to speak. At the same time, the history of development of American geography has been such that there is less agreement among us over the nature of our field than there is concerning substantive questions within the field.

That we regard methodological questions as important is attested by the extent to which we discuss them. Yet our discussions of methodology tend easily to argument and fruitless controversy. Recognizing this, we criticise each other for the way in which we discuss and argue these questions about geography. Such criticisms imply the existence of rules of proper attitudes and conduct, unwritten rules more or less generally accepted by American geographers.

Recent experience, both as critic and as target for criticism, caused me to ponder on what are the mores of American geographers as a social group in discussions of this nature. Concrete evidence of our reactions was available to some extent in published literature, in larger part in my memory of discussions at annual meetings of this Association, and particularly in many specific comments that have been made

* Presented at the Annual Meeting, December, 1947.

to me orally and in correspondence concerning my own discussions of the nature of geography.[1]

The purpose of this paper is to induce from this evidence the specific attitudes and rules that represent current *mores* of methodological discussion among American geographers, and to analyze and determine the validity of each of these.

I. WHY TALK ABOUT GEOGRAPHY?

The first conclusion of fact is that there is considerable doubt among American geographers whether there is any value in methodological discussions. Superficial examination of the evidence might lead one to conclude that all geographers could be divided into two groups; those who argue about what geography is and those who argue that such talk is waste of time. But we find that individuals who have been prominent in discussions about geography have at other times indicated that such discussions are not important. Many stand on the side-lines and accept in part the viewpoint of both groups. Others feel that the discussions are of value only in earlier years of training, but are fruitless after a worker has established his objectives and methods.

The existence of these amphibians suggests that the issue involved is not realistic. Surely anyone who likes to be called a geographer—and don't we all?—has some concern over what the word means. Some may say that they prefer to demonstrate their concept of geography by their substantive works and wish that others would be content to do likewise. But I found that you need only to scratch a geographer to discover that he has a fairly definitely formulated concept of what geography is, what it should do, and how it should do it, and that he does not regard this concept as uniquely valuable to himself alone. If he writes a book, he will state his concept in the introduction if not in the text and he constantly presents it to the students who are under his influence.

In short, none of us would really accept the popular idea that it doesn't matter where you are going as long as you are on the way. We all recognize, in fact, that each geographer must somehow find answers to the questions: "What is geography?", "What is its purpose?", and "How should it be studied?"

The real issue is not over the importance of the questions, but whether any

[1] No small part of the evidence on which this paper relies is drawn from criticism directed at the author's study of *The Nature of Geography* both in published reviews and still more in personal letters, many of which have been frank and penetrating though always cordial. In using evidence of this character it does not seem appropriate to specify the sources, but to limit myself to a general expression of appreciation to all colleagues who have thus contributed to the thinking in this paper. Particular appreciation however should be expressed of Lester E. Klimm's "Commentary" on the Second Printing of the work (*Geographical Review*, July 1947, pp. 486–490). In addition to the information he took the trouble to gather on the use to which the work has been put, he has summarized with considerable effectiveness, I believe, the reactions of American geographers not only to that work, but to methodological discussion in general.

progress can be expected from discussion of them. Some hold that such discussion produces only argument, yielding heat but no light. Can we accept this form of intellectual defeatism?—that men of similar training, engaged in the same field, but holding different concepts of that field and how one should work in it, cannot by discussion arrive at any greater degree of agreement, or at least of mutual understanding of why they disagree?

This question is answered I think by the reaction of geographers to serious effective studies in methodology; many who have published such studies will confirm—often with regret—that those have aroused more attention and response than any or all of their substantive works. Such popularity may not be deserved, but it is not without reason. I suggest that the reason is that American Geographers, by and large, no matter what they say about it, are in fact seriously concerned with methodological questions.

If this judgment of fact is correct, we need only to recognize it. We can then agree to treat methodological questions as among the important matters that we have to study and discuss in common.

II. TO TALK OR TO STUDY?

How should methodological questions be studied and discussed? It is a very common attitude among American geographers that the fundamental questions of what and how you do in geography are suitable for oral discussions whenever geographers gather together, but are not appropriate for serious scholarly writings and work; that such discussions are not productive. It is only a minor modification of this to add that on occasion, notably in a presidential address, it is appropriate for a geographer to present to his colleagues, as a "confession of faith," the personal views he happens at the moment to hold on these questions. Such addresses, it is held, are to be considered as little more than well-formulated statements for oral presentation and should not be subjected to critical examination as though they were scholarly writings.

If the validity of methods may be judged by the results obtained, the weight of evidence completely reverses this rule. All of us have witnessed lengthy oral arguments on these questions and observed little or no influence on the thought of either the participants or their hearers. Yet no one familiar with the development of geographic thought during the past twenty years can question that scholarly studies, presenting not merely the thought of the authors but that of large numbers of their colleagues and predecessors, had a pronounced and developmental effect on the thinking of a very large number of research geographers. In affecting their thinking it has influenced their substantive work; it has been productive in its effects.

We might therefore agree with those who say that discussion of such questions is fruitless in so far as such discussion consists of oral argument. But to say that scholarly discussion, based on critical examination of previous writings and cool logical analysis, was without effect, would be to deny the facts of experience.

These considerations lead to the following conclusions:

1. The most rapid progress toward a common understanding of sound methodology in geography is to be expected by the application of responsible scholarship to the problem concerned.

2. Oral discussion has value only in a secondary phase, in which it can utilize the results of such scholarship rather than depend merely on what the individuals assembled happen to think.

As a corollary we may add that such oral discussion is likely to be of most value if held in seminar form and least likely to be productive at formal meetings of geographers' associations.[2]

III. IS SERIOUS CRITICISM DESIRABLE?

The use of the words "responsible" and "scholarship" in the previous conclusion brings us to a third attitude, namely, that it is not *fair practice* to take "too seriously" what a geographer happens to write and publish on methodological questions. One should realize that the authors of methodological studies, in writing them, did not take them as seriously as they do their "real scientific work." Why not? Have they any reason to suppose that their substantive work will have a greater influence in geography than their methodological writings? Whatever the answer they might give to that question, they must have hoped that their methodological writings would have some significant influence on readers; why else offer them for publication? Geographers presumably do not write and publish merely for self-expression; like all other serious writers, their purpose is to influence people.

It is true that at times one writes with the intention of influencing merely the thought of the moment, whereas in other studies one hopes to produce something of lasting influence. But which will prove to be the case cannot be determined by the author. The novelist Blackmore is said to have been indignant that the public would not recognize what he regarded as his more important works, but insisted on acclaiming and reading only one of the lesser ones—*Lorna Doone*. After more than a century, Julius Fröbel is known in geography almost exclusively for two rather ill-informed, methodological essays written in his youth. A much more thorough essay in the same field, as well as substantive works that he regarded as much more important, were apparently never of much influence and indeed have been mentioned in our time only because current interest had been aroused again in his early im-

[2] It was largely because of the conclusions stated above that *The Nature of Geography* came to be written. There was need, it seemed, for a thorough, even exhaustive, examination of what our fellow-workers in previous periods and other countries had learned from serious study of the problems which we were debating orally. Such a study might, it was hoped, make unnecessary much of the seemingly endless argumentation into which our method of discussion had let us and, for questions still unresolved, would prove a background of learning that might make future discussions productive. In view of the use to which that study is being put in graduate training, according to Klimm's report, a number of references to it are included in this paper, not for personal defense, but in order to explain misunderstandings and to prevent misuse.

mature articles.³ The critic of current thought in geography may agree with Fröbel's judgment of the relative merit of his different works and still concern himself only with those two articles as the only work of Fröbel that is significant in current thought.

Whatever a writer may think of the relative importance of his different works, it is difficult to take with any seriousness the claim that one has taken any of his serious writings too seriously. I know of no geographer who has objected because students who accepted what he wrote had taken his precepts too seriously. As long as a geographer's methodological writings are taken as gospel, no objection from the author is heard; seriousness becomes excessive only when adversely critical.

If we assume that the purpose of methodological discussion is to seek truth, to find the most reliable method of approach to our subject, we should suppose that critical discussion of a man's methodological writing was the only kind of discussion that could be of value to him for his own thinking. The questions of methodology are involved and logically complicated, seldom permitting of a clear and certain answer. They deal with ideas that cannot be expressed in relatively precise forms of statistics, map distribution, or clearly defined descriptive terms, but only in words commonly ill-defined and slippery in meaning. Consequently there is always great chance that the writer deceives himself and his readers. Fortunately, on the other hand, most of the problems of geographic methodology have been studied in published writings by a large number of geographers during the past two centuries. These provide a multitude of checks on the thinking of any who attempt to consider the same problems today.

It is therefore amazing to learn that an author of a methodological study is not interested in the comparison of his views with those of others who have studied the same problem, or that he is not interested in critical examination of the logical reasoning of his own presentation. Is it really of no value to a geographer holding a particular concept to have his ideas subjected to the test of comparison with those of others, to the test of logical analysis by others? Is any of us so certain that he is on the right track that an examination of that track which other geographers have made before or make now is of no significance to him?

One may seek to dismiss these questions by simply saying, "I do not like dialectics," but to do so is simply to say he does not like logical argumentation, that is, that he is not interested in having his writings subjected to the only check by which we can make sure of the soundness of conclusions reached by the process of reasoning.

In short, anyone who says that he is not interested in a critical analysis of the logic of his writing is saying that it is of no concern to him whether his conclusions are sound, dubious, or completely erroneous. If that be true of any writer—if it can be true—it cannot be true of serious students who will read his writings.

³ *The Nature of Geography,* pp. 72–75, 102–106.

In place of the attitude that methodological writings should not be taken too seriously—that methodology is a field in which any geographer may dabble freely when the spirit moves him—we are led to the following conclusions:

1. The measure by which we judge the seriousness with which any writings are to be considered is not the degree of importance given them by the author, but the degree of importance they have for readers.

2. Methodological studies require standards of scholarship no less exacting than those expected in substantive work.

3. It is both appropriate and desirable that any methodological writing worthy of publication should be subjected to critical evaluation in terms of its basic assumptions, the reliability of its evidence, the soundness of its logical reasoning, and extensive comparison with the findings of other students of the same problem.[4]

IV. MAY ONE ASK "IS IT GEOGRAPHY?"?

The best known of all our unwritten rules is at the same time the least understood. This is: that while it may be desirable and important to seek an answer to the question, "What is geography?", it is undesirable if not reprehensible to apply the answer arrived at to any specific study by a geographer—to raise the question "Is it geography?" Perhaps the rule is best known because of the frequency with which it is violated and the vigorous reactions of disapproval which such instances provoke. One such expression deserves to be lifted from its relative obscurity: "I am at a loss to understand this pernicious anemia that has seized upon some geographers, expressed in a weary shaking of the head at sight of a fellow student happily productive and in the monotonous, sepulchral query: 'But is it geography?' . . . whence do they derive the authority to correct or even read out of the party those who concern themselves also with" certain topics that the geographers in question have decided are not included in the field?[5] I presume all of us would join in the chorus of disapproval. And yet which one among us has not, in listening to the papers presented at our annual meetings or reading papers published in geographical journals, asked himself, "Is this also to be included in the field that I should recognize as geography?" Evidently there is confusion in our thinking concerning the proper purpose of formulating a definition of geography and the usage to which it should be put.

We may make a major step toward dissolving this confusion, and, I would hope, be able to eliminate much discussion that is not pertinent and often offensive, if we would agree upon the following three principles.

[4] For an excellent example of critical analysis of previous methodological writings, see Edward A. Ackerman, "Geographic Training, Wartime Research, and Immediate Professional Objectives," *Annals of the Association of American Geographers*, XXXV, No. 4 (December 1945), 121–143.

[5] Carl Sauer: Correspondence with the Editor, *Geographical Review*, XXII (October 1932), 528.

A. Is the Question Relevant?

A judgment classifying the relation of a particular research study to the field of geography does not constitute a significant criticism of that study: it cannot in any way reflect on the value of that study; it is merely a statement of classification devoid of value judgment. I know of no basis for assuming that a geological study or an historical study is of less value than a geographic one; a study that combined the techniques of two or more disciplines to such an extent as to be difficult to classify might well produce a superior contribution.

It follows therefore that, when a substantive work is presented for consideration, our proper concern is with the work itself, its thoroughness, soundness, and significance for learning. Any question concerning its relation to the field of geography diverts the consideration away from the work itself into a discussion of another topic, namely methodology. Like any other diversion it is out of order. The question of classification might well be left to the cataloguers.

B. Who Shall Set the Limits?

The second principle is that there is not and cannot be any basis on which geographers as a group may decide that any particular topic is not to be included in the field of geography.

In view of the wide expanse of a field that literally covers the earth and in that coverage must include, in any man's definition, such a great variety of different categories of phenomena, it is entirely natural that geographers should wish somehow to limit the scope of study in which they are expected to be competent. Each geographer must decide for himself in how wide a field he will endeavor to attain and maintain competence. If he decides that a particular research work is so remote from his own field of study that he finds no desire or need to incorporate its findings into his own scope of knowledge, that judgment is of concern only to himself; ordinary courtesy should prevent him from giving public expression to his lack of interest in the work.

It is however a common and proper desire of most geographers to include within their field of general interest and competence all of the field that they regard as geography. Any methodological study of geography is of value to them therefore if it provides them with a basis for measuring the scope of that field and for orienting in relation to it substantive studies which they hear or read. If some of us present definite statements of what geography is or ought to be, any who are impressed by what we write will tend to apply the conclusions to specific substantive papers. Since the purpose of any attempt to find an answer to the question "What is geography?" is to provide a systematic orientation, such use of the conclusions is appropriate and to be expected. In concrete terms, when any geographer who has developed or acquired some definite concept of the field of geography reads or hears a substantive paper, he may find it useful to ask and answer to himself the question: "Where does this study fit into the field of geography as I understand it?"

If however he gives public expression to a conclusion that a particular work does not belong in the field of geography, he is merely challenging one personal opinion—that of the author who presumably thought his work was included in geography—with another personal opinion of no greater validity. For any judgment classifying a particular work in relation to the field of geography is valid only in terms of the particular concept of the field on which it is based and there is no license for assuming that any one concept carries any more authority than any other conceivable concept of geography.

Likewise we may not seek to establish any authority. Though the proper purpose of methodological study of the question "What is geography?" is to seek the maximum degree of mutual understanding in the field, even if we should attain to virtual unanimity, we do not desire and may not require conformance. In the field of learning we can recognize no orthodoxy and therefore no heterodoxies. In studying the question "What is geography?" therefore, we may not seek and cannot produce a formula for determining whether specific works by geographers are admissible or inadmissible in the field of geography. The word "but" in the question "But is it geography?" is always out of order.

C. Purposes for Which the Question Is Relevant

In a methodological discussion it may be appropriate or even necessary to illustrate or to test theoretical conclusions by applying them to specific substantive works and in doing so to discuss the relation of those substantive works to the field of geography.

To take a simple example, one may need to clarify a difficult theoretical discussion of the logical distinction between geography and some other field by applying the theoretical concept to specific borderline studies. Any conclusion placing those studies on one side or the other of a theoretical borderline has, as we have seen, no significance for those studies; significant only is the fundamental theoretical distinction thus illustrated.[6]

A more important case occurs when a writer presents a statement of the scope of geography which clearly would not include certain types of work hitherto cultivated by geographers. To justify this, to defend his concept against the obvious charge of being inadequate, the writer must show that works of the type omitted are more properly included in some discipline other than geography. He cannot do that in terms of theory; for that would be merely to argue in a logical circle. He must show by judgments of concrete cases that studies of the type in question lack characteristics—of purpose, organization and techniques—generally recognized as

[6] Examples may be found in *The Nature of Geography, pp.* 408, 419. In both cases any possibility of protest from authors was eliminated by selecting examples from the writer's own publications, though at the risk of appearing to regard them as unduly important.

essential to geographic work and are in these respects more akin to work in another discipline.[7]

A similar problem arises when a methodological writer attempts to test the validity of precepts for geographic work by examining the fruits of those precepts as presented in substantive works produced under their influence.   The lack of such concrete tests seems to me to be a major weakness in much of our methodological discussion.   In considering a particular concept of the field of geography, or specific precepts for work in geography, it is directly pertinent and significant to determine in fact whether the works produced under the influence of those theoretical directions are in their organization, purposes, and techniques of a character generally recognized as geographic or whether they are not more akin to other fields.[8]

In all of these cases, it should be noted, the judgments made of substantive works are appropriate only as contributing to the methodological discussion (and are in order only if the topic of discussion is methodological).   In presenting such judgments, the writer cannot prevent the reader from regarding them as criticisms of the substantive studies used as examples—even though he insists that no criticism is involved.   Neither can he prevent the reader from treating them as judgments determining the admissibility or inadmissibility of those studies in geography.   But in either case it is the reader who is responsible for the conclusion which is in conflict with either the first or second principles stated above.

D. Is the Question an Attack on the Author?

If it be agreed that it is no criticism of a study made by a geographer to say that it is not geography, is it not a criticism of the geographer who made this study? This common attitude is based on the assumption that those who call themselves geographers should be working at geography.   Regardless of the significance or value of a particular study, if it cannot be classified as geography, the geographer should not have made the study.   Why not?   The only requirement for work in any field of learning is competence, and if our colleague is said to be working outside the field of geography, geographers are not the ones to judge his competence in that outside field.   Furthermore the final test of competence for any particular study is the work accomplished on that study.   If the finished work is found to be original and sound, how can its author be judged incompetent to do it?

The fact is that the scope of the field of learning in which individual scholars are competent to work cannot be defined in terms of the specific discipline in which they were trained.   Each of my readers, I have no doubt, could name certain topics outside the field of geography on which he would be more competent than he is on certain topics within the field.

[7] For example see Carl Sauer's discussions of specific works on geographic influences in history and of works in political geography in "Recent Developments in Cultural Geography," *Recent Developments in the Social Sciences* (1927), pp. 199 f., 207–211.

[8] Examples of this use of substantive works to test theoretical conclusions may be found in *The Nature of Geography,* pp. 122–125, 177 f., and 219.   These include the two examples to which Klimm points an accusing finger.   *Op. cit.,* 488.

It is not therefore a proper purpose in seeking to determine what is the nature and scope of geography to find limits that individual geographers may not overstep without criticism. The proper purpose is to provide orientation—to recognize a central core within and around which geographers work, radiating out in diverse directions, though ever conscious of where they are. If we observe that any of our colleagues, whether because of their personal inclinations or in pursuit of particular topics, are drawn so far from the central core that we think they have crossed into fields logically included in other disciplines, that conclusion carries no criticism. On the contrary, in the anticipation of gains that may result from cross-fertilization, we may well regard such expeditions with positive approval. We only ask that a worker on the periphery of our field shall not lose his sense of orientation so that he feels called upon to shout back to his colleagues: "Here is the proper center of our field; here is where all geographers should work."

Granting the value of recognizing complete freedom of the individual geographer to operate on due occasion as widely as the spirit moves him and his competence can safely carry him, is it not his business, the reader may ask, to devote most of his work to the field of geography? I presume that as a matter of intellectual and academic honesty, anyone who calls himself a geographer will wish to devote most of his attention to what he believes is geography. But since, as noted above, there can be no authority to determine what is geography, any statement by others that implied that they found his work to be outside geography would not constitute a criticism of his work, but only of his concept of geography. No one will question that any geographer's concept of geography, whether stated in methodological writings or expressed in his substantive works, is open to critical consideration by his colleagues.

We may summarize our discussion of the problems involved in putting the question "Is it geography?" by saying that the question is pertinent only for methodological purposes. Insofar as it arises from an individual's need to classify and organize in his thinking the particular piece of substantive knowledge included in a particular study, he is concerned only with his own answer to the question and therefore should have no reason to suppose anyone else is interested in his answer. For general discussion the question is in order only if the topic of discussion is a methodological one. In that case classification of substantive works may lead to critical reflections on methodological theories; in no case can they be considered as casting critical reflections on the substantive studies themselves or on their authors.

V. THE PERSONAL PROBLEM

The discussion of the geographer in the previous section brings us to what is perhaps the cardinal difficulty underlying all the problems discussed in this paper— the *personal* problem.

Throughout the sections preceding the last one, I have referred to writings by geographers as though they were things in themselves, existing independent of their authors. The justification for this assumption is that it is a fact. When a man publishes something he has written, he has turned his child loose on the world to live or die, to influence readers, independent of anything he can do about it. Whether or not he continues to think well of it, even if he would totally denounce it, it continues its independent existence, coming to life whenever a reader in a library reacts intellectually to what *it* says—perhaps long after the original author is dead and possibly forgotten as a person.

If this were the whole of the picture we would have no problem. The problem exists because while articles once published, like the off-spring of the lower animals, are no longer attached to their authors, the authors are very like human parents in that they continue to be attached to what they have published. The attachment is not merely one of feelings; in major degree a man's brain-children determine his reputation in the field of learning.

The universal rule in learned critical discussions is that the critic considers not the writer but the writing. Though we know that criticism of the writing may hurt the person of the writer, we read critical reviews of substantive works without regarding legitimate criticism of the work as attack on the writer.

Methodological writings however seem much more closely associated with the person of the writer than are substantive works. The facts that constitute much of the content of substantive work obviously belong to no one; ideas appear as the cherished possession of him who expresses them. We accentuate this impression by a convention that appears unfortunate but unavoidable; though our concern is properly with writings rather than with writers, in methodological discussion particularly we refer to the writings not in terms of their titles, but in terms of the names of their authors. While substantive works are commonly distinct units whose titles are often convenient handles, sometimes better known than the names of the authors, the opposite is generally the case for methodological writings. If we refer to a work as "The Geography of the World's Agriculture," "The California Raisin Industry," or "Süddeutschland," the informed reader knows of what we speak. Even if it were feasible to refer repeatedly to such titles as "The Inductive Study of the Content of Geography," "Some Comments on Contemporary Geographic Methods," or "Die Wesen und die Methoden der Geographie," few readers would recognize that we were referring to studies familiar to them as the writings of Davis, Leighly, and Hettner. Further, in many cases, methodological studies are not distinct units; an essentially unified viewpoint may be presented in the course of several years in a number of articles each with an independent title. The only label that ties them together is the name of the author.

As a consequence of this tendency to think of methodological discussion in personal terms, a critic who subjects a methodological thesis to logical analysis is regarded as dissecting the writer who presented the thesis. If the original writer replies to explain or defend the thesis, he is regarded as defending himself and the discussion is considered as a personal conflict.

This is not because the theoretical issues involved are of such intense concern to us; on the contrary since we are not sufficiently serious about them our attention is caught by the extraneous, but more exciting personal aspects. We have not yet attained the degree of objectivity that is presumed to underly a scientific discussion. An additional factor is the erroneous assumption that trained men of eminent standing in their field are expected to be free of any errors of logical reasoning. Consequently the discovery of any such error instead of being received as a stimulus to correct the thinking is taken as damaging criticism.

We do not view methodological discussion as an earnest and humble effort of fallible human minds to seek by interchange of ideas the closest approximation of truth. Rather the discussion comes to be considered as a *debate,* in which points of agreement are regarded as cancelled out rather than as underscored, points of difference magnified, the prize not truth but personal prestige.

If we could escape from this concern about persons, if in our discussions we could focus our interest on the writings rather than the writers, many of the difficulties previously noted would be eliminated. Whether a writer regarded his methodological papers as less important than his substantive works would then be recognized as irrelevant; relevant only is the question of how important the writings are in influencing the thought of readers. Whether a study had just appeared in print, was twenty years old, or even centuries old, would be seen to be equally irrelevant. Any work that influences the current thought of students may properly be discussed as of current interest. Whether it represents the present view of its author is not the concern of the critic of geographic writings. If the author no longer holds the views once expressed, but they continue to influence readers, it is his concern, not the critic's, to inform readers of the change in his thinking. The critic can properly be expected to take account of "current thought" in geography only as that is expressed in geographic literature.[9]

In sum, the universal rule in all branches of learning applies to discussions of methodology; scholarship is not concerned with persons but with writings. There is a continuing obligation to focus methodological criticism on writings rather than on persons, and to maintain the tone of the discussion on the impersonal level. If the present writer has at any time failed to adhere to these standards, he expresses his regrets to any who may have been offended. But it is only fair to remind the reader again that the obligation applies to both parties in reading a paper which thereby becomes a discussion between writer and reader.

Since much of our discussion of the mores in methodological discussion has been critical of the present customs and standards of American geographers as a social group, it is appropriate to close with a recognition of one established custom to which we may point with pride. I refer to what is, I hope, so well-established as a practice as to take on the nature of a compelling social law. This is, that even when a methodological discussion appears to produce conflicts between persons rather than between ideas, the engagements are regarded, at most, as those of a

[9] Cf. the discussion on this point in Klimm's review, *op. cit.,* 488.

tournament, not of a battle, and certainly not of a war. It should harm no one to be thrown from his intellectual horse—however high the horse—on the contrary, the experience may be taken as a part of a continuing education. Over a goodly number of years, American geographers have consistently adhered to the principle that methodological controversy is not a basis for personal enmity or schisms in the field. In the writing of *The Nature of Geography* therefore, great effort—greater than the reader is likely to realize—was made by the author and his editorial critics to avoid provocation that could lead to disruption within the geographic fellowship. This effort, whether completely successful in itself, has met with full cooperation, so far as I can learn, from all concerned. American geographers, whether they agree or disagree, maintain their standard of cordial intellectual relations.

12144